Progress in Mathematical Physics

Volume 22

Progress in Mathematical Physics

Volume 22

Carlo Cercignani
Gilberto Medeiros Kremer

The Relativistic Boltzmann Equation: Theory and Applications

Springer Basel AG

Authors:

Carlo Cercignani
Dipartimento di Matematica
Politecnico di Milano
Piazza Leonardo da Vinci 32
20133 Milano
Italy
e-mail: carcer@mate.polimi.it

Gilberto Medeiros Kremer
Departamento de Física
Universidade Federal do Paraná
Caixa Postal 19044
81531-990 Curitiba
Brazil
kremer@fisica.ufpr.br

2000 Mathematics Subject Classification 76L05, 82B05, 82B40, 83-01, 83F05

A CIP catalogue record for this book is available from the Library of Congress, Washington D.C., USA

Deutsche Bibliothek Cataloging-in-Publication Data

Cercignani, Carlo:
The relativistic Boltzmann equation : theory and application / Carlo Cercignani ; Gilberto Medeiros Kremer. – Boston ; Basel ; Berlin : Birkhäuser, 2002
(Progress in mathematical physics ; Vol. 22)
ISBN 978-3-0348-9463-0 ISBN 978-3-0348-8165-4 (eBook)
DOI 10.1007/978-3-0348-8165-4

ISBN 978-3-0348-9463-0

Originally published by Birkhäuser Verlag Basel in 2002
Softcover reprint of the hardcover 1st edition 2002

Printed on acid-free paper produced of chlorine-free pulp. TCF ∞

ISBN 978-3-0348-9463-0

9 8 7 6 5 4 3 2 1

www.birkhauser-science.com

Contents

Preface

The development of the relativistic kinetic theory started in 1911 when Jüttner derived an equilibrium distribution function which was a generalization of the Maxwell distribution function for a relativistic gas. He also succeeded, in 1928, to derive a distribution function for relativistic gases whose particles obey the Fermi–Dirac and Bose–Einstein statistics.

The next step towards a statistical description of a relativistic gas was the generalization of the Boltzmann equation in a covariant formulation given by Lichnerowicz and Marrot in the forties.

Beginning in the 1960s, rapid progress was made in relativistic kinetic theory, in particular when Israel and Kelly calculated from the Boltzmann equation the transport coefficients of a relativistic gas. The result that a relativistic gas has a bulk viscosity – unlike a non-relativistic monatomic gas – called the attention of many researchers to a number of applications of the results of this theory: the effect of neutrino viscosity on the evolution of the universe, the study of galaxy formation, neutron stars, controlled thermonuclear fusion, etc.

The aim of this book is to present the theory and applications of the relativistic Boltzmann equation in a self-contained manner, even for those readers who have no familiarity with special and general relativity. Though an attempt has been made to present the basic concepts in a complete fashion, the style of presentation is chosen to be appealing to readers who want to understand how kinetic theory is used for explicit calculations.

The book is organized as follows. In the first chapter we introduce the basic principles of special relativity theory and tensor analysis in a Minkowski space. The relativistic Boltzmann and transport equations for a single non-degenerate and for a single degenerate gas are derived in the second chapter where the macroscopic description, the equilibrium and the trend to equilibrium of a relativistic gas are analyzed. In the third chapter the fields in equilibrium of a relativistic gas, i.e., particle number density, pressure and temperature, are studied and applied to a Fermi–Dirac system where the Chandrasekhar limit of the mass of a white dwarf star is derived and to a Bose–Einstein system where the relativistic Bose–Einstein condensation is discussed. Chapter Four presents a phenomenological theory for a single relativistic gas in a non-equilibrium state based on the linear thermodynamics of irreversible processes. The determination of the constitutive equations from the Boltzmann equation by using the method of Chapman and Enskog for a viscous, heat conducting relativistic gas – which correspond to the laws of Navier–Stokes and Fourier – is the subject of the fifth chapter. Chapter Six deals with development of the method of moments of Grad and with derivation of the Burnett linear constitutive equations. Mixtures of relativistic gases in which chemical or nuclear reactions occur are analyzed in chapter Seven where the Onsager reciprocity relations between coefficients of cross-effects are discussed. In chapter Eight we analyze the model equations of Marle and Anderson and Witting for the relativistic Boltzmann equation and apply these models to determine the transport coefficients of relativistic quantum gases and relativistic ionized gases. The problems concerning the propagation of shock and sound waves in a relativistic gas is the subject of Chapter Nine. Chapter Ten deals with the basic principles of the tensor calculus in general coordinates, while Chapter Eleven introduces tensors in Riemannian space and the principles of the theory of general relativity. In Chapter Twelve we derive the Boltzmann and transfer equation for a relativistic gas in a gravitational field and analyze the equilibrium state of the gas. The last chapter of the book is dedicated to the study of the Vlasov–Einstein system.

We expect that this book can be helpful not only as a textbook for an advanced course on relativistic kinetic theory but also as a reference for physicists, astrophysicists and applied mathematicians who are interested in the theory and application of the relativistic Boltzmann equation.

Milano and Curitiba C.C. and G.M.K
June 2001

Chapter 1

Special Relativity

1.1 Introduction

Before entering the subject of the book, relativistic kinetic theory, we introduce in this chapter the basic principles of special relativity theory and tensor analysis in Minkowski spaces. This chapter, as well as those on tensor analysis in general coordinates and Riemannian spaces, has the purpose of making the book self-contained. It can be skipped by readers thoroughly familiar with the subject, who may just give a glance at it, to be sure that the notation is the same the reader is used to.

The notation used throughout this book is:

a) Latin indices $i, j, k, \dots$ denote the three-dimensional space coordinate system x, y, z running from 1, 2, 3;

b) Greek indices $\alpha, \beta, \gamma, \dots$ are used to specify the four-dimensional space-time coordinate system ct, x, y, z running from 0, 1, 2, 3;

c) The Einstein summation convention over repeated indices is also used. In other words, we write, e.g., $T_{\alpha\beta}v^\beta$ for $\sum_{\beta=0}^{3} T_{\alpha\beta}v^\beta$;

d) We use the convention of writing the number of the equation that is being used in a passage over the equality sign, e.g., $\overset{(1.23)}{=}$ means that we have used equation (1.23).

1.2 Lorentz transformations

The special theory of relativity was developed by Einstein (see the collection of papers in [3]) in 1905 and accepts the usual Euclidean structure of space (homogeneous and isotropic) and in particular the possibility of using Cartesian coordinates. Observers moving with respect to each other can measure distances and times as in classical mechanics. Time is homogeneous in the sense that the laws are invariant with respect to a change in the origin of time. Any method of assigning coordinates and measuring time provides a reference frame, exactly as in classical mechanics. Differences from classical mechanics arise only when measurements of two such observers are compared.

The first basic difference from classical mechanics concerns the fact that, whereas in the non-relativistic case one can easily split the subject into kinematics and dynamics (statics being a limiting case of the latter) and imagine that in kinematics any observer is as good as any other one, in relativistic mechanics we must talk of inertial frames (which in classical mechanics belong to dynamics) to start with. To be sure, one can never easily split physics into its parts. Even in classical mechanics one uses yardsticks and clocks from the beginning, and the existence and behavior of these tools should be discussed. However, it is easy to postulate that the lengths and time intervals measured by these instruments have a conventionally accepted significance, since they are not influenced by motion, which is the only thing kinematics is concerned with. In relativistic mechanics motion influences the results of length and time measurements. In addition the propagation of light (an electromagnetic phenomenon) plays a privileged role. Thus not only can we not easily split kinematics and dynamics apart; we cannot even split mechanics and electromagnetism.

Here we could repeat the trick of classical mechanics and assume that we only have clocks and light-beams (yardsticks are made unnecessary, by the fact that a length can be measured by the time a light beam takes to travel through it), postulate rules on their behavior and proceed to describe relativistic kinematics. However, we need also another concept, that of an inertial frame of reference. This is a frame with respect to which a point-like body at rest removed from the action of other bodies remains at rest. On the basis of experience (and without any logical contradiction) these frames turn out to be infinitely many and to translate at constant velocity with respect to each other; this has the consequence that a point-like body, removed from the action of other bodies and moving with a given velocity with respect to an inertial frame at a certain time instant, will continue to move with the same velocity at subsequent times. This is of course true in classical mechanics and remains true in special relativity.

The special theory of relativity originates from the necessity, shown by experiments, of extending the Galilei invariance of the laws of mechanics to electromagnetism.

The departing structure is based on two postulates:

1) The laws of physics should be invariant in all inertial frames of reference (in other words, it should be impossible for an observer to detect uniform translational motion by measurements made by him);
2) The speed of light in free space has the same value for all observers that are in inertial frames.

The first postulate states that the physical laws are identical in all inertial frames of reference, i.e., they are invariant with respect to space-time transformations between inertial systems. From the second postulate it follows that the velocity of propagation of signals is the same in all inertial systems of reference. Hence the velocity of propagation of signals is a universal constant which is equal to the speed of light in free space, $c = 299\,792\,458$ m/s. Another consequence is that the time interval is not absolute and that simultaneous events in one frame of reference are not simultaneous in another frame of reference. An event is characterized by the point occupied at the time it has occurred, i.e., it is a point in a space-time coordinate system. The set of all events forms a four-dimensional space, the events are represented by points called world points and the trajectory of a particle in space-time is called a world line. All this can be introduced in classical mechanics as well, but it becomes the simplest way of describing things in relativistic mechanics, as we shall presently show.

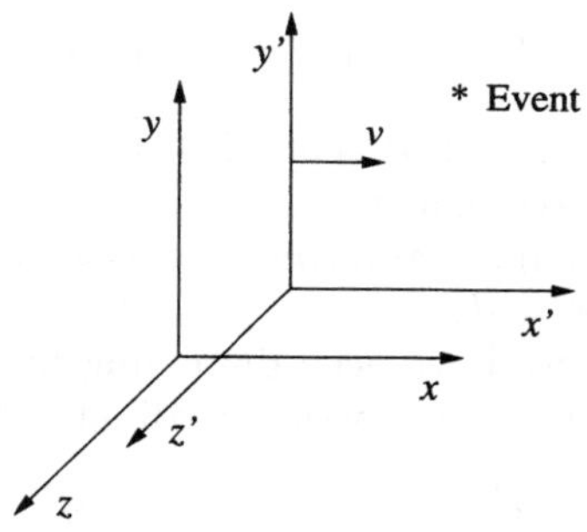

Figure 1.1: Representation of two inertial systems

We consider two reference systems K and K' such that K' is moving with uniform velocity v directed along the x axis (see Figure 1.1). The times in K and K' are denoted by t and t', respectively. If in the reference frame K we send at time t_0 a signal that propagates with the speed of light from a point x_0, y_0, z_0, the equation for the wave front of this signal is given by:

$$c^2(t-t_0)^2 = (x-x_0)^2 + (y-y_0)^2 + (z-z_0)^2. \tag{1.1}$$

According to the postulates above an identical equation must hold in the frame of reference K',

$$c^2(t'-t'_0)^2 = (x'-x'_0)^2 + (y'-y'_0)^2 + (z'-z'_0)^2. \tag{1.2}$$

Let x_1, y_1, z_1, t_1 and x_2, y_2, z_2, t_2 be the coordinates of two events in the reference frame K. It is convenient to define the interval s_{12} between these two events by

$$s_{12}^2 = c^2(t_2-t_1)^2 - (x_2-x_1)^2 - (y_2-y_1)^2 - (z_2-z_1)^2. \tag{1.3}$$

When the two events are infinitesimally close to each other the interval between them will be denoted by ds and we have that

$$ds^2 = c^2dt^2 - dx^2 - dy^2 - dz^2. \tag{1.4}$$

If the interval between two events in an inertial frame is zero it will be zero in all inertial systems, hence $ds^2 = 0$ implies that $ds'^2 = 0$. Since the intervals ds^2 and ds'^2 are infinitesimals of the same order, we conclude that they must be proportional to each other, i.e.,

$$ds^2 = a\, ds'^2. \tag{1.5}$$

Due to the hypothesis of homogeneity of space-time and of the isotropy of space, a can at most depend on the relative velocity and it is an even function of v, i.e.,

$$a = a(v), \qquad \text{with} \qquad a(v) = a(-v). \tag{1.6}$$

From the first postulate we have also that

$$ds'^2 = a(-v)ds^2 = a(v)ds^2. \tag{1.7}$$

Now it follows from (1.5) and (1.6) that $a(v) = \pm 1$. The negative value is not considered since we must have that $a = 1$ when $v = 0$ and we expect a continuous dependence on v. Hence the interval between two events is the same in all inertial frames of reference, i.e., $ds^2 = ds'^2$.

We can formalize what we have been discussing by introducing a four-dimensional space with the so-called Minkowski coordinates defined by:

$$x^0 = ct, \qquad x^1 = x, \qquad x^2 = y, \qquad x^3 = z. \tag{1.8}$$

Hence we can write equation (1.4) as

$$ds^2 = \eta_{\alpha\beta} dx^\alpha dx^\beta, \tag{1.9}$$

where $\eta_{\alpha\beta}$ are the components of the metric tensor in the four-dimensional space characterized by the element ds. This space is called Minkowski space and the metric tensor is given by:

$$(\eta_{\alpha\beta}) = \begin{pmatrix} +1 & 0 & 0 & 0 \\ 0 & -1 & 0 & 0 \\ 0 & 0 & -1 & 0 \\ 0 & 0 & 0 & -1 \end{pmatrix}. \tag{1.10}$$

It is clear that this matrix is its own inverse. If we denote the elements of the inverse by $\eta^{\alpha\beta}$ (the convenience of this notation will be clear in the next section), we have

$$(\eta^{\alpha\beta}) = \begin{pmatrix} +1 & 0 & 0 & 0 \\ 0 & -1 & 0 & 0 \\ 0 & 0 & -1 & 0 \\ 0 & 0 & 0 & -1 \end{pmatrix}, \tag{1.11}$$

and

$$\eta_{\alpha\delta}\eta^{\delta\beta} = \delta_\alpha^\beta \tag{1.12}$$

where δ_α^β is the Kronecker symbol defined by

$$\delta_\alpha^\gamma = \begin{cases} 1 & \text{if } \alpha = \gamma, \\ 0 & \text{if } \alpha \neq \gamma, \end{cases} \qquad \text{with} \qquad \delta_\alpha^\alpha = 4. \tag{1.13}$$

Let us determine the transformation law between two inertial systems x'^α and x^β that leaves invariant the interval between the two inertial systems: $ds'^2 = ds^2$. We first discard the trivial solution offered by the translation $x'^\alpha = x^\alpha + c^\alpha$, where c^α are four arbitrary constants. This transformation can be used to transform any point in Minkowski space into the origin. Then if we are at the origin, we can use the property of space-time of admitting straight lines to see that the invariance of $\eta_{\gamma\delta} dx^\gamma dx^\delta$ is equivalent to the invariance of $\eta_{\gamma\delta} x^\gamma x^\delta$ (each infinitesimal displacement can be replaced by a finite one along the straight line having the same direction). It is then easy but lengthy to show that the most general transformation which leaves the latter invariant must be linear and homogeneous (Problem 1.2.2). We denote by Λ_β^α the transformation matrix between the two inertial systems and write[1]

$$x'^\alpha = \Lambda_\beta^\alpha x^\beta. \tag{1.14}$$

From the above equation it follows that

$$ds'^2 = \eta_{\alpha\beta} dx'^\alpha dx'^\beta = \eta_{\alpha\beta} \Lambda_\gamma^\alpha \Lambda_\delta^\beta dx^\gamma dx^\delta = ds^2 = \eta_{\gamma\delta} dx^\gamma dx^\delta. \tag{1.15}$$

Hence we get the relationship

$$\eta_{\gamma\delta} = \eta_{\alpha\beta} \Lambda_\gamma^\alpha \Lambda_\delta^\beta, \qquad \text{or} \qquad \boldsymbol{\eta} = \boldsymbol{\Lambda}^T \boldsymbol{\eta} \boldsymbol{\Lambda}. \tag{1.16}$$

If the two inertial systems are the systems K and K' of the Figure 1.1, we have that $x^2 = x'^2$, $x^3 = x'^3$ and the transformation matrix $\boldsymbol{\Lambda}$ reduces to

$$(\Lambda_\beta^\alpha) = \begin{pmatrix} \Lambda_0^0 & \Lambda_1^0 & 0 & 0 \\ \Lambda_0^1 & \Lambda_1^1 & 0 & 0 \\ 0 & 0 & 1 & 0 \\ 0 & 0 & 0 & 1 \end{pmatrix}. \tag{1.17}$$

If we insert (1.17) into (1.16), we get the following system of equations for the components:

$$\begin{cases} (\Lambda_0^0)^2 - (\Lambda_0^1)^2 = 1, \\ (\Lambda_1^0)^2 - (\Lambda_1^1)^2 = -1, \\ \Lambda_0^0 \Lambda_1^0 - \Lambda_0^1 \Lambda_1^1 = 0. \end{cases} \tag{1.18}$$

To solve this system of equations we write, without loss of generality, $\Lambda_1^0 = -\sinh\phi$, $\Lambda_0^1 = -\sinh\psi$ and get from $(1.18)_{1,2}$:

$$\left(\Lambda_0^0\right)^2 = 1 + \sinh^2\psi = \cosh^2\psi, \qquad \left(\Lambda_1^1\right)^2 = 1 + \sinh^2\phi = \cosh^2\phi. \tag{1.19}$$

[1]The set of all transformations (1.14) is called a homogeneous Lorentz group.

Now the substitution of the above equations into $(1.18)_3$ yields:

$$\tanh\psi = \tanh\phi. \tag{1.20}$$

We conclude from (1.20) that $\phi = \psi$ and transformation matrix (1.17) can be written as

$$(\Lambda^\alpha_\beta) = \begin{pmatrix} \cosh\psi & -\sinh\psi & 0 & 0 \\ -\sinh\psi & \cosh\psi & 0 & 0 \\ 0 & 0 & 1 & 0 \\ 0 & 0 & 0 & 1 \end{pmatrix}. \tag{1.21}$$

For the determination of ψ let us analyze the motion of the origin of the system K'. In this case we have $x'^1 = 0$, $x^0 = ct$, $x^1 = vt$ and it follows from (1.14) that

$$x'^1 = 0 = \Lambda^1_0 x^0 + \Lambda^1_1 x^1 = -ct\,\sinh\psi + vt\,\cosh\psi. \tag{1.22}$$

As a consequence ψ is given by the relationship

$$\tanh\psi = \frac{v}{c}, \tag{1.23}$$

and the transformation matrix (1.21) has its final form given by

$$(\Lambda^\alpha_\beta) = \begin{pmatrix} \frac{1}{\sqrt{1-v^2/c^2}} & \frac{-v/c}{\sqrt{1-v^2/c^2}} & 0 & 0 \\ \frac{-v/c}{\sqrt{1-v^2/c^2}} & \frac{1}{\sqrt{1-v^2/c^2}} & 0 & 0 \\ 0 & 0 & 1 & 0 \\ 0 & 0 & 0 & 1 \end{pmatrix}. \tag{1.24}$$

The Lorentz transformations for the two systems K and K' are obtained from (1.14) and (1.24):

$$x'^1 = \frac{x^1 - vt}{\sqrt{1-v^2/c^2}}, \qquad x'^2 = x^2, \qquad x'^3 = x^3, \qquad t' = \frac{t - x^1 v/c^2}{\sqrt{1-v^2/c^2}}. \tag{1.25}$$

The Galilean transformations can be obtained as a particular case of the Lorentz transformations (1.25) when one considers that the velocities are small relative to the light speed i.e., $v \ll c$. In this case we have

$$x'^1 = x^1 - vt, \qquad x'^2 = x^2, \qquad x'^3 = x^3, \qquad t' = t. \tag{1.26}$$

Let us introduce the concept of proper time of a particle which is the time measured by a clock that moves with the particle. As usual the proper time will be denoted by τ and in the system K' we have that $t' = \tau$. Since in this system the clock is at rest it follows that $dx'^i = 0$ for $i = 1, 2, 3$ and one can get from (1.4) that the interval between the events is given by $ds = ds' = cd\tau$. In this case we have from (1.25) that

$$dx'^1 = 0 = \frac{dx^1 - v\,dt}{\sqrt{1-v^2/c^2}} \qquad \text{and} \qquad d\tau = dt' = \frac{dt - dx^1\,v/c^2}{\sqrt{1-v^2/c^2}}. \tag{1.27}$$

The first of these equations provides that $dx^1 = vdt$, which is introduced in the second one yielding

$$d\tau = dt' = dt\sqrt{1 - v^2/c^2}. \tag{1.28}$$

Hence the time interval in the system K at rest is longer than the time interval in the system K' that is moving with velocity v. Moreover, integration of the above equation leads to

$$\tau_2 - \tau_1 = \int_{t_1}^{t_2} \sqrt{1 - v^2/c^2}dt = (t_2 - t_1)\sqrt{1 - v^2/c^2}. \tag{1.29}$$

From the Lorentz transformation (1.25) it follows that the inverse transformation is given by

$$x^1 = \frac{x'^1 + vt'}{\sqrt{1 - v^2/c^2}}, \qquad x^2 = x'^2, \qquad x^3 = x'^3, \qquad t = \frac{t' + x'^1 v/c^2}{\sqrt{1 - v^2/c^2}}. \tag{1.30}$$

In the next section we shall show that the transformation matrix $\bar{\boldsymbol{\Lambda}}$ of $x^\alpha = \bar{\Lambda}^\alpha_\beta x'^\beta$ is the inverse of the matrix $\boldsymbol{\Lambda}$ such that $\bar{\boldsymbol{\Lambda}}(\mathbf{v}) = \boldsymbol{\Lambda}(-\mathbf{v})$ i.e., the matrix $\bar{\boldsymbol{\Lambda}}$ can be obtained from the matrix $\boldsymbol{\Lambda}$ by interchanging $\mathbf{v}$ into $-\mathbf{v}$.

Another consequence of the Lorentz transformation is the so-called Lorentz contraction which we proceed to analyze. We consider a rod placed parallel to the x axis. In the frame of reference K the rod has a proper length that is given by $l_0 \equiv \Delta x = x^1_A - x^1_B$ where A and B denote the end-points of the rod. The motion of the end-points obtained from $(1.30)_1$ is:

$$x^1_A = \frac{x'^1_A + vt'}{\sqrt{1 - v^2/c^2}}, \qquad x^1_B = \frac{x'^1_B + vt'}{\sqrt{1 - v^2/c^2}}. \tag{1.31}$$

If we denote by $l \equiv \Delta x' = x'^1_A - x'^1_B$ the length of the rod in the frame of reference K', it follows by subtracting the two above equations that

$$l = l_0\sqrt{1 - v^2/c^2}. \tag{1.32}$$

This equation represents the Lorentz contraction and shows that a rod has its largest length in a system at rest.

Let V_0 and V denote the volumes of a body in the frames K and K', respectively, which are connected by the Lorentz transformations (1.30). The volume contraction of the proper volume V_0 is then given by:

$$V = V_0\sqrt{1 - v^2/c^2}. \tag{1.33}$$

The general transformation law (1.24) for two inertial systems in which one of them is moving with a relative velocity with respect to the other that is not parallel to the x axis but has an arbitrary direction, is given by

$$(\Lambda^\alpha_\beta) = \begin{pmatrix} \gamma & -\gamma\frac{v^1}{c} & -\gamma\frac{v^2}{c} & -\gamma\frac{v^3}{c} \\ -\gamma\frac{v^1}{c} & 1+\frac{(\gamma-1)v^1v^1}{|\mathbf{v}|^2} & \frac{(\gamma-1)v^1v^2}{|\mathbf{v}|^2} & \frac{(\gamma-1)v^1v^3}{|\mathbf{v}|^2} \\ -\gamma\frac{v^2}{c} & \frac{(\gamma-1)v^2v^1}{|\mathbf{v}|^2} & 1+\frac{(\gamma-1)v^2v^2}{|\mathbf{v}|^2} & \frac{(\gamma-1)v^2v^3}{|\mathbf{v}|^2} \\ -\gamma\frac{v^3}{c} & \frac{(\gamma-1)v^3v^1}{|\mathbf{v}|^2} & \frac{(\gamma-1)v^3v^2}{|\mathbf{v}|^2} & 1+\frac{(\gamma-1)v^3v^3}{|\mathbf{v}|^2} \end{pmatrix}, \tag{1.34}$$

where we have introduced the usual abbreviation

$$\gamma = \frac{1}{\sqrt{1-|\mathbf{v}|^2/c^2}}. \tag{1.35}$$

The components of Λ^α_β can be written in the compact form:

$$\Lambda^0_0 = \gamma, \quad \Lambda^i_0 = -\gamma\frac{v^i}{c}, \quad \Lambda^0_i = -\gamma\frac{v^j}{c}\delta_{ij}, \quad \Lambda^i_j = \delta^i_j + (\gamma-1)\frac{v^iv^k}{|\mathbf{v}|^2}\delta_{jk}. \tag{1.36}$$

Problems

1.2.1 Prove that if a transformation law between two inertial systems x'^α and x^β leaves invariant $\eta_{\alpha\beta}x'^\alpha x'^\beta = \eta_{\gamma\delta}x^\gamma x^\delta$ for any choice of x^δ, then it leaves $\eta_{\gamma\delta}x^\gamma y^\delta$ also invariant for any choice of x^δ and y^δ. (Hint: apply the assumed property to $x^\delta + y^\delta$, x^δ, and $y^\delta \ldots$).

1.2.2 Show that if a transformation law between two inertial systems x'^α and x^β leaves invariant $\eta_{\alpha\beta}x'^\alpha x'^\beta = \eta_{\gamma\delta}x^\gamma x^\delta$ for any choice of x^δ, then it must be linear and homogeneous and hence given by $x'^\alpha = \Lambda^\alpha_\beta x^\beta$. (Hint: take four quadruplets $y^{\alpha(\beta)}$ where β is fixed for any quadruplet and takes the values 0,1,2,3. Then apply the result of the previous problem to each quadruplet to obtain $\eta_{\gamma\delta}x'^\gamma y'^{\delta(\beta)} = \eta_{\gamma\delta}x^\gamma y^{\delta(\beta)}$; then choose $y'^{\delta(\beta)} = \eta^{\delta\beta}$ and denote by Λ^β_γ the combination $\eta_{\gamma\delta}y^{\delta(\beta)}$).

1.2.3 Show that the transformation matrix $\mathbf{\Lambda}$ reduces to (1.17) when $x^2 = x'^2$ and $x^3 = x'^3$.

1.3 Tensors in Minkowski spaces

A four-vector A is a quantity which can be described by four components A^α with respect to a given reference frame, that transform according to

$$A'^\alpha = \Lambda^\alpha_\beta A^\beta, \tag{1.37}$$

when the transformation law between two coordinate systems is given by (1.14). By using the transformation matrix (1.24) it follows that:

$$A'^0 = \frac{A^0 - A^1v/c}{\sqrt{1-v^2/c^2}}, \quad A'^1 = \frac{A^1 - A^0v/c}{\sqrt{1-v^2/c^2}}, \quad A'^2 = A^2, \quad A'^3 = A^3. \tag{1.38}$$

In general one introduces two representations to the components of a four-vector: the contravariant components denoted by A^α, and the covariant components denoted by A_α. The covariant components of a four-vector are related to the contravariant components through

$$A_\alpha = \eta_{\alpha\beta} A^\beta, \qquad \text{that is} \quad A_0 = A^0, \quad A_i = -A^i. \tag{1.39}$$

The contravariant and covariant components can also be represented as

$$(A^\alpha) = (A^0, \mathbf{A}) = (A_0, \mathbf{A}), \qquad \text{and} \quad (A_\alpha) = (A_0, -\mathbf{A}). \tag{1.40}$$

The inverse of the metric tensor matrix $(\eta_{\alpha\beta})$ is the matrix $(\eta^{\alpha\beta})$ which we have already met in the previous section such that

$$\eta_{\alpha\beta}\eta^{\beta\gamma} = \delta_\alpha^\gamma. \tag{1.41}$$

In the above equation δ_α^γ is the Kronecker symbol defined by (1.13). It is easy to verify from (1.39) and (1.41) that the following relationships hold (the first one has been given in a more explicit form in the previous section):

$$((\eta^{\alpha\beta})) = ((\eta_{\alpha\beta})) \qquad \text{and} \qquad A^\alpha = \eta^{\alpha\beta} A_\beta. \tag{1.42}$$

The transformation law of a covariant four-vector is

$$A'_\alpha = \eta_{\alpha\beta} A'^\beta \stackrel{(1.37)}{=} \eta_{\alpha\beta}\Lambda^\beta_\gamma A^\gamma \stackrel{(1.42)}{=} \eta_{\alpha\beta}\Lambda^\beta_\gamma \eta^{\gamma\delta} A_\delta = \bar{\Lambda}^\delta_\alpha A_\delta, \tag{1.43}$$

where we have introduced the transformation matrix $\bar{\mathbf{\Lambda}}$ of $x^\alpha = \bar{\Lambda}^\alpha_\beta x'^\beta$ that is given by

$$\bar{\Lambda}^\delta_\alpha = \eta_{\alpha\beta}\Lambda^\beta_\gamma \eta^{\gamma\delta}. \tag{1.44}$$

The matrix $\bar{\mathbf{\Lambda}}$ is the inverse matrix of $\mathbf{\Lambda}$ since

$$\bar{\Lambda}^\alpha_\beta \Lambda^\beta_\gamma = \eta_{\beta\delta}\Lambda^\delta_\epsilon \eta^{\epsilon\alpha}\Lambda^\beta_\gamma \stackrel{(1.16)}{=} \eta_{\epsilon\gamma}\eta^{\epsilon\alpha} = \delta^\alpha_\gamma. \tag{1.45}$$

Further the matrices $\bar{\mathbf{\Lambda}}$ and $\mathbf{\Lambda}$ are symmetric and

$$\Lambda^\alpha_\beta \bar{\Lambda}^\beta_\gamma = \delta^\alpha_\gamma. \tag{1.46}$$

As a consequence we have that

$$\bar{\mathbf{\Lambda}}(\mathbf{v}) = \mathbf{\Lambda}^{-1}(\mathbf{v}) = \mathbf{\Lambda}(-\mathbf{v}). \tag{1.47}$$

Hence the transformation matrix of the contravariant four-vectors is $\mathbf{\Lambda}$ while the transformation matrix of the covariant four-vectors is $\bar{\mathbf{\Lambda}}$. The matrix $\bar{\mathbf{\Lambda}}$ is also the matrix of the inverse transformation from A'^α to A^α since the multiplication of (1.37) by $\bar{\Lambda}^\gamma_\alpha$ leads to

$$\bar{\Lambda}^\gamma_\alpha A'^\alpha = \bar{\Lambda}^\gamma_\alpha \Lambda^\alpha_\beta A^\beta = \delta^\gamma_\beta A^\beta = A^\gamma. \tag{1.48}$$

The scalar product of two four-vectors A^α and B_α is defined by

$$A^\alpha B_\alpha = A^0 B_0 + A^i B_i = A^0 B_0 - \mathbf{A}\cdot\mathbf{B}, \tag{1.49}$$

where $\mathbf{A}\cdot\mathbf{B}$ is the usual scalar product of two vectors in a three-dimensional space. Further the scalar product is an invariant:

$$A'^\alpha B'_\alpha = \Lambda^\alpha_\beta A^\beta \bar{\Lambda}^\gamma_\alpha B_\gamma = \delta^\gamma_\beta A^\beta B_\gamma = A^\alpha B_\alpha. \tag{1.50}$$

The gradient $\partial/\partial x^\alpha$ with respect to the contravariant coordinates transforms like a covariant four-vector:

$$\frac{\partial}{\partial x'^\alpha} = \frac{\partial x^\beta}{\partial x'^\alpha}\frac{\partial}{\partial x^\beta} = \bar{\Lambda}^\beta_\alpha \frac{\partial}{\partial x^\beta}. \tag{1.51}$$

On the other hand, the gradient $\partial/\partial x_\alpha$ with respect to the covariant coordinates transforms like a contravariant four-vector:

$$\frac{\partial}{\partial x'_\alpha} = \frac{\partial x^\beta}{\partial x'_\alpha}\frac{\partial}{\partial x_\beta} = \Lambda^\alpha_\beta \frac{\partial}{\partial x_\beta}. \tag{1.52}$$

The components of the two above gradients are:

$$\left(\frac{\partial}{\partial x^\alpha}\right) = \left(\frac{\partial}{\partial x^0}, \vec{\nabla}\right), \qquad \text{and} \qquad \left(\frac{\partial}{\partial x_\alpha}\right) = \left(\frac{\partial}{\partial x_0}, -\vec{\nabla}\right), \tag{1.53}$$

where $\vec{\nabla}$ is the usual gradient in a three-dimensional space. We shall also use the notation

$$\frac{\partial}{\partial x^\alpha} \equiv \partial_\alpha, \qquad \text{and} \qquad \frac{\partial}{\partial x_\alpha} \equiv \partial^\alpha. \tag{1.54}$$

The d'Alembertian is also an invariant:

$$\Box \equiv \partial_\alpha \partial^\alpha = \eta^{\alpha\beta}\partial_\alpha\partial_\beta = \frac{1}{c^2}\frac{\partial^2}{\partial t^2} - \nabla^2, \tag{1.55}$$

where $\nabla^2 = \partial^2/\partial x_i \partial x_i$ denotes the Laplacian.

A quantity $\mathbf{T}$ which has p contravariant and q covariant components is called a tensor if its components transform according to

$$T'^{\alpha_1\ldots\alpha_p}_{\beta_1\ldots\beta_q} = (\text{sgn}(\det\mathbf{\Lambda}))^n \Lambda^{\alpha_1}_{\gamma_1}\ldots\Lambda^{\alpha_p}_{\gamma_p}\bar{\Lambda}^{\epsilon_1}_{\beta_1}\ldots\bar{\Lambda}^{\epsilon_q}_{\beta_q} T^{\gamma_1\ldots\gamma_p}_{\epsilon_1\ldots\epsilon_q}, \quad (n = 0, 1) \tag{1.56}$$

for a change of the system of coordinates given by $x'^\alpha = \Lambda^\alpha_\beta x^\beta$. In the above equation sgn(det$\mathbf{\Lambda}$) is the sign of the determinant of the matrix $\mathbf{\Lambda}$. If $n = 0$ the tensor is called absolute, whereas if $n = 1$ it is called axial or pseudo-tensor. The components of a pseudo-tensor transform like an absolute tensor for proper rotations but not for reflections of the coordinate axis.

For simplicity absolute tensors are just called tensors. The contravariant and the covariant components of a second order tensor are transformed according to

$$T'^{\alpha\beta} = \Lambda^{\alpha}_{\gamma}\Lambda^{\beta}_{\epsilon}T^{\gamma\epsilon} \qquad \text{and} \qquad T'_{\alpha\beta} = \bar{\Lambda}^{\gamma}_{\alpha}\bar{\Lambda}^{\epsilon}_{\beta}T_{\gamma\epsilon}, \tag{1.57}$$

while their inverse transformations are given by

$$T^{\alpha\beta} = \bar{\Lambda}^{\alpha}_{\gamma}\bar{\Lambda}^{\beta}_{\epsilon}T'^{\gamma\epsilon} \qquad \text{and} \qquad T_{\alpha\beta} = \Lambda^{\gamma}_{\alpha}\Lambda^{\epsilon}_{\beta}T'_{\gamma\epsilon}. \tag{1.58}$$

A second order tensor $T_{\alpha\beta}$ can be decomposed into the sum of a symmetric tensor $T^{S}_{\alpha\beta}$ and an antisymmetric tensor $T^{A}_{\alpha\beta}$:

$$T_{\alpha\beta} = T^{S}_{\alpha\beta} + T^{A}_{\alpha\beta}, \tag{1.59}$$

where the symmetric tensor and the antisymmetric tensor are given by

$$T^{S}_{\alpha\beta} = \frac{1}{2}\left(T_{\alpha\beta} + T_{\beta\alpha}\right), \qquad T^{A}_{\alpha\beta} = \frac{1}{2}\left(T_{\alpha\beta} - T_{\beta\alpha}\right). \tag{1.60}$$

Further it is usual to introduce a traceless symmetric tensor

$$T^{TR}_{\alpha\beta} = T^{S}_{\alpha\beta} - \frac{1}{4}T^{\gamma}_{\gamma}\eta_{\alpha\beta}. \tag{1.61}$$

The components of the tensors $\eta_{\alpha\beta}$, $\eta^{\alpha\beta}$ and δ^{α}_{β} are the same in all coordinate systems, i.e., $\eta'^{\alpha\beta} = \eta^{\alpha\beta}$ and $\delta'^{\alpha}_{\beta} = \delta^{\alpha}_{\beta}$.

The Levi–Civita tensor defined by

$$\epsilon^{\alpha\beta\gamma\delta} = \begin{cases} +1 & \text{if } (\alpha,\, \beta,\, \gamma,\, \delta) \text{ is an even permutation of } (0,1,2,3), \\ -1 & \text{if } (\alpha,\, \beta,\, \gamma,\, \delta) \text{ is an odd permutation of } (0,1,2,3), \\ 0 & \text{otherwise,} \end{cases} \tag{1.62}$$

is a pseudo-tensor with $\epsilon^{0123} = 1$ and $\epsilon_{0123} = -\epsilon^{0123}$. However the product $\epsilon^{\alpha\beta\gamma\delta}\epsilon_{\mu\nu\kappa\lambda}$ is an absolute tensor of eighth order whose components can be expressed as combinations of the unit tensor δ^{β}_{α}:

$$\epsilon^{\alpha\beta\gamma\delta}\epsilon_{\mu\nu\kappa\lambda} = -\begin{vmatrix} \delta^{\alpha}_{\mu} & \delta^{\alpha}_{\nu} & \delta^{\alpha}_{\kappa} & \delta^{\alpha}_{\lambda} \\ \delta^{\beta}_{\mu} & \delta^{\beta}_{\nu} & \delta^{\beta}_{\kappa} & \delta^{\beta}_{\lambda} \\ \delta^{\gamma}_{\mu} & \delta^{\gamma}_{\nu} & \delta^{\gamma}_{\kappa} & \delta^{\gamma}_{\lambda} \\ \delta^{\delta}_{\mu} & \delta^{\delta}_{\nu} & \delta^{\delta}_{\kappa} & \delta^{\delta}_{\lambda} \end{vmatrix}. \tag{1.63}$$

The successive contractions of two indices of (1.63) lead to

$$\epsilon^{\alpha\beta\gamma\delta}\epsilon_{\mu\nu\kappa\delta} = -\begin{vmatrix} \delta^{\alpha}_{\mu} & \delta^{\alpha}_{\nu} & \delta^{\alpha}_{\kappa} \\ \delta^{\beta}_{\mu} & \delta^{\beta}_{\nu} & \delta^{\beta}_{\kappa} \\ \delta^{\gamma}_{\mu} & \delta^{\gamma}_{\nu} & \delta^{\gamma}_{\kappa} \end{vmatrix}, \qquad \epsilon^{\alpha\beta\gamma\delta}\epsilon_{\mu\nu\gamma\delta} = -2\left(\delta^{\alpha}_{\mu}\delta^{\beta}_{\nu} - \delta^{\alpha}_{\nu}\delta^{\beta}_{\mu}\right), \tag{1.64}$$

$$\epsilon^{\alpha\beta\gamma\delta}\epsilon_{\mu\beta\gamma\delta} = -6\delta^{\alpha}_{\mu}, \qquad \epsilon^{\alpha\beta\gamma\delta}\epsilon_{\alpha\beta\gamma\delta} = -24. \tag{1.65}$$

The determinant of the covariant components of a tensor $|T_{\sigma\tau}| \equiv \det\mathbf{T}$ is expressed in terms of the Levi–Civita tensor as

$$\begin{cases} \epsilon^{\alpha\beta\gamma\delta}T_{\mu\alpha}T_{\nu\beta}T_{\kappa\gamma}T_{\lambda\delta} = -(\det\mathbf{T})\epsilon_{\mu\nu\kappa\lambda}, \\ \epsilon^{\alpha\beta\gamma\delta}\epsilon^{\mu\nu\kappa\lambda}T_{\mu\alpha}T_{\nu\beta}T_{\kappa\gamma}T_{\lambda\delta} = 24(\det\mathbf{T}). \end{cases} \tag{1.66}$$

The corresponding pseudo-tensor $\boldsymbol{\epsilon}$ in a three-dimensional space is ϵ_{ijk}. The product $\epsilon_{ijk}\epsilon^{lmn}$ is also an absolute tensor whose components, given in terms of the unit tensor δ_i^j, are:

$$\epsilon_{ijk}\epsilon^{lmn} = \begin{vmatrix} \delta_i^l & \delta_i^m & \delta_i^n \\ \delta_j^l & \delta_j^m & \delta_j^n \\ \delta_k^l & \delta_k^m & \delta_k^n \end{vmatrix}. \tag{1.67}$$

The successive contractions of (1.67) are

$$\epsilon_{ijk}\epsilon^{lmk} = \left(\delta_i^l\delta_j^m - \delta_i^m\delta_j^l\right), \qquad \epsilon_{ijk}\epsilon^{ljk} = 2\delta_i^l, \qquad \epsilon_{ijk}\epsilon^{ijk} = 6. \tag{1.68}$$

An absolute vector in the three-dimensional space is called a polar vector, whereas a pseudo-vector is an axial vector. If $\boldsymbol{\omega}$ is an axial vector it is dual to an antisymmetric tensor $\mathbf{W}$:

$$\omega_i = -\frac{1}{2}\epsilon_{ijk}W^{jk}, \qquad W_{ij} = \epsilon_{ikj}\omega^k. \tag{1.69}$$

An antisymmetric tensor $A^{\alpha\beta} = -A^{\beta\alpha}$ in four dimensions can be represented by a polar vector $\mathbf{p}$ and an axial vector $\mathbf{a}$ in three dimensions as follows:

$$\left(A^{\alpha\beta}\right) = \begin{pmatrix} 0 & p^1 & p^2 & p^3 \\ -p^1 & 0 & -a^3 & a^2 \\ -p^2 & a^3 & 0 & -a^1 \\ -p^3 & -a^2 & a^1 & 0 \end{pmatrix}, \tag{1.70}$$

$$\left(A_{\alpha\beta}\right) = \begin{pmatrix} 0 & -p^1 & -p^2 & -p^3 \\ p^1 & 0 & -a^3 & a^2 \\ p^2 & a^3 & 0 & -a^1 \\ p^3 & -a^2 & a^1 & 0 \end{pmatrix}, \tag{1.71}$$

or in a compact form

$$\left(A^{\alpha\beta}\right) = (\mathbf{p}, \mathbf{a}), \qquad \left(A_{\alpha\beta}\right) = (-\mathbf{p}, \mathbf{a}). \tag{1.72}$$

We proceed now to analyze the invariance property of volume elements that will be needed in the next chapters. Let A^α be the components of a four-vector, i.e., four numbers satisfying the transformation law (1.37). The volume element $dA^0dA^1dA^2dA^3$ in four-dimensional space spanned by $\mathbf{A}$ is a scalar invariant. Indeed, it is easy to show that the modulus of the Jacobian

$$J = \frac{\partial(A'^0, A'^1, A'^2, A'^3)}{\partial(A^0, A^1, A^2, A^3)} \tag{1.73}$$

is equal to one. Hence it follows that

$$dA'^0dA'^1dA'^2dA'^3 = |J|dA^0dA^1dA^2dA^3 = dA^0dA^1dA^2dA^3. \tag{1.74}$$

In particular we have that

$$cdtdV = dx^0dx^1dx^2dx^3 = dx'^0dx'^1dx'^2dx'^3. \tag{1.75}$$

The elements

$$dA^0dA^1dA^2, \quad dA^0dA^1dA^3, \quad dA^0dA^2dA^3, \quad dA^1dA^2dA^3 \tag{1.76}$$

are not scalar invariants, but the ratios

$$\frac{dA^0dA^1dA^2}{A_3}, \quad \frac{dA^0dA^1dA^3}{A_2}, \quad \frac{dA^0dA^2dA^3}{A_1}, \quad \frac{dA^1dA^2dA^3}{A_0}, \tag{1.77}$$

are scalar invariants provided that $A^\alpha A_\alpha$ = constant. We proceed to show that $dA^1dA^2dA^3/A_0$ is a scalar invariant if the latter condition is satisfied. If we take into account the fact that, because of that condition, A_0 is implicitly a function of A_i (i=1,2,3), the transformation law between the elements (1.76) is given by

$$dA'^1dA'^2dA'^3 = |J|dA^1dA^2dA^3 \tag{1.78}$$

where according to (1.38)

$$J = \frac{\partial(A'^1, A'^2, A'^3)}{\partial(A^1, A^2, A^3)} = \begin{vmatrix} \gamma\left(1 - \frac{v}{c}\frac{\partial A^0}{\partial A^1}\right) & -\gamma\frac{v}{c}\frac{\partial A^0}{\partial A^2} & -\gamma\frac{v}{c}\frac{\partial A^0}{\partial A^3} \\ 0 & 1 & 0 \\ 0 & 0 & 1 \end{vmatrix}$$

$$= \gamma\left(1 - \frac{v}{c}\frac{\partial A^0}{\partial A^1}\right). \tag{1.79}$$

By considering $A^\alpha A_\alpha$ = constant and by using the property

$$0 = A^\alpha\frac{\partial A_\alpha}{\partial A^i} = A_\alpha\frac{\partial A^\alpha}{\partial A^i} = A_0\frac{\partial A^0}{\partial A^i} + A_k\frac{\partial A^k}{\partial A^i} = A_0\frac{\partial A^0}{\partial A^i} + A_k\delta_i^k$$

$$= A_0\frac{\partial A^0}{\partial A^i} + A_i \quad (i = 1, 2, 3), \tag{1.80}$$

it follows that

$$\frac{\partial A^0}{\partial A^i} = -\frac{A_i}{A_0} \quad (i = 1, 2, 3). \tag{1.81}$$

Now the substitution of (1.81) into (1.79) yields

$$J = \frac{\gamma}{A_0}\left(A_0 + \frac{v}{c}A_1\right) = \frac{A_0'}{A_0}. \tag{1.82}$$

This last equation is introduced in (1.78) and it follows that

$$\frac{dA'^1dA'^2dA'^3}{A_0'} = \frac{dA^1dA^2dA^3}{A_0}. \tag{1.83}$$

The above equation shows that the ratios (1.77) are scalar invariants.

Problems

1.3.1 Show by direct construction that the determinant of the mixed components of a second order tensor in a three-dimensional space $|T^\alpha_\beta| \equiv \det \mathbf{T}$ can be expressed by

$$\det \mathbf{T} = \epsilon^{ijk} T^1_i T^2_j T^3_k = \frac{1}{3!}\epsilon^{ijk}\epsilon_{lmn} T^l_i T^m_j T^n_k, \qquad \text{or}$$
$$\epsilon^{lmn} \det \mathbf{T} = \epsilon^{ijk} T^l_i T^m_j T^n_k.$$

1.3.2 By successive differentiation of the first relationship of problem 1.3.1 show that

$$\frac{\partial^3 \det \mathbf{T}}{\partial T^l_i \partial T^m_j \partial T^n_k} = \epsilon^{ijk}\epsilon_{lmn}.$$

Further show that the successive differentiation of the second relationship of problem 1.3.1 together with the relationship just found leads to (1.67).

1.3.3 By differentiating the second relationship of problem 1.3.1 show that

$$\frac{\partial \det \mathbf{T}}{\partial T^j_i} = \det \mathbf{T}(T^{-1})^i_j,$$

where $(T^{-1})^i_j$ is the inverse of $\mathbf{T}$, i.e., $(T^{-1})^i_j T^j_k = \delta^i_k$.

1.3.4 By using the relationship $(T^{-1})^i_j T^j_k = \delta^i_k$ show that

$$\frac{\partial (T^{-1})^i_j}{\partial T^k_l} = -(T^{-1})^i_k (T^{-1})^l_j.$$

1.4 Relativistic mechanics

1.4.1 Four-velocity

The reason why four-tensors (in particular four-vectors, i.e., tensors of rank 1) are useful, is that we can associate such quantities with typical classical quantities and obtain relations that automatically transform correctly under a Lorentz transformation. The first example is velocity, as we proceed to show.

Let us first analyze the relativistic transformation of the velocities when two frames of reference are connected by the Lorentz transformations (1.25). In this case we obtain the differentials

$$dx'^1 = \frac{dx^1 - v\,dt}{\sqrt{1-v^2/c^2}}, \quad dx'^2 = dx^2, \quad dx'^3 = dx^3, \quad dt' = \frac{dt - dx^1 v/c^2}{\sqrt{1-v^2/c^2}}. \tag{1.84}$$

Since the velocities are defined by

$$\mathbf{v} = \frac{d\mathbf{x}}{dt}, \qquad \mathbf{v}' = \frac{d\mathbf{x}'}{dt'}, \tag{1.85}$$

the transformation laws of the components of the velocities are obtained by dividing the first of the three equations (1.84) by the fourth yielding

$$v'^1 = \frac{v^1 - v}{1 - v^1 v/c^2}, \qquad v'^2 = \frac{v^2\sqrt{1-(v/c)^2}}{1 - v^1 v/c^2}, \qquad v'^3 = \frac{v^3\sqrt{1-(v/c)^2}}{1 - v^1 v/c^2}. \quad (1.86)$$

For small velocities $v \ll c$ we get the classical expression for the transformation of velocities:

$$v'^1 = v^1 - v, \qquad v'^2 = v^2, \qquad v'^3 = v^3. \quad (1.87)$$

When working in space-time it is convenient to define the four-velocity of a particle U^α as

$$U^\alpha = \frac{dx^\alpha}{d\tau}, \quad (1.88)$$

where $d\tau = ds/c = dt\sqrt{1 - v^2/c^2}$ is the proper time, which is a scalar invariant. Since x^α is a four-vector, U^α is called a four-velocity. The contravariants and covariants components of the four-velocity are given respectively by:

$$(U^\alpha) = \left(\frac{c}{\sqrt{1-v^2/c^2}}, \frac{\mathbf{v}}{\sqrt{1-v^2/c^2}}\right), \quad (1.89)$$

$$(U_\alpha) = \left(\frac{c}{\sqrt{1-v^2/c^2}}, \frac{-\mathbf{v}}{\sqrt{1-v^2/c^2}}\right). \quad (1.90)$$

From the above equations it is easy to verify that $U^\alpha U_\alpha = c^2$.

Problem

1.4.1.1 Show that the transformation laws of the components of the velocity are given by (1.86).

1.4.2 Minkowski force

The four-acceleration ω^α is defined in the same manner as the four-velocity, that is in terms of the derivatives of the four-velocity with respect to the proper time:

$$\omega^\alpha = \frac{dU^\alpha}{d\tau}. \quad (1.91)$$

The contravariant components of the four-acceleration ω^α are:

$$(\omega^\alpha) = \left(\frac{\mathbf{a}\cdot\mathbf{v}}{c(1-v^2/c^2)^2}, \frac{\mathbf{a} + \mathbf{v}\times(\mathbf{v}\times\mathbf{a})/c^2}{(1-v^2/c^2)^2}\right), \quad (1.92)$$

where $a^i = dv^i/dt$ is the usual acceleration in the three-dimensional space.

The four-velocity and the four-acceleration are perpendicular to each other, as is the case for any four-vector A^α such that $A^\alpha A_\alpha = \text{const.}$, and its derivative. One can also use (1.90) and (1.92) to verify that

$$U_\alpha \omega^\alpha = 0. \tag{1.93}$$

The rest mass of a particle m is a scalar invariant; hence its product with the four-velocity

$$p^\alpha = mU^\alpha \tag{1.94}$$

defines a momentum four-vector. The contravariant components of the momentum four-vector p^α are

$$(p^\alpha) = \left(\frac{mc}{\sqrt{1 - v^2/c^2}}, \frac{m\mathbf{v}}{\sqrt{1 - v^2/c^2}} \right) = \left(\frac{E}{c}, \mathbf{p} \right) = (p^0, \mathbf{p}), \tag{1.95}$$

where the energy of a particle is given by

$$E = cp^0 = \frac{mc^2}{\sqrt{1 - v^2/c^2}}. \tag{1.96}$$

The condition that the velocity of the particle is zero $\mathbf{v} = \mathbf{0}$ implies that the rest energy of a particle is $E = mc^2$. This is the energy equivalent of a mass m, the famous Einstein formula. The difference between the general value of E given by (1.96) and the rest energy mc^2 is the kinetic energy. Further for small velocities $v \ll c$ we have

$$E - mc^2 \approx \frac{1}{2} mv^2, \tag{1.97}$$

which is the classical expression for the kinetic energy of a particle.

The transformation laws for the components of the momentum four-vector can be obtained from (1.24), (1.37) and (1.95) yielding

$$E = \frac{E' + vp'^1}{\sqrt{1 - v^2/c^2}}, \quad p^1 = \frac{p'^1 + E'v/c^2}{\sqrt{1 - v^2/c^2}}, \quad p^2 = p'^2, \quad p^3 = p'^3. \tag{1.98}$$

The condition $U_\alpha U^\alpha = c^2$ and the definition (1.94) give the result that the scalar product of the momentum four-vector with itself is

$$p_\alpha p^\alpha = m^2 c^2. \tag{1.99}$$

We remark that the concept of force has not such a clear status in relativity, since there are no such general properties as the principle of action and reaction (in this sense, the forces exerted by magnetic fields are non-classical). Thus there is a certain amount of arbitrariness when defining the relativistic counterpart of the classical force. Usually one defines the Minkowski force through

$$K^\alpha = \frac{dp^\alpha}{d\tau} = m\frac{dU^\alpha}{d\tau} = m\omega^\alpha. \tag{1.100}$$

From (1.100), (1.92) and (1.95) it follows that the contravariant components of the Minkowski force are

$$(K^\alpha) = \left(\frac{m\mathbf{a}\cdot\mathbf{v}}{c(1-v^2/c^2)^2}, \frac{m[\mathbf{a}+\mathbf{v}\times(\mathbf{v}\times\mathbf{a})/c^2]}{(1-v^2/c^2)^2}\right)$$

$$= \left(\frac{\mathbf{F}\cdot\mathbf{v}}{c(1-v^2/c^2)^{\frac{1}{2}}}, \frac{\mathbf{F}}{(1-v^2/c^2)^{\frac{1}{2}}}\right), \tag{1.101}$$

where $F^i = dp^i/dt$ is the non-relativistic force. It is clear that the condition $K^\alpha U_\alpha = 0$ holds and this makes the concept of Minkowski force occasionally useful, as we shall see in the next chapter.

Problems

1.4.2.1 Show that the contravariant components of the four-acceleration are given by (1.92).

1.4.2.2 Show that any four-vector A^α and its derivative are perpendicular to each other provided that $A^\alpha A_\alpha$ =constant.

1.4.2.3 Obtain the transformations laws (1.98) for the components of the momentum four-vector.

1.4.3 Elastic collisions

We consider an elastic collision between two identical particles of rest mass m, one of which is labeled by an index $*$ to distinguish it from the other (unlabeled). Before the collision the two particles have momenta $\mathbf{p}$, $\mathbf{p}_*$ and energies cp^0, cp^0_*, where according to (1.99)

$$p^0 = \sqrt{|\mathbf{p}|^2 + m^2c^2} \qquad \text{and} \qquad p^0_* = \sqrt{|\mathbf{p}_*|^2 + m^2c^2}. \tag{1.102}$$

The conservation laws of momentum and energy can be written in terms of the momentum four-vectors:

$$p^\alpha + p^\alpha_* = p'^\alpha + p'^\alpha_*, \tag{1.103}$$

where the quantities p'^α, p'^α_* denote the values of the momentum four-vectors after the collision.

Equation (1.103) represents a collisional invariant and one can further define the invariants:

$$s = (p^\alpha + p^\alpha_*)(p_\alpha + p_{*\alpha}) = (p'^\alpha + p'^\alpha_*)(p'_\alpha + p'_{*\alpha}), \tag{1.104}$$

$$t = (p^\alpha - p'^\alpha)(p_\alpha - p'_\alpha) = (p^\alpha_* - p'^\alpha_*)(p_{*\alpha} - p'_{*\alpha}), \tag{1.105}$$

$$u = (p^\alpha - p'^\alpha_*)(p_\alpha - p'_{*\alpha}) = (p^\alpha_* - p'^\alpha)(p_{*\alpha} - p'_\alpha), \tag{1.106}$$

which are called the Mandelstam variables. Only two of these invariants are linearly independent since the relationship holds that

$$s+t+u=4m^2c^2. \tag{1.107}$$

The invariants s and t have the following physical interpretation. We consider a Lorentz frame where the spatial components of the total momentum four-vector vanishes, i.e.,

$$\mathbf{p}+\mathbf{p}_*=\mathbf{p}'+\mathbf{p}'_*=\mathbf{0}. \tag{1.108}$$

A system where (1.108) holds is called a center-of-mass system. From (1.102) and (1.108) we have that

$$p^0=p_*^0=p'^0=p_*'^0. \tag{1.109}$$

Now by using the definition of the invariant s we get

$$\begin{aligned} s&=(p^\alpha+p_*^\alpha)(p_\alpha+p_{*\alpha})=2m^2c^2+2p_*^\alpha p_\alpha=(2p^0)^2\\ &=(p^0+p_*^0)^2=\frac{1}{c^2}(E+E_*)^2. \end{aligned} \tag{1.110}$$

Hence the Mandelstam variable s is the square of the energy in the center-of-mass system divided by c^2. On the other hand it follows from the definition of the invariant t (1.105) that

$$t=(p^\alpha-p'^\alpha)(p_\alpha-p'_\alpha)=2m^2c^2-2p^\alpha p'_\alpha. \tag{1.111}$$

If we use the relation

$$\begin{aligned} p^\alpha p'_\alpha&=p^0p'_0-\mathbf{p}\cdot\mathbf{p}'=(p^0)^2-|\mathbf{p}|^2\cos\Theta\\ &\overset{(1.102)}{=}(p^0)^2-[(p^0)^2-m^2c^2]\cos\Theta, \end{aligned} \tag{1.112}$$

where Θ is the scattering angle in the center-of-mass system, (1.111) reduces to

$$\cos\Theta=1+\frac{2t}{s-4m^2c^2}. \tag{1.113}$$

Hence t is related to the scattering angle Θ in the center-of-mass system.

The representation of the free parameters in a collision by the invariants s, t, u is by no means unique. One might, e.g., introduce a scalar C and a vector with components N^α, such that

$$p'_\alpha=p_\alpha+CN_\alpha\,;\quad p'_{*\alpha}=p_{*\alpha}-CN_\alpha\,;\quad N_\alpha N^\alpha=-1. \tag{1.114}$$

Then, since $p'_\alpha p'^\alpha=p_\alpha p^\alpha=p'_{*\alpha}p'^\alpha_*=p_{*\alpha}p_*^\alpha=m^2c^2$:

$$C=2N^\alpha p_\alpha=-2N^\alpha p_{*\alpha}, \tag{1.115}$$

or, equivalently,

$$C = N^{\alpha}(p_{\alpha} - p_{*\alpha}), \quad N^{\alpha}(p_{\alpha} + p_{*\alpha}) = 0. \tag{1.116}$$

If we introduce the time and space components of N_{α} we obtain:

$$N^0(p_0 + p_{*0}) = \mathbf{N} \cdot (\mathbf{p} + \mathbf{p}_*), \tag{1.117}$$

or, in terms of the components of $\overline{p}_{\alpha} = (p_{\alpha} + p_{*\alpha})/2$:

$$N^0 = \frac{\mathbf{N} \cdot \overline{\mathbf{p}}}{\overline{p_0}}. \tag{1.118}$$

This relation gives the time component of N_{α} in terms of the space components. If we let $\mathbf{N} = N\mathbf{n}$, where $\mathbf{n}$ is a unit vector in ordinary space, we have from $(1.114)_3$:

$$N^2 \left(\frac{\mathbf{n} \cdot \overline{\mathbf{p}}}{\overline{p_0}}\right)^2 - N^2 = -1 \tag{1.119}$$

or

$$N = \left[\frac{1}{1 - (\mathbf{n} \cdot \overline{\mathbf{p}}/\overline{p_0})^2}\right]^{\frac{1}{2}}. \tag{1.120}$$

Thus giving the unit vector $\mathbf{n}$ in ordinary space it determines completely the state after a collision from the state before the collision. Formulas simplify if we use the center-of-mass system. Then $N = 1$, $N^0 = 0$, $C = -\mathbf{n} \cdot (\mathbf{p} - \mathbf{p}_*)$ and

$$p'_0 = p_0 = p'_{*0} = p_{*0} = mc, \tag{1.121}$$

$$\mathbf{p}' = \mathbf{p} - \mathbf{n}[\mathbf{n} \cdot (\mathbf{p} - \mathbf{p}_*)], \quad \mathbf{p}'_* = \mathbf{p}_* + \mathbf{n}[\mathbf{n} \cdot (\mathbf{p} - \mathbf{p}_*)]. \tag{1.122}$$

Problems

1.4.3.1 Show that the relationship (1.107) between the Mandelstam variables holds.

1.4.3.2 Show that the scattering angle in the center-of-mass system is given by (1.113).

1.4.3.3 Show that in the center-of-mass system (1.121) and (1.122) hold.

1.4.4 Relative velocity

We shall now determine the relative velocity of two identical particles with rest mass m and velocities $\mathbf{v}$ and $\mathbf{v}_*$ with respect to a rest frame. Let x^{α} be the space-time coordinates of the particle with velocity $\mathbf{v}_*$ in the rest frame, while x'^{α} denotes its space-time coordinates in the frame of the particle which is moving with velocity $\mathbf{v}$. According to the general transformation law (1.14) together with (1.36) we have that

$$\mathbf{x}' = \mathbf{x} - \mathbf{v}t + (\gamma - 1)\frac{\mathbf{v}}{|\mathbf{v}|^2}\left(\mathbf{v}\cdot\mathbf{x} - |\mathbf{v}|^2 t\right), \tag{1.123}$$

$$t' = \gamma\left(t - \frac{\mathbf{v}\cdot\mathbf{x}}{c^2}\right). \tag{1.124}$$

Since $d\mathbf{x}/dt = \mathbf{v}_*$ is the velocity of the particle with label $*$ with respect to a rest frame, we have that $d\mathbf{x}'/dt' = \mathbf{v}_{\rm rel}$ is the velocity of this particle in the frame of the particle which is moving with velocity $\mathbf{v}$. Hence we get from (1.123) and (1.124) the relative velocity:

$$\mathbf{v}_{\rm rel} = \frac{1}{\gamma(1 - \mathbf{v}\cdot\mathbf{v}_*/c^2)}\left[\mathbf{v}_* - \mathbf{v} + (\gamma - 1)\frac{\mathbf{v}}{|\mathbf{v}|^2}\left(\mathbf{v}\cdot\mathbf{v}_* - |\mathbf{v}|^2\right)\right], \tag{1.125}$$

which for small velocities $v/c \ll 1$ and $v_*/c \ll 1$ reduces to the classical definition of the relative velocity $\mathbf{v}_{\rm rel} = \mathbf{v}_* - \mathbf{v}$.

From (1.125) it is easy to obtain the modulus of the relative velocity but here we shall determine it by using another method. We consider a reference frame in which the particle without the label $*$ is at rest ($\mathbf{v} = \mathbf{0}$). In this case we have according to (1.95)

$$p_\alpha p_*^\alpha = \frac{m^2c^2}{\sqrt{1 - v_{\rm rel}^2/c^2}}, \tag{1.126}$$

where the modulus of the relative velocity $v_{\rm rel}$ is equal to the modulus of the velocity of the particle labeled by $*$ in the reference frame where the particle without label is at rest. From (1.126) it is easy to express $v_{\rm rel}$ in terms of $p_\alpha p_*^\alpha$:

$$v_{\rm rel} = c\sqrt{1 - \frac{m^4c^4}{(p_\alpha p_*^\alpha)^2}}. \tag{1.127}$$

On the other hand, the product $p_\alpha p_*^\alpha$ when $\mathbf{v}$ is not zero is given by

$$p_\alpha p_*^\alpha = \frac{m^2c^2 - m^2\mathbf{v}\cdot\mathbf{v}_*}{\sqrt{1 - v^2/c^2}\sqrt{1 - v_*^2/c^2}} = \left(1 - \frac{\mathbf{v}}{c}\cdot\frac{\mathbf{v}_*}{c}\right)p^0 p_*^0. \tag{1.128}$$

Since $p_\alpha p_*^\alpha$ is a scalar invariant we can substitute (1.128) into (1.127) and get the final form of the relative speed in terms of $\mathbf{v}$ and $\mathbf{v}_*$:

$$v_{\rm rel} = \frac{1}{1 - (\mathbf{v}\cdot\mathbf{v}_*)/c^2}\sqrt{(\mathbf{v} - \mathbf{v}_*)^2 - \frac{1}{c^2}(\mathbf{v}\times\mathbf{v}_*)^2}. \tag{1.129}$$

From the above equation one can verify that for small velocities $v/c \ll 1$ and $v_*/c \ll 1$ it follows the classical definition of the relative speed. Further (1.129) is symmetric with respect to $\mathbf{v}$ and $\mathbf{v}_*$; hence the relative speed is independent of the choice of the particle used to define it.

We remark that

$$p_\alpha p_*^\alpha = \frac{1}{2}(p_\alpha + p_{*\alpha})(p^\alpha + p_*^\alpha) - m^2c^2 = \frac{1}{2}(p'_\alpha + p'_{*\alpha})(p'^\alpha + p'^\alpha_*) - m^2c^2 = p'_\alpha p'^\alpha_*. \tag{1.130}$$

Thus $p_\alpha p_*^\alpha$ and hence the relative speed are the same before and after the collision.

In the next chapters we shall use another form of the relative speed known as the Møller relative speed:

$$g_\o = \sqrt{(\mathbf{v} - \mathbf{v}_*)^2 - \frac{1}{c^2}(\mathbf{v} \times \mathbf{v}_*)^2} = v_{\text{rel}} \frac{p_\alpha p_*^\alpha}{p^0 p_*^0}, \tag{1.131}$$

which has a simpler expression but is not Lorentz invariant. It does not preserve its value through a collision (Problem 1.4.4.2) and is only convenient because it simplifies formulas.

Problems

1.4.4.1 By using (1.129) show that $v_{\text{rel}}^2 < c^2$. (Hint: Use the fact that $|\mathbf{v}|^2 < c^2$, $|\mathbf{v}_*|^2 < c^2$ and build the relationship $1 - v_{\text{rel}}^2/c^2$.)

1.4.4.2 Prove that Møller relative speed is not invariant through a collision. (Hint: Use the fact that v_{rel} is invariant and $p^0 p_*^0$ is not, the second statement following from the relation between $p_0 p_*^0$ and $p_\alpha p_*^\alpha$.)

1.5 Electrodynamics in free space

1.5.1 Maxwell equations

Here we assume that the reader is familiar with Maxwell's equations. In the unlikely case that he/she is not, he/she should consider these equations, to be introduced below, as the basic (partial differential) equations describing how the electromagnetic field changes in space and time. They condense the information that charges at rest produce electric fields, and moving charges also produce magnetic fields, whereas there are no magnetic charges (or "mono-poles") that do the same thing with an interchange between the words "electric" and "magnetic".

Let $\mathbf{E}(\mathbf{x},t)$ be the electric field intensity, $\mathbf{B}(\mathbf{x},t)$ the magnetic flux density (or magnetic induction), $\mathbf{D}(\mathbf{x},t)$ the electric displacement, $\mathbf{H}(\mathbf{x},t)$ the magnetic field intensity, $\varrho_q(\mathbf{x},t)$ the volumetric charge density and $\mathbf{I}(\mathbf{x},t)$ the electric current density. The Maxwell equations for these fields are given by:

$$\operatorname{div} \mathbf{D} = \varrho_q, \qquad \operatorname{rot} \mathbf{H} = \mathbf{I} + \frac{\partial \mathbf{D}}{\partial t}, \tag{1.132}$$

$$\operatorname{rot} \mathbf{E} + \frac{\partial \mathbf{B}}{\partial t} = 0, \qquad \operatorname{div} \mathbf{B} = 0. \tag{1.133}$$

In free space the following relationships between the fields **D**, **H** and **E**, **B** hold:

$$\mathbf{D} = \epsilon_0 \mathbf{E}, \qquad \mathbf{H} = \frac{1}{\mu_0}\mathbf{B}. \tag{1.134}$$

In (1.134) ϵ_0 is the permittivity of free space and μ_0 the permeability of free space. The values of these constants, in the usual MKSA system, are:

$$\epsilon_0 = 8.854 \times 10^{-12}\frac{\mathrm{C}^2}{\mathrm{N\,m}^2}, \quad \mu_0 = 4\pi \times 10^{-7}\frac{\mathrm{N}}{\mathrm{A}^2}, \quad \text{with} \quad c^2\epsilon_0\mu_0 = 1. \tag{1.135}$$

Hence in free space the Maxwell equations (1.132) and (1.133) together with (1.134) can be written as:

$$\operatorname{div}\mathbf{E} = \frac{1}{\epsilon_0}\varrho_q = c^2\mu_0\varrho_q, \qquad \operatorname{rot}\mathbf{B} = \mu_0\mathbf{I} + \frac{1}{c^2}\frac{\partial\mathbf{E}}{\partial t}, \tag{1.136}$$

$$\operatorname{rot}\mathbf{E} + \frac{\partial\mathbf{B}}{\partial t} = 0, \qquad \operatorname{div}\mathbf{B} = 0. \tag{1.137}$$

If we differentiate $(1.136)_1$ with respect to the time, take the divergence of $(1.136)_2$ and combine both resulting equations, we get the continuity equation for the charge density:

$$\frac{\partial\varrho_q}{\partial t} + \operatorname{div}\mathbf{I} = 0. \tag{1.138}$$

The electric charge in a volume element dV, given by $dq = \varrho_q dV$, is a scalar invariant. Hence the product $dq\,dx^\alpha$ is a four-vector and we can write:

$$dq\,dx^\alpha = \varrho_q dV\,dx^\alpha = \varrho_q dV\,dt\frac{dx^\alpha}{dt}. \tag{1.139}$$

Since $dV\,c\,dt$ is a scalar invariant (see (1.75)) it follows that $\varrho_q dx^\alpha/dt$ is a four-vector called the current four-vector

$$J^\alpha = \varrho_q\frac{dx^\alpha}{dt}, \tag{1.140}$$

whose contravariant components are

$$(J^\alpha) = (c\varrho_q, \mathbf{I}), \qquad \text{where} \qquad \mathbf{I} = \varrho_q\mathbf{v}. \tag{1.141}$$

The continuity equation for the charge density (1.138) written in terms of the current four-vector is:

$$\frac{\partial J^\alpha}{\partial x^\alpha} = \partial_\alpha J^\alpha = 0. \tag{1.142}$$

The electric field intensity **E** is a polar vector while the magnetic induction **B** is an axial vector. According to (1.70) and (1.71) we can introduce an antisymmetric tensor $F^{\alpha\beta}$ called the electromagnetic field tensor and defined through:

$$(F^{\alpha\beta}) = \begin{pmatrix} 0 & -E^1 & -E^2 & -E^3 \\ E^1 & 0 & -cB^3 & cB^2 \\ E^2 & cB^3 & 0 & -cB^1 \\ E^3 & -cB^2 & cB^1 & 0 \end{pmatrix}, \quad (F^{\alpha\beta}) = (-\mathbf{E}, c\mathbf{B}), \tag{1.143}$$

$$(F_{\alpha\beta}) = \begin{pmatrix} 0 & E^1 & E^2 & E^3 \\ -E^1 & 0 & -cB^3 & cB^2 \\ -E^2 & cB^3 & 0 & -cB^1 \\ -E^3 & -cB^2 & cB^1 & 0 \end{pmatrix}, \qquad (F_{\alpha\beta}) = (\mathbf{E}, c\mathbf{B}). \tag{1.144}$$

In terms of the electromagnetic field tensor $F^{\alpha\beta}$ the Maxwell equations (1.136) and (1.137) can be written, respectively, as:

$$\frac{\partial F^{\alpha\beta}}{\partial x^\beta} = \partial_\beta F^{\alpha\beta} = -\mu_0 c J^\alpha, \tag{1.145}$$

$$\epsilon_{\alpha\beta\gamma\delta}\frac{\partial F^{\gamma\delta}}{\partial x_\beta} = \epsilon_{\alpha\beta\gamma\delta}\partial^\beta F^{\gamma\delta} = 0, \quad \text{or} \quad \partial_\gamma F_{\alpha\beta} + \partial_\alpha F_{\beta\gamma} + \partial_\beta F_{\gamma\alpha} = 0. \tag{1.146}$$

The transformation law of the electromagnetic field tensor is:

$$F'^{\alpha\beta} = \Lambda^\alpha_\gamma \Lambda^\beta_\delta F^{\gamma\delta}, \qquad \text{or} \qquad F^{\alpha\beta} = \bar{\Lambda}^\alpha_\gamma \bar{\Lambda}^\beta_\delta F'^{\gamma\delta}. \tag{1.147}$$

Hence the transformation law for the components of $F^{\alpha\beta}$ follows from (1.36) and (1.147) yielding

$$\mathbf{E}' = \gamma\mathbf{E} - (\gamma - 1)\frac{\mathbf{v}}{|\mathbf{v}|^2}(\mathbf{v}\cdot\mathbf{E}) + \gamma(\mathbf{v}\times\mathbf{B}), \tag{1.148}$$

$$\mathbf{B}' = \gamma\mathbf{B} - (\gamma - 1)\frac{\mathbf{v}}{|\mathbf{v}|^2}(\mathbf{v}\cdot\mathbf{B}) - \frac{\gamma}{c^2}(\mathbf{v}\times\mathbf{E}). \tag{1.149}$$

From the above equations we infer that the electric field intensity and the magnetic induction are not invariant with respect to a change of the frame of reference. In particular **E** and **B** can vanish in one reference frame but be non-zero in another frame of reference.

It is easy to verify that a solution of the Maxwell equation (1.146) is of the form

$$F^{\alpha\beta} = \partial^\alpha A^\beta - \partial^\beta A^\alpha. \tag{1.150}$$

The four-vector A^α is called four-potential and its contravariant components are given by

$$(A^\alpha) = (\phi, c\mathbf{A}), \tag{1.151}$$

where ϕ is the scalar potential and **A** the vector potential. Further one can get from (1.143), (1.150) and (1.151) that

$$\mathbf{E} = -\operatorname{grad}\phi - \frac{\partial\mathbf{A}}{\partial t}, \qquad \mathbf{B} = \operatorname{rot}\mathbf{A}. \tag{1.152}$$

The four-potential A^α is not uniquely determined by (1.150) since the gauge transformation

$$A^\alpha \to A^\alpha + \partial^\alpha\psi, \tag{1.153}$$

where $\psi \equiv \psi(x^\alpha)$ is a scalar function, produces another potential which also satisfies (1.150).

The equation of motion of a particle with electric charge q in the electromagnetic field can be obtained from the equation of motion for a particle in an electric field in the local Lorentz rest frame where $\mathbf{v} = \mathbf{0}$, i.e., $d\mathbf{p}/dt = q\mathbf{E}$. The vector corresponding to $\mathbf{E}$ in Minkowski space has components $F^{\alpha\beta}U_\beta/c = (0, \mathbf{E})$ in the local Lorentz rest frame. Thus the equation of a charged particle is given by

$$\frac{dp^\alpha}{d\tau} = \frac{q}{c} F^{\alpha\beta}U_\beta. \tag{1.154}$$

The spatial components reduce to the equations of motion of a charge under the action of the well-known Lorentz force (which turns out to be a consequence of the expression of the force of a purely electric field and Lorentz invariance):

$$\frac{d\mathbf{p}}{dt} = q\left(\mathbf{E} + \mathbf{v} \times \mathbf{B}\right). \tag{1.155}$$

The time component gives the energy equation:

$$\frac{dE}{dt} = q\mathbf{E} \cdot \mathbf{v}. \tag{1.156}$$

Problems

1.5.1.1 Obtain $(1.146)_2$ from $(1.146)_1$ and show that the Maxwell equations (1.136) and (1.137) are equivalent to (1.145) and (1.146).

1.5.1.2 Show that the transformation laws for the electric field intensity and for the magnetic induction are given by (1.148) and (1.149), respectively.

1.5.2 Energy-momentum tensor

To obtain the expression of the energy-momentum tensor of the electromagnetic field we multiply the Maxwell equation (1.145) by $F_{\gamma\alpha}$ and get

$$-\mu_0 c\, F_{\gamma\alpha}J^\alpha = F_{\gamma\alpha}\partial_\beta F^{\alpha\beta} = \partial_\beta(F_{\gamma\alpha}F^{\alpha\beta}) - F^{\alpha\beta}\partial_\beta F_{\gamma\alpha}. \tag{1.157}$$

Further (1.146) gives

$$\begin{aligned} F^{\alpha\beta}\partial_\beta F_{\gamma\alpha} &= F^{\alpha\beta}\left(-\partial_\gamma F_{\alpha\beta} - \partial_\alpha F_{\beta\gamma}\right) = -F^{\alpha\beta}\partial_\gamma F_{\alpha\beta} - F^{\beta\alpha}\partial_\beta F_{\alpha\gamma} \\ &= -F^{\alpha\beta}\partial_\gamma F_{\alpha\beta} - F^{\alpha\beta}\partial_\beta F_{\gamma\alpha}, \end{aligned} \tag{1.158}$$

by renaming the dummy indices. Hence it follows from (1.158) that

$$F^{\alpha\beta}\partial_\beta F_{\gamma\alpha} = -\frac{1}{2}F^{\alpha\beta}\partial_\gamma F_{\alpha\beta} = -\frac{1}{4}\partial_\gamma(F^{\alpha\beta}F_{\alpha\beta}), \tag{1.159}$$

and we can write (1.157) as

$$-\mu_0 c F^{\gamma}_{\ \alpha} J^{\alpha} = \partial_{\beta}\left(F^{\gamma}_{\ \alpha} F^{\alpha\beta} + \frac{1}{4}\eta^{\beta\gamma} F^{\alpha\delta} F_{\alpha\delta}\right), \tag{1.160}$$

or

$$\partial_{\beta} T^{\alpha\beta}_{\text{em}} = -\frac{1}{c} F^{\alpha\beta} J_{\beta}. \tag{1.161}$$

Here $T^{\alpha\beta}_{\text{em}}$ is the energy-momentum tensor of the electromagnetic field, which is not uniquely defined. From (1.161) we see that a convenient definition is

$$T^{\alpha\beta}_{\text{em}} = \epsilon_0 \left(F^{\alpha}_{\ \gamma} F^{\gamma\beta} + \frac{1}{4}\eta^{\alpha\beta} F^{\gamma\delta} F_{\gamma\delta}\right). \tag{1.162}$$

The energy-momentum tensor of the electromagnetic field, defined in this way, is a symmetric tensor with vanishing trace. Indeed one can verify from (1.162) that

$$T^{\alpha\beta}_{\text{em}} = T^{\beta\alpha}_{\text{em}}, \qquad \eta_{\alpha\beta} T^{\alpha\beta}_{\text{em}} = 0. \tag{1.163}$$

The components of the energy-momentum tensor of the electromagnetic field are:

a) The energy density of the electromagnetic field W,

$$T^{00}_{\text{em}} = \frac{\epsilon_0}{2}(E^2 + c^2 B^2) = W; \tag{1.164}$$

b) The Poynting vector $\mathbf{S}$,

$$T^{0i}_{\text{em}} = \frac{1}{\mu_0 c}(\mathbf{E} \times \mathbf{B})^i = \frac{1}{c} S^i; \tag{1.165}$$

c) The Maxwell stress tensor t^{ij},

$$T^{ij}_{\text{em}} = \epsilon_0 \left[-E^i E^j - c^2 B^i B^j + \frac{1}{2}(E^2 + c^2 B^2)\delta^{ij}\right] = -t^{ij}. \tag{1.166}$$

From the above identifications we can write the matrix of the energy-momentum tensor of the electromagnetic field as:

$$(T^{\alpha\beta}_{\text{em}}) = \begin{pmatrix} W & S^1/c & S^2/c & S^3/c \\ S^1/c & -t^{11} & -t^{12} & -t^{13} \\ S^2/c & -t^{21} & -t^{22} & -t^{23} \\ S^3/c & -t^{31} & -t^{32} & -t^{33} \end{pmatrix}. \tag{1.167}$$

For $\alpha = 0$, (1.161) reduces to

$$\frac{\partial W}{\partial t} + \operatorname{div} \mathbf{S} + \mathbf{E} \cdot \mathbf{I} = 0, \tag{1.168}$$

which is the balance equation for the energy density of the electromagnetic field. For $\alpha = i$ we get the balance equation for the momentum density of the electromagnetic field,

$$\frac{\partial}{\partial t}\left(\frac{\mathbf{S}}{c^2}\right) - \operatorname{div}\mathsf{t} = -\varrho_q(\mathbf{E} + \mathbf{v} \times \mathbf{B}). \tag{1.169}$$

Hence $\mathbf{S}/c^2$ is interpreted as a momentum density of the electromagnetic field and t is the momentum flux density tensor (Maxwell's stresses).

Problem

1.5.2.1 Obtain the balance equations for the energy density (1.168) and for the momentum density (1.169) of the electromagnetic field.

1.5.3 Retarded potentials

The gauge transformation (1.153) implies that we can choose a suitable function ψ such that the so-called Lorentz gauge fixing

$$\partial_\alpha A^\alpha = 0 \tag{1.170}$$

is satisfied (Problem 1.5.3.1). Hence it follows from the insertion of (1.150) into the Maxwell equation (1.145) that

$$\Box A^\alpha = \mu_0\, c\, J^\alpha, \tag{1.171}$$

which represents an inhomogeneous wave equation.

The formal solution of the inhomogeneous wave equation (1.171) is

$$A^\alpha(x) = \mu_0 c \int G(x|x')J^\alpha(x')d^4x' + \mathcal{A}^\alpha(x), \tag{1.172}$$

where x and x' denote the components x^α and x'^α, respectively. $\mathcal{A}^\alpha(x)$ is a solution of the homogeneous wave equation

$$\Box \mathcal{A}^\alpha = 0, \tag{1.173}$$

and $G(x|x')$ is a Green function that satisfies the equation

$$\Box_x G(x|x') = \delta^4(x - x'). \tag{1.174}$$

In (1.174) we have introduced the four-dimensional delta function

$$\delta^4(x - x') = \delta(x^0 - x'^0)\delta(x^1 - x'^1)\delta(x^2 - x'^2)\delta(x^3 - x'^3). \tag{1.175}$$

We proceed to determine from (1.174) the Green function $G(x|x')$. First we note that the equation (1.174) and the Green function $G(x|x')$ must be invariant with respect to space-time translations, hence it follows that $G(x|x')$ must depend only on the difference $z^\alpha = x^\alpha - x'^\alpha$, i.e., $G(x|x') = G(z)$, and we can rewrite (1.174) as:

$$\Box_z G(z) = \delta^4(z). \tag{1.176}$$

We introduce the Fourier transform of $G(z)$ and its inverse through

$$\tilde{G}(k) = \frac{1}{(\sqrt{2\pi})^4}\int G(z)e^{ik\cdot z}d^4z, \qquad G(z) = \frac{1}{(\sqrt{2\pi})^4}\int \tilde{G}(k)e^{-ik\cdot z}d^4k, \tag{1.177}$$

while the four-dimensional delta function has the representation

$$\delta^4(z) = \frac{1}{(2\pi)^4}\int e^{-ik\cdot z}d^4k. \tag{1.178}$$

In (1.177) and (1.178) $k \cdot z = k_0(x^0 - x'^0) - \mathbf{k}\cdot(\mathbf{x}-\mathbf{x}')$.

The four-fold Fourier transform of (1.176) gives

$$\tilde{G}(k) = -\frac{1}{(\sqrt{2\pi})^4}\frac{1}{k_0^2 - |\mathbf{k}|^2}\,. \tag{1.179}$$

The singularity at $k_0^2 = |\mathbf{k}|^2$ shows that the calculation must be performed with some care, since it turns out that it is meaningful only if we think of generalized functions or distributions and not of ordinary functions. Thus (1.179) does not give the most general solution of the Fourier-transformed form of (1.176). We might add another term proportional to $\delta(k_0^2 - |\mathbf{k}|^2)$, the coefficient being an arbitrary function of $\mathbf{k}$ (this is the Fourier transform of the general solution of the homogeneous equation $\Box G = 0$)(Problem 1.5.3.2). This indicates that there is not just one Green function. Additional conditions are thus required. We shall make use of this freedom when choosing the integration path to calculate the Green function from its Fourier transform. If we accept (1.179) for the moment, the Green function $(1.177)_2$ reduces to

$$G(z) = -\frac{1}{(2\pi)^4}\int \frac{e^{i\mathbf{k}\cdot(\mathbf{x}-\mathbf{x}')}e^{-ik_0(x^0-x'^0)}}{k_0^2 - |\mathbf{k}|^2}dk_0d^3k, \tag{1.180}$$

where $d^4k = dk_0dk_1dk_2dk_3 = dk_0d^3k$.

To solve the integral (1.180) we introduce the abbreviations

$$\mathbf{r} = \mathbf{x} - \mathbf{x}', \qquad \tau = x^0 - x'^0, \qquad |\mathbf{k}| = \kappa, \tag{1.181}$$

and write (1.180) as

$$G(z) = -\frac{1}{(2\pi)^4}\int\left\{\int_{-\infty}^{+\infty}\frac{e^{-ik_0\tau}}{(k_0-\kappa)(k_0+\kappa)}dk_0\right\}e^{i\mathbf{k}\cdot\mathbf{r}}d^3k. \tag{1.182}$$

It is clear that the integration over k_0 exhibits a singularity at $k_0 = \kappa$. Thus we must choose a *recipe* to avoid the singularity. One possibility would be to take the

Cauchy principal value for the integral. This mathematically interesting choice should, however, be justified from a physical viewpoint. There are many other possibilities, which are offered by choosing different paths of integration in the complex plane of the variable k_0. From a physical viewpoint we shall impose that $G(z) = 0$ for all $\tau < 0$, since this means that a perturbation appears only after the source generated it. Since the integral (1.182) in k_0 has two simple poles in $k_0 = \pm\kappa$ we integrate in the contour indicated in Figure 1.2 and get by the use of the residue theorem[2]

$$\int_{-\infty}^{+\infty} \frac{e^{-ik_0\tau}}{(k_0-\kappa)(k_0+\kappa)} dk_0 = -2\pi i \left\{ \frac{e^{i\kappa\tau}}{-2\kappa} + \frac{e^{-i\kappa\tau}}{2\kappa} \right\} = -\frac{2\pi}{\kappa} \sin\kappa\tau, \qquad (1.183)$$

where the minus sign is due to a clockwise integration and $\tau > 0$. From (1.182)

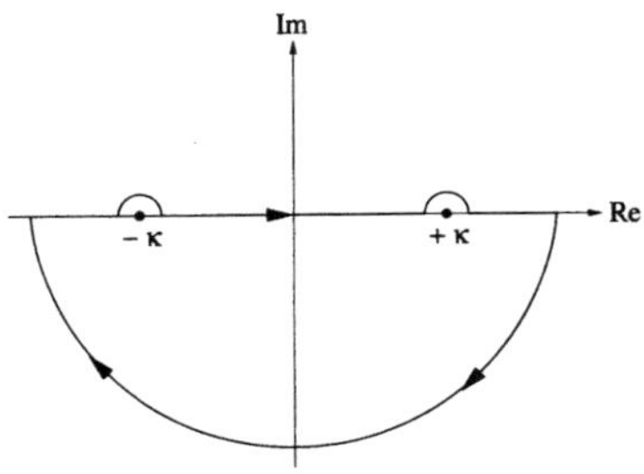

Figure 1.2: k_0 complex plane

and (1.183) it follows that the Green function $G(z)$ reduces to

$$G(z) = \frac{1}{(2\pi)^3}\theta(\tau) \int \frac{e^{i\mathbf{k}\cdot\mathbf{r}}}{\kappa} \sin\kappa\tau \, d^3k, \qquad (1.184)$$

where $\theta(\tau)$ is the Heaviside step function

$$\theta(\tau) = \begin{cases} 0 & \text{if } \tau < 0, \\ 1 & \text{if } \tau > 0. \end{cases} \qquad (1.185)$$

For the integration in d^3k we use spherical coordinates and choose $\mathbf{r}$ in the direction of k_3. Hence

$$d^3k = \kappa^2 \sin\vartheta \, d\vartheta \, d\phi \, d\kappa, \qquad \mathbf{k}\cdot\mathbf{r} = r\kappa\cos\vartheta, \qquad (1.186)$$

[2]If we write $k_0 = \Re\{k_0\} + i\Im\{k_0\}$, then $e^{-ik_0\tau} = e^{-i\Re\{k_0\}\tau}e^{\Im\{k_0\}\tau}$. Since $|e^{-i\Re\{k_0\}\tau}| = 1$, the factor involving the imaginary part dictates the choice of the path: for $\tau < 0$, we must choose the upper-half contour in order to have bounded values for $e^{-ik_0\tau}$, and this gives zero as required, since there are no poles inside the contour. For $\tau > 0$, we must choose the lower-half contour in order to have bounded values for $e^{-ik_0\tau}$ and this gives the result indicated in the text.

and we have from (1.184) that

$$G(z) = \frac{1}{(2\pi)^3}\theta(\tau) \int_0^{\infty} \int_0^{\pi} \int_0^{2\pi} e^{ir\kappa \cos\vartheta} \kappa \sin \kappa\tau \, \sin\vartheta \, d\phi \, d\vartheta d\kappa. \tag{1.187}$$

By performing the integrations in the angles ϕ and ϑ we get

$$G(z) = \frac{1}{2\pi^2} \frac{\theta(\tau)}{r} \int_0^{\infty} \sin \kappa\tau \sin \kappa r \, d\kappa. \tag{1.188}$$

We remark that this integral, like all the calculations concerning the Green function which have been performed so far, is meaningless for ordinary functions, but has a precise meaning for distributions. The above equation can be written as

$$G(z) = \frac{1}{2\pi^2} \frac{\theta(\tau)}{r} \frac{1}{2} \int_{-\infty}^{+\infty} \left[\frac{e^{i\kappa(\tau - r)} - e^{i\kappa(\tau + r)}}{2} \right] d\kappa, \tag{1.189}$$

since the integrals of the products of the sines with the cosines vanish. Now we use the integral form of the delta function to get:

$$G(z) = \frac{1}{4\pi} \frac{\theta(\tau)}{r} \left[\delta(\tau - r) - \delta(\tau + r) \right]. \tag{1.190}$$

The delta function $\delta(\tau + r)$ is always zero because $\tau > 0$ and $r > 0$. If we return now to the variables t and $\mathbf{x}$ we have that

$$\begin{aligned} G(z) &= \frac{1}{4\pi} \frac{\theta(x^0 - x'^0)}{|\mathbf{x} - \mathbf{x}'|} \delta(c(t - t') - |\mathbf{x} - \mathbf{x}'|) \\ &= \frac{1}{4\pi} \frac{\theta(x^0 - x'^0)}{c|\mathbf{x} - \mathbf{x}'|} \delta\left((t - t') - \frac{|\mathbf{x} - \mathbf{x}'|}{c} \right), \end{aligned} \tag{1.191}$$

by the use of the relationship $\delta(ax) = \delta(x)/a$ which is valid for all $a > 0$.

Finally we insert (1.191) into (1.172) and obtain, by performing the time integration

$$A^\alpha(\mathbf{x}, t) = \frac{\mu_0 c}{4\pi} \int \frac{J^\alpha(\mathbf{x}', t - |\mathbf{x} - \mathbf{x}'|/c)}{|\mathbf{x} - \mathbf{x}'|} dx'^1 dx'^2 dx'^3 + \mathcal{A}(\mathbf{x}, t). \tag{1.192}$$

The potential $A^\alpha(\mathbf{x}, t)$ which is a solution of the inhomogeneous wave equation (1.171) and has the form (1.192) is called retarded potential. There is a completely analogous procedure which produces advanced potentials, for which a perturbation appears only before the source generated it (Problem 1.5.3.3). Though this may seem paradoxical, it is a consequence of the invariance of the Maxwell equations with respect to time reversal, first discovered by Boltzmann [2]. We should hasten to add that if one solves the initial value problem for the Maxwell equations in the presence of given sources for $t > 0$, only retarded potentials occur. Advanced potentials may be regarded as a mathematical curiosity, but serious attempts have been made to use them to produce a time-symmetric form of the solution (this requires cosmological assumptions on the behavior of emitters and absorbers of radiation).

Problems

1.5.3.1 Prove that we can choose the gauge in such a way that the Lorentz condition (1.170) is satisfied. (Hint: After having chosen an arbitrary vector potential giving the correct electromagnetic field, the divergence $\partial_\alpha A^\alpha$ will be, in general, nonzero. If we change the gauge the divergence will change by the addition of a term proportional to $\Box\psi$ and hence ψ can be chosen by solving an inhomogeneous wave equation for ψ to make $\partial_\alpha A^\alpha = 0$ in the new gauge).

1.5.3.2 Prove that the Fourier transform of the general solution of the homogeneous wave equation $\Box G = 0$ is $A(\mathbf{k})\delta(k_0^2 - |\mathbf{k}|^2)$, where the coefficient A is an arbitrary function of $\mathbf{k}$. (Hint: The four-fold Fourier transform of $\Box G = 0$ is $(k_0^2 - |\mathbf{k}|^2)\tilde{G}(k) = 0$).

1.5.3.3 Repeat the calculation of the vector potential, but choose the path *under* the poles in Figure 1.2. Note that now you obtain the so-called advanced potentials, which exhibit the effect of a source before the source appears.

References

[1] J. Aharoni, *The special theory of relativity*, 2nd. ed. (Oxford Univ. Press, London, 1965).

[2] C. Cercignani, *Ludwig Boltzmann. The man who trusted atoms*, (Oxford University Press, Oxford, 1998).

[3] H.A. Lorentz, A. Einstein, H. Minkowski and H. Weyl, *The principle of relativity* (Dover, New York, 1952).

[4] S. R. de Groot, W. A. van Leeuwen and Ch. G. van Weert, *Relativistic kinetic theory* (North-Holland, Amsterdam, 1980).

[5] T. Fliessbach, *Allgemeine Relativitätstheorie* (B. I. Wissenschaftsverlag, Mannheim, 1990).

[6] V. Fock, *The theory of space time and gravitation* (Pergamon Press, London, 1959).

[7] J. D. Jackson, *Classical electrodynamics*, 2nd. ed. (John Wiley, New York, 1975).

[8] G. M. Kremer, *Uma introdução à teoria cinética relativística* (Editora da UFPR, Curitiba, 1998).

[9] L. D. Landau and E. M. Lifshitz, *The classical theory of fields*, 4th ed. (Pergamon Press, Oxford, 1980).

[10] C. Møller, *The theory of relativity*, (Oxford University Press, London, 1952).

Chapter 2

Relativistic Boltzmann Equation

2.1 Single non-degenerate gas

The purpose of this chapter is to introduce the basic concepts of relativistic kinetic theory and the relativistic Boltzmann equation, which rules the time evolution of the distribution function.

In this section we shall consider a single non-degenerate relativistic gas, i.e., a gas where quantum effects are not taken into account.

A gas particle of rest mass m is characterized by the space-time coordinates $(x^\alpha) = (ct, \mathbf{x})$ and by the momentum four-vector $(p^\alpha) = (p^0, \mathbf{p})$. Due to the constraint that the length of the momentum four-vector is mc, p^0 is given in terms of $\mathbf{p}$ by (1.102), i.e., $p^0 = \sqrt{|\mathbf{p}| + m^2c^2}$.

The one-particle distribution function, defined in terms of the space-time and momentum coordinates $f(x^\alpha, p^\alpha) = f(\mathbf{x}, \mathbf{p}, t)$, is such that

$$f(\mathbf{x}, \mathbf{p}, t)d^3x\, d^3p = f(\mathbf{x}, \mathbf{p}, t)dx^1dx^2dx^3dp^1dp^2dp^3 \tag{2.1}$$

gives at time t the number of particles in the volume element d^3x about $\mathbf{x}$ and with momenta in a range d^3p about $\mathbf{p}$.

The number of particles in a volume element is a scalar invariant since all observers will count the same particles. Let us examine the invariance of the volume element d^3xd^3p. According to (1.83)

$$\frac{d^3p'}{p_0'} = \frac{d^3p}{p_0} \tag{2.2}$$

is a scalar invariant. Let us choose the primed frame of reference as a rest frame, i.e., where $\mathbf{p}' = \mathbf{0}$. In this case d^3x' is the proper volume and from (1.33) we have that

$$d^3x = \sqrt{1 - v^2/c^2}d^3x' = \frac{1}{\gamma}d^3x'. \tag{2.3}$$

On the other hand, from the transformation of the components of the momentum four-vector $(1.98)_1$ we have that

$$p_0' = \frac{1}{\gamma} p_0, \tag{2.4}$$

since $p^0 = p_0$. Now we build the product

$$d^3x d^3p = \frac{1}{\gamma} d^3x' \frac{p^0}{p'^0} d^3p' = d^3x' d^3p' \tag{2.5}$$

and conclude that d^3xd^3p is a scalar invariant. Since the number of the particles in the volume element $d^3x\, d^3p$ is also a scalar invariant we conclude that the one-particle distribution function $f(\mathbf{x}, \mathbf{p}, t)$ is a scalar invariant.

Let us denote the volume element at time t by

$$d\mu(t) = d^3x d^3p. \tag{2.6}$$

The number of particles in this volume element at time t is

$$N(t) = f(\mathbf{x}, \mathbf{p}, t) d\mu(t). \tag{2.7}$$

Further, the number of particles in the volume element $d\mu(t + \Delta t)$ at time $t + \Delta t$ is

$$N(t + \Delta t) = f(\mathbf{x} + \Delta\mathbf{x}, \mathbf{p} + \Delta\mathbf{p}, t + \Delta t) d\mu(t + \Delta t). \tag{2.8}$$

The collisions between the particles imply that $N(t)$ is not equal to $N(t+\Delta t)$ and the change in the number of particles is given by

$$\begin{aligned} \Delta N &= N(t + \Delta t) - N(t) \\ &= f(\mathbf{x} + \Delta\mathbf{x}, \mathbf{p} + \Delta\mathbf{p}, t + \Delta t)\, d\mu(t + \Delta t) - f(\mathbf{x}, \mathbf{p}, t)\, d\mu(t), \end{aligned} \tag{2.9}$$

where the increments in the position and in the momentum read

$$\Delta\mathbf{x} = \mathbf{v}\Delta t, \qquad \Delta\mathbf{p} = \mathbf{F}\Delta t. \tag{2.10}$$

$\mathbf{F}(\mathbf{x}, \mathbf{p}, t)$ denotes the external force that acts on the particles and $\mathbf{v} = c\mathbf{p}/p^0$ is the velocity of the particle with momentum $\mathbf{p}$.

The relationship between $d\mu(t + \Delta t)$ and $d\mu(t)$ is given by:

$$d\mu(t + \Delta t) = |J| d\mu(t), \tag{2.11}$$

with J denoting the Jacobian of the transformation

$$J = \frac{\partial(x^1(t + \Delta t), x^2(t + \Delta t), \ldots, p^3(t + \Delta t))}{\partial(x^1(t), x^2(t), \ldots, p^3(t))}. \tag{2.12}$$

If we consider up to linear terms in Δt we get from (2.12) that the Jacobian reduces to

$$J = 1 + \frac{\partial F^i}{\partial p^i}\Delta t + \mathcal{O}[(\Delta t)^2]. \tag{2.13}$$

Now by expanding $f(\mathbf{x}+\Delta\mathbf{x}, \mathbf{p}+\Delta\mathbf{p}, t+\Delta t)$ in Taylor series about the point $(\mathbf{x}, \mathbf{p}, t)$ and by considering only linear terms in Δt it follows that

$$f(\mathbf{x}+\Delta\mathbf{x}, \mathbf{p}+\Delta\mathbf{p}, t+\Delta t) \approx f(\mathbf{x}, \mathbf{p}, t) + \frac{\partial f}{\partial t}\Delta t + \frac{\partial f}{\partial x^i}\Delta x^i + \frac{\partial f}{\partial p^i}\Delta p^i + \mathcal{O}[(\Delta t)^2]. \tag{2.14}$$

We combine equations (2.9) through (2.14) and get the total change in the number of particles per unit of time interval:

$$\frac{\Delta N}{\Delta t} = \left[\frac{\partial f}{\partial t} + v^i\frac{\partial f}{\partial x^i} + F^i\frac{\partial f}{\partial p^i} + f\frac{\partial F^i}{\partial p^i}\right] d\mu(t)$$

$$= \left[\frac{\partial f}{\partial t} + v^i\frac{\partial f}{\partial x^i} + \frac{\partial f F^i}{\partial p^i}\right] d\mu(t). \tag{2.15}$$

ΔN is a scalar invariant as well as the proper time $\Delta\tau = \Delta t/\gamma$, hence

$$\gamma\frac{\Delta N}{\Delta t} = \frac{\Delta N}{\Delta \tau} = \gamma\left[\frac{\partial f}{\partial t} + v^i\frac{\partial f}{\partial x^i} + \frac{\partial f F^i}{\partial p^i}\right] d\mu(t) \tag{2.16}$$

is also a scalar invariant. We have shown in (2.5) that $d\mu = d^3x\, d^3p$ is a scalar invariant, and as a consequence the expression multiplying it in (2.16) must have the same property as we shall show. We first consider the term

$$\gamma\left[\frac{\partial f}{\partial t} + v^i\frac{\partial f}{\partial x^i}\right] = \gamma\left[\frac{\partial f}{\partial t} + \frac{cp^i}{p^0}\frac{\partial f}{\partial x^i}\right] = \frac{c\gamma}{p^0}p^\alpha\frac{\partial f}{\partial x^\alpha} = \frac{1}{m}p^\alpha\frac{\partial f}{\partial x^\alpha}. \tag{2.17}$$

Due to the fact that f is a scalar invariant, $\partial f/\partial x^\alpha$ is a four-vector and the scalar product $p^\alpha \partial f/\partial x^\alpha$ is a scalar invariant. We need only to prove that $\gamma \partial f F^i/\partial p^i$ has the same property. For the proof we consider the Minkowski force K^α defined by (1.101) that satisfies

$$K^\alpha p_\alpha = K^0 p_0 - \mathbf{K}\cdot\mathbf{p} = 0, \tag{2.18}$$

and the relationship

$$\mathbf{F} = \frac{\mathbf{K}}{\gamma} = \frac{mc\mathbf{K}}{p^0}, \tag{2.19}$$

where m is the rest mass. If we consider p^0 as an independent variable and make use of the chain rule:

$$\frac{\partial}{\partial p^i} \rightarrow \frac{\partial p^0}{\partial \mathbf{p}}\frac{\partial}{\partial p^0} + \frac{\partial}{\partial \mathbf{p}} = \frac{\mathbf{p}}{p^0}\frac{\partial}{\partial p^0} + \frac{\partial}{\partial \mathbf{p}}, \tag{2.20}$$

we can write the following expression

$$\gamma\frac{\partial fF^i}{\partial p^i} = \gamma mc\left[\frac{\mathbf{p}}{p^0}\cdot\frac{\partial}{\partial p^0}\left(\frac{f\mathbf{K}}{p^0}\right)+\frac{\partial}{\partial \mathbf{p}}\cdot\left(\frac{f\mathbf{K}}{p^0}\right)\right]. \tag{2.21}$$

Since p^0 and $\mathbf{p}$ are treated as independent variables the above equation reduces to

$$\begin{aligned}\gamma\frac{\partial fF^i}{\partial p^i} &= \gamma mc\left[\frac{1}{p^0}\frac{\partial}{\partial p^0}\left(\frac{f\mathbf{K}\cdot\mathbf{p}}{p^0}\right)+\frac{1}{p^0}\frac{\partial}{\partial \mathbf{p}}\cdot(f\mathbf{K})\right]\\ &\overset{(2.18)}{=}\frac{\gamma mc}{p^0}\left[\frac{\partial fK^0}{\partial p^0}+\frac{\partial}{\partial \mathbf{p}}\cdot(f\mathbf{K})\right]\\ &=\frac{\gamma mc}{p^0}\frac{\partial fK^\alpha}{\partial p^\alpha}=\frac{\partial fK^\alpha}{\partial p^\alpha},\end{aligned} \tag{2.22}$$

which is a scalar invariant.

Now according to (2.17) and (2.22) equation (2.15) reads

$$\frac{\Delta N}{\Delta t} = \frac{c}{p^0}\left[p^\alpha\frac{\partial f}{\partial x^\alpha}+m\frac{\partial fK^\alpha}{\partial p^\alpha}\right]d\mu(t). \tag{2.23}$$

To determine $\Delta N/\Delta t$ we decompose it in two terms

$$\frac{\Delta N}{\Delta t} = \frac{(\Delta N)^+}{\Delta t} - \frac{(\Delta N)^-}{\Delta t}, \tag{2.24}$$

where $(\Delta N)^-/\Delta t$ corresponds to the particles that leave the volume element $d^3x d^3p$, whereas $(\Delta N)^+/\Delta t$ corresponds to those particles that enter in the same volume element. Further we assume the following:

a) Only collisions between pairs of particles are taken into account, i.e., only binary collisions are considered (this is reasonable if the gas is dilute, i.e., if the volume occupied by the molecules is much smaller than the volume of the gas);

b) if $\mathbf{p}$ and $\mathbf{p}_*$ denote the momenta of two particles before collision they are not correlated. This will be applied to the momenta $\mathbf{p}$ of the particle that we are following, and $\mathbf{p}_*$ of its collision partner, as well as to two momenta $\mathbf{p}'$ and $\mathbf{p}'_*$ possessed by two particles before a collision that will transform them into particles with momenta $\mathbf{p}$ and $\mathbf{p}_*$ after collision. This hypothesis is the so-called molecular chaos assumption;

c) the one-particle distribution function $f(\mathbf{x},\mathbf{p},t)$ does not vary very much over a time interval which is larger than the duration of a collision but smaller than the time between collisions. The same applies to the change of f over a distance of the order of the interaction range.

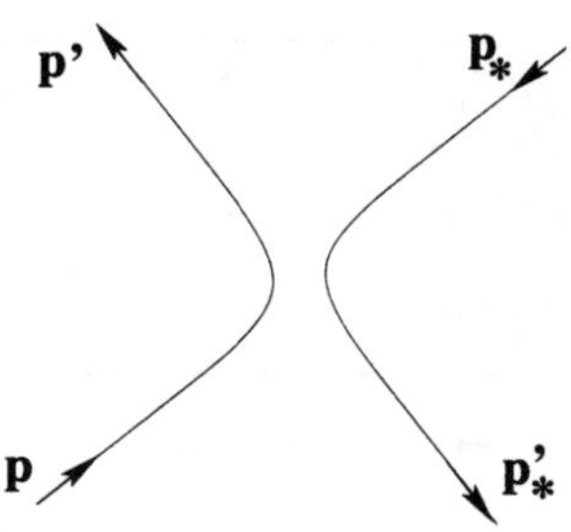

Figure 2.1: Representation of a binary collision

The German word *Stoßzahlansatz*, which means supposition of number of collisions, is frequently employed in the literature to indicate this set of assumptions that will be presently used to determine $\Delta N/\Delta t$.

We consider a collision between two beams of particles with velocities $\mathbf{v} = c\mathbf{p}/p^0$ and $\mathbf{v}_* = c\mathbf{p}_*/p_*^0$. The particle number densities of these two beams in their own frames are denoted by dn and dn_*. The d in front of n and n_* indicates that these number densities are infinitesimal because they refer to volume elements d^3p and d^3p_* of momentum space ($dn = fd^3p$ and $dn_* = f_*d^3p_*$). We first consider a reference frame where the particles without label are at rest, i.e., $\mathbf{v} = \mathbf{0}$. The total number of these particles about $\mathbf{x}$ is dnd^3x. The total number of particles that will collide with the former and are in a volume element dV_* will be $dn_*dV_* = dn_*dV/\sqrt{1-v_{\rm rel}^2/c^2}$, where $v_{\rm rel}$ is the relative speed and $dV/\sqrt{1-v_{\rm rel}^2/c^2}$ is a proper volume.

The particles with density dn_* in the volume dV are differently scattered by their partners in the collision through different angles. Each collision will occur in a plane with some scattering angle Θ; another angle is needed to single out the plane (which must contain the relative velocity) and two infinitesimal neighborhoods of the two angles together single out a solid angle element $d\Omega$. The volume element dV can be written in terms of the so-called collision cylinder of base $\sigma d\Omega$ and height $v_{\rm rel}\Delta t$. Δt is identified with the differential of the proper time, because of the choice of the reference frame. The factor σ has clearly the dimensions of an area and is called the differential cross-section of the scattering process corresponding to the relative speed $v_{\rm rel}$ and the scattering angle Θ. In another reference system where $\mathbf{v} \neq \mathbf{0}$, $d^3x\Delta t$, σ, $d\Omega$ and $v_{\rm rel}$ are scalar invariants.

The total number of collisions will be given then by the product of the particle numbers corresponding to the velocities $\mathbf{v}$ and $\mathbf{v}_*$:

$$dnd^3x\frac{dn_*}{\sqrt{1-v_{\rm rel}^2/c^2}}dV = dnd^3x\frac{dn_*}{\sqrt{1-v_{\rm rel}^2/c^2}}(\sigma d\Omega v_{\rm rel}\Delta t), \tag{2.25}$$

where we have rewritten the volume element dV in terms of the collision cylinder as discussed above.

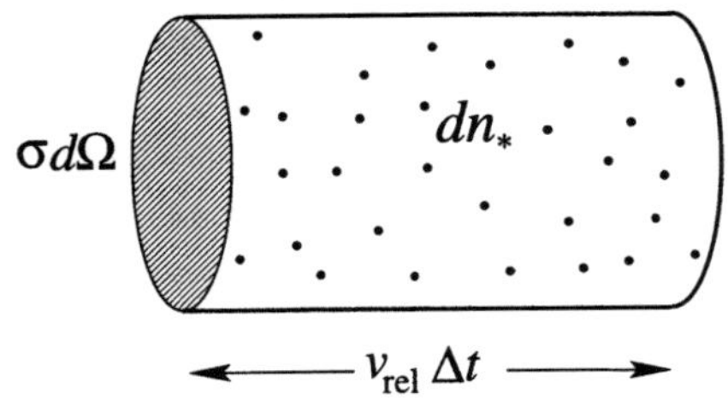

Figure 2.2: Representation of the collision cylinder

Let us consider the product of the particle number densities in a system where $\mathbf{v} \neq \mathbf{0}$:

$$\frac{dn\,dn_*}{\sqrt{1-v_{\rm rel}^2/c^2}} = \frac{dn}{\sqrt{1-v^2/c^2}}\frac{dn_*}{\sqrt{1-v_*^2/c^2}}\frac{\sqrt{1-v^2/c^2}\sqrt{1-v_*^2/c^2}}{\sqrt{1-v_{\rm rel}^2/c^2}}$$

$$= f(\mathbf{x},\mathbf{p},t)d^3p f(\mathbf{x},\mathbf{p}_*,t)d^3p_*\frac{\sqrt{1-v^2/c^2}\sqrt{1-v_*^2/c^2}}{\sqrt{1-v_{\rm rel}^2/c^2}}$$

$$\overset{(1.126)}{=} f(\mathbf{x},\mathbf{p},t)d^3p f(\mathbf{x},\mathbf{p}_*,t)d^3p_*\frac{p_\alpha p_*^\alpha}{p^0p_*^0}. \tag{2.26}$$

The particle number densities above were written in terms of the one-particle distribution functions

$$\frac{dn}{\sqrt{1-v^2/c^2}} \equiv f(\mathbf{x},\mathbf{p},t)d^3p, \qquad \frac{dn_*}{\sqrt{1-v_*^2/c^2}} \equiv f(\mathbf{x},\mathbf{p}_*,t)d^3p_*. \tag{2.27}$$

Hence instead of (2.25) we have that the total number of collisions reads

$$f(\mathbf{x},\mathbf{p},t)d^3p f(\mathbf{x},\mathbf{p}_*,t)d^3p_* v_{\rm rel}\frac{p_\alpha p_*^\alpha}{p^0p_*^0}\sigma d\Omega d^3x\Delta t$$

$$\overset{(1.131)}{=} f(\mathbf{x},\mathbf{p},t)d^3p f(\mathbf{x},\mathbf{p}_*,t)d^3p_* g_\o \sigma d\Omega d^3x\Delta t \tag{2.28}$$

where we have introduced Møller's relative speed $g_\o$.

Now the total number of particles that leave the volume element d^3xd^3p is obtained from (2.28) by integrating it over all momenta $\mathbf{p}_*$ and over all solid angle $d\Omega$, yielding

$$(\Delta N)^- = \int_\Omega\int_{\mathbf{p}_*} f(\mathbf{x},\mathbf{p},t)f(\mathbf{x},\mathbf{p}_*,t)\, g_\o\, \sigma\, d\Omega\, d^3p_*\, d^3x\, d^3p\, \Delta t. \tag{2.29}$$

This is frequently called the loss term because it describes the loss of particles in the volume element d^3xd^3p in phase space, due to collisions.

We remark that sometimes this equation is written with a factor $1/2$ in front of the above integral. This is clearly related to the definition of the cross-section. The fact is that we are dealing with identical particles. Whereas in non-quantum mechanics identical particles can be regarded as distinguishable (because we can follow their motion in a continuous way), this is not the case in quantum mechanics. Thus if we start with two states for the colliding particles we find for each pair of final states a number which is the double of what we should expect from an analogy with a non-quantum calculation. This is due to the fact that we compute the two scattering processes leading the particles from the states $(\mathbf{p}', \mathbf{p}'_*)$ to the states $(\mathbf{p}, \mathbf{p}_*)$ and from the states $(\mathbf{p}', \mathbf{p}'_*)$ to the states $(\mathbf{p}_*, \mathbf{p})$, respectively, as the same process, due to the indistinguishability of the particles involved. The definition of cross-section in quantum mechanics thus leads to a result which is twice as much the result expected from an analogy with classical mechanics. Thus if we use the quantum cross-section, we must divide by two the result for the loss term obtained above. We shall mainly deal with non-quantum effects and write the collision terms without the factor $1/2$; the reader is advised to check what convention each author is using when comparing equations in different publications. We remark that, when considering a mixture, one deals with collisions of distinguishable particles and then our convention agrees with the opposite one; this means that authors using the latter have a factor $1/2$ in front of collision terms associated with particles of the same species, whereas this factor is absent in the collision terms referring to different species.

The same reasoning and comments apply to the computation of the total number of particles that leave the volume element $d^3x'\,d^3p'$ and enter the volume element $d^3x\,d^3p$. We consider a collision between two beams of particles with velocities $\mathbf{v}'_* = c\mathbf{p}'_*/p'^0_*$ and $\mathbf{v}' = c\mathbf{p}'/p'^0$ and by taking into account (2.29) we write the total number of particles that leave the volume element $d^3x'\,d^3p'$ as

$$(\Delta N)^+ = \int_{\Omega'}\int_{\mathbf{p}'_*} f(\mathbf{x}, \mathbf{p}', t) f(\mathbf{x}, \mathbf{p}'_*, t)\, g'_\varnothing\, \sigma'\, d\Omega'\, d^3p'_*\, d^3x'\, d^3p'\, \Delta t', \tag{2.30}$$

which is called the gain term since it describes the gain of particles in the volume element d^3xd^3p.

For relativistic particles we have that $g_\varnothing \neq g'_\varnothing$. However, Liouville's theorem asserts that if we follow the evolution of a volume element in phase space its volume does not change in the course of time. Here we have that

$$g_\varnothing\, \Delta t\, \sigma\, d\Omega\, d^3p_* d^3x\, d^3p = g'_\varnothing\, \Delta t'\, \sigma'\, d\Omega'\, d^3p'_* d^3x'\, d^3p'. \tag{2.31}$$

Since $d^3x\,\Delta t = d^3x'\,\Delta t'$ is an invariant it follows from (2.31) that

$$\int_\Omega g_\varnothing \sigma\, d\Omega\, d^3p_*\, d^3p = \int_{\Omega'} g'_\varnothing \sigma'\, d\Omega'\, d^3p'_*\, d^3p'. \tag{2.32}$$

Now we get from (2.23) together with (2.24), (2.29), (2.30) and (2.32):

$$\frac{c}{p^0}\left[p^\alpha\frac{\partial f}{\partial x^\alpha}+m\frac{\partial fK^\alpha}{\partial p^\alpha}\right]d^3x\,d^3p=\frac{(\Delta N)^+-(\Delta N)^-}{\Delta t}$$

$$=\int (f'_*f'-f_*f)\,g_\varnothing\,\sigma\,d\Omega\,d^3p_*\,d^3x\,d^3p, \tag{2.33}$$

where we have introduced the abbreviations

$$f'_*\equiv f(\mathbf{x},\mathbf{p}'_*,t),\quad f'\equiv f(\mathbf{x},\mathbf{p}',t),\quad f_*\equiv f(\mathbf{x},\mathbf{p}_*,t),\quad f\equiv f(\mathbf{x},\mathbf{p},t). \tag{2.34}$$

If we denote by F the invariant flux

$$F=\frac{p^0p^0_*}{c}g_\varnothing=\frac{p^0p^0_*}{c}\sqrt{(\mathbf{v}-\mathbf{v}_*)^2-\frac{1}{c^2}(\mathbf{v}\times\mathbf{v}_*)^2}=\sqrt{(p^\alpha_*p_\alpha)^2-m^4c^4}, \tag{2.35}$$

equation (2.33) reduces to

$$p^\alpha\frac{\partial f}{\partial x^\alpha}+m\frac{\partial fK^\alpha}{\partial p^\alpha}=\int (f'_*f'-f_*f)\,F\,\sigma\,d\Omega\frac{d^3p_*}{p_{*0}}, \tag{2.36}$$

which is the final form of the relativistic Boltzmann equation for a single non-degenerate relativistic gas. In (2.36) we have denoted by only one symbol the integrals over Ω and $\mathbf{p}_*$.

Another expression for the Boltzmann equation (2.36) is obtained by the combination of (2.15), (2.23) and (2.33), yielding

$$\frac{\partial f}{\partial t}+v^i\frac{\partial f}{\partial x^i}+\frac{\partial fF^i}{\partial p^i}=\int (f'_*f'-f_*f)\,g_\varnothing\,\sigma\,d\Omega d^3p_*. \tag{2.37}$$

The above equation has the same expression as that of the classical Boltzmann equation.

Problems

2.1.1 Show that the Jacobian J of the transformation between the two volume elements $d\mu(t)$ and $d\mu(t+\Delta t)$ reduces to (2.13), when one considers up to linear terms in Δt.

2.1.2 Show that F defined in (2.35) is indeed a scalar invariant.

2.1.3 Show that the Boltzmann equation (2.37) reduces to

$$\frac{\partial f}{\partial t}+v^i\frac{\partial f}{\partial x^i}+F^i\frac{\partial f}{\partial p^i}=\int (f'_*f'-f_*f)\,g_\varnothing\,\sigma\,d\Omega d^3p_*,$$

when the external forces that act on the particles do not depend on the momentum $\mathbf{p}$ or depend on it through the Lorentz force $F^i=q(\mathbf{v}\times\mathbf{B})^i$. The Boltzmann equation given above is discussed in books on non-relativistic kinetic theory (see for example Cercignani [8]).

2.2 Single degenerate gas

In this section we are interested only in the derivation of the relativistic Uehling–Uhlenbeck equation that is a quasi-classical Boltzmann equation [28], which incorporates modifications in the collision term of the Boltzmann equation since the particles obey quantum statistics. For a complete quantum mechanical description based on the Wigner distribution function one is referred to the book of de Groot, van Leeuwen and van Weert [17].

We note that according to (2.5) the volume element in phase space is a scalar invariant. When quantum effects are taken into account in a semi-classical description, one frequently divides the volume element by h^3, where $h = 6.626\times\ 10^{-34}$ Js is the Planck constant. Thus we write

$$\frac{d^3xd^3p}{h^3} = \frac{d^3x'd^3p'}{h^3} \tag{2.38}$$

which, of course, is also a scalar invariant. Here d^3xd^3p/h^3 can be interpreted as the number of available states in the volume element d^3xd^3p. For particles with spin s there are more states, corresponding to the values that the spin component on a given axis can take and we have to introduce the degeneracy factor g_s. Hence we write the number of available states as

$$g_\mathrm{s}\frac{d^3xd^3p}{h^3} \qquad \text{where} \qquad g_\mathrm{s} = \begin{cases} 2\mathrm{s}+1 & \text{for} \quad m \neq 0, \\ 2\mathrm{s} & \text{for} \quad m = 0, \end{cases} \tag{2.39}$$

where m is the rest mass of a particle.

According to quantum mechanics a system of identical particles is described by two kinds of particles called bosons and fermions. Bosons have integral spin, obey the Bose–Einstein statistics and include mesons (pions and kaons), photons, gluons and nuclei of even mass number like helium-4. Fermions have half-integral spin, obey the Fermi–Dirac statistics and include leptons (electrons and muons), baryons (neutrons, protons and lambda particles) and nuclei of odd mass number like helium-3.

From our viewpoint the main difference between bosons and fermions concerns the occupation number of a state. Any number of bosons may occupy the same state, while fermions obey the Pauli exclusion principle and at most one particle may occupy each state.

Let us now turn to the modifications we have to introduce in the collision term of the Boltzmann equation in order to incorporate the statistics of bosons and fermions. First we note that due to the fact that fermions obey the Pauli exclusion principle we could infer that the phase space is completely occupied if the number of the particles in d^3xd^3p is equal to the number of available states:

$$fd^3xd^3p = g_\mathrm{s}\frac{d^3xd^3p}{h^3}, \qquad \text{i.e.,} \qquad f = \frac{g_\mathrm{s}}{h^3}. \tag{2.40}$$

As a consequence $(1 - fh^3/g_s)$ gives the number of vacant states in the phase space. If the number of particles that enter the volume element d^3xd^3p in phase space as a consequence of a collision is proportional to $f'f'_*$ we must multiply it by the number of vacant states which is proportional to $(1-fh^3/g_s)(1-f_*h^3/g_s)$. In this sense we have to make the following substitution in the collision term of the Boltzmann equation

$$f'f'_* \to f'f'_* \left(1 - \frac{fh^3}{g_s}\right)\left(1 - \frac{f_*h^3}{g_s}\right). \tag{2.41}$$

Based on the same reasoning for the particles that leave the volume element d^3xd^3p in phase space, we have to substitute

$$ff_* \to ff_* \left(1 - \frac{f'h^3}{g_s}\right)\left(1 - \frac{f'_*h^3}{g_s}\right). \tag{2.42}$$

For bosons the factor $(1 - fh^3/g_s)$ must be replaced by $(1 + fh^3/g_s)$ in order to include the apparent attraction between the particles (due to the statistics of indistinguishable particles with no restrictions on the occupation of a state). Now we can infer from the Boltzmann equation (2.36) and from the above conclusions the relativistic Uehling–Uhlenbeck equation which reads:

$$p^\alpha \frac{\partial f}{\partial x^\alpha} + m\frac{\partial f K^\alpha}{\partial p^\alpha} = \int \left[f'_* f' \left(1 + \varepsilon\frac{fh^3}{g_s}\right)\left(1 + \varepsilon\frac{f_*h^3}{g_s}\right)\right.$$
$$\left. - f_* f \left(1 + \varepsilon\frac{f'h^3}{g_s}\right)\left(1 + \varepsilon\frac{f'_*h^3}{g_s}\right)\right] F\,\sigma\, d\Omega \frac{d^3p_*}{p_{*0}}, \tag{2.43}$$

where ε is defined through

$$\varepsilon = \begin{cases} +1 & \text{for Bose–Einstein statistics,} \\ -1 & \text{for Fermi–Dirac statistics, and} \\ 0 & \text{for Maxwell–Boltzmann statistics.} \end{cases} \tag{2.44}$$

We remark that frequently fh^3/g_s is replaced by f; then f becomes a probability (that a state is occupied) rather than a probability density in phase space.

Problems

2.2.1 Show that if we let $\overline{f} = g_s/h^3 - f$ in the case of a gas of Fermions ($\varepsilon = -1$), then $\overline{f}$ satisfies the same kinetic equation as f.

2.2.2 Give an interpretation of the result of the previous problem. (Hint: g_s/h^3 is the number of states available per unit volume in phase space. Then the number of unoccupied states per unit volume in phase space (*holes*) is . . . .)

2.3 General equation of transfer

Here we are interested only in external forces that do not depend on the momentum four-vector p^α or are of electromagnetic nature of the type (1.154). In both cases we have that $\partial K^\alpha/\partial p^\alpha = 0$ and the Boltzmann equation (2.36) for a single non-degenerate gas reduces to

$$p^\alpha \frac{\partial f}{\partial x^\alpha} + mK^\alpha \frac{\partial f}{\partial p^\alpha} = \int (f'_* f' - f_* f)\, F\, \sigma\, d\Omega \frac{d^3 p_*}{p_{*0}}. \tag{2.45}$$

In order to obtain the transfer equation we multiply (2.45) by an arbitrary function $\psi(x^\beta, p^\beta)$ and integrate the resulting equation with respect to d^3p/p_0, yielding

$$\int \psi \left[p^\alpha \frac{\partial f}{\partial x^\alpha} + mK^\alpha \frac{\partial f}{\partial p^\alpha} \right] \frac{d^3p}{p_0} = \int \psi (f'_* f' - f_* f) F\, \sigma\, d\Omega \frac{d^3 p_*}{p_{*0}} \frac{d^3 p}{p_0}. \tag{2.46}$$

The first term on the left-hand side of (2.46) can be written as

$$\int \psi p^\alpha \frac{\partial f}{\partial x^\alpha} \frac{d^3p}{p_0} = \frac{\partial}{\partial x^\alpha} \int \psi p^\alpha f \frac{d^3p}{p_0} - \int p^\alpha f \frac{\partial \psi}{\partial x^\alpha} \frac{d^3p}{p_0}, \tag{2.47}$$

while for the second term on the left-hand side of (2.46) we have that

$$m \int \psi K^\alpha \frac{\partial f}{\partial p^\alpha} \frac{d^3p}{p_0} = m \int \frac{\partial \psi K^\alpha f}{\partial p^\alpha} \frac{d^3p}{p_0} - m \int K^\alpha f \frac{\partial \psi}{\partial p^\alpha} \frac{d^3p}{p_0}. \tag{2.48}$$

For the first term on the right-hand side of (2.48) we use the divergence theorem in order to transform the volume integral in momentum space into an integral over an infinitely far surface. Since the one-particle distribution function tends to zero for large values of the momentum variables, this integral vanishes.

The term on the right-hand side of (2.46) can be transformed as follows. First we write

$$\int \psi f'_* f' F\, \sigma\, d\Omega \frac{d^3 p_*}{p_{*0}} \frac{d^3 p}{p_0} = \int \psi' f_* f\, F' \sigma' d\Omega' \frac{d^3 p'_*}{p'_{*0}} \frac{d^3 p'}{p'_0}$$
$$\overset{(2.32)}{=} \int \psi' f_* f\, F\, \sigma\, d\Omega \frac{d^3 p_*}{p_{*0}} \frac{d^3 p}{p_0}, \tag{2.49}$$

by interchanging the role of the particles before and after collision i.e., we replace (p^α, p_*^α) by $(p'^\alpha, p_*'^\alpha)$. Hence, we write the right-hand side of (2.46) as

$$\int \psi (f'_* f' - f_* f)\, F\, \sigma\, d\Omega \frac{d^3 p_*}{p_{*0}} \frac{d^3 p}{p_0} = \int (\psi' - \psi)\, f_*\, f\, F\, \sigma\, d\Omega \frac{d^3 p_*}{p_{*0}} \frac{d^3 p}{p_0}. \tag{2.50}$$

Further, if we rename the two colliding particles in the above equation and interchange (p^α, p'^α) by $(p_*^\alpha, p_*'^\alpha)$, it follows that

$$\int (\psi' - \psi)\, f_*\, f\, F\, \sigma\, d\Omega \frac{d^3 p_*}{p_{*0}} \frac{d^3 p}{p_0} = \int (\psi'_* - \psi_*)\, f_*\, f\, F\, \sigma\, d\Omega \frac{d^3 p_*}{p_{*0}} \frac{d^3 p}{p_0}, \tag{2.51}$$

and (2.50) reads

$$\int \psi(f'_*f' - f_*f)F\sigma d\Omega \frac{d^3p_*}{p_{*0}}\frac{d^3p}{p_0} = \frac{1}{2}\int(\psi' + \psi'_* - \psi - \psi_*)f_*fF\sigma d\Omega \frac{d^3p_*}{p_{*0}}\frac{d^3p}{p_0}$$

$$= \frac{1}{4}\int(\psi + \psi_* - \psi' - \psi'_*)\,(f'_*f' - f_*f)\,F\,\sigma\,d\Omega \frac{d^3p_*}{p_{*0}}\frac{d^3p}{p_0}. \qquad (2.52)$$

The last term on the right-hand side of (2.52) was obtained by replacing (p^α, p^α_*) by (p'^α, p'^α_*).

Now, based on (2.46), (2.47), (2.48) and (2.52), we get the general equation of transfer for a non-degenerate relativistic gas:

$$\frac{\partial}{\partial x^\alpha}\int \psi p^\alpha f\frac{d^3p}{p_0} - \int\left[p^\alpha\frac{\partial\psi}{\partial x^\alpha} + mK^\alpha\frac{\partial\psi}{\partial p^\alpha}\right]f\frac{d^3p}{p_0}$$

$$= \frac{1}{2}\int(\psi' + \psi'_* - \psi - \psi_*)\,f_*f\,F\,\sigma\,d\Omega\frac{d^3p_*}{p_{*0}}\frac{d^3p}{p_0} = \mathcal{P}$$

$$= \frac{1}{4}\int(\psi + \psi_* - \psi' - \psi'_*)\,(f'_*f' - f_*f)\,F\,\sigma\,d\Omega\frac{d^3p_*}{p_{*0}}\frac{d^3p}{p_0}. \qquad (2.53)$$

For a single degenerate gas the left-hand side of (2.53) remains unchanged while the production term due to collisions $\mathcal{P}$ is written as

$$\mathcal{P} = \frac{1}{4}\int(\psi + \psi_* - \psi' - \psi'_*)\left[f'_*f'\left(1 + \varepsilon\frac{fh^3}{g_s}\right)\left(1 + \varepsilon\frac{f_*h^3}{g_s}\right)\right.$$

$$\left. -f_*f\left(1 + \varepsilon\frac{f'h^3}{g_s}\right)\left(1 + \varepsilon\frac{f'_*h^3}{g_s}\right)\right]F\,\sigma\,d\Omega\frac{d^3p_*}{p_{*0}}\frac{d^3p}{p_0}. \qquad (2.54)$$

Problem

2.3.1 Show that the production term due to collisions of a single degenerate gas is given by (2.54).

2.4 Summational invariants

The concept of summational invariants plays an important role in the study of the non-relativistic and relativistic Boltzmann equation since it is related to the equilibrium distribution function. Let us proceed to study it.

The right-hand side of (2.53) or (2.54) represent a production term due to collisions and it vanishes if

$$\psi + \psi_* = \psi' + \psi'_*. \qquad (2.55)$$

A function that satisfies (2.55) is called a summational invariant. For summational invariants the following theorem holds:

Theorem: *A continuous and differentiable function of class C^2 $\psi(p^\alpha)$ is a summational invariant if and only if it is given by:*

$$\psi(p^\alpha) = A + B_\alpha p^\alpha, \tag{2.56}$$

where A is an arbitrary scalar and B_α an arbitrary four-vector that do not depend on p^α.

Proof: Assume that $\psi(p^\alpha)$ is given by (2.56). Due to the energy-momentum conservation law $p^\alpha + p_*^\alpha = p'^\alpha + p_*'^\alpha$, the relationship (2.56) satisfies (2.55) identically.

Next suppose that (2.55) holds. Due to the constraint $p^\alpha p_\alpha = m^2c^2$ or $p^0 = \sqrt{|\mathbf{p}|^2 + m^2c^2}$ we have that $\psi(p^\alpha) = \psi(\mathbf{p})$. Further the energy-momentum conservation law implies that there exists a function Ψ such that

$$\psi(\mathbf{p}) + \psi(\mathbf{p}_*) = \Psi(\boldsymbol{\pi}, u), \tag{2.57}$$

where

$$\boldsymbol{\pi} = \mathbf{p} + \mathbf{p}_*, \qquad u = p_0 + p_{*0} = \sqrt{|\mathbf{p}|^2 + m^2c^2} + \sqrt{|\mathbf{p}_*|^2 + m^2c^2}. \tag{2.58}$$

In fact the first and last expression in (2.57) must be equal for all the four-moment vectors for which the energy-momentum conservation law holds and hence they must be both functions of the quantities which are invariant when we pass from the unprimed to the primed variables according to the definition of this transformation.

In the following we shall need to know the derivatives of p_0 with respect to $\mathbf{p}$ which are given by

$$\frac{\partial p_0}{\partial p^i} = -\frac{p_i}{p_0}, \qquad \frac{\partial^2 p_0}{\partial p^i \partial p^j} = -\left(\frac{p_i p_j}{p_0^3} + \frac{\eta_{ij}}{p_0}\right). \tag{2.59}$$

First we differentiate (2.57) with respect to $\mathbf{p}$ and get

$$\frac{\partial \psi}{\partial p^i} = \frac{\partial \Psi}{\partial \pi^i} - \frac{\partial \Psi}{\partial u}\frac{p_i}{p_0}. \tag{2.60}$$

By applying the same procedure we differentiate (2.57) with respect to $\mathbf{p}_*$ and write the difference

$$\frac{\partial \psi}{\partial p^i} - \frac{\partial \psi}{\partial p_*^i} = -\frac{\partial \Psi}{\partial u}\left(\frac{p_i}{p_0} - \frac{p_{*i}}{p_{*0}}\right), \tag{2.61}$$

which implies that

$$\left(\frac{\partial \psi}{\partial p^i} - \frac{\partial \psi}{\partial p_*^i}\right)\left(\frac{p_j}{p_0} - \frac{p_{*j}}{p_{*0}}\right) = \left(\frac{\partial \psi}{\partial p^j} - \frac{\partial \psi}{\partial p_*^j}\right)\left(\frac{p_i}{p_0} - \frac{p_{*i}}{p_{*0}}\right). \tag{2.62}$$

Next the differentiation of the above equation with respect to p^k, yields

$$\frac{\partial^2\psi}{\partial p^i\partial p^k}\left(\frac{p_j}{p_0}-\frac{p_{*j}}{p_{*0}}\right)+\left(\frac{\partial\psi}{\partial p^i}-\frac{\partial\psi}{\partial p_*^i}\right)\left(\frac{p_jp_k}{p_0^3}+\frac{\eta_{jk}}{p_0}\right)$$
$$=\frac{\partial^2\psi}{\partial p^j\partial p^k}\left(\frac{p_i}{p_0}-\frac{p_{*i}}{p_{*0}}\right)+\left(\frac{\partial\psi}{\partial p^j}-\frac{\partial\psi}{\partial p_*^j}\right)\left(\frac{p_ip_k}{p_0^3}+\frac{\eta_{ik}}{p_0}\right). \tag{2.63}$$

Further the differentiation of (2.63) with respect to p_*^l reads

$$\frac{\partial^2\psi}{\partial p^i\partial p^k}\left(\frac{p_{*j}p_{*l}}{p_{*0}^3}+\frac{\eta_{jl}}{p_{*0}}\right)+\frac{\partial^2\psi}{\partial p_*^i\partial p_*^l}\left(\frac{p_jp_k}{p_0^3}+\frac{\eta_{jk}}{p_0}\right)$$
$$=\frac{\partial^2\psi}{\partial p^j\partial p^k}\left(\frac{p_{*i}p_{*l}}{p_{*0}^3}+\frac{\eta_{il}}{p_{*0}}\right)+\frac{\partial^2\psi}{\partial p_*^j\partial p_*^l}\left(\frac{p_ip_k}{p_0^3}+\frac{\eta_{ik}}{p_0}\right). \tag{2.64}$$

Equation (2.64) has the form

$$C_{ik}(\mathbf{p})D_{jl}(\mathbf{p}_*)+C_{il}(\mathbf{p}_*)D_{jk}(\mathbf{p})=C_{jk}(\mathbf{p})D_{il}(\mathbf{p}_*)+C_{jl}(\mathbf{p}_*)D_{ik}(\mathbf{p}). \tag{2.65}$$

Equation (2.65) is a tensorial equation of fourth order, the fourth order tensors being products of second order tensors that depend on different variables $\mathbf{p}$ and $\mathbf{p}_*$. It will be satisfied only if

$$C_{ik}(\mathbf{p})=-B^0D_{ik}(\mathbf{p}),\quad\text{or}\quad\frac{\partial^2\psi}{\partial p^i\partial p^k}=-B^0\left(\frac{p_ip_k}{p_0^3}+\frac{\eta_{ik}}{p_0}\right)=B^0\frac{\partial^2p_0}{\partial p^i\partial p^k}, \tag{2.66}$$

where B^0 is a scalar. Now the integration of $(2.66)_2$ leads to

$$\psi=A+B^ip_i+B^0p_0,\quad\text{or}\quad\psi=A+B^\alpha p_\alpha, \tag{2.67}$$

where A is another scalar and B^i a vector that do not depend on p^α. Hence we have proved the above theorem.

Other proofs of (2.67) can be found in Chernikov [14], Bichteler [1], Boyer [4], Marle [23], Ehlers [16] and Dijkstra [15]. The proof given above is based on that given by Cercignani [9] for the non-relativistic case and was first presented in a paper by the authors [12], where a slightly different proof is also given (Problem 2.4.1).

We remark that the assumption $\psi\in C^2$ might seem too strong. However, as remarked in [12] thanks to an idea of Wennberg [30], this is not a strong restriction, because one can assume that the collision invariants are distributions and the derivatives are taken in the sense of distributions. The advantage of the proof is that the property of being a collision invariant is shown to be a local one, and this is retained in the case of distributions; the cross-section might vanish for sets of non-zero measure (provided the complement is not of zero measure), but the set of collision invariants would remain the same. Thus, as a corollary, using Wennberg's argument [30], we also prove (Problem 2.4.2).

Corollary. *A generalized function, or distribution, $\psi(p^\alpha)$ is a summational invariant if and only if it is given by (2.56), where A is an arbitrary scalar and B^α an arbitrary four-vector that do not depend on p^α.*

Problems

2.4.1 Prove the theorem given in this section by the following different method: assume that the constraint $p^0 = \sqrt{|\mathbf{p}|^2 + m^2c^2}$ is replaced by the weaker constraint

$$p^\alpha p_\alpha + p_*^\alpha p_{*\alpha} = p'^\alpha p'_\alpha + p'^\alpha_* p'_{*\alpha}$$

and use the same method used in the classical case [9], the only difference being the space dimensions.

2.4.2 Prove the corollary given above when ψ is assumed to be a distribution rather than a function (see [30] for the classical case).

2.5 Macroscopic description

The macroscopic description of the relativistic gas is based on the moments of the one-particle distribution function and defined by

$$T^{\alpha\beta\dots\gamma\delta} = c\int p^\alpha p^\beta \dots p^\gamma p^\delta f \frac{d^3p}{p_0}. \tag{2.68}$$

The first moment is called particle four-flow

$$N^\alpha = c\int p^\alpha f \frac{d^3p}{p_0}. \tag{2.69}$$

The second moment is the energy-momentum tensor

$$T^{\alpha\beta} = c\int p^\alpha p^\beta f \frac{d^3p}{p_0}, \tag{2.70}$$

while the moments of higher order do not have proper names.

The contraction of the third-order moment

$$T^{\alpha\beta\gamma} = c\int p^\alpha p^\beta p^\gamma f \frac{d^3p}{p_0} \tag{2.71}$$

reduces to the particle four-flow, i.e.,

$$T^{\alpha\beta}_{\beta} = c\int m^2c^2 p^\alpha f \frac{d^3p}{p_0} = m^2c^2N^\alpha. \tag{2.72}$$

Here we are interested in the case where the relativistic gas is free of external forces, i.e., $K^\alpha = 0$ and in the following we derive the balance equations for the moments N^α, $T^{\alpha\beta}$ and $T^{\alpha\beta\gamma}$ from the general equation of transfer (2.53) by choosing $\psi = c, cp^\alpha, cp^\alpha p^\beta$, respectively.

- Balance of the particle four-flow: $\psi = c$,

$$\partial_\alpha N^\alpha = 0; \tag{2.73}$$

- Balance of the energy-momentum tensor: $\psi = cp^\alpha$,

$$\partial_\beta T^{\alpha\beta} = 0; \tag{2.74}$$

- Balance of the third-order moment: $\psi = cp^\alpha p^\beta$,

$$\partial_\gamma T^{\alpha\beta\gamma} = P^{\alpha\beta}. \tag{2.75}$$

In (2.75) $P^{\alpha\beta}$ is the production term of the third-order moment defined by

$$P^{\alpha\beta} = \frac{c}{2}\int (p'^\alpha p'^\beta + p'^\alpha_* p'^\beta_* - p^\alpha p^\beta - p^\alpha_* p^\beta_*)\, f_* f\, F\,\sigma\, d\Omega \frac{d^3p_*}{p_{*0}}\frac{d^3p}{p_0} \tag{2.76}$$

for single non-degenerate gases, whereas for single degenerate gases it reads

$$P^{\alpha\beta} = \frac{c}{4}\int (p'^\alpha p'^\beta + p'^\alpha_* p'^\beta_* - p^\alpha p^\beta - p^\alpha_* p^\beta_*)\left[f'_* f'\left(1+\varepsilon\frac{fh^3}{g_s}\right)\left(1+\varepsilon\frac{f_*h^3}{g_s}\right)\right.$$
$$\left. - f_* f\left(1+\varepsilon\frac{f'h^3}{g_s}\right)\left(1+\varepsilon\frac{f'_*h^3}{g_s}\right)\right] F\,\sigma\, d\Omega\frac{d^3p_*}{p_{*0}}\frac{d^3p}{p_0}. \tag{2.77}$$

We note that the contraction of (2.75) leads to (2.73) since

$$P^\alpha_\alpha = 0. \tag{2.78}$$

Another interesting choice for the arbitrary function ψ is

$$\psi = -kc\left[\ln\left(\frac{fh^3}{g_s}\right) - \left(1+\frac{g_s}{\varepsilon fh^3}\right)\ln\left(1+\frac{\varepsilon fh^3}{g_s}\right)\right] \tag{2.79}$$

where $k = 1.38 \times 10^{-23}$J/K is the Boltzmann constant. In this case we get from the general equation of transfer (2.53), with the production term due to collisions $\mathcal{P}$ given by (2.54), a balance equation that, after some manipulations, reads:

$$\partial_\alpha S^\alpha = \varsigma, \tag{2.80}$$

where the quantities S^α and ς are defined by

$$S^\alpha = \int p^\alpha f\left[-kc\ln\left(\frac{fh^3}{g_s}\right) + kc\left(1+\frac{g_s}{\varepsilon fh^3}\right)\ln\left(1+\frac{\varepsilon fh^3}{g_s}\right)\right]\frac{d^3p}{p_0}, \tag{2.81}$$

$$\varsigma = \frac{kc}{4}\int f_* f \ln\left[\frac{f'_* f'\,(1+\varepsilon f h^3/g_s)\,(1+\varepsilon f_* h^3/g_s)}{f_* f\,(1+\varepsilon f' h^3/g_s)\,(1+\varepsilon f'_* h^3/g_s)}\right]\left(1+\varepsilon\frac{f' h^3}{g_s}\right)$$
$$\times\left(1+\varepsilon\frac{f'_* h^3}{g_s}\right)\left[\frac{f'_* f'\,(1+\varepsilon f h^3/g_s)\,(1+\varepsilon f_* h^3/g_s)}{f_* f\,(1+\varepsilon f' h^3/g_s)\,(1+\varepsilon f'_* h^3/g_s)}-1\right] F\,\sigma\,d\Omega\frac{d^3p_*}{p_{*0}}\frac{d^3p}{p_0}. \tag{2.82}$$

Since the distribution functions are non-negative quantities we use the relationships

$$\begin{cases}(1-x)\ln x < 0 & \text{for all} \quad x>0 \quad \text{and} \quad x\neq 1,\\ (1-x)\ln x = 0 & \text{for} \quad x=1,\end{cases} \tag{2.83}$$

to infer that the right-hand side of (2.80) is non-negative and vanishes if and only if

$$f'_* f'\left(1+\varepsilon\frac{f h^3}{g_s}\right)\left(1+\varepsilon\frac{f_* h^3}{g_s}\right) = f_* f\left(1+\varepsilon\frac{f' h^3}{g_s}\right)\left(1+\varepsilon\frac{f'_* h^3}{g_s}\right). \tag{2.84}$$

Eq. (2.80) represents the balance equation of the entropy four-flow S^α and we identify ς as the entropy production rate which is non-negative. The balance equation for the entropy four-flow is then written as

$$\partial_\alpha S^\alpha = \varsigma \geq 0. \tag{2.85}$$

It is interesting to have the entropy four-flow for a gas that obeys the Maxwell–Boltzmann statistics. In this case, taking the limit when $\varepsilon \to 0$ in (2.81) yields

$$S^\alpha = -kc\int p^\alpha f \ln\left(\frac{f h^3}{e g_s}\right)\frac{d^3p}{p_0}. \tag{2.86}$$

Problems

2.5.1 Show that the choice of ψ given by (2.79) in the general equation of transfer (2.53) with $\mathcal{P}$ expressed by (2.54) leads to the equation (2.80) where S^α and ς are represented by (2.81) and (2.82), respectively.

2.5.2 Show that the entropy four-flow of a gas that obeys the Maxwell–Boltzmann statistics is given by (2.86), while the entropy production rate reduces to

$$\varsigma = \frac{kc}{4}\int f_* f \ln\left(\frac{f'_* f'}{f_* f}\right)\left[\frac{f'_* f'}{f_* f}-1\right] F\,\sigma\,d\Omega\frac{d^3p_*}{p_{*0}}\frac{d^3p}{p_0}.$$

2.6 Local Lorentz rest frame

We introduce a frame, called local Lorentz rest frame and denoted by an index R, in which the gas is seen as an isotropic body for an observer that moves with the gas velocity $\mathbf{v}$. In this frame the four-velocity of the gas

$$(U^\alpha) = (c\gamma, \gamma\mathbf{v}), \quad \text{such that} \quad U^\alpha U_\alpha = c^2, \tag{2.87}$$

reduces to

$$(U_R^\alpha) = (c, 0, 0, 0) = (c, \mathbf{0}). \tag{2.88}$$

In the local Lorentz rest frame we introduce the quantities:

$$\begin{cases} n \text{ - particle number density,} \\ p \text{ - isotropic pressure,} \\ e \text{ - internal energy per particle,} \\ T \text{ - temperature,} \\ s \text{ - entropy per particle.} \end{cases} \tag{2.89}$$

The particle number density is related to the number of particles minus the number of anti-particles per unit proper volume.

The flow of the particles of a gas, described by the particle four-flow N^α, has components identified as

$$\begin{cases} N^0 \text{ - } c\times \text{ particle number density,} \\ N^i \text{ - particle flux density.} \end{cases} \tag{2.90}$$

In a local Lorentz rest frame the equilibrium particle four-flow has components

$$(N_R^\alpha) = (cn, \mathbf{0}), \quad \text{or} \quad N_R^0 = cn, \quad N_R^i = 0. \tag{2.91}$$

Let x'^α denote the coordinates of the local Lorentz rest frame and x^α the coordinates of an arbitrary Lorentz frame that are connected by the transformation law $x^\alpha = \bar{\Lambda}^\alpha_\beta x'^\beta$, where $\bar{\mathbf{\Lambda}}(\mathbf{v}) = \mathbf{\Lambda}(-\mathbf{v})$ is given by (1.36). Hence the equilibrium particle four-flow in an arbitrary Lorentz frame is:

$$N_E^\alpha = \bar{\Lambda}^\alpha_\beta N_R^\beta, \quad \text{or} \quad N_E^0 = \gamma nc, \quad N_E^i = \gamma n v^i, \tag{2.92}$$

where the index E denotes the equilibrium value of the particle four-flow. In terms of the four-velocity of the gas U^α the equilibrium particle four-flow in an arbitrary Lorentz frame reads

$$N_E^\alpha = nU^\alpha. \tag{2.93}$$

The energy-momentum tensor of a relativistic gas has its components identified as

$$\begin{cases} T^{00} \text{ - internal energy density,} \\ T^{0i} \text{ - energy flux density}/c, \\ T^{i0} \text{ - } c\times \text{ momentum density,} \\ T^{ij} \text{ - momentum flux density.} \end{cases} \tag{2.94}$$

In a local Lorentz rest frame the equilibrium energy-momentum tensor has the form

$$\left(T_R^{\alpha\beta}\right) = \begin{pmatrix} ne & 0 & 0 & 0 \\ 0 & p & 0 & 0 \\ 0 & 0 & p & 0 \\ 0 & 0 & 0 & p \end{pmatrix}, \tag{2.95}$$

i.e., in this frame the components of $T_R^{\alpha\beta}$ are $T_R^{00} = ne$, $T_R^{0i} = T_R^{i0} = 0$ and $T_R^{ij} = -p\eta^{ij}$.

In an arbitrary Lorentz frame the equilibrium energy-momentum tensor is given by

$$T_E^{\alpha\beta} = \bar{\Lambda}_\gamma^\alpha \bar{\Lambda}_\delta^\beta T_R^{\gamma\delta}, \tag{2.96}$$

which by the use of (1.36) has components

$$\begin{cases} T_E^{00} = (ne+p)\gamma^2 - p, \qquad T_E^{i0} = T_E^{0i} = (ne+p)\gamma^2 v^i/c, \\ T_E^{ij} = (ne+p)\gamma^2 v^i v^j/c^2 - p\eta^{ij}. \end{cases} \tag{2.97}$$

Hence the equilibrium energy-momentum tensor can be written in terms of the four-velocity of the gas U^α as

$$T_E^{\alpha\beta} = (ne+p)\frac{U^\alpha U^\beta}{c^2} - p\eta^{\alpha\beta}. \tag{2.98}$$

Another important equation in the equilibrium theory of a relativistic gas is the Gibbs equation. This equation identifies the potential of the Pfaffian form $de - pdn/n^2$ – whose integrating factor is the reciprocal of the temperature T – with the equilibrium entropy per particle s_E, i.e.,

$$ds_E = \frac{1}{T}\left(de - \frac{p}{n^2}dn\right). \tag{2.99}$$

Problems

2.6.1 Obtain the expressions for the equilibrium values of the particle four-flow $(2.92)_{2,3}$ and of the energy-momentum tensor (2.97) in an arbitrary Lorentz frame.

2.6.2 The equilibrium value of the entropy four-flow in a local Lorentz rest frame is given by $(S_R^\alpha) = (cns_E, \mathbf{0})$. Show that in an arbitrary Lorentz frame it reads $S_E^\alpha = ns_E U^\alpha$.

2.7 Equilibrium distribution function

In equilibrium the entropy production rate must vanish; this, as we know, implies that (2.84) holds. We write this equation again, for convenience:

$$f_*'^{(0)} f'^{(0)} \left(1 + \varepsilon\frac{f^{(0)}h^3}{g_s}\right)\left(1 + \varepsilon\frac{f_*^{(0)}h^3}{g_s}\right)$$

$$= f_*^{(0)} f^{(0)} \left(1 + \varepsilon\frac{f'^{(0)}h^3}{g_s}\right)\left(1 + \varepsilon\frac{f_*'^{(0)}h^3}{g_s}\right) \tag{2.100}$$

where the superscript (0) denotes the equilibrium value of the one-particle distribution function.

Thus we see that the right-hand side of (2.43) must also vanish; this means that the number of particles that enter the volume element d^3xd^3p in phase space must equal the number of particles that leave it.

By taking the logarithms of both sides of (2.100) it follows that

$$\ln\left(\frac{f^{(0)}}{1+\varepsilon f^{(0)}h^3/g_s}\right)+\ln\left(\frac{f_*^{(0)}}{1+\varepsilon f_*^{(0)}h^3/g_s}\right)$$

$$=\ln\left(\frac{f'^{(0)}}{1+\varepsilon f'^{(0)}h^3/g_s}\right)+\ln\left(\frac{f_*'^{(0)}}{1+\varepsilon f_*'^{(0)}h^3/g_s}\right). \tag{2.101}$$

Hence $\ln[f^{(0)}/(1+\varepsilon f^{(0)}h^3/g_s)]$ is a summational invariant and according to (2.56) it must be a linear combination of the momentum four-vector p^α:

$$\ln\left(\frac{f^{(0)}}{1+\varepsilon f^{(0)}h^3/g_s}\right)=-(A+B^\alpha p_\alpha),\quad\text{or}\quad f^{(0)}=\frac{g_s/h^3}{e^{-a+B^\alpha p_\alpha}-\varepsilon}, \tag{2.102}$$

where $a=-A-\ln(g_s/h^3)$.

We shall now determine the two parameters a and B^α of the equilibrium distribution function $(2.102)_2$. First we assume that in the local Lorentz rest frame B^α has only temporal coordinates which we write as

$$(B_R^\alpha)=\left(\frac{\zeta}{mc},\mathbf{0}\right), \tag{2.103}$$

where ζ is a parameter which we shall identify later. Since B^α is a four-vector we have that

$$B_R^\alpha B_{R\alpha}=B^\alpha B_\alpha=\frac{\zeta^2}{(mc)^2},\quad\text{and}\quad\frac{\partial\zeta}{\partial B_\alpha}=\frac{(mc)^2}{\zeta}B^\alpha. \tag{2.104}$$

Inserting the equilibrium distribution function $(2.102)_2$ into the definition of the particle four-flow (2.69) yields

$$N_E^\alpha=c\int p^\alpha\frac{g_s/h^3}{e^{-a+B^\alpha p_\alpha}-\varepsilon}\frac{d^3p}{p_0}. \tag{2.105}$$

Let $\mathcal{I}$ be the integral

$$\mathcal{I}=\int\frac{g_s/h^3}{e^{-a+B^\alpha p_\alpha}-\varepsilon}\frac{d^3p}{p_0}. \tag{2.106}$$

If we differentiate N_E^α with respect to a and $\mathcal{I}$ with respect to B_α we get that

$$\frac{\partial N_E^\alpha}{\partial a}=-c\frac{\partial\mathcal{I}}{\partial B_\alpha}. \tag{2.107}$$

On the other hand, in a local Lorentz rest frame we can use spherical coordinates

$$d^3p = |\mathbf{p}|^2 \sin\theta d|\mathbf{p}|d\theta d\varphi, \tag{2.108}$$

and change the variable of integration by introducing ϑ such that

$$B_R^\alpha p_\alpha = \frac{p_0\zeta}{mc} = \zeta\cosh\vartheta, \quad |\mathbf{p}|^2 = p_0^2 - m^2c^2 = m^2c^2\sinh^2\vartheta, \quad \frac{d|\mathbf{p}|}{p_0} = d\vartheta. \tag{2.109}$$

In this case all integrals in d^3p/p_0 can be written in terms of the integrals

$$J_{nm}(\zeta, a) = \int_0^\infty \frac{\sinh^n\vartheta\cosh^m\vartheta}{e^{\zeta\cosh\vartheta - a} - \varepsilon} d\vartheta. \tag{2.110}$$

Hence in a local Lorentz rest frame we get that the expressions for the particle number density n and for the integral $\mathcal{I}$ in terms of the integrals $J_{nm}(\zeta, a)$ are

$$n = 4\pi(mc)^3\frac{g_s}{h^3}J_{21}, \qquad \mathcal{I} = 4\pi(mc)^2\frac{g_s}{h^3}J_{20}, \tag{2.111}$$

where we have performed the integrations in the angles $0 \le \theta \le \pi$ and $0 \le \varphi \le 2\pi$ which furnish 4π.

By differentiating n with respect to a and $\mathcal{I}$ with respect to ζ and by using the relationships

$$\frac{\partial J_{nm}}{\partial a} = \frac{n-1}{\zeta}J_{n-2,m+1} + \frac{m}{\zeta}J_{n,m-1}, \tag{2.112}$$

$$\frac{\partial J_{nm}}{\partial \zeta} = -\frac{n-1}{\zeta}J_{n-2,m+2} - \frac{m+1}{\zeta}J_{n,m} = -\frac{\partial J_{n,m+1}}{\partial a}, \tag{2.113}$$

it follows that

$$\frac{\partial n}{\partial a} = -\frac{1}{mc}\frac{\partial \mathcal{I}}{\partial \zeta}. \tag{2.114}$$

Now by combining equations (2.93), $(2.104)_2$, (2.107) and (2.114) we get that B^α is given by

$$B^\alpha = \frac{\zeta}{mc^2}U^\alpha. \tag{2.115}$$

Hence we have identified B^α, and we proceed to identifying ζ and a. If we insert the equilibrium distribution function $(2.102)_2$ into the definition of the energy-momentum tensor (2.70), and consider a local Lorentz rest frame, we get

$$ne = 4\pi m^4c^5\frac{g_s}{h^3}J_{22}, \qquad p = \frac{4\pi}{3}m^4c^5\frac{g_s}{h^3}(J_{22} - J_{20}) = \frac{4\pi}{3}m^4c^5\frac{g_s}{h^3}J_{40}, \tag{2.116}$$

by using the relationships

$$ne = \frac{1}{c^2}T_E^{\alpha\beta}U_\alpha U_\beta, \qquad -3p + ne = T_E^{\alpha\beta}\eta_{\alpha\beta}. \tag{2.117}$$

Since it is not possible to obtain from (2.111) and (2.116) explicit expressions for a and ζ, we calculate equilibrium value of the entropy per particle s which is given by (see problem 2.6.2)

$$s_E = \frac{1}{nc^2} S_E^\alpha U_\alpha. \tag{2.118}$$

Hence we obtain, from (2.81), $(2.102)_2$ and (2.118),

$$s_E = k\left(\frac{\zeta}{mc^2}e - a + \frac{4\pi\zeta}{3}\frac{m^3c^3}{n}\frac{g_s}{h^3}J_{40}\right). \tag{2.119}$$

Now we take the differential of the above equation and get

$$ds_E = \frac{k\zeta}{mc^2}\left(de - \frac{p}{n^2}dn\right), \tag{2.120}$$

by using the relationships (2.112), (2.113) and (2.116). We compare (2.120) with the Gibbs equation (2.99), and identify

$$\zeta = \frac{mc^2}{kT}, \tag{2.121}$$

i.e., ζ is the ratio between the rest energy of a particle mc^2 and kT, which gives the order of magnitude of the thermal energy of the gas.

Further (2.119) can be written in terms of the chemical potential in equilibrium, which is identified with the Gibbs function per particle $\mu_E = e - Ts_E + p/n$:

$$a = \frac{1}{kT}\left(e - Ts_E + \frac{p}{n}\right) = \frac{\mu_E}{kT}. \tag{2.122}$$

That is a is identified as the ratio between the chemical potential in equilibrium and kT which gives the order of magnitude of the thermal energy of the gas.

Hence we have identified a, ζ and B^α, and the equilibrium distribution function $(2.102)_2$ can be written as:

- relativistic Maxwell–Boltzmann statistics:

$$f^{(0)} = \frac{g_s}{h^3}e^{\frac{\mu_E}{kT} - \frac{U^\alpha p_\alpha}{kT}}, \tag{2.123}$$

- relativistic Fermi–Dirac (+) and Bose–Einstein (−) statistics:

$$f^{(0)} = \frac{g_s/h^3}{e^{-\frac{\mu_E}{kT} + \frac{U^\alpha p_\alpha}{kT}} \pm 1}. \tag{2.124}$$

The expression (2.123) was obtained by Jüttner [18] in 1911 and the expression (2.124) was also obtained by him [19] in 1928. The distribution function (2.123) is also known in the literature as the Maxwell–Jüttner distribution function.

We remark that the relation between the temperature and the internal energy per particle is not linear and thus one cannot speak of a "temperature" for a non-equilibrium gas, as one frequently does in the non-relativistic case.

Problems

2.7.1 Prove the relationships (2.112) and (2.113).

2.7.2 Obtain the two equations (2.119) and (2.120) for the entropy per particle in equilibrium s_E.

2.8 Trend to equilibrium. $\mathcal{H}$-theorem

Solutions which depend on just one space-time coordinate are simpler than others. A frame-invariant definition is

$$f = f(a_\alpha x^\alpha, p^\alpha) \tag{2.125}$$

where a_α are the components of a constant four-vector.

There are two important sub-cases, according to whether a_α is time-like or space-like. In the first case $|a_0| > |\mathbf{a}|$, where $\mathbf{a}$ is the vector made out of the three-dimensional components, in the second $|a_0| < |\mathbf{a}|$. In the first case we can always find a frame where the solution is space homogeneous, i.e., it depends on just the time coordinate. This frame is obtained by taking the time axis aligned with the constant four-vector. In the second case, the solution is steady (and dependent on just one space coordinate) in a reference frame which is obtained by taking the first space axis aligned with the constant four-vector. An example of the latter case is offered by the discussion of the structure of a shock wave in Chapter 9.

Here we discuss the space homogeneous case. Before we proceed the analysis of the trend to equilibrium of a relativistic gas – which refers to the so-called $\mathcal{H}$-theorem – we shall introduce some inequalities that will be used in this section. Let y be a positive real variable; then it is easy to show that the inequalities hold:

$$(y-1) - \ln y \geq 0, \qquad \text{or with} \quad y = \frac{1}{x}, \qquad x\ln x + 1 - x \geq 0. \tag{2.126}$$

Further one can always find a constant $\mathcal{C}$ such that the following inequality holds (Problem 2.7.1),

$$x\ln x + 1 - x - \mathcal{C}\mathcal{G}(|x-1|)|x-1| \geq 0, \tag{2.127}$$

where

$$\mathcal{G}(|x-1|) = \begin{cases} |x-1| & \text{if} \quad 0 \leq |x-1| \leq 1, \\ 1 & \text{if} \quad |x-1| \geq 1. \end{cases} \tag{2.128}$$

We return to the balance equation for the entropy four-flow (2.85) and write it as

$$\partial_\alpha \mathcal{H}^\alpha = \mathcal{S} \leq 0, \tag{2.129}$$

where $\mathcal{H}^\alpha = -S^\alpha/k$ and $\mathcal{S} = -\varsigma/k$. By considering that the one-particle distribution function does not depend on the space coordinates and introducing the scalar invariant in a local Lorentz rest frame

$$\mathcal{H} = \frac{1}{c^2} U_R^\alpha \mathcal{H}_\alpha = \frac{1}{c}\mathcal{H}^0$$

$$= \int p^0 f \left[\ln\left(\frac{fh^3}{g_s}\right) - \left(1 + \frac{\varepsilon f h^3}{g_s}\right) \ln\left(1 + \frac{\varepsilon f h^3}{g_s}\right)\right] \frac{d^3p}{p_0}, \tag{2.130}$$

we get that the inequality (2.129) reduces to

$$\frac{d\mathcal{H}}{dt} = \mathcal{S} \leq 0. \tag{2.131}$$

From the above equation one infers that $\mathcal{H}$ decreases. Further the time derivative of $\mathcal{H}$ vanishes when the one-particle distribution function is the equilibrium distribution function, which will be denoted by $\mathcal{H}_E \equiv \mathcal{H}(f^{(0)})$.

Now we choose in the inequalities (2.126)

$$y = \frac{1 + \varepsilon f h^3/g_s}{1 + \varepsilon f^{(0)} h^3/g_s}, \quad \text{and} \quad x = \frac{fh^3/g_s}{1 + \varepsilon f h^3/g_s} \frac{1 + \varepsilon f^{(0)} h^3/g_s}{f^{(0)} h^3/g_s}, \tag{2.132}$$

and get by adding the two resulting inequalities:

$$f \left[\ln \frac{fh^3/g_s}{1 + \varepsilon f h^3/g_s} - \ln \frac{f^{(0)} h^3/g_s}{1 + \varepsilon f^{(0)} h^3/g_s}\right] - \frac{g_s}{\varepsilon h^3} \ln\left(\frac{1 + \varepsilon f h^3/g_s}{1 + \varepsilon f^{(0)} h^3/g_s}\right) \geq 0. \tag{2.133}$$

Next we multiply (2.133) by p^0 and integrate the resulting equation over all values of d^3p/p_0 and obtain

$$\mathcal{H} - \mathcal{H}_E \geq \int p^0 (f - f^{(0)}) \ln \frac{f^{(0)} h^3/g_s}{1 + \varepsilon f^{(0)} h^3/g_s} \frac{d^3p}{p_0}$$

$$\stackrel{(2.102)_1}{=} -\frac{U_R^\alpha}{c} \int p_\alpha (f - f^{(0)}) [A + B^\beta p_\beta] \frac{d^3p}{p_0}. \tag{2.134}$$

The right-hand side of the above inequality vanishes if we impose that

$$U_\alpha N^\alpha = U_\alpha N_E^\alpha, \qquad U_\alpha U_\beta T^{\alpha\beta} = U_\alpha U_\beta T_E^{\alpha\beta}, \tag{2.135}$$

and by recalling that B^α is proportional to U^α. The conditions (2.135) imply that

$$\mathcal{H} \geq \mathcal{H}_E, \tag{2.136}$$

and we conclude from (2.136) and (2.131) that $\mathcal{H}$ is bounded from below by $\mathcal{H}_E$, its derivative being negative and vanishing when the one-particle distribution function is the equilibrium one. Although these conditions do not assure that $\mathcal{H}$ tends to $\mathcal{H}_E$ when $t \to \infty$, we shall assume it in order to prove the following

Theorem 1: *If $\mathcal{H}$ tends to $\mathcal{H}_E$ when $t \to \infty$, then f tends strongly to $f^{(0)}$ in L^1.*

Proof: We begin by writing instead of (2.133) the inequality

$$f\left[\ln\frac{fh^3/g_s}{1+\varepsilon fh^3/g_s}-\ln\frac{f^{(0)}h^3/g_s}{1+\varepsilon f^{(0)}h^3/g_s}\right]-\frac{g_s}{\varepsilon h^3}\ln\left(\frac{1+\varepsilon fh^3/g_s}{1+\varepsilon f^{(0)}h^3/g_s}\right)$$

$$-\mathcal{C}\mathcal{G}\left(\frac{|f^{(0)}-f|}{f^{(0)}(1+\varepsilon fh^3/g_s)}\right)\frac{|f^{(0)}-f|}{1+\varepsilon f^{(0)}h^3/g_s}\geq 0, \tag{2.137}$$

which is a consequence of (2.127). Following the same procedure as above in order to derive (2.134) we get

$$\mathcal{H}-\mathcal{H}_E\geq\mathcal{C}\left[\int_{L_t}p^0\frac{|f^{(0)}-f|}{1+\varepsilon f^{(0)}h^3/g_s}\frac{d^3p}{p_0}\right.$$

$$\left.+\int_{S_t}p^0\frac{|f^{(0)}-f|^2}{f^{(0)}(1+\varepsilon fh^3/g_s)(1+\varepsilon f^{(0)}h^3/g_s)}\frac{d^3p}{p_0}\right]. \tag{2.138}$$

In the above inequality L_t and S_t denote the integration domains where $|f^{(0)}-f|$ is larger or smaller than $f^{(0)}$, respectively. Since we have assumed that $\mathcal{H}$ tends to $\mathcal{H}_E$ when $t\to\infty$ and the integrands are positive, both integrals on the right-hand side of (2.138) must tend to zero in this limit. Further by using Schwarz's inequality one can show that:

$$\int_{S_t}p^0\frac{|f^{(0)}-f|}{1+\varepsilon f^{(0)}h^3/g_s}\frac{d^3p}{p_0}\leq\left[\int_{S_t}p^0\frac{f^{(0)}(1+\varepsilon fh^3/g_s)}{(1+\varepsilon f^{(0)}h^3/g_s)}\frac{d^3p}{p_0}\right]^{\frac{1}{2}}$$

$$\times\left[\int_{S_t}p^0\frac{|f^{(0)}-f|^2}{f^{(0)}(1+\varepsilon fh^3/g_s)(1+\varepsilon f^{(0)}h^3/g_s)}\frac{d^3p}{p_0}\right]^{\frac{1}{2}}\to 0, \tag{2.139}$$

when $t\to\infty$. Hence the following integral over the entire domain of integration tends to zero when $t\to\infty$,

$$\int p^0\frac{|f^{(0)}-f|}{1+\varepsilon f^{(0)}h^3/g_s}\frac{d^3p}{p_0}=\int_{L_t}p^0\frac{|f^{(0)}-f|}{1+\varepsilon f^{(0)}h^3/g_s}\frac{d^3p}{p_0}$$

$$+\int_{S_t}p^0\frac{|f^{(0)}-f|}{1+\varepsilon f^{(0)}h^3/g_s}\frac{d^3p}{p_0}\to 0. \tag{2.140}$$

From the above equation we conclude that f tends strongly to $f^{(0)}$ in L^1, provided that $\mathcal{H}$ tends to $\mathcal{H}_E$ when $t\to\infty$, thus the theorem is proved.

This proof of the theorem is less technical than the following one and hence more appealing to a physicist than that of the following theorem. There is the strong assumption, however, that $\mathcal{H}$ tends to $\mathcal{H}_E$. A simple proof of this result is needed; to this aim, one of the authors conjectured (in the classical case) an inequality on the entropy source [10] which would lead to an exponential decay of the entropy to its equilibrium value. This inequality has been disproved by several counterexamples if only mass, energy and entropy are assumed to exist at $t = 0$ [2, 31]. An example where the entropy dissipation rate is arbitrarily low was recently supplied by Bobylev and Cercignani [3]. A modified form of the inequality, which still serves the purpose, has been proved by Toscani and Villani [27]. Entropic convergence to equilibria for general initial data was, however, first discussed by Carlen and Carvalho [7] and inequalities showing that entropy converges in broad generality were established later by Marten and Carvalho [24].

It is not easy to prove that $\mathcal{H}$ tends to $\mathcal{H}_E$ when $t \to \infty$. All the available proofs are rather technical. The proof given here is based on a paper by the authors [13], which in turn is based on the corresponding result in the non-relativistic case given by Carleman [5, 6] and Cercignani [11]. Thus we prove the following theorem.

Theorem 2: *If $\mathcal{H}(t)$ is a continuous and differentiable function of t that satisfies (2.131) and (2.136) and f is uniformly bounded and equicontinuous in p_α, and $\int p_0^{1+\eta} f d^3p$ ($\eta > 0$) is uniformly bounded, then $\mathcal{H}$ tends to $\mathcal{H}_E$ when $t \to \infty$.*

Proof: From (2.136) and (2.131) we have inferred that $\mathcal{H}$ is bounded from below by $\mathcal{H}_E$, its derivative being negative and vanishing when the one-particle distribution function is the equilibrium one. Hence it is possible to find a sequence of instants of time $t_1, t_2, \ldots, t_n, \ldots$ such that

$$\lim_{n\to\infty} \frac{d\mathcal{H}}{dt}(t_n) = 0. \tag{2.141}$$

Further, because of Ascoli–Arzelà's theorem, there exists a uniformly converging sequence $f(t_n) \equiv f_n$ such that on any compact set $\mathcal{D}$

$$\lim_{n\to\infty} f_n = f_\infty. \tag{2.142}$$

If we prove that

$$f'_{*\infty} f'_\infty \left(1 + \varepsilon\frac{f_\infty h^3}{g_s}\right)\left(1 + \varepsilon\frac{f_{*\infty} h^3}{g_s}\right) = f_{*\infty} f_\infty \left(1 + \varepsilon\frac{f'_\infty h^3}{g_s}\right)\left(1 + \varepsilon\frac{f'_{*\infty} h^3}{g_s}\right), \tag{2.143}$$

then according to (2.100) f_∞ is an equilibrium distribution function and, as we shall prove, $\mathcal{H}$ tends to $\mathcal{H}_E$ when $t \to \infty$.

In order to prove (2.143) we suppose that there exists a domain of positive measure $\mathcal{D}$ such that

$$\left| f'_{*\infty} f'_\infty \left(1 + \varepsilon\frac{f_\infty h^3}{g_s}\right)\left(1 + \varepsilon\frac{f_{*\infty} h^3}{g_s}\right) \right.$$
$$\left. - f_{*\infty} f_\infty \left(1 + \varepsilon\frac{f'_\infty h^3}{g_s}\right)\left(1 + \varepsilon\frac{f'_{*\infty} h^3}{g_s}\right)\right| \geq R > 0, \tag{2.144}$$

where R is a constant. The uniform convergence of f_n to f_∞ implies that it is possible to find an n_0 such that

$$\left| f'_{*n} f'_n \left(1 + \varepsilon \frac{f_n h^3}{g_s}\right)\left(1 + \varepsilon \frac{f_{*n} h^3}{g_s}\right)\right.$$

$$\left. - f_{*n} f_n \left(1 + \varepsilon \frac{f'_n h^3}{g_s}\right)\left(1 + \varepsilon \frac{f'_{*n} h^3}{g_s}\right)\right| \geq \frac{R}{2} > 0, \tag{2.145}$$

for $n > n_0$ in $\mathcal{D}$. Hence it follows that

$$\left| f'_{*n} f'_n \left(1 + \varepsilon \frac{f_n h^3}{g_s}\right)\left(1 + \varepsilon \frac{f_{*n} h^3}{g_s}\right) - f_{*n} f_n \left(1 + \varepsilon \frac{f'_n h^3}{g_s}\right)\left(1 + \varepsilon \frac{f'_{*n} h^3}{g_s}\right)\right|$$

$$\times \left| \ln \frac{f'_{*n} f'_n \left(1 + \varepsilon f_n h^3/g_s\right)\left(1 + \varepsilon f_{*n} h^3/g_s\right)}{f_{*n} f_n \left(1 + \varepsilon f'_n h^3/g_s\right)\left(1 + \varepsilon f'_{*n} h^3/g_s\right)} \right|$$

$$\geq \frac{R}{2} \ln\left(1 + \frac{R}{2M^2(1 + \varepsilon M h^3/g_s)^2}\right), \tag{2.146}$$

by considering that the one-particle distribution function f is bounded by a constant M.

If we multiply (2.146) by $(c/4) F \sigma d\Omega (d^3 p_*/p_{*0})(d^3 p/p_0)$ and integrate the resulting equation over all values of $d^3 p_*/p_{*0}$ and $d^3 p/p_0$ we get

$$-\frac{d\mathcal{H}}{dt}(t_n) \geq \frac{Rc}{8} \ln\left(1 + \frac{R}{2M^2(1 + \varepsilon M h^3/g_s)^2}\right) \int_{\mathcal{D}} F \sigma d\Omega \frac{d^3 p_*}{p_{*0}} \frac{d^3 p}{p_0}, \tag{2.147}$$

by the use of (2.82) and (2.129) through (2.131). Taking the limit of the above expression when $n \to \infty$ yields

$$0 \leq -\frac{Rc}{8} \ln\left(1 + \frac{R}{2M^2(1 + \varepsilon M h^3/g_s)^2}\right) \int_{\mathcal{D}} F \sigma d\Omega \frac{d^3 p_*}{p_{*0}} \frac{d^3 p}{p_0}, \tag{2.148}$$

which contradicts the premise that $\mathcal{D}$ has a non-zero measure. Hence f_∞ is an equilibrium distribution function; in order to prove the theorem we need to show that f_∞ is the equilibrium distribution function $f^{(0)}$ determined by the initial data and the conservation laws; this is easy because we can pass to the limit under the integral sign in the expression of the conserved moments thanks to the assumption that $\int p_0^{1+\eta} f d^3 p$ $(\eta > 0)$ is uniformly bounded. Then $\mathcal{H}$ tends to $\mathcal{H}_E$ and f tends to $f^{(0)}$ when $t \to \infty$ along an arbitrary sequence. This is based on the well-known fact that under the constraints provided by the conservation laws, $\mathcal{H}$ is a convex functional with $\mathcal{H}_E$ as its minimum (attained for $f = f^{(0)}$). The theorem is proved.

Problems

2.7.1 Prove (2.127).

2.7.2 Show that the inequalities (2.137) and (2.146) hold.

2.9 The projector $\Delta_{\alpha\beta}$

Before we analyze the equilibrium states of a relativistic gas we shall introduce a projector that will be useful to interpret physically the consequences dictated by an equilibrium distribution function.

From the definition of the four-velocity of the fluid U^α we introduce a symmetric tensor

$$\Delta^{\alpha\beta} = \eta^{\alpha\beta} - \frac{1}{c^2}U^\alpha U^\beta, \tag{2.149}$$

that projects an arbitrary four-vector into another four-vector perpendicular to U^α since

$$\Delta^{\alpha\beta}U_\beta = 0. \tag{2.150}$$

The tensor $\Delta^{\alpha\beta}$ is called a projector and it has the properties

$$\Delta^{\alpha\beta}\Delta_{\beta\gamma} = \Delta^\alpha_\gamma, \qquad \Delta^\alpha_\beta\Delta^{\beta\gamma} = \Delta^{\alpha\gamma}, \qquad \Delta^\alpha_\alpha = 3. \tag{2.151}$$

In a local Lorentz rest frame the projector has the form:

$$(\Delta_R^{\alpha\beta}) = (\Delta_{R\alpha\beta}) = \begin{pmatrix} 0 & 0 & 0 & 0 \\ 0 & -1 & 0 & 0 \\ 0 & 0 & -1 & 0 \\ 0 & 0 & 0 & -1 \end{pmatrix}, \qquad (\Delta^\alpha_{R\beta}) = \begin{pmatrix} 0 & 0 & 0 & 0 \\ 0 & 1 & 0 & 0 \\ 0 & 0 & 1 & 0 \\ 0 & 0 & 0 & 1 \end{pmatrix}. \tag{2.152}$$

If A^α is a four-vector and $T^{\alpha\beta}$ a tensor,

$$A^{(\alpha)} = \Delta^\alpha_\beta A^\beta, \tag{2.153}$$

$$T^{(\alpha\beta)} = \frac{1}{2}\left(\Delta^\alpha_\gamma\Delta^\beta_\delta + \Delta^\beta_\gamma\Delta^\alpha_\delta\right)T^{\gamma\delta} \tag{2.154}$$

$$T^{[\alpha\beta]} = \frac{1}{2}\left(\Delta^\alpha_\gamma\Delta^\beta_\delta - \Delta^\beta_\gamma\Delta^\alpha_\delta\right)T^{\gamma\delta} \tag{2.155}$$

represent a four-vector, a symmetric tensor and an antisymmetric tensor that have only the spatial components in a local Lorentz rest frame. Further

$$T^{\langle\alpha\beta\rangle} = T^{(\alpha\beta)} - \frac{1}{3}\Delta^{\alpha\beta}\Delta_{\gamma\delta}T^{(\gamma\delta)} \tag{2.156}$$

is a traceless symmetric tensor since $\eta_{\alpha\beta}T^{\langle\alpha\beta\rangle} = 0$, and it is easy to show that we have also $\Delta_{\alpha\beta}T^{\langle\alpha\beta\rangle} = 0$.

We shall decompose the contravariant components of the gradient ∂^α into two parts, i.e., one parallel to the four-velocity U^α and another perpendicular to it:

$$\partial^\alpha = \frac{1}{c^2}U^\alpha U^\beta \partial_\beta + \left(\eta^{\alpha\beta} - \frac{1}{c^2}U^\alpha U^\beta\right)\partial_\beta = \frac{1}{c^2}U^\alpha D + \nabla^\alpha, \tag{2.157}$$

where

$$D = U^\alpha \partial_\alpha, \qquad \nabla^\alpha = \left(\eta^{\alpha\beta} - \frac{1}{c^2}U^\alpha U^\beta\right)\partial_\beta = \Delta^{\alpha\beta}\partial_\beta. \tag{2.158}$$

The differential operator D is called a convective time derivative and in a local Lorentz rest frame it reduces to a partial differentiation with respect to time,

$$D_R = \frac{\partial}{\partial t}. \tag{2.159}$$

The gradient operator ∇^α in a local Lorentz rest frame reduces to

$$(\nabla_{R\alpha}) = (0, \vec{\nabla}), \tag{2.160}$$

that is it has only spatial components. Further it is easy to verify from $(2.158)_2$ that $U_\alpha \nabla^\alpha = 0$.

Problems

2.9.1 Show that the properties of the projector $\Delta^{\alpha\beta}$ given by (2.151) are valid.

2.9.2 Show that $U^\alpha U^\beta / c^2 - \eta^{\alpha\beta}$ is not a projector. (Hint: a projector P is such that $P^2 = P$.)

2.9.3 What is the form of the projector $\Delta^{\alpha\beta}$ in a Minkowski space with signature $(-1, 1, 1, 1)$?

2.9.4 Show that the relationship between $T_{(\alpha\beta)}$ and the symmetric part of a tensor $T^S_{\alpha\beta} = (T_{\alpha\beta} + T_{\beta\alpha})/2$ is given by

$$T_{(\alpha\beta)} = T^S_{\alpha\beta} - \frac{1}{c^2}U_\alpha U^\gamma T^S_{\gamma\beta} - \frac{1}{c^2}U_\beta U^\gamma T^S_{\gamma\alpha} + \frac{1}{c^4}U_\alpha U_\beta U^\gamma U^\delta T^S_{\gamma\delta}.$$

Further show that $\nabla^{(\alpha}U^{\beta)}$ is equal to the symmetric part of $\nabla^\alpha U^\beta$.

2.10 Equilibrium states

As we know the equilibrium distribution function (2.123) or (2.124) imposes that the Boltzmann equation (2.43) has a vanishing right-hand side. A question that arises refers to the conditions imposed on the left-hand side of the Boltzmann equation when the distribution function is the equilibrium one with parameters depending on space-time coordinates. We proceed to determine these conditions in the case of a relativistic gas with particles of non-vanishing rest mass and without external forces $K^\alpha = 0$. We begin by inserting (2.123) or (2.124) into (2.43), yielding

$$(\mu_E \partial_\alpha T - T\partial_\alpha \mu_E)\, p^\alpha + (T\partial_\alpha U_\beta - U_\beta \partial_\alpha T)\, p^\alpha p^\beta = 0. \tag{2.161}$$

Equation (2.161) can be transformed into a polynomial equation for the spatial components of the momentum four-vector p^α. This last equation must be valid for all values of p^i and it is satisfied only if all coefficients of the polynomial equation vanish, i.e.,

$$\mu_E \partial_\alpha T = T\partial_\alpha \mu_E, \qquad T\,(\partial_\alpha U_\beta + \partial_\beta U_\alpha) = U_\beta \partial_\alpha T + U_\alpha \partial_\beta T. \tag{2.162}$$

If we multiply $(2.162)_2$ by $U^\alpha U^\beta$, we get by using the definition of the convective time derivative $(2.158)_1$ and of the relationship which follows from the constraint $U^\alpha U_\alpha = c^2$, i.e., $U_\alpha \partial_\beta U^\alpha = 0$:

$$DT = 0. \tag{2.163}$$

Hence in equilibrium the convective time derivative of the temperature T is zero, and in particular in a local Lorentz rest frame the temperature may not depend on the time since $\partial T/\partial t = 0$. The same conclusion can be drawn for the chemical potential μ_E through multiplication of $(2.162)_1$ by U^α.

Next we multiply $(2.162)_2$ by U^β only and get

$$DU^\alpha = \frac{c^2}{T}\partial^\alpha T + U^\beta U_\alpha \partial_\beta T = \frac{c^2}{T}\nabla^\alpha T, \tag{2.164}$$

through the use of (2.157) and (2.163). This equation dictates that the existence of a temperature gradient in equilibrium must be compensated for by an acceleration.

On the other hand, we can write $(2.162)_2$ as

$$\frac{\partial}{\partial x^\alpha}\left(\frac{U^\beta}{kT}\right) + \frac{\partial}{\partial x^\beta}\left(\frac{U^\alpha}{kT}\right) = 0. \tag{2.165}$$

The above equation, known as a Killing equation (for more details see Weinberg [29]), has a general solution given by

$$\frac{U^\alpha}{kT} = \Omega^{\alpha\beta} x_\beta + \Omega^\alpha. \tag{2.166}$$

The four-vector $U^\alpha/(kT)$ is called a Killing vector; $\Omega^{\alpha\beta} = -\Omega^{\beta\alpha}$ denotes an antisymmetrical constant tensor and Ω^α a constant four-vector. Hence the general solution for the four-velocity of a relativistic gas in equilibrium is the superposition of a translation and a Lorentz transformation.

We collect all results on the equilibrium conditions for a relativistic gas dictated by the Boltzmann equation:

a) the convective time derivatives of the temperature and of the chemical potential in equilibrium must vanish;
b) the existence of a temperature gradient must be compensated for by an acceleration;
c) the four-velocity is the superposition of a translation and a Lorentz transformation.

Problems

2.10.1 Show that if $DT = 0$ and $D\mu_E = 0$, then we have also that $Dp = 0$. (Hint: Use the definition of the chemical potential in equilibrium and of the Gibbs equation.)

2.10.2 Show that (2.164) can be written as

$$\frac{nh_E}{c^2} DU^\alpha = \nabla^\alpha p,$$

where $h_E = e + p/n$ is the enthalpy per particle. (Hint: Use $(2.162)_1$ and the Gibbs equation.)

References

[1] K. Bichteler, Bemerkungen über relativistische Stoßinvarianten, *Z. Physik* **182**, 521–523 (1965).

[2] A. V. Bobylev, Exact solutions of the Boltzmann equation and the theory of relaxation of Maxwell gas, *USSR Theor. Math Phys.* **60**, 280–310 (1984).

[3] A.V. Bobylev and C. Cercignani, On the rate of entropy production for the Boltzmann equation *J. Stat. Phys.* **94**, 603–618 (1999).

[4] R. H. Boyer, Some uses of hyperbolic velocity space, *Am. J. Phys.* **33**, 910–916 (1965).

[5] T. Carleman, Sur la théorie de l'équation intégro-differentielle de Boltzmann, *Acta Mathematica* **60**, 91–146 (1933).

[6] T. Carleman, *Problèmes mathématiques dans la théorie cinétique des gaz* (Almqvist & Wiksells, Uppsala, 1957).

[7] E. A. Carlen and M. C. Carvalho, The central limit theorem, entropy production and stability of the rate of convergence for the Boltzmann equation, in *Probabilistic methods in mathematical physics*, pp. 140–154, eds. F. Gerra, M. I. Loffredo and C. Marchioro, (World Scientific, Singapore, 1992).

[8] C. Cercignani, *The Boltzmann equation and its applications* (Springer, New York, 1988).

[9] C. Cercignani, Are there more than five linearly-independent collision invariants for the Boltzmann equation?, *J. Stat. Phys.* **58**, 817–823 (1990).

[10] C. Cercignani, H-theorem and trend to equilibrium in the kinetic theory of gases, *Arch. Mech.* **34**, 231–241 (1982).

[11] C. Cercignani, *Ludwig Boltzmann. The man who trusted atoms*, (Oxford Univ. Press, Oxford, 1998).

[12] C. Cercignani and G. M. Kremer, On relativistic collisional invariants, *J. Stat. Phys.* **96**, 439–445 (1999).

[13] C. Cercignani and G. M. Kremer, Trend to equilibrium of a degenerate relativistic gas, *J. Stat. Phys.* **98**, 441–456 (2000).

[14] N. A. Chernikov, Equilibrium distribution of the relativistic gas, *Acta Phys. Polon.* **26**, 1069–1092 (1964).

[15] J. J. Dijkstra, Mathematical aspects of relativistic kinetic theory II. The summational invariants, *Proc. Kon. Ned. Akad. Wetensch.* **81B**, 265–289 (1978).

[16] J. Ehlers, General relativity and kinetic theory, in *Proceedings of the international school of physics, Enrico Fermi, course 47* , pp. 1–70, ed. R. K. Sachs (Academic Press, New York, 1971).

[17] S. R. de Groot, W. A. van Leeuwen and Ch. G. van Weert, *Relativistic kinetic theory* (North-Holland, Amsterdam, 1980).

[18] F. Jüttner, Das Maxwellsche Gesetz der Geschwindigkeitsverteilung in der Relativtheorie, *Ann. Physik und Chemie* **34**, 856–882 (1911).

[19] F. Jüttner, Die relativistische Quantentheorie des idealen Gases, *Zeitschr. Physik* **47**, 542–566 (1928).

[20] D. C. Kelly, *The kinetic theory of a relativistic gas*, unpublished report (Miami University, Oxford, 1963).

[21] G. M. Kremer, *Uma introdução à teoria cinética relativística* (Editora da UFPR, Curitiba, 1998).

[22] L. D. Landau and E. M. Lifshitz, *The classical theory of fields*, 4th ed. (Pergamon Press, Oxford, 1980).

[23] C. Marle, Sur l'établissement des équations de l'hydrodynamique des fluides relativistiques dissipatifs I. - L'équation de Boltzmann relativiste, *Ann. Inst. Henri Poincaré* **10**, 67–126 (1969).

[24] E. A. Marten and M. C. Carvalho, Entropy production estimates for Boltzmann equations with physically realistic collisions kernels, *J. Stat. Phys.* **74**, 743–782 (1994).

[25] J. M. Stewart, *Non-equilibrium relativistic kinetic theory*, Lecture Notes in Physics, vol. 10 (Springer, Heidelberg, 1971).

[26] J. L. Synge, *The relativistic gas* (North-Holland, Amsterdam, 1957).

[27] G. Toscani and C. Villani, Sharp entropy dissipation bounds and explicit rate of trend to equilibrium for the spatially homogeneous Boltzmann equation, *Comm. Math. Phys.* **203**, 667–706 (1999).

[28] E. A. Uehling, and G. E. Uhlenbeck, Transport phenomena in Einstein–Bose and Fermi–Dirac gases. I, *Phys. Rev.* **43**, 552–561 (1933).

[29] S. Weinberg, *Gravitation and cosmology. Principles and applications of the theory of relativity* (Wiley, New York, 1972).

[30] B. Wennberg, On an entropy dissipation inequality for the Boltzmann equation, *C. R. A. S. (Paris)* **315**, 1441–1446 (1992).

[31] B. Wennberg, Entropy dissipation and and moment production for the Boltzmann equation, *J. Stat. Phys.* **86**, 1053–1066 (1997).

Chapter 3

Fields in Equilibrium

3.1 The general case

The study of the fields particle number density, pressure and internal energy per particle – or simply energy per particle – in equilibrium gives precious preliminary information on the behavior of a gas described by an equilibrium distribution function. This information is useful not only when the gas actually is in a state of homogeneous equilibrium, but also when the main behavior is described in a sufficiently accurate way by an equilibrium distribution function; we shall see in the next few chapters several instances of this situation and the conditions under which it occurs. This topic is usually paid a lip-service in the non-relativistic case, because the fields to be studied are simply related; in the relativistic case rather complicated functions of at least one parameter occur (in the simplest case only modified Bessel functions show up).

It is thus required to investigate the various cases that may occur with particular attention to the general behavior of the relations between the fields and the parameters, as well as to the limiting form of these relations when the parameters become very small or very large. As a simple application we shall indicate how to use the relations to be developed in the case of white dwarf stars. We shall also discuss relativistic Bose–Einstein condensation. A typical quantity that is studied and plotted in the various cases is the heat capacity (at constant volume), which is a constant ($3k/2$) in the non-relativistic case.

In Section 2.7 we have written the equilibrium quantities[1] in terms of the integrals $J_{nm}(\zeta, \mu_E)$ given in (2.110) which we reproduce below

$$J_{nm}(\zeta, \mu_E) = \int_0^\infty \frac{\sinh^n \vartheta \cosh^m \vartheta}{e^{-\mu_E/kT} e^{\zeta \cosh \vartheta} - \varepsilon} d\vartheta. \tag{3.1}$$

[1] Recall that we have identified $\zeta = mc^2/(kT)$ and $a = \mu_E/(kT)$.

Here we introduce another integral $I_n(\zeta,\mu_E)$ defined by

$$I_n(\zeta,\mu_E) = \int_0^\infty \frac{\cosh(n\vartheta)}{e^{-\mu_E/kT}e^{\zeta\cosh\vartheta}-\varepsilon}d\vartheta, \tag{3.2}$$

such that the integrals $J_{nm}(\zeta,\mu_E)$ reduce to a combination of the $I_n(\zeta,\mu_E)$ integrals by the use of simple relationships between hyperbolic functions. Here we give only the equations we shall use in this chapter:

$$J_{21} = \frac{1}{4}(I_3 - I_1), \qquad J_{22} = \frac{1}{8}(I_4 - I_0), \qquad J_{40} = \frac{1}{8}(I_4 - 4I_2 + 3I_0). \tag{3.3}$$

The above relationships are used to rewrite the particle number density $(2.111)_1$, the energy per particle $(2.116)_1$ and the pressure $(2.116)_2$ as

$$n = 4\pi(mc)^3\frac{g_s}{h^3}J_{21} = \pi(mc)^3\frac{g_s}{h^3}(I_3 - I_1), \tag{3.4}$$

$$ne = 4\pi m^4c^5\frac{g_s}{h^3}J_{22} = \frac{\pi}{2}m^4c^5\frac{g_s}{h^3}(I_4 - I_0), \quad e = mc^2\frac{J_{22}}{J_{21}} = \frac{1}{2}mc^2\frac{I_4 - I_0}{I_3 - I_1}, \tag{3.5}$$

$$p = \frac{4\pi}{3}m^4c^5\frac{g_s}{h^3}J_{40} = \frac{4\pi}{3}m^4c^5\frac{g_s}{h^3}(J_{22} - J_{20}) = \frac{\pi}{6}m^4c^5\frac{g_s}{h^3}(I_4 - 4I_2 + 3I_0). \tag{3.6}$$

The above constitutive equations, that are functions of (T,μ_E), represent the general case of a relativistic gas with particles of non-vanishing rest mass. For relativistic gases with particles of vanishing rest mass we have to use another change of the variable of integration and instead of (2.109) we write:

$$B_R^\alpha p_\alpha = \frac{cp_0}{kT}, \qquad |\mathbf{p}|^2 = p_0^2, \qquad x = \frac{cp_0}{kT}. \tag{3.7}$$

With this change of the variable of integration, instead of the integrals $J_{nm}(\zeta,\mu_E)$ we have another type of integrals which read:

$$i_n(\mu_E) = \int_0^\infty \frac{x^n}{e^{-\mu_E/kT}e^x - \varepsilon}dx. \tag{3.8}$$

In a local Lorentz rest frame the particle number density, the energy per particle and the pressure for a relativistic gas with particles of vanishing rest mass – which is the case of photons – read as

$$n = 4\pi\frac{g_s}{h^3}\left(\frac{kT}{c}\right)^3 i_2, \tag{3.9}$$

$$ne = 4\pi c\frac{g_s}{h^3}\left(\frac{kT}{c}\right)^4 i_3, \quad \text{or} \quad e = kT\frac{i_3}{i_2}, \tag{3.10}$$

$$p = \frac{1}{3}ne. \tag{3.11}$$

In the next sections we shall analyze the limiting cases of the above expressions.

Problems

3.1.1 Show that the relationships (3.3) between the integrals $J_{nm}(\zeta,\mu_E)$ and $I_n(\zeta,\mu_E)$ hold.

3.1.2 Obtain the expressions for the particle number density (3.9), energy per particle (3.10) and pressure (3.11) of a relativistic gas whose particles have a vanishing rest mass.

3.2 Non-degenerate gas

Non-degenerate gases obey Maxwell–Boltzmann statistics so that the one-particle distribution function is the Maxwell–Jüttner distribution function (2.123). The fields in equilibrium for non-degenerate gases are represented in terms of modified Bessel functions of the second kind, so that we begin this section with a brief review of the properties of these functions. After introducing the general expressions for the fields in equilibrium, we study the limiting cases of a non-degenerate gas that correspond to a non-relativistic gas and to an ultra-relativistic gas.

3.2.1 Modified Bessel function of second kind

A particularity of the integrals $I_n(\zeta,\mu_E)$ given in (3.2) is that they tend to a modified Bessel function of the second kind $K_n(\zeta)$ in the limits $e^{-\mu_E/kT}\gg 1$ (when $\varepsilon\neq 0$) or $\varepsilon=0$ that refer to a non-degenerate relativistic gas characterized by the Maxwell–Jüttner distribution function (2.123). Hence we have

$$I_n(\zeta,\mu_E)\longrightarrow e^{\mu_E/kT}K_n(\zeta),\quad \text{for}\quad e^{-\mu_E/kT}\gg 1\quad\text{or}\quad \varepsilon=0, \tag{3.12}$$

where the modified Bessel function of the second kind is defined by

$$K_n(\zeta)=\int_0^\infty e^{-\zeta\cosh\vartheta}\cosh(n\vartheta)d\vartheta. \tag{3.13}$$

The modified Bessel function satisfies the linear differential equation of second order for $\omega=\omega(\zeta)$:

$$\zeta^2\frac{d^2\omega}{d\zeta^2}+\zeta\frac{d\omega}{d\zeta}-(\zeta^2+n^2)\omega=0, \tag{3.14}$$

which reduces to the Bessel differential equation of order n by the change of variable $z=i\zeta$. To check that (3.13) verifies the differential equation (3.14) one has to make use of the recurrence relation

$$K_{n+1}(\zeta)=K_{n-1}(\zeta)+\frac{2n}{\zeta}K_n(\zeta), \tag{3.15}$$

which is a direct consequence of the definition (3.13).

Another integral representation of $K_n(\zeta)$ can be obtained by changing the variable of integration as follows. First we write (3.13) as

$$K_n(\zeta) = \frac{1}{2}\int_{-\infty}^{+\infty} e^{-\zeta\cosh\vartheta} e^{n\vartheta}\, d\vartheta, \tag{3.16}$$

and introduce the variable η through $e^{\vartheta} = 1/(2\eta^2\zeta)$, yielding

$$\begin{aligned} K_n(\zeta) &= \frac{1}{(2\zeta)^n}\int_0^\infty \frac{e^{-(\frac{1}{4\eta^2}+\eta^2\zeta^2)}}{\eta^{(2n+1)}}\, d\eta \\ &= \frac{1}{(2\zeta)^n}\frac{1}{\Gamma(n+\frac{1}{2})}\int_0^\infty e^{-(\frac{1}{4\eta^2}+\eta^2\zeta^2)}\, d\eta \int_0^\infty e^{-\tau\eta^2}\tau^{n-\frac{1}{2}}\, d\tau \\ &= \frac{1}{(2\zeta)^n}\frac{\Gamma(\frac{1}{2})}{\Gamma(n+\frac{1}{2})}\int_0^\infty \frac{1}{2\sqrt{\zeta^2+\tau}} e^{-\sqrt{\zeta^2+\tau}}\tau^{n-\frac{1}{2}}\, d\tau. \end{aligned} \tag{3.17}$$

Here we have used the integrals

$$\int_0^\infty e^{-\tau\eta^2}\tau^{n-\frac{1}{2}}\, d\tau = \frac{\Gamma(n+\frac{1}{2})}{\eta^{2n+1}}, \qquad \int_0^\infty e^{-(a^2\xi^2+\frac{b^2}{\xi^2})}\, d\xi = \frac{\Gamma(\frac{1}{2})}{2a} e^{-2ab}, \tag{3.18}$$

where $\Gamma(n+1) = n\Gamma(n)$ with $\Gamma(1) = 1$ and $\Gamma(\frac{1}{2}) = \sqrt{\pi}$ denotes the gamma function. Next we introduce another variable of integration $y = \sqrt{1+\tau/\zeta^2}$ and get an integral representation of the modified Bessel function of the second kind that reads

$$K_n(\zeta) = \left(\frac{\zeta}{2}\right)^n \frac{\Gamma(\frac{1}{2})}{\Gamma(n+\frac{1}{2})}\int_1^\infty e^{-\zeta y}(y^2-1)^{n-\frac{1}{2}}\, dy. \tag{3.19}$$

From the above representation one can easily obtain the recurrence relations

$$\frac{d}{d\zeta}\left(\frac{K_n(\zeta)}{\zeta^n}\right) = -\frac{K_{n+1}(\zeta)}{\zeta^n}, \qquad \frac{d}{d\zeta}\left(\zeta^n K_n(\zeta)\right) = -\zeta^n K_{n-1}(\zeta). \tag{3.20}$$

Another representation of the modified Bessel function of the second kind is obtained by replacing the variable of integration by $y = \cosh t$, yielding

$$K_n(\zeta) = \left(\frac{\zeta}{2}\right)^n \frac{\Gamma(\frac{1}{2})}{\Gamma(n+\frac{1}{2})}\int_0^\infty e^{-\zeta\cosh t}\sinh^{2n} t\, dt. \tag{3.21}$$

The asymptotic expansion of $K_n(\zeta)$ for large values of ζ, i.e., $\zeta \gg 1$, is given by [1, 4]

$$\begin{aligned} K_n(\zeta) = \sqrt{\frac{\pi}{2\zeta}}\frac{1}{e^\zeta}\Big[1 &+ \frac{4n^2-1}{8\zeta} + \frac{(4n^2-1)(4n^2-9)}{2!(8\zeta)^2} \\ &+ \frac{(4n^2-1)(4n^2-9)(4n^2-25)}{3!(8\zeta)^3} + \cdots\Big]. \end{aligned} \tag{3.22}$$

For small values of ζ, i.e., $\zeta \ll 1$, the following expansion of $K_n(\zeta)$ is useful [1, 4],

$$\begin{aligned} K_n(\zeta) &= (-1)^{n+1}\sum_{k=0}^{\infty}\frac{\left(\frac{\zeta}{2}\right)^{n+2k}}{k!(n+k)!}\times\left[\ln\frac{\zeta}{2}-\frac{1}{2}\psi(k+1)-\frac{1}{2}\psi(n+k+1)\right] \\ &\quad +\frac{1}{2}\sum_{k=0}^{n-1}(-1)^k\frac{(n-k-1)!}{k!\left(\frac{\zeta}{2}\right)^{n-2k}}. \end{aligned} \tag{3.23}$$

Here the function $\psi(n)$ is defined by

$$\psi(n+1) = -\gamma + \sum_{k=1}^{n}\frac{1}{k}, \qquad \psi(1) = -\gamma \tag{3.24}$$

with $\gamma = 0.577\,215\,664\ldots$ being the Euler constant.

Problems

3.2.1.1 Show that the recurrence relation (3.15) holds and check that (3.13) verifies the differential equation (3.14).

3.2.1.2 Show that

$$\lim_{\zeta\to 0}(K_n(\zeta)\zeta^n) = 2^{n-1}(n-1)!.$$

(Hint: Change the variable of integration of (3.19) by introducing $z = \zeta y$.)

3.2.1.3 Introduce the abbreviation $G(\zeta) = K_3(\zeta)/K_2(\zeta)$ and show from the recurrence relation (3.15) that

$$\frac{K_0(\zeta)}{K_2(\zeta)} = 1+\frac{8}{\zeta^2}-2\frac{G(\zeta)}{\zeta}, \qquad \frac{K_1(\zeta)}{K_2(\zeta)} = G(\zeta)-\frac{4}{\zeta},$$

$$\frac{K_4(\zeta)}{K_2(\zeta)} = 1+6\frac{G(\zeta)}{\zeta}, \qquad \frac{K_5(\zeta)}{K_2(\zeta)} = G(\zeta)+\frac{8}{\zeta}+48\frac{G(\zeta)}{\zeta^2},$$

$$\frac{K_6(\zeta)}{K_2(\zeta)} = 1+\frac{80}{\zeta^2}+16\frac{G(\zeta)}{\zeta}+480\frac{G(\zeta)}{\zeta^3}.$$

These relationships will be used later in this chapter and in the subsequent chapters.

3.2.2 Expressions for n, e and p

Here we give general expressions for the fields in equilibrium of non-degenerate gases, i.e., relativistic gases that obey the Maxwell–Boltzmann statistics. We begin with the particle number density (3.4) which for a non-degenerate gas reduces to

$$n = 4\pi m^2 ckT\frac{g_s}{h^3}K_2(\zeta)e^{\mu_E/kT}, \tag{3.25}$$

by the use of (3.12) and of the relationships given in problem 3.2.1.3. From (3.25) one can express the equilibrium chemical potential μ_E in terms of the particle number density n and of the temperature T:

$$\mu_E = kT \ln\left[\frac{nh^3}{4\pi g_s m^2 ckT K_2(\zeta)}\right]. \tag{3.26}$$

We can eliminate also the chemical potential from the equilibrium distribution function in the Maxwell–Boltzmann statistics, i.e., from the Maxwell–Jüttner distribution function (2.123), yielding

$$f^{(0)} = \frac{n}{4\pi m^2 ckT K_2(\zeta)} e^{-\frac{U^\alpha p_\alpha}{kT}}. \tag{3.27}$$

The energy per particle (3.5) for a non-degenerate relativistic gas is written as

$$e = mc^2\left[\frac{K_3(\zeta)}{K_2(\zeta)} - \frac{kT}{mc^2}\right] = mc^2\left[G(\zeta) - \frac{1}{\zeta}\right], \tag{3.28}$$

if we make use again of (3.12) and of the relationships given in Problem 3.2.1.3.

For a non-degenerate relativistic gas the pressure (3.6) is given by

$$p = 4\pi m^2 c(kT)^2 \frac{g_s}{h^3} K_2(\zeta) e^{\mu_E/kT} \overset{(3.25)}{=} nkT, \tag{3.29}$$

by taking into account the same relationships we have used to obtain the particle number density and the energy per particle.

The entropy per particle in equilibrium for a non-degenerate gas can be obtained from (2.122), (3.26), (3.28) and (3.29), yielding

$$s_E = -k\left\{\ln\left[\frac{nh^3\zeta}{4\pi g_s (mc)^3 K_2(\zeta)}\right] - \zeta G(\zeta)\right\}. \tag{3.30}$$

Another important quantity is the thermodynamic potential called enthalpy per particle defined by

$$h_E = e + \frac{p}{n}, \tag{3.31}$$

which for a non-degenerate relativistic gas can be written as

$$h_E = mc^2 G(\zeta), \tag{3.32}$$

thanks to (3.28) and (3.29).

The heat capacities per particle at constant volume and at constant pressure are defined, respectively, by

$$c_v = \left(\frac{\partial e}{\partial T}\right)_v, \qquad c_p = \left(\frac{\partial h_E}{\partial T}\right)_p. \tag{3.33}$$

For a non-degenerate relativistic gas their expressions follow from (3.28) and (3.32) and read

$$c_v = k\left(\zeta^2 + 5G\zeta - G^2\zeta^2 - 1\right), \qquad c_p = k\left(\zeta^2 + 5G\zeta - G^2\zeta^2\right). \tag{3.34}$$

Problems

3.2.2.1 Obtain (3.30) through insertion of the Maxwell–Jüttner distribution function (3.27) into the definition of the entropy four-flow for a gas that obeys the Maxwell–Boltzmann statistics, given in (2.86), and integration of the resulting equation. Recall that $s_E = S_E^\alpha U_\alpha/(nc^2)$, and that the calculation must be done in a local Lorentz rest frame.

3.2.2.2 Obtain the expressions (3.34) for the heat capacities. (Hint: Show first that the derivative of $G(\zeta)$ with respect to ζ, denoted by G', is given by $G' = G^2 - 5G/\zeta - 1$.)

3.2.3 Non-relativistic limit

The equations of the last section are functions of a parameter which represents the ratio between the rest energy and the thermal energy of a particle, i.e., $\zeta = mc^2/(kT)$. For electrons (say) whose rest mass is $m_e \approx 9.11 \times 10^{-31}$ kg, we have that $\zeta = 5.93 \times 10^9/T$. Hence by considering low temperatures where $\zeta \gg 1$, we can use the approximation (3.22) for the modified Bessel function of the second kind and write the energy per particle (3.28) in the non-relativistic limit as

$$e = mc^2 + \frac{3}{2}kT\left(1 + \frac{5}{4\zeta} + \cdots\right). \tag{3.35}$$

The first term on the right-hand side of (3.35) is the rest energy, the second term corresponds to the non-relativistic value of the energy per particle, while the third one is its first relativistic correction. Note that if we consider only the first and the second terms on the right-hand side of (3.35) we get that the difference between the energy per particle and the rest energy, denoted by $e^\star$, is proportional to the pressure, i.e.,

$$n(e - mc^2) = ne^\star \approx \frac{3}{2}p. \tag{3.36}$$

The entropy per particle in equilibrium (3.30) in the non-relativistic limit reduces to

$$s_E = \underline{k\left\{\ln\frac{T^{\frac{3}{2}}}{n} - \ln\left[\frac{h^3}{g_s(2\pi mk)^{\frac{3}{2}}}\right] + \frac{5}{2}\right\}} + k\left\{\frac{15}{4\zeta} + \cdots\right\}, \tag{3.37}$$

where the underlined term corresponds to the classical Sackur–Tetrode formula, while the other terms are its relativistic corrections.

In the non-relativistic limit the enthalpy per particle (3.31) and the heat capacities per particle at constant volume and constant pressure (3.34) are given by

$$h_E = mc^2 + \frac{5}{2}kT\left(1 + \frac{3}{4\zeta} + \cdots\right), \tag{3.38}$$

$$c_v = \frac{3}{2}k\left(1+\frac{5}{2\zeta}+\cdots\right), \qquad c_p = \frac{5}{2}k\left(1+\frac{3}{2\zeta}+\cdots\right). \tag{3.39}$$

From (3.39) one can build the ratio between the heat capacities per particle, yielding

$$\frac{c_p}{c_v} = \frac{5}{3}\left(1-\frac{1}{\zeta}+\cdots\right). \tag{3.40}$$

The first term on the right-hand side of (3.39) and of (3.40) correspond to the non-relativistic values.

Since a non-degenerate gas is characterized by $e^{-\mu_E/kT} \gg 1$ while a non-relativistic gas by $\zeta \gg 1$, we may say that a non-degenerate *and* non-relativistic gas should be characterized by $e^{-\mu_E/kT+\zeta} \gg 1$. This last condition implies together with (3.26) and the approximation (3.22) $K_2(\zeta) \approx \sqrt{\pi/(2\zeta)}/e^{\zeta}$ that

$$\frac{n}{g_s}\left(\frac{h^2}{2\pi mkT}\right)^{\frac{3}{2}} \ll 1. \tag{3.41}$$

This is a well-known condition in non-relativistic statistical mechanics, that for a non-degenerate gas the thermal wavelength $\lambda = h/(2\pi mkT)^{\frac{1}{2}}$ is much smaller than the average interparticle distance $l = (1/n)^{\frac{1}{3}}$, i.e., $\lambda \ll l$.

Problems

3.2.3.1 Obtain the expressions in the limit of a non-relativistic gas for the energy per particle (3.35), entropy per particle (3.37) and heat capacities (3.39).

3.2.3.2 Consider a local Lorentz rest frame where $(U^\alpha) = (c, \mathbf{0})$ and that the modulus of the velocity $\mathbf{v} = c\mathbf{p}/p^0$ of the gas particles is small when compared to the speed of light, i.e., $v \ll c$. Show that the Maxwell–Jüttner distribution function (3.27) reduces to the Maxwell distribution function for the moments:

$$f^{(0)} = \frac{n}{(2\pi kT)^{3/2}} e^{-\frac{|\mathbf{p}|^2}{2mkT}},$$

in the non-relativistic limiting case where $\zeta \gg 1$. Note that the distribution function tends rapidly to zero for large speeds.

3.2.4 Ultra-relativistic limit

The ultra-relativistic limit is characterized by the condition that the parameter $\zeta = mc^2/(kT) \ll 1$. This condition is attained if the temperature is very high, or if the rest mass is very small which is the case of neutrinos and, in the limiting case, photons since their rest mass is zero. When $\zeta \ll 1$ one can use the asymptotic expansion for $K_n(\zeta)$, given in (3.23), and get that the leading terms of $K_2(\zeta)$ and $K_3(\zeta)$ are

$$K_2(\zeta) \approx \frac{2}{\zeta^2}, \qquad K_3(\zeta) \approx \frac{8}{\zeta^3}. \tag{3.42}$$

Based on the above asymptotic values it is easy to get that the ultra-relativistic expressions for the particle number density (3.25), energy per particle (3.28), pressure (3.29) and enthalpy per particle (3.32) reduce to

$$n = 8\pi\frac{g_s}{h^3}\left(\frac{kT}{c}\right)^3 e^{\mu_E/kT}, \qquad \text{or} \qquad \mu_E = kT\ln\left[\frac{n}{8\pi g_s}\left(\frac{hc}{kT}\right)^3\right], \tag{3.43}$$

$$e = 3kT, \qquad p = \frac{1}{3}ne = nkT, \qquad h_E = 4kT. \tag{3.44}$$

Note that the pressure is one-third of the value of the energy density in the ultra-relativistic limiting case, whereas its value in the non-relativistic limiting case is two-thirds of the energy density (see (3.36)).

The heat capacities per particle at constant volume and at constant pressure in the ultra-relativistic limit – obtained through the derivatives of the energy per particle and of the entropy per particle with respect to the temperature, respectively – are given by

$$c_v = 3k, \qquad c_p = 4k, \qquad \text{such that} \qquad \frac{c_p}{c_v} = \frac{4}{3}. \tag{3.45}$$

In Figure 3.1 we have plotted the ratio c_v/k as a function of the parameter ζ. We note that in the non-relativistic limit it tends to 3/2 while in the ultra-relativistic limit to 3.

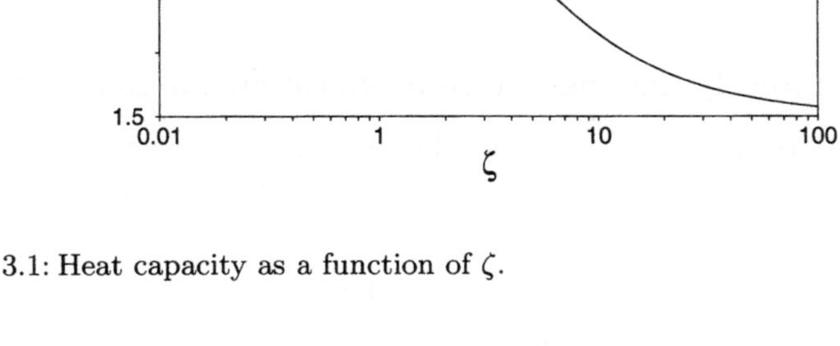

Figure 3.1: Heat capacity as a function of ζ.

Problems

3.2.4.1 Show that in the case when $kT \gg mc^2$, the integrals $i_n(\mu_E)$ in (3.8) for $e^{-\mu_E/kT} \gg 1$ (when $\varepsilon \neq 0$) or $\varepsilon = 0$ reduce to

$$i_n(\mu_E) \longrightarrow e^{\mu_E/kT}\Gamma(n+1), \quad \text{for} \quad e^{-\mu_E/kT} \gg 1 \quad \text{or} \quad \varepsilon = 0.$$

3.2.4.2 Show that the expressions for the particle number density (3.43) and for the energy per particle (3.44) follow also by using the above result and equations (3.9) and (3.10), respectively.

3.2.4.3 Show that the entropy per particle of an ultra-relativistic gas is given by:

$$s_E = -k\left\{\ln\left[\frac{n}{8\pi g_s}\left(\frac{hc}{kT}\right)^3\right] - 4\right\}.$$

3.3 Degenerate relativistic Fermi gas

In this section we shall study a relativistic gas that obeys the Fermi–Dirac statistics. First we shall deal with the case of a completely degenerate relativistic Fermi gas. In this case the distribution function has a constant value, equal to one, up to a momentum which is known as the Fermi momentum. The distribution function vanishes for all values which are larger than the Fermi momentum. The Fermi momentum is associated with a Fermi energy which is the lowest energy where all levels of energy are occupied by one particle. The equilibrium fields of particle number density, energy per particle and pressure are calculated for this case and the non-relativistic and ultra-relativistic limits are determined. The results obtained for the completely degenerate relativistic Fermi gas are used to study white dwarf stars, and to show that there exists no white dwarf star with mass larger than a limiting mass, the so-called Chandrasekhar limit. We finish this section on relativistic Fermi gas with an analysis of the case of a strongly degenerate Fermi gas in which a less restrictive assumption on the distribution function is made so that the fields may also depend on the temperature and not only on the particle number density. As in the case of a complete degeneracy, the equilibrium fields of particle number density, energy per particle and pressure are calculated and the non-relativistic and ultra-relativistic limits are determined.

3.3.1 Completely degenerate relativistic Fermi gas

Let us first write the particle number density (3.4), the energy density (3.5) and the pressure (3.6) for a degenerate relativistic Fermi gas, explicitly in terms of the integrals (3.2):

$$n = 4\pi(mc)^3\frac{g_s}{h^3}\int_0^\infty \frac{\sinh^2\vartheta\cosh\vartheta}{e^{-u_0}e^{\zeta(\cosh\vartheta-1)}+1}d\vartheta, \tag{3.46}$$

$$ne = 4\pi m^4 c^5 \frac{g_s}{h^3} \int_0^\infty \frac{\sinh^2 \vartheta \cosh^2 \vartheta}{e^{-u_0} e^{\zeta(\cosh\vartheta - 1)} + 1} d\vartheta, \tag{3.47}$$

$$p = \frac{4}{3}\pi m^4 c^5 \frac{g_s}{h^3} \int_0^\infty \frac{\sinh^4 \vartheta}{e^{-u_0} e^{\zeta(\cosh\vartheta - 1)} + 1} d\vartheta, \tag{3.48}$$

where $u_0 = \mu_E/kT - \zeta$.

In Section 3.2 we stated that a non-degenerate and non-relativistic gas is characterized by $e^{-\mu_E/kT+\zeta} = e^{-u_0} \gg 1$. Hence the condition $e^{-u_0} \ll 1$ should describe a degenerate relativistic Fermi gas.

The gas is called completely degenerate if u_0 is so large that we can take the limit $u_0 \to \infty$. In the case of complete degeneracy the denominators of the integrands in (3.46) through (3.48) tend to unity up to a fixed value of ϑ whereas from this value the denominators tend to infinity. Hence we get:

$$n = 4\pi (mc)^3 \frac{g_s}{h^3} \int_0^{\vartheta_F} \sinh^2 \vartheta \cosh \vartheta d\vartheta, \tag{3.49}$$

$$ne = 4\pi m^4 c^5 \frac{g_s}{h^3} \int_0^{\vartheta_F} \sinh^2 \vartheta \cosh^2 \vartheta d\vartheta, \tag{3.50}$$

$$p = \frac{4}{3}\pi m^4 c^5 \frac{g_s}{h^3} \int_0^{\vartheta_F} \sinh^4 \vartheta d\vartheta, \tag{3.51}$$

where ϑ_F is the largest value of ϑ where the denominators of the integrands in (3.46) through (3.48) tend to unity. Since $\cosh\vartheta = p_0/mc$ we have that

$$\begin{aligned} \vartheta_F &= \operatorname{arcosh} \frac{p_0^F}{mc} = \operatorname{arcosh} \sqrt{1 + \frac{p_F^2}{m^2 c^2}} \\ &= \operatorname{arsinh} \frac{p_F}{mc} = \operatorname{arcosh} \left(\frac{E_F}{mc^2} + 1 \right). \end{aligned} \tag{3.52}$$

The quantity $p_F = |\mathbf{p}_F|$ is the so-called Fermi momentum and E_F is the Fermi energy which is the lowest energy where all levels of energy are occupied by one particle. Note that we have excluded the rest energy from the definition of the Fermi energy.

Now we integrate (3.49) to (3.51) and get

$$n = \frac{4\pi}{3}(mc)^3 \frac{g_s}{h^3} \sinh^3 \vartheta_F = \frac{4\pi}{3} \frac{g_s}{h^3} p_F^3 = \frac{4\pi}{3}(mc)^3 \frac{g_s}{h^3} x^3, \tag{3.53}$$

$$ne = \frac{\pi}{2} m^4 c^5 \frac{g_s}{h^3} \left[\frac{\sinh 4\vartheta_F}{4} - \vartheta_F \right] = \frac{\pi}{6} m^4 c^5 \frac{g_s}{h^3} \mathcal{G}(x), \tag{3.54}$$

$$p = \frac{\pi}{6} m^4 c^5 \frac{g_s}{h^3} \left[\frac{\sinh 4\vartheta_F}{4} - 2\sinh 2\vartheta_F + 3\vartheta_F \right] = \frac{\pi}{6} m^4 c^5 \frac{g_s}{h^3} \mathcal{F}(x). \tag{3.55}$$

Here we have followed Chandrasekhar [2] and introduced the abbreviations:

$$x = \frac{p_F}{mc}, \tag{3.56}$$

$$\mathcal{G}(x) = 3x(2x^2+1)\sqrt{1+x^2} - 3\,\mathrm{arsinh}\,x, \tag{3.57}$$

$$\mathcal{F}(x) = x(2x^2-3)\sqrt{1+x^2} + 3\,\mathrm{arsinh}\,x. \tag{3.58}$$

The asymptotic forms of $\mathcal{G}(x)$ and $\mathcal{F}(x)$ for $x \ll 1$ and $x \gg 1$ read:

$$\mathcal{G}(x) = 8x^3 + \frac{12}{5}x^5 - \frac{3}{7}x^7 + \frac{1}{6}x^9 + \cdots, \qquad x \ll 1, \tag{3.59}$$

$$\mathcal{F}(x) = \frac{8}{5}x^5 - \frac{4}{7}x^7 + \frac{1}{3}x^9 + \cdots, \qquad x \ll 1, \tag{3.60}$$

$$\mathcal{G}(x) = 6x^4 + 6x^2 - 3\ln(2x) + \frac{3}{4} - \frac{3}{4}\frac{1}{x^2} + \cdots, \qquad x \gg 1, \tag{3.61}$$

$$\mathcal{F}(x) = 2x^4 - 2x^2 + 3\ln(2x) - \frac{7}{4} + \frac{5}{4}\frac{1}{x^2} + \cdots, \qquad x \gg 1. \tag{3.62}$$

The condition $x \ll 1$ or $p_F \ll mc$ corresponds to a completely degenerate non-relativistic Fermi gas and in this case we have from (3.53) through (3.55) together with (3.59) and (3.60):

$$p_F = \left(\frac{3nh^3}{4\pi g_s}\right)^{\frac{1}{3}}, \tag{3.63}$$

$$e = mc^2 + \frac{3}{10}\left(\frac{3h^3}{4\pi m^{\frac{3}{2}} g_s}\right)^{\frac{2}{3}} n^{\frac{2}{3}} \left[1 - \frac{5}{28}\left(\frac{3h^3}{4\pi g_s}\right)^{\frac{2}{3}} \frac{n^{\frac{2}{3}}}{m^2c^2} + \cdots\right], \tag{3.64}$$

$$p = \frac{1}{5}\left(\frac{3h^3}{4\pi m^{\frac{3}{2}} g_s}\right)^{\frac{2}{3}} n^{\frac{5}{3}} \left[1 - \frac{5}{14}\left(\frac{3h^3}{4\pi g_s}\right)^{\frac{2}{3}} \frac{n^{\frac{2}{3}}}{m^2c^2} + \cdots\right]. \tag{3.65}$$

From (3.64) and (3.65) we conclude that for a completely degenerate non-relativistic gas the energy per particle e and the pressure p depend only on the particle number density n. Further if we consider only the first approximation in the brackets of equations (3.64) and (3.65) we infer that the pressure is proportional to $n^{\frac{5}{3}}$ and that the difference between the energy per particle e and the rest energy mc^2 is related to the pressure, i.e.,

$$p \approx \frac{1}{5}\left(\frac{3h^3}{4\pi m^{3/2} g_s}\right)^{\frac{2}{3}} n^{\frac{5}{3}}, \qquad n(e - mc^2) = ne^\star \approx \frac{3}{2}p. \tag{3.66}$$

The result (3.66) is the same as we have found for a non-degenerate and non-relativistic gas (3.36), i.e., the pressure is two-thirds of the energy density.

For $x \gg 1$ or $p_F \gg mc$ we have a completely degenerate ultra-relativistic Fermi gas so that (3.54) and (3.55) reduce to

$$e = \frac{3}{4}\left(\frac{3c^3h^3}{4\pi g_s}\right)^{\frac{1}{3}} n^{\frac{1}{3}} \left[1 + \left(\frac{4\pi g_s}{3h^3}\right)^{\frac{2}{3}} \frac{m^2c^2}{n^{\frac{2}{3}}} + \cdots\right], \tag{3.67}$$

$$p = \frac{1}{4}\left(\frac{3c^3h^3}{4\pi g_s}\right)^{\frac{1}{3}} n^{\frac{4}{3}} \left[1 - \left(\frac{4\pi g_s}{3h^3}\right)^{\frac{2}{3}} \frac{m^2c^2}{n^{\frac{2}{3}}} + \cdots\right]. \tag{3.68}$$

We infer from (3.67) and (3.68) that the energy per particle and the pressure of a completely degenerate ultra-relativistic Fermi gas depend only on the particle number density. Again by considering only the first term in the brackets of equations (3.67) and (3.68) we have

$$p \approx \frac{1}{4}\left(\frac{3c^3h^3}{4\pi g_s}\right)^{\frac{1}{3}} n^{\frac{4}{3}}, \qquad ne = 3p, \tag{3.69}$$

that is the pressure is proportional to $n^{\frac{4}{3}}$ and the energy density ne is proportional to the pressure. This last result is the same as we have found for a non-degenerate and ultra-relativistic gas $(3.44)_2$, i.e., the pressure is one-third of the energy density.

Problems

3.3.1.1 Show that for a completely degenerate Fermi gas the particle number density, the energy density and the pressure are given by (3.53), (3.54) and (3.55), respectively.

3.3.1.2 Obtain the asymptotic expansions for $\mathcal{G}(x)$ and $\mathcal{F}(x)$ for $x \ll 1$ and $x \gg 1$ which are given by (3.59) through (3.62).

3.3.1.3 Show that the heat capacity at constant volume of a completely degenerate Fermi gas vanishes.

3.3.2 White dwarf stars

An interesting application of the theory of a completely degenerate relativistic Fermi gas is related to the study of white dwarf stars. A white dwarf star is characterized by a low luminosity since it has used up wholly all its nuclear fuel, its mass is about that of the Sun ($M_\odot \approx 1.99 \times 10^{30}$ kg) and its radius of order of that of the Earth ($R_\oplus \approx 6.38 \times 10^6$m). The temperature at the center of a white dwarf star is about 10^7 K, and at this temperature all atoms are completely ionized.

According to Cox and Giuli [3] the probable composition of a white dwarf star is:

a) $M \lesssim 0.08 M_\odot$ predominantly H^1;
b) $0.08 M_\odot \lesssim M \lesssim 0.35 M_\odot$ predominantly He^4;
c) $0.35 M_\odot \lesssim M \lesssim 1.00 M_\odot$ predominantly a mixture of C^{12} and O^{16};
d) $1.00 M_\odot \lesssim M \lesssim 1.4 M_\odot$ predominantly mixtures whose elements have atomic masses in the range 20 to 28.

Let us consider a white dwarf star with mass $M \approx 1.02 M_\odot$ and radius $R \approx 5.4 \times 10^6\,\text{m}$, which could represent Sirius B, the companion that orbits around the star Sirius. The mass density of Sirius B is then about $\varrho = M/(4\pi R^3/3) \approx 3 \times 10^9\,\text{kg/m}^3$ which is 2×10^6 times larger than the mass density of the Sun whose radius is about $R_\odot \approx 6.96 \times 10^8\,\text{m}$.

In a completely ionized mixture the relationship between the particle number density of the electrons n and the mass density of the mixture ϱ is given by

$$n = \varrho \sum_a \frac{X_a}{m_u} \frac{Z_a}{M_a}, \tag{3.70}$$

where $m_u = 1.6605402 \times 10^{-27}\text{kg}$ denotes the unified atomic mass unity, while X_a is the mass fraction, Z_a the atomic number and M_a the relative atomic mass of constituent a. We can write (3.70) as

$$n \approx \frac{\varrho}{m_u} \left[X_H + \frac{1}{2} \sum_{a \neq H} X_a \right], \tag{3.71}$$

where the index H denotes the constituent hydrogen and it was considered that $Z_a \approx \frac{1}{2} M_a$ for the constituents of relative atomic mass $M_a \geq 4$, which is the case of helium, carbon, oxygen and the elements with atomic masses in the range 20 to 26. We use now the constraint $\sum_{a \neq H} X_a = 1 - X_H$ and get from (3.71) the desired relationship between the mass density of the white dwarf star ϱ and the particle number density of the electrons n:

$$\varrho = \frac{2 m_u}{1 + X_H} n. \tag{3.72}$$

Since $0 \leq X_H \leq 1$ the particle number density of the electrons may vary in the range

$$\frac{\varrho}{2 m_u} \leq n \leq \frac{\varrho}{m_u}, \tag{3.73}$$

the lower limit describing a star with no hydrogen content, the upper one a hydrogen star.

The particle number density of the electrons for the lower limit of (3.73) is

$$n = \frac{\varrho}{2 m_u} = \frac{3 \times 10^9}{2 \times 1,66 \times 10^{-27}} \frac{\text{electrons}}{\text{m}^3} \approx 9 \times 10^{35} \frac{\text{electrons}}{\text{m}^3}, \tag{3.74}$$

by considering the mass density of Sirius B. The ratio of the Fermi momentum for the electrons (3.63) and the momentum $m_e c$ is

$$\frac{p_F}{m_e c} = \frac{1}{m_e c}\left(\frac{3nh^3}{8\pi}\right)^{\frac{1}{3}} \approx 1.15, \tag{3.75}$$

since for electrons the degeneracy factor is $g_s = 2$. Hence the electrons may be considered as a completely degenerate relativistic Fermi gas.

Let us estimate the ratios between the two parameters $\zeta_e = m_e c^2/(kT)$ of the electrons and $\zeta_n = \bar{m}_n c^2/(kT)$ of the nuclei:

$$\frac{\zeta_n}{\zeta_e} = \frac{\bar{m}_n}{m_e} \approx 4.4 \times 10^4, \tag{3.76}$$

where $\bar{m}_n = (m_{C^{12}} + m_{O^{16}})/2 \approx 3.98 \times 10^{-26}$kg is the mean mass of the nuclei. We conclude from the above estimates that the nuclei may be treated non-relativistically. Further it is possible to obtain from $\varrho = m_e n + \bar{m}_n n_n$ the particle number density of the nuclei which is about $n_n \approx 7.5 \times 10^{34}$ nuclei/m^3. Moreover the condition (3.41) – that indicates if the gas is degenerate or non-degenerate – implies that

$$n_n\left(\frac{h^2}{2\pi\bar{m}_n kT}\right)^{\frac{3}{2}} \approx 1.08 \times 10^{-4} \ll 1, \tag{3.77}$$

by considering the degeneracy factor of the nuclei equal to one. Hence we infer that the nuclei may be considered as a non-degenerate and non-relativistic gas.[2] The pressure of the nuclei, calculated from the equation of state $p_n = n_n kT$, gives

$$p_n \approx 1.04 \times 10^{19}\,\text{N/m}^2. \tag{3.78}$$

On the other hand the pressure of the electrons, calculated by using the equation of state of a completely degenerate relativistic Fermi gas (3.55), is

$$p \approx 1.37 \times 10^{22}\,\text{N/m}^2, \tag{3.79}$$

since from (3.58) and (3.75) we have that $\mathcal{F}(1.15) \approx 2.29$. Hence the pressure of the electrons is about 1317 times larger than the pressure of the nuclei, and we may say that a white dwarf star is supported against its own gravitation by the pressure of the degenerated relativistic electron gas in its interior. We proceed to relate the electron pressure with the force of gravitational attraction per unit of area.

We consider a spherically symmetrical distribution of matter in the interior of a white dwarf star, so that the mass density is only a function of the modulus of the radius vector r measured from the center of the star. If $M(r)$ denotes the mass which is enclosed inside r, we have that

$$M(r) = \int_0^r \varrho(r) 4\pi r^2 dr. \tag{3.80}$$

[2] This is a rough assumption, but here we are interested only in orders of magnitudes.

Further we consider a spherical shell of radius r and thickness dr in the interior of the star. If p is the pressure at r and $p+dp$ the corresponding pressure at $r+dr$, the force on the spherical shell in the direction of increasing r is $-4\pi r^2 dp$. This force is equal to the gravitational force that acts on the mass of the spherical shell $dM(r) = 4\pi r^2 \varrho(r)dr$ by the mass $M(r)$ which is enclosed inside r, i.e.,

$$-4\pi r^2 dp = \frac{GM(r)dM(r)}{r^2} = \frac{GM(r)4\pi r^2 \varrho(r)dr}{r^2}, \tag{3.81}$$

where $G = 6.673 \times 10^{-11}\,\mathrm{N\,m^2/kg^2}$ is the gravitational constant. Here we have ignored gravitational effects which go beyond the Newtonian approximation, which is reasonable in the present case. For consideration of gravitational effects one is referred to Chapter 11 on general relativity.

In order to show that there exists an upper limit for the mass of possible white dwarf stars, the so-called Chandrasekhar limit, we shall first analyze a simple model based on the mean pressure in the interior of the star which is defined by

$$\bar{p} = \frac{1}{M(R)} \int_0^R p(r)dM(r), \tag{3.82}$$

where R is the radius of the white dwarf star and it is assumed that the pressure at the surface of the star vanishes, i.e., $p(R) = 0$.

We integrate by parts (3.82) and get

$$\bar{p} = -\frac{1}{M(R)} \int_0^R M(r)dp(r) \overset{(3.81)}{=} \frac{G}{4\pi M(R)} \int_0^R \frac{M(r)^2 dM(r)}{r^4}. \tag{3.83}$$

If $\bar{\varrho}$ denotes a mean mass density of the star, the mass in the interior of a radius r is given by

$$M(r) = \bar{\varrho}\frac{4\pi}{3}r^3, \tag{3.84}$$

so that (3.83) can be integrated, yielding

$$\bar{p} = \frac{3}{20\pi}\frac{GM(R)^2}{R^4}. \tag{3.85}$$

In the book by Chandrasekhar [2] this expression is the lowest bound for the mean pressure in a star, and for more details one is referred to that book.

We shall apply the formula for the mean pressure (3.85) to a completely degenerate electron gas, since we have concluded that in a white dwarf star the pressure of the degenerate relativistic gas is larger than that of the nuclei. According to (3.53), (3.55) and (3.72) the mass density of the white dwarf star ϱ and the mean pressure $\bar{p}$ of the electron in a white dwarf star are given by

$$\varrho = Bx^3, \qquad \bar{p} = A\mathcal{F}(x), \tag{3.86}$$

where B and A are the constants:

$$B = \frac{16\pi}{3}\frac{m_u}{1+X_H}\left(\frac{m_e c}{h}\right)^3, \qquad A = \frac{\pi}{3}m_e c^2\left(\frac{m_e c}{h}\right)^3. \tag{3.87}$$

Now we shall analyze the two limiting cases which correspond to a non-relativistic and an ultra-relativistic gas. In the limiting case of a completely degenerate non-relativistic Fermi gas we may use the approximation (3.60), write $\mathcal{F}(x) \approx 8x^5/5$ and get, by the use of (3.84) through (3.86),

$$\bar{p} \approx \frac{8}{5}Ax^5 = \frac{8}{5}A\left(\frac{\varrho}{B}\right)^{\frac{5}{3}} = \frac{8}{5}A\left[\frac{1}{B}\frac{3M(R)}{4\pi R^3}\right]^{\frac{5}{3}} = \frac{3}{20\pi}\frac{GM(R)^2}{R^4}. \tag{3.88}$$

From the above equation one can easily obtain the following relationship between the radius R of a white dwarf star and its mass $M(R)$:

$$R = \frac{A}{G}\left(\frac{3}{\pi}\right)^{\frac{2}{3}}\left(\frac{2}{B}\right)^{\frac{5}{3}}\frac{1}{M(R)^{\frac{1}{3}}}. \tag{3.89}$$

We conclude that in the limiting case of a completely degenerate non-relativistic Fermi gas the radius decreases with increasing mass, since R is proportional to $M(R)^{-\frac{1}{3}}$. Further by considering a white dwarf star devoid of hydrogen, i.e., $X_{\mathrm{H}} = 0$, it follows from (3.87) and (3.89) through the substitution of numerical values that

$$\frac{M(R)}{M_\odot} \approx 1.13 \times 10^{-6}\left(\frac{R_\odot}{R}\right)^3. \tag{3.90}$$

For the limiting case of a completely degenerate ultra-relativistic Fermi gas we approximate according to (3.62), $\mathcal{F}(x) \approx 2(x^4 - x^2)$ and get from (3.85) and (3.86)

$$\bar{p} \approx 2A(x^4 - x^2) = 2A\left[\left(\frac{\varrho}{B}\right)^{\frac{4}{3}} - \left(\frac{\varrho}{B}\right)^{\frac{2}{3}}\right]$$

$$= 2A\left\{\left[\frac{1}{B}\frac{3M(R)}{4\pi R^3}\right]^{\frac{4}{3}} - \left[\frac{1}{B}\frac{3M(R)}{4\pi R^3}\right]^{\frac{2}{3}}\right\} = \frac{3}{20\pi}\frac{GM(R)^2}{R^4}. \tag{3.91}$$

By solving (3.91) for R it follows that

$$R = \left(\frac{3M(R)}{4\pi B}\right)^{\frac{1}{3}}\left[1 - \left(\frac{M(R)}{M_{\mathrm{Ch}}}\right)^{\frac{2}{3}}\right]^{\frac{1}{2}}. \tag{3.92}$$

We note from (3.92) that R has no real solutions for $M(R) > M_{\mathrm{Ch}}$. Hence there exists no white dwarf star with mass larger than M_{Ch}. The limiting mass M_{Ch}, called the Chandrasekhar limit, is given here by

$$M_{\mathrm{Ch}} = \frac{3}{512\pi}\left(\frac{10ch}{G}\right)^{\frac{3}{2}}\left(\frac{1+X_{\mathrm{H}}}{m_u}\right)^2. \tag{3.93}$$

If we insert numerical values in (3.93) and choose a white dwarf star devoid of hydrogen ($X_{\rm H} = 0$) we find that $M_{\rm Ch} \approx 1.75 M_\odot$.

In the following we shall find another value for the limiting mass $M_{\rm Ch}$ by using a more elaborate method based on the differential equation that follows from (3.81):

$$\frac{dp}{dr} = -\frac{GM(r)\varrho(r)}{r^2}, \quad \text{or} \quad \frac{d}{dr}\left(\frac{r^2}{\varrho(r)}\frac{dp}{dr}\right) = -G\frac{dM(r)}{dr} = -G4\pi r^2 \varrho(r). \quad (3.94)$$

We are interested in the so-called polytropic equations of state, which are of the form

$$p = \kappa \varrho^{\frac{(n+1)}{n}}, \quad (3.95)$$

where κ and n are constants and n is called the polytropic index. We shall deal with two polytropic equations of state for:

a) a completely degenerate non-relativistic Fermi gas:

$$p = \frac{8}{5}Ax^5 \stackrel{(3.86)_1}{=} \frac{8}{5}A\left(\frac{\varrho}{B}\right)^{\frac{5}{3}}, \qquad \text{where} \qquad n = \frac{3}{2}, \quad \kappa = \frac{8}{5}\frac{A}{B^{\frac{5}{3}}}; \quad (3.96)$$

b) a completely degenerate ultra-relativistic Fermi gas:

$$p = 2Ax^4 \stackrel{(3.86)_1}{=} 2A\left(\frac{\varrho}{B}\right)^{\frac{4}{3}}, \qquad \text{where} \qquad n = 3, \quad \kappa = 2\frac{A}{B^{\frac{4}{3}}}. \quad (3.97)$$

If we insert the polytropic equation of state (3.95) into $(3.94)_2$ and introduce the dimensionless quantities

$$\Theta = \left(\frac{\varrho}{\varrho_c}\right)^{\frac{1}{n}}, \qquad \xi = \frac{r}{a}, \quad (3.98)$$

we get the so-called Lane–Emden equation of index n:

$$\frac{1}{\xi^2}\frac{d}{d\xi}\left(\xi^2\frac{d\Theta}{d\xi}\right) = -\Theta^n. \quad (3.99)$$

In (3.98) ϱ_c is the central mass density, i.e., $\varrho_c = \varrho(0)$ and a is the constant

$$a = \left[\frac{(n+1)\kappa\varrho_c^{\frac{1-n}{n}}}{4\pi G}\right]^{\frac{1}{2}}. \quad (3.100)$$

The boundary conditions at the center $r = 0$ for the Lane–Emden equation (3.99) are

$$\Theta(0) = 1, \qquad \text{and} \qquad \frac{d\Theta}{d\xi}(0) = 0. \quad (3.101)$$

The first condition is a direct consequence of $(3.98)_1$ and of the definition of the central mass density ϱ_c. The boundary condition $(3.101)_2$ can be obtained by using the approximation $M(r) \approx 4\pi r^3 \varrho_c/3$ near the center of the white dwarf star, so that the limiting value of $(3.94)_1$ approaching zero is

$$\lim_{r\to 0}\frac{dp}{dr}=0,\quad \text{or by (3.95)}\ \lim_{r\to 0}\frac{d\varrho}{dr}=0,\quad \text{or by (3.98)}\ \lim_{\xi\to 0}\frac{d\Theta}{d\xi}=0. \tag{3.102}$$

The Lane–Emden equation (3.99) with the boundary conditions (3.101) can be integrated numerically. For $n = 3/2$ and $n = 3$, Θ is a monotonically decreasing function of ξ, vanishing at a finite value $\xi = \xi_{\mathrm{R}}$ (see Figure 3.2). Since $\Theta(\xi_{\mathrm{R}}) = 0$ implies that the mass density and the pressure also vanish, the point ξ_{R} corresponds to the surface of the white dwarf star. Further from (3.98) and (3.100) we can write the radius of the white dwarf star as

$$R = a\xi_{\mathrm{R}} = \left[\frac{(n+1)\kappa\varrho_c^{\frac{1-n}{n}}}{4\pi G}\right]^{\frac{1}{2}}\xi_{\mathrm{R}}. \tag{3.103}$$

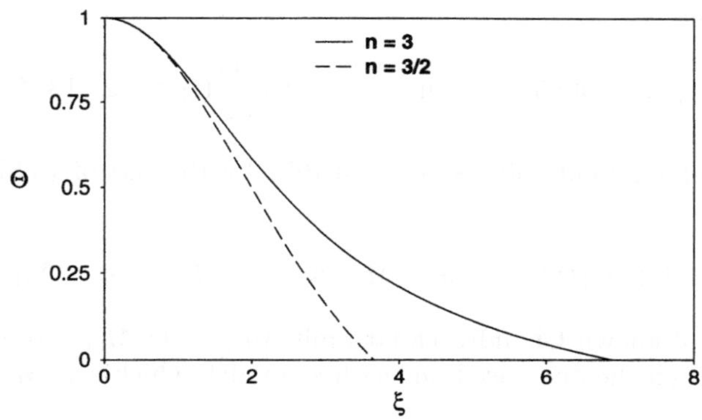

Figure 3.2: Θ as a function of ξ.

In order to calculate the mass of a white dwarf star we apply (3.80) and obtain

$$M(R) = \int_0^R 4\pi r^2\varrho(r)dr \overset{(3.98)}{=} 4\pi a^3\varrho_c\int_0^{\xi_{\mathrm{R}}}\xi^2\Theta^n d\xi$$

$$= -4\pi a^3\varrho_c\int_0^{\xi_{\mathrm{R}}}\frac{d}{d\xi}\left(\xi^2\frac{d\Theta}{d\xi}\right)d\xi = -4\pi a^3\varrho_c\xi_{\mathrm{R}}^2\frac{d\Theta}{d\xi}(\xi_{\mathrm{R}}). \tag{3.104}$$

Now we eliminate a and ϱ_c from (3.104) by using (3.100) and (3.103), yielding

$$M(R) = -4\pi\left[\frac{(n+1)\kappa}{4\pi G}\right]^{\frac{n}{n-1}}\left(\frac{R}{\xi_{\mathrm{R}}}\right)^{\frac{n-3}{n-1}}\xi_{\mathrm{R}}^2\frac{d\Theta}{d\xi}(\xi_{\mathrm{R}}). \tag{3.105}$$

For a completely degenerate non-relativistic Fermi gas, where $n = 3/2$ we have

$$\xi_{\rm R} \approx 3.65375, \qquad \text{and} \qquad -\xi_{\rm R}^2 \frac{d\Theta}{d\xi}(\xi_{\rm R}) \approx 2.71406. \tag{3.106}$$

Hence it follows from (3.105) together with (3.87) and (3.96) for a white dwarf star devoid of hydrogen ($X_{\rm H} = 0$) that

$$\frac{M(R)}{M_\odot} \approx 2.09 \times 10^{-6} \left(\frac{R_\odot}{R}\right)^3. \tag{3.107}$$

The above estimate is about 1.85 times larger than (3.90), which was based on a simpler model. Here again, we have that the radius decreases with increasing mass.

From (3.105) we could infer that the mass of a white dwarf star does not depend on its radius for $n = 3$ which is the limiting case that corresponds to the Chandrasekhar limit. This is the case of a completely degenerate ultra-relativistic Fermi gas where

$$\xi_{\rm R} \approx 6.89685, \qquad \text{and} \qquad -\xi_{\rm R}^2 \frac{d\Theta}{d\xi}(\xi_{\rm R}) \approx 2.01824. \tag{3.108}$$

The Chandrasekhar limit follows from (3.105) by the use of (3.87) and (3.97), yielding

$$M_{\rm Ch} \approx 1.459(1 + X_{\rm H})^2 M_\odot \qquad \text{or} \qquad M_{\rm Ch} \approx 1.459 M_\odot \qquad \text{for} \qquad X_{\rm H} = 0, \tag{3.109}$$

which is the well-known Chandrasekhar limit $M_{\rm Ch} \approx 1.4 M_\odot$. Note that this last value is lower than the one based on the first model, which was $M_{\rm Ch} \approx 1.75 M_\odot$.

Problems

3.3.2.1 Solve (3.91) for R and show that it leads to (3.92).

3.3.2.2 Obtain the Lane–Emden equation (3.99) and show that a is given by (3.100).

3.3.2.3 Obtain the numerical values for $\xi_{\rm R}$ and $-\xi_{\rm R}^2 (d\Theta/d\xi)|_{\xi_{\rm R}}$ given in (3.106) and (3.108).

3.3.3 Strongly degenerate relativistic Fermi gas

In Section 3.3.1 we calculated the expression for the energy per particle e and for the pressure p and have found that in a completely degenerate relativistic Fermi gas they are only functions of the particle number density n but not of the temperature T. Here we shall consider a less restrictive assumption on the denominators of the integrands of (3.46) through (3.48) so that we can write the integrals in (3.2) according to an asymptotic formula due to Sommerfeld [12]. Here we draw on Chandrasekhar [2] to state the following lemma:

Sommerfeld's Lemma: Let $\varphi(u)$ be a continuous and differentiable function of exponential order[3] u_0 and such that $\varphi(0) = 0$, then the following integral has the asymptotic formula

$$\int_0^\infty \frac{du}{e^{-u_0}e^u + 1}\frac{d\varphi(u)}{du} = \varphi(u_0) + 2\sum_{n=1}^{\infty} C_{2n}\frac{d^{2n}\varphi}{du^{2n}}(u_0), \tag{3.110}$$

provided that we neglect quantities of order e^{-u_0}. In (3.110) the coefficients C_{2n} are given by

$$C_{2n} = 1 - \frac{1}{2^{2n}} + \frac{1}{3^{2n}} - \frac{1}{4^{2n}} + \ldots = \frac{\pi^{2n}(2^{2n-1}-1)}{(2n)!}B_n, \tag{3.111}$$

with B_n denoting Bernoulli numbers: $B_1 = 1/6$, $B_2 = 1/30$, and so on.

Proof: We begin by writing the integral on the left-hand side of (3.110) as:

$$\int_0^\infty \frac{du}{e^{-u_0}e^u + 1}\frac{d\varphi(u)}{du} = \int_0^{u_0}\frac{d\varphi(u)}{du}du + \int_0^{u_0}\left(\frac{1}{e^{-u_0}e^u+1} - 1\right)\frac{d\varphi(u)}{du}du$$
$$+\int_{u_0}^\infty \frac{du}{e^{-u_0}e^u + 1}\frac{d\varphi(u)}{du} = \varphi(u_0) - \int_0^{u_0}\frac{du}{1+e^{u_0}e^{-u}}\frac{d\varphi(u)}{du}$$
$$+\int_{u_0}^\infty \frac{du}{e^{-u_0}e^u + 1}\frac{d\varphi(u)}{du}. \tag{3.112}$$

If in the first integral on the right-hand side of (3.112) we change the variables of integration by writing $u = u_0(1-t)$, while for the second integral we put $u = u_0(1+t)$, the expression (3.112) reduces to:

$$\int_0^\infty \frac{du}{e^{-u_0}e^u + 1}\frac{d\varphi(u)}{du} = \varphi(u_0) - u_0\int_0^1 \frac{dt}{e^{u_0t}+1}\frac{d\varphi}{du}(u_0(1-t))$$
$$+u_0\int_0^\infty \frac{dt}{e^{u_0t}+1}\frac{d\varphi}{du}(u_0(1+t)) = \varphi(u_0) + u_0\int_0^\infty \frac{dt}{e^{u_0t}+1}\left[\frac{d\varphi}{du}(u_0(1+t))\right.$$
$$\left.-\frac{d\varphi}{du}(u_0(1-t))\right] + u_0\int_1^\infty \frac{dt}{e^{u_0t}+1}\frac{d\varphi}{du}(u_0(1-t)). \tag{3.113}$$

One can show that the last integral on the right-hand side of (3.113) is of order e^{-u_0} provided that $\varphi(u)$ is of exponential order u_0. Hence we can neglect this integral

[3] A function $\varphi(t)$ is of exponential order u_0 on the interval $a \leq t \leq \infty$ if it increases not faster than e^{u_0t} when $t \to \infty$, i.e., $|e^{-u_0t}\varphi(t)| \leq M$ where M is a real positive constant.

due to the condition stated in the lemma. Now we expand $d\varphi/du$ in (3.113) in Taylor series about u_0 and get

$$\int_0^\infty \frac{du}{e^{-u_0}e^u+1}\frac{d\varphi(u)}{du} \simeq \varphi(u_0) + 2\sum_{n=1}^\infty \frac{u_0^{2n}}{(2n-1)!}\frac{d^{2n}\varphi}{du^{2n}}(u_0)\int_0^\infty \frac{t^{2n-1}}{e^{u_0 t}+1}dt. \tag{3.114}$$

The integral on the right-hand side of (3.114) can be evaluated since we can write it as

$$\int_0^\infty \frac{t^{2n-1}}{e^{u_0 t}+1}dt = \int_0^\infty t^{2n-1}e^{-u_0 t}\left(1 - e^{-u_0 t} + e^{-2u_0 t} - e^{-3u_0 t} + \cdots\right)dt$$
$$= \frac{(2n-1)!}{u_0^{2n}}\left[1 - \frac{1}{2^{2n}} + \frac{1}{3^{2n}} - \frac{1}{4^{2n}} + \cdots\right] = \frac{(2n-1)!}{u_0^{2n}}C_{2n}. \tag{3.115}$$

By inserting (3.115) into (3.114) we get (3.110) and we have proved Sommerfeld's lemma.

For a Fermi gas the integral (3.2) is written as

$$I_n(\zeta,\mu_E) = \int_0^\infty \frac{\cosh(n\vartheta)}{e^{\zeta\cosh\vartheta}e^{-\mu_E/kT}+1}d\vartheta, \tag{3.116}$$

so that if we introduce a new variable of integration

$$u = \zeta(\cosh\vartheta - 1), \qquad u_0 = \frac{\mu_E}{kT} - \zeta, \tag{3.117}$$

we can express (3.116) as

$$I_n(\zeta,\mu_E) = \int_0^\infty \frac{du}{e^{-u_0}e^u+1}\frac{d}{du}\left\{\frac{1}{n}\sinh\left[n\,\mathrm{arcosh}\left(\frac{u+\zeta}{\zeta}\right)\right]\right\}. \tag{3.118}$$

Let us apply Sommerfeld's lemma to the integral (3.118). First we identify

$$\varphi(u) = \frac{1}{n}\sinh\left[n\,\mathrm{arcosh}\left(\frac{u+\zeta}{\zeta}\right)\right], \tag{3.119}$$

and get by taking into account the asymptotic formula (3.110):

$$I_n(\zeta,\mu_E) = \frac{1}{n}\sinh n\xi\left[1 + \frac{\pi^2}{6}\frac{n}{\zeta^2}\frac{n\sinh\xi\sinh n\xi - \cosh\xi\cosh n\xi}{\sinh^3\xi\sinh n\xi} + \cdots\right]. \tag{3.120}$$

In this section we shall use the abbreviations

$$\xi = \mathrm{arcosh}\left(\frac{\mu_E}{mc^2}\right), \qquad y = \sinh\xi. \tag{3.121}$$

From (3.120) we can obtain the asymptotic formulae for the integrals that appear in equations (3.4) through (3.6) in terms of y:

$$I_0(\zeta,\mu_E) = \mathrm{arsinh}\, y - \frac{\pi^2}{6\zeta^2}\frac{\sqrt{1+y^2}}{y^3} + \cdots, \tag{3.122}$$

$$I_1(\zeta, \mu_E) = y\left[1 - \frac{\pi^2}{6\zeta^2}\frac{1}{y^4} + \cdots\right], \tag{3.123}$$

$$I_2(\zeta, \mu_E) = y\sqrt{1+y^2}\left[1 + \frac{\pi^2}{6\zeta^2}\frac{2y^2-1}{y^4} + \cdots\right], \tag{3.124}$$

$$I_3(\zeta, \mu_E) = \frac{y}{3}(3+4y^2)\left[1 + \frac{\pi^2}{2\zeta^2}\frac{8y^4+4y^2-1}{y^4(3+4y^2)} + \cdots\right], \tag{3.125}$$

$$I_4(\zeta, \mu_E) = y\sqrt{1+y^2}(2y^2+1)\left[1 + \frac{\pi^2}{6\zeta^2}\frac{24y^4+8y^2-1}{y^4(2y^2+1)} + \cdots\right]. \tag{3.126}$$

We insert (3.122) through (3.126) into (3.4) through (3.6) and get the expressions for the particle number density n, energy density ne and pressure p for a strongly degenerate relativistic Fermi gas, that read:

$$n = \frac{4\pi}{3}(mc)^3\frac{g_s}{h^3}y^3\left[1 + \frac{\pi^2}{2\zeta^2}\frac{2y^2+1}{y^4} + \cdots\right], \tag{3.127}$$

$$ne = \frac{\pi}{6}m^4c^5\frac{g_s}{h^3}\mathcal{G}(y)\left[1 + \frac{4\pi^2}{\zeta^2}\frac{(3y^2+1)\sqrt{1+y^2}}{\mathcal{G}(y)y} + \cdots\right], \tag{3.128}$$

$$p = \frac{\pi}{6}m^4c^5\frac{g_s}{h^3}\mathcal{F}(y)\left[1 + \frac{4\pi^2}{\zeta^2}\frac{y\sqrt{1+y^2}}{\mathcal{F}(y)} + \cdots\right], \tag{3.129}$$

where the functions $\mathcal{G}(y)$ and $\mathcal{F}(y)$ are defined in (3.57) and (3.58), respectively.

One may also obtain from (3.127) and (3.128) the difference between the energy per particle e and the rest energy mc^2:

$$ne^\star = \frac{\pi}{6}m^4c^5\frac{g_s}{h^3}\mathcal{G}^\star(y)\left[1 + \frac{4\pi^2}{\zeta^2}\frac{(3y^2+1)\sqrt{1+y^2}-(2y^2+1)}{\mathcal{G}^\star(y)y} + \cdots\right], \tag{3.130}$$

where $\mathcal{G}^\star(y)$ is given in terms of $\mathcal{G}(y)$ by

$$\mathcal{G}^\star(y) = \mathcal{G}(y) - 8y^3. \tag{3.131}$$

We shall now calculate the heat capacity per particle at constant volume of a strongly degenerate relativistic Fermi gas. First we note that according to (3.127) we have $n = n(y,T)$, or by inverting the power series we can obtain $y = y(n,T)$. Hence if we differentiate y with respect to T we get

$$\left(\frac{\partial y}{\partial T}\right)_n = -\frac{\pi^2}{3\zeta^2}\frac{2y^2+1}{y^3}\frac{1}{T} + \cdots. \tag{3.132}$$

Now taking the derivative of the energy per particle (3.128) with respect to T yields

$$c_v = \frac{\pi^2k^2}{mc^2}\frac{\sqrt{1+y^2}}{y^2}T + \cdots, \tag{3.133}$$

by the use of (3.127), (3.132) and of the relationship

$$\frac{d\mathcal{G}(y)}{dy} = 24y^2\sqrt{1+y^2}. \tag{3.134}$$

The above equations were first obtained by Chandrasekhar (see [2] pp. 388–394), and in the following we proceed to analyze some of these equations in order to get a better physical interpretation. First we note that the particle number density for a completely and for a strongly degenerate relativistic Fermi gas should be the same so that we can equate the two expressions (3.53) and (3.127) and obtain a relationship between x and y:

$$x^3 = y^3\left[1+\frac{\pi^2}{2\zeta^2}\frac{2y^2+1}{y^4}+\cdots\right], \tag{3.135}$$

or by inverting the power series

$$y = x\left[1-\frac{\pi^2}{6\zeta^2}\frac{2x^2+1}{x^4}+\cdots\right]. \tag{3.136}$$

The abbreviation x and y are related to the Fermi momentum p_F and to the chemical potential in equilibrium μ_E, respectively, through (3.53) and (3.121):

$$x = \frac{p_F}{mc}, \qquad y = \sqrt{\left(\frac{\mu_E}{mc^2}\right)^2 - 1}. \tag{3.137}$$

Let us analyze first a non-relativistic strongly degenerate Fermi gas. In this case we have $p_F \ll mc$ so that we get from (3.52) the relationship between the Fermi momentum p_F and the Fermi energy E_F that reads

$$E_F \approx \frac{p_F^2}{2m}. \tag{3.138}$$

If we introduce $\mu_E^\star = \mu_E - mc^2$, which is defined as the difference between the equilibrium chemical potential μ_E and the rest mass mc^2, we get from (3.136) and (3.137) the relationship

$$\mu_E^\star = E_F\left[1-\frac{\pi^2}{12}\left(\frac{kT}{E_F}\right)^2+\cdots\right]. \tag{3.139}$$

Further it follows from (3.129), (3.130) and (3.133) by use of the approximations $\mathcal{G}^\star(y) \approx 12y^5/5$ and $\mathcal{F}(y) \approx 8y^5/5$ (see (3.59), (3.60) and (3.131)) that

$$e^\star = \frac{3}{5}E_F\left[1+\frac{5\pi^2}{12}\left(\frac{kT}{E_F}\right)^2+\cdots\right], \tag{3.140}$$

$$\frac{p}{n} = \frac{2}{5} E_F \left[1 + \frac{5\pi^2}{12} \left(\frac{kT}{E_F} \right)^2 + \cdots \right], \quad \text{i.e.,} \quad p = \frac{3}{2} n e^\star, \tag{3.141}$$

$$c_v = \frac{\pi^2}{2} \frac{k^2 T}{E_F} + \cdots. \tag{3.142}$$

The expressions for the equilibrium chemical potential (3.139), energy per particle (3.140) and pressure (3.141) show the first corrections, which depend on the temperature, to their corresponding values of a completely degenerate non-relativistic Fermi gas. The entropy per particle in equilibrium can also be obtained in this limit. Indeed from (2.122) and (3.139) through (3.141) we get

$$s_E = \frac{\pi^2}{3} \frac{k^2 T}{E_F} + \cdots. \tag{3.143}$$

We note first that the heat capacity at constant volume of a strongly degenerate Fermi gas is non-zero. Further we infer from (3.142) and (3.143) that s_E and c_v tend to zero when $T \to 0$ which agrees with the third law of thermodynamics. Equation (3.138) to (3.143) are well-known approximations in non-relativistic statistical mechanics (see for example Pathria [10] pp. 200–201).

The limiting case of an ultra-relativistic strongly degenerate Fermi gas ($p_F \gg mc$) can be obtained in the same manner. The equations that correspond to (3.138) through (3.143) in this case are:

$$E_F = c p_F, \tag{3.144}$$

$$\mu_E = E_F \left[1 - \frac{\pi^2}{3} \left(\frac{kT}{E_F} \right)^2 + \cdots \right], \tag{3.145}$$

$$e = \frac{3}{4} E_F \left[1 + \frac{2\pi^2}{3} \left(\frac{kT}{E_F} \right)^2 + \cdots \right], \tag{3.146}$$

$$\frac{p}{n} = \frac{1}{4} E_F \left[1 + \frac{2\pi^2}{3} \left(\frac{kT}{E_F} \right)^2 + \cdots \right], \quad \text{i.e.,} \quad p = \frac{1}{3} n e, \tag{3.147}$$

$$c_v = \pi^2 \frac{k^2 T}{E_F} + \cdots, \qquad s_E = \pi^2 \frac{k^2 T}{E_F} + \cdots. \tag{3.148}$$

The above equations show the dependence on the temperature of the first corrections to the corresponding values of a completely degenerate ultra-relativistic Fermi gas.

Problems

3.3.3.1 Show that the last integral of the right-hand side of (3.113) is of order e^{-u_0} provided that $\varphi(u)$ is of exponential order u_0.

3.3.3.2 Obtain the asymptotic formulae (3.122) through (3.126).

3.3.3.3 Obtain the expressions (3.144) through (3.148) for a strongly degenerate ultra-relativistic Fermi gas.

3.4 Degenerate relativistic Bose gas

Degenerate Bose gases have a remarkable feature, which is the so-called Bose–Einstein condensation, where some part of the particles of the gas are concentrated in the state of zero energy below a critical temperature known as the condensation temperature. The heat capacity at constant volume has also a remarkable behavior at the condensation temperature, being discontinuous at this point for the case of a degenerate ultra-relativistic Bose gas. We begin this section by giving some relationships between the integrals defined in terms of the Bose–Einstein distribution function which will be useful for study of the relativistic Bose–Einstein condensation.

3.4.1 Some useful integrals

We follow the work by Landsberg and Dunning-Davies [8] and introduce the integrals

$$\mathcal{J}_n(\zeta,\mu_E^\star)=\int_0^\infty \frac{(u^2+2\zeta u)^{\frac{1}{2}}(u^n+\zeta u^{n-1})}{e^{-\mu_E^\star/kT}e^u-1}du, \tag{3.149}$$

$$\mathcal{K}_n(\zeta,\mu_E^\star)=\int_0^\infty \frac{(u^2+2\zeta u)^{\frac{1}{2}}(u^n+\zeta u^{n-1})}{\left(e^{-\mu_E^\star/kT}e^u-1\right)^2}e^{-\mu_E^\star/kT}e^u du, \tag{3.150}$$

$$\mathcal{I}_n(\mu_E^\star)=\int_0^\infty \frac{u^n}{e^{-\mu_E^\star/kT}e^u-1}du, \tag{3.151}$$

where $\mu_E^\star=\mu_E-mc^2$ is the difference between the chemical potential in equilibrium and the rest energy.

The relationship between the integrals $\mathcal{J}_n(\zeta,\mu_E^\star)$ and $\mathcal{K}_n(\zeta,\mu_E^\star)$ is given by

$$\frac{d}{d\zeta}\left[\frac{1}{\zeta^{n+2}}\mathcal{J}_n(\zeta,\mu_E^\star)\right]=-\frac{1}{\zeta^{n+3}}\mathcal{K}_{n+1}(\zeta,\mu_E^\star)+\frac{1}{\zeta^{n+2}}\mathcal{K}_n(\zeta,\mu_E^\star)\frac{d}{d\zeta}\left(\frac{\mu_E^\star}{kT}\right). \tag{3.152}$$

The integral $\mathcal{I}_n(\mu_E^\star)$ can be evaluated when $\mu_E^\star=0$, yielding

$$\mathcal{I}_n(0)=\int_0^\infty \frac{u^n}{e^u-1}du=\Gamma(n+1)\zeta_{\mathcal{R}}(n+1). \tag{3.153}$$

In the above equation $\zeta_{\mathcal{R}}(n)$ is the Riemann zeta function[4]

$$\zeta_{\mathcal{R}}(n) = \sum_{j=1}^{\infty} \frac{1}{j^n} = 1 + \frac{1}{2^n} + \frac{1}{3^n} + \cdots, \tag{3.154}$$

and some values of the Riemann zeta function are given below:

$$\zeta_{\mathcal{R}}(1) \to \infty, \qquad \zeta_{\mathcal{R}}(3/2) \approx 2.61238, \qquad \zeta_{\mathcal{R}}(2) = \pi^2/6, \tag{3.155}$$

$$\zeta_{\mathcal{R}}(5/2) \approx 1.34149, \qquad \zeta_{\mathcal{R}}(3) \approx 1.20206, \qquad \zeta_{\mathcal{R}}(4) = \pi^4/90. \tag{3.156}$$

In the limiting case where $e^{-\mu_E^\star/kT} \gg 1$, which corresponds to a non-degenerate gas, we have that

$$\mathcal{I}_n(\mu_E^\star) \to e^{\mu_E^\star/kT}\Gamma(n+1). \tag{3.157}$$

In the non-relativistic and ultra-relativistic limiting cases we obtain that the integrals $\mathcal{J}_n(\zeta, \mu_E^\star)$ and $\mathcal{K}_n(\zeta, \mu_E^\star)$ tend to the integral $\mathcal{I}_n(\mu_E^\star)$, i.e.

a) non-relativistic $\zeta \gg 1$,

$$\mathcal{J}_n(\zeta, \mu_E^\star) \approx \sqrt{2}\zeta^{\frac{3}{2}}\mathcal{I}_{n-\frac{1}{2}}(\mu_E^\star), \tag{3.158}$$

$$\mathcal{K}_n(\zeta, \mu_E^\star) \approx \sqrt{2}\zeta^{\frac{3}{2}}\left(n - \frac{1}{2}\right)\mathcal{I}_{n-\frac{3}{2}}(\mu_E^\star), \quad \text{for} \quad n \geq \frac{1}{2}; \tag{3.159}$$

b) ultra-relativistic $\zeta \ll 1$,

$$\mathcal{J}_n(\zeta, \mu_E^\star) \approx \mathcal{I}_{n+1}(\mu_E^\star), \tag{3.160}$$

$$\mathcal{K}_n(\zeta, \mu_E^\star) \approx (n+1)\mathcal{I}_n(\mu_E^\star), \quad \text{for} \quad n \geq -1. \tag{3.161}$$

Problems

3.4.1.1 Show that the relationship between the integrals $\mathcal{J}_n(\zeta, \mu_E^\star)$ and $\mathcal{K}_n(\zeta, \mu_E^\star)$ is given by (3.152). (Hint: change the variable of integration to $t = u/\zeta$.)

3.4.1.2 Obtain the non-relativistic and the ultra-relativistic limits of the integrals $\mathcal{J}_n(\zeta, \mu_E^\star)$ and $\mathcal{K}_n(\zeta, \mu_E^\star)$ given in (3.158) through (3.161).

---

[4] Here the Riemann zeta function is written as $\zeta_{\mathcal{R}}$ in order to distinguish it from the parameter $\zeta = mc^2/(kT)$.

3.4.2 Relativistic Bose–Einstein condensation

According to (2.124) the equilibrium distribution function for a Bose gas is given by

$$f^{(0)} = \frac{g_s/h^3}{e^{-\mu_E^\star/kT}e^{E/kT} - 1}. \tag{3.162}$$

In (3.162) we have considered a local Lorentz rest frame where $(U^\alpha) = (c, \mathbf{0})$ and excluded from the energy the rest energy, i.e., $E = cp^0 - mc^2$.

If the volume and the number of particles are large enough we may consider that the energy levels vary continuously and the lowest energy level can be chosen to be zero. This last condition implies that we must have $\mu_E^\star \leq 0$ so that the positiveness of the equilibrium distribution function is not violated.

The particle number density n and the difference between the energy per particle and the rest energy $e^\star = e - mc^2$ of a Bose gas can be written in terms of the integral $\mathcal{J}_n(\zeta, \mu_E^\star)$, given in (3.149), as

$$n = 4\pi \frac{(mc)^3}{\zeta^3}\frac{g_s}{h^3}\mathcal{J}_1(\zeta, \mu_E^\star), \tag{3.163}$$

$$ne^\star = 4\pi \frac{m^4c^5}{\zeta^4}\frac{g_s}{h^3}\mathcal{J}_2(\zeta, \mu_E^\star). \tag{3.164}$$

The above expressions follow from $(2.111)_1$ and $(2.116)_1$ together with (2.110), that is

$$n = 4\pi(mc)^3\frac{g_s}{h^3}\int_0^\infty \frac{\sinh^2\vartheta\cosh\vartheta}{e^{-\mu_E^\star/kT}e^{\zeta(\cosh\vartheta - 1)} - 1}d\vartheta, \tag{3.165}$$

$$ne^\star = 4\pi m^4c^5\frac{g_s}{h^3}\int_0^\infty \frac{\sinh^2\vartheta\cosh\vartheta(\cosh\vartheta - 1)}{e^{-\mu_E^\star/kT}e^{\zeta(\cosh\vartheta - 1)} - 1}d\vartheta, \tag{3.166}$$

by considering

$$\vartheta = \operatorname{arcosh}\left(\frac{u}{\zeta} + 1\right) = \operatorname{arsinh}\left(\frac{u^2}{\zeta^2} + 2\frac{u}{\zeta}\right)^{\frac{1}{2}}. \tag{3.167}$$

Let us decrease the temperature of the Bose gas by maintaining the number of particles and the volume fixed, i.e., we keep n fixed. When the temperature decreases, $|\mu_E^\star|$ must also decrease so that there exists a minimum temperature T_c for which the chemical potential must vanish, i.e., for $T = T_c$ we have that $\mu_E^\star = 0$. The minimum value of the temperature T_c is obtained from (3.163):

$$n = 4\pi \frac{(mc)^3}{\zeta_c^3}\frac{g_s}{h^3}\mathcal{J}_1(\zeta_c, 0), \tag{3.168}$$

where $\zeta_c = mc^2/(kT_c)$. In the limiting cases we obtain from (3.168) together with (3.153), (3.158) and (3.160):

a) non-relativistic case $\zeta_c \gg 1$,

$$T_c = \frac{h^2}{2\pi mk}\left(\frac{n}{g_s\zeta_{\mathcal{R}}(3/2)}\right)^{\frac{2}{3}}; \tag{3.169}$$

b) ultra-relativistic case $\zeta_c \ll 1$,

$$T_c = \frac{hc}{2k}\left(\frac{n}{\pi g_s\zeta_{\mathcal{R}}(3)}\right)^{\frac{1}{3}}. \tag{3.170}$$

Since the temperature of the gas can decrease below T_c the expression for the particle number density (3.163) must be modified for $T < T_c$. When the temperature decreases below T_c the number of particles with zero energy, denoted by N_0, will increase and this contribution was neglected in the integral (3.163) since it has a weight zero. Hence instead of (3.163) we write

$$N = N_0 + 4\pi V\frac{(mc)^3}{\zeta^3}\frac{g_s}{h^3}\mathcal{J}_1(\zeta, 0), \qquad \text{for} \qquad T < T_c, \tag{3.171}$$

where V denotes the volume of the gas. We have also the condition that

$$N_0 \ll N, \qquad \text{for} \qquad T \geq T_c. \tag{3.172}$$

In the two limiting cases we get from (3.171) together with (3.153), (3.158), (3.160), (3.169) and (3.170):

a) non-relativistic case $\zeta_c \gg 1$,

$$\frac{N_0}{N} = 1 - \left(\frac{T}{T_c}\right)^{\frac{3}{2}}, \qquad \text{for} \qquad T \leq T_c; \tag{3.173}$$

b) ultra-relativistic case $\zeta_c \ll 1$,

$$\frac{N_0}{N} = 1 - \left(\frac{T}{T_c}\right)^{3}, \qquad \text{for} \qquad T \leq T_c. \tag{3.174}$$

Hence we have particles in states with energy different from zero and particles in the state of zero energy. The process of concentrating particles in the state of zero energy as the temperature decreases below T_c is known as Bose–Einstein condensation and the temperature T_c is called the condensation temperature. Note that above the condensation temperature the number of particles in the state of zero energy N_0 is negligible in comparison to the total number of the particles N (see Figure 3.3).

One of the main interesting features of Bose–Einstein condensation concerns the behavior of the heat capacity as the temperature increases, which we proceed to analyze. We begin by writing the energy per particle (3.164) as

$$ne^\star = 4\pi\frac{m^4c^5}{\zeta^4}\frac{g_s}{h^3}\mathcal{J}_2(\zeta, \mu_E^\star), \qquad \text{for} \qquad T > T_c, \tag{3.175}$$

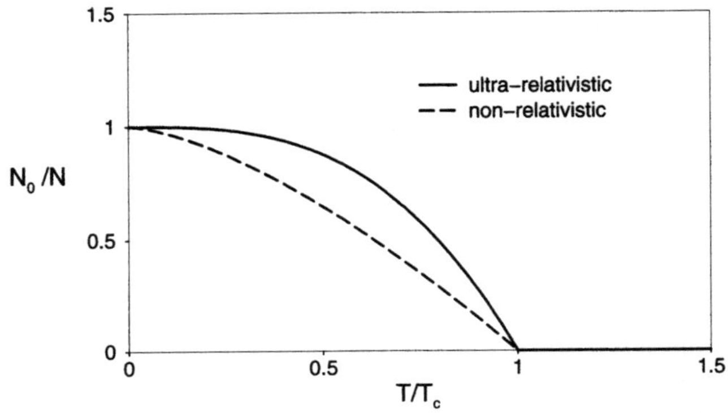

Figure 3.3: Number of particles with zero energy as function of T/T_c.

$$ne^\star = 4\pi \frac{m^4c^5}{\zeta^4}\frac{g_s}{h^3}\mathcal{J}_2(\zeta, 0) \qquad \text{for} \qquad T \le T_c. \tag{3.176}$$

Next we differentiate n, given by (3.163), with respect to T, equate the resulting equation to zero and obtain the relationship

$$\left[\frac{\partial}{\partial T}\left(\frac{\mu_E^\star}{kT}\right)\right]_v = -\frac{1}{T}\frac{\mathcal{K}_2(\zeta, \mu_E^\star)}{\mathcal{K}_1(\zeta, \mu_E^\star)}, \tag{3.177}$$

where (3.152) has been used. Now by differentiating $ne^\star$ with respect to T, there follows from (3.175) through (3.177) the expressions for the heat capacity per particle at constant volume:

$$nc_v = \frac{4\pi k^4T^3 g_s}{c^3h^3}\left[\mathcal{K}_3(\zeta, \mu_E^\star) - \frac{\mathcal{K}_2(\zeta, \mu_E^\star)^2}{\mathcal{K}_1(\zeta, \mu_E^\star)}\right], \qquad \text{for} \qquad T > T_c, \tag{3.178}$$

$$nc_v = \frac{4\pi k^4T^3 g_s}{c^3h^3}\mathcal{K}_3(\zeta, 0), \qquad \text{for} \qquad T \le T_c. \tag{3.179}$$

From (3.178) and (3.179) we can build the difference between the heat capacities at $T = T_c$:

$$\Delta = c_v(T_c^-) - c_v(T_c^+) = k\frac{\mathcal{K}_2(\zeta_c, 0)^2}{\mathcal{J}_1(\zeta_c, 0)\mathcal{K}_1(\zeta_c, 0)}, \tag{3.180}$$

by the use of (3.168).

We shall analyze the above expressions in the two limiting cases. The first case is the non-relativistic case characterized by $\zeta \gg 1$. For temperatures lower than the condensation temperature $T \le T_c$ the heat capacity per particle at constant volume c_v, the energy per particle $e^\star$ and the pressure $p = 2ne^\star/3$ read

$$c_v = \frac{15\zeta_{\mathcal{R}}(5/2)}{4\zeta_{\mathcal{R}}(3/2)}k\left(\frac{T}{T_c}\right)^{\frac{3}{2}} \approx 1.9257k\left(\frac{T}{T_c}\right)^{\frac{3}{2}}, \tag{3.181}$$

$$e^{\star} = \frac{3\zeta_{\mathcal{R}}(5/2)}{2\zeta_{\mathcal{R}}(3/2)} k \frac{T^{\frac{5}{2}}}{T_c^{\frac{3}{2}}} \approx 0.7703 k \frac{T^{\frac{5}{2}}}{T_c^{\frac{3}{2}}}, \tag{3.182}$$

$$p = \zeta_{\mathcal{R}}(5/2) \left(\frac{2\pi m k}{h^2} \right)^{\frac{3}{2}} g_s k T^{\frac{5}{2}}. \tag{3.183}$$

Equations (3.181) and (3.182) were obtained from (3.179) and (3.176) by taking into account the expression for the particle number density (3.168). Moreover we have used the expressions (3.158) and (3.159) together with (3.153).

The difference between the heat capacities at $T = T_c$, given by (3.180) is zero, i.e., $\Delta = 0$, since the integral $\mathcal{K}_1(\zeta_c, 0)$ can be transformed into

$$\mathcal{K}_1(\zeta_c, 0) = \int_0^{\infty} \frac{1}{e^u - 1} \frac{2u^2 + 4\zeta_c u + \zeta_c^2}{(u^2 + 2\zeta_c u)^{\frac{1}{2}}} du \tag{3.184}$$

by integration by parts and the integral (3.184) tends to infinity[5] for all $\zeta_c \neq 0$.

Further for temperatures larger than the condensation temperature i.e., $T > T_c$, the heat capacity per particle at constant volume (3.178) can be written, thanks to (3.158), (3.159) and (3.163), as:

$$\frac{c_v}{k} = \frac{5\mathcal{I}_{3/2}(\mu_E^{\star})}{2\mathcal{I}_{1/2}(\mu_E^{\star})} - \frac{9\mathcal{I}_{1/2}(\mu_E^{\star})}{2\mathcal{I}_{-1/2}(\mu_E^{\star})}. \tag{3.185}$$

In the limiting case of a non-degenerate gas $e^{-\mu_E^{\star}/kT} \gg 1$ we may use (3.157) to get from (3.185)

$$c_v = \frac{3}{2} k. \tag{3.186}$$

We close the analysis of the non-relativistic Bose gas by concluding as follows:

i) the heat capacity per particle at constant volume c_v is proportional to $T^{3/2}$ for $T \leq T_c$ and has a maximum value $c_v \approx 1.9257k$ at $T = T_c$. It is continuous at $T = T_c$ and decreases to $c_v = 1.5k$ for $T \gg T_c$ (see Figure 3.4);

ii) for $T \leq T_c$ the energy per particle $e^{\star}$ and the pressure p do not depend on the particle number density but only on the temperature and are proportional to $T^{5/2}$.

The second limiting case is the ultra-relativistic case characterized by $\zeta \ll 1$. For $T \leq T_c$ we have from (3.168), (3.179) and (3.176) and from the relationship $p = ne/3$:

$$c_v = \frac{12\zeta_{\mathcal{R}}(4)}{\zeta_{\mathcal{R}}(3)} k \left(\frac{T}{T_c} \right)^3 \approx 10.8047 k \left(\frac{T}{T_c} \right)^3, \tag{3.187}$$

$$e = \frac{3\zeta_{\mathcal{R}}(4)}{\zeta_{\mathcal{R}}(3)} k \frac{T^4}{T_c^3} \approx 2.7012 k \frac{T^4}{T_c^3}, \tag{3.188}$$

[5]Note that for $\zeta_c \neq 0$ the integrand is positive everywhere and can be approximated by $\zeta_c^{3/2}/(\sqrt{2}u^{3/2})$ near $u = 0$, and the integral of the latter function diverges.

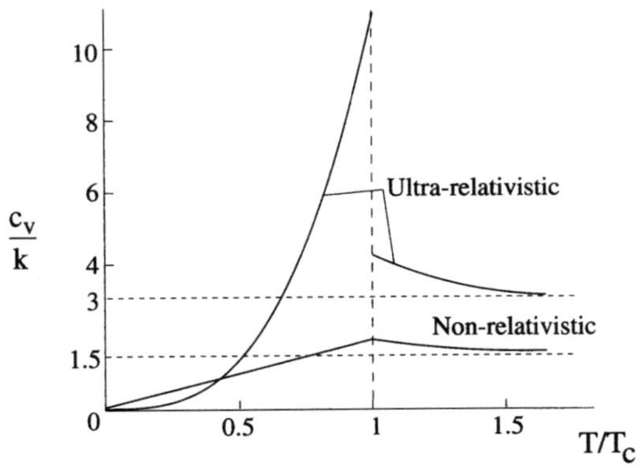

Figure 3.4: Heat capacity as a function of T.

$$p = 8\pi\zeta_{\mathcal{R}}(4)\left(\frac{k}{hc}\right)^3 g_s kT^4. \tag{3.189}$$

To obtain the above equations we have used (3.160) and (3.161) together with (3.153). We note that for the ultra-relativistic case $e/(kT) = (e^\star + mc^2)/(kT) \approx e^\star/(kT)$ since $mc^2/(kT) \ll 1$.

At $T = T_c$ the difference between the heat capacities follows from (3.180) together with (3.153), (3.160) and (3.161), yielding

$$\Delta = \frac{9\zeta_{\mathcal{R}}(3)}{\zeta_{\mathcal{R}}(2)}k \approx 6.5769k. \tag{3.190}$$

For $T > T_c$ the heat capacity per particle at constant volume (3.178) reads

$$\frac{c_v}{k} = \frac{4\mathcal{I}_3(\mu_E^\star)}{\mathcal{I}_2(\mu_E^\star)} - \frac{9\mathcal{I}_2(\mu_E^\star)}{2\mathcal{I}_1(\mu_E^\star)}, \tag{3.191}$$

thanks to (3.163), (3.160) and (3.161). The above equation has the limits

$$\frac{c_v}{k} = \frac{12\zeta_{\mathcal{R}}(4)}{\zeta_{\mathcal{R}}(3)} - \frac{9\zeta_{\mathcal{R}}(3)}{\zeta_{\mathcal{R}}(2)} \approx 4.2278, \quad \text{at} \quad T = T_c \quad \text{when} \quad \mu_E^\star = 0, \tag{3.192}$$

$$c_v = 3k, \qquad \text{when} \quad e^{-\mu_E^\star/kT} \gg 1, \tag{3.193}$$

where we have used the approximations (3.153) and (3.157), respectively.

The conclusions for the ultra-relativistic Bose gas are:

i) the heat capacity per particle at constant volume c_v is proportional to T^3 for $T \leq T_c$ having a maximum value $c_v \approx 10.8047k$ at $T = T_c$. It is discontinuous at $T = T_c$ having a jump of $\Delta \approx 6.5768k$ and decreases to $c_v = 3k$ for $T > T_c$ (see Figure 3.4). The discontinuity of the heat capacity at $T = T_c$ was first reported by Landsberg and Dunning-Davis [9];

ii) the energy per particle and the pressure for $T \le T_c$ depend only on the temperature being proportional to T^4.

We note that in both limiting cases – i.e., non-relativistic and ultra-relativistic – the heat capacities per particle at constant volume tend to zero when $T \to 0$ which agree with the third law of thermodynamics.

Problems

3.4.2.1 Obtain the expressions for the condensation temperature in the non-relativistic limiting case (3.169) and in the ultra-relativistic limiting case (3.170).

3.4.2.2 Show that the derivative of $\mu_E^\star/(kT)$ with respect to T at constant volume is given by (3.177) and obtain the expressions (3.178) and (3.179) for the heat capacity at constant volume when $T > T_c$ and $T < T_c$, respectively.

3.4.2.3 Show that for $T \le T_c$: a) the heat capacity at constant volume, the energy per particle and the pressure of a non-relativistic Bose gas are given by (3.181), (3.182) and (3.183), respectively; b) the heat capacity at constant volume, the energy per particle and the pressure of an ultra-relativistic Bose gas are given by (3.187), (3.188) and (3.189), respectively.

References

[1] M. Abramowitz and I. A. Stegun, *Handbook of mathematical functions* (Dover, New York, 1968).

[2] S. Chandrasekhar, *An introduction to the study of stellar structure* (Dover, New York, 1958).

[3] J. P. Cox and R. T. Giuli, *Principles of stellar structure*, Vols. I and II, (Gordon and Breach, New York, 1968).

[4] I. S. Gradshteyn and I. M. Ryzhik, *Tables of integrals, series and products* 5th ed. (Academic Press, San Diego, 1994).

[5] S. R. de Groot, W. A. van Leeuwen and Ch. G. van Weert, *Relativistic kinetic theory* (North-Holland, Amsterdam, 1980).

[6] D. C. Kelly, *The kinetic theory of a relativistic gas*, unpublished report (Miami University, Oxford, 1963).

[7] G. M. Kremer, *Uma introdução à teoria cinética relativística* (Editora da UFPR, Curitiba, 1998).

[8] P. T. Landsberg and J. Dunning-Davis, Statistical thermodynamics of the ideal relativistic quantum gas, in *Statistical mechanics of equilibrium and non-equilibrium* pp. 36–51, ed. J. Meixner (North-Holland, Amsterdam, 1961).

[9] P. T. Landsberg and J. Dunning-Davis, Ideal relativistic Bose condensation *Phys. Rev. A* **138**, 1049–1052 (1965).

[10] R. K. Pathria, *Statistical mechanics* 2nd. ed. (Butterworth-Heinemann, Oxford, 1996).

[11] S. L. Shapiro and S. A. Teukolsky, *Black holes, white dwarfs and neutron stars*, (John Wiley and Sons, New York, 1983).

[12] A. Sommerfeld, Zur Elektronentheorie der Metalle auf Grund der Fermischen Statistik. I. Teil: Allgemeines, Strömungs- und Austrittsvorgänge *Z. Physik* **47**, 1–82 (1928).

[13] J. M. Stewart, *Non-equilibrium relativistic kinetic theory*, Lecture Notes in Physics, vol. 10 (Springer, Heidelberg, 1971).

[14] J. L. Synge, *The relativistic gas* (North-Holland, Amsterdam, 1957).

Chapter 4

Thermomechanics of Relativistic Fluids

4.1 Introduction

Before we begin the study of the kinetic theory of a non-equilibrium relativistic gas we shall present a phenomenological theory based on the linear thermodynamics of irreversible processes. The constitutive equations obtained in this chapter will be derived in the next chapters on the basis of the relativistic Boltzmann equation and explicit expressions for the transport coefficients will be given in terms of the interaction potential between the relativistic particles. In this chapter we shall deal only with a single relativistic gas. The reader who is interested in a more elaborated thermodynamic theory based on rational extended thermodynamics should consult the book by Müller and Ruggeri [5].

4.2 Thermodynamics of perfect fluids

In this section we shall deal only with perfect fluids that are fluids where dissipative effects like viscous forces and heat conduction are neglected.

The particle four-flow is a conservative quantity whose balance equation is given, according to (2.73) and (2.92), by

$$\partial_\alpha N^\alpha = 0, \qquad \text{or} \qquad \frac{\partial \gamma n}{\partial t} + \frac{\partial \gamma n v^i}{\partial x^i} = 0. \tag{4.1}$$

The above equation reduces to the non-relativistic continuity equation

$$\frac{\partial n}{\partial t} + \frac{\partial n v^i}{\partial x^i} = 0. \tag{4.2}$$

for small velocities $v \ll c$ since in this case we have that $\gamma \to 1$.

Further according to (2.74) the conservation law of the energy-momentum tensor reads

$$\partial_\beta T^{\alpha\beta} = 0. \tag{4.3}$$

If we introduce the expression of the energy-momentum tensor for a perfect fluid (2.98)

$$T^{\alpha\beta} = (ne+p)\frac{U^\alpha U^\beta}{c^2} - p\eta^{\alpha\beta}, \tag{4.4}$$

in the above equation we get

$$\frac{\partial p}{\partial x_\alpha} = \frac{\partial}{\partial x_\beta}\left[(ne+p)\frac{U^\alpha U^\beta}{c^2}\right], \tag{4.5}$$

where the temporal ($\alpha = 0$) and spatial ($\alpha = i$) components are given by

$$\frac{\partial p}{\partial ct} = \frac{\partial}{\partial x_\beta}\left[(ne+p)\frac{U^0 U^\beta}{c^2}\right], \qquad \frac{\partial p}{\partial x_i} = \frac{\partial}{\partial x_\beta}\left[(ne+p)\frac{U^0 U^\beta}{c^2}\frac{v^i}{c}\right]. \tag{4.6}$$

In the last equation we have used the relationship $U^i = \gamma v^i = U^0 v^i/c$. Now we eliminate from $(4.6)_2$ the term that corresponds to the left-hand side of $(4.6)_1$ yielding

$$\frac{\partial p}{\partial x_i} = \frac{\partial p}{\partial ct}\frac{v^i}{c} + (ne+p)\frac{U^0 U^\beta}{c^2}\frac{\partial}{\partial x_\beta}\left(\frac{v^i}{c}\right), \qquad \text{or} \tag{4.7}$$

$$n\frac{d\mathbf{v}}{dt} = \frac{-c^2}{\gamma^2 h_E}\left[\vec{\nabla}p - \frac{\mathbf{v}}{c^2}\frac{\partial p}{\partial t}\right]. \tag{4.8}$$

In (4.8) we introduced the enthalpy per particle $h_E = e + p/n$ and the material time derivative d/dt which is defined by

$$\frac{d}{dt} = \frac{\partial}{\partial t} + v^i\frac{\partial}{\partial x^i}. \tag{4.9}$$

Equation (4.8) reduces to the non-relativistic balance equation for the momentum density of a perfect fluid

$$\varrho\frac{d\mathbf{v}}{dt} + \vec{\nabla}p = 0, \tag{4.10}$$

since in the non-relativistic limiting case we have $h_E \to mc^2$ (by taking into account only the rest energy, see (3.38)), $v \ll c$ and $\gamma \to 1$. In (4.10) $\varrho = mn$ denotes the mass density of the fluid.

On the other hand, the scalar product of (4.3) by U_α leads to

$$U_\alpha\partial_\beta T^{\alpha\beta} = \partial_\beta(U_\alpha T^{\alpha\beta}) - T^{\alpha\beta}\partial_\beta U_\alpha \overset{(4.4)}{=} \partial_\beta(neU_\beta) + p\partial_\beta U^\beta, \tag{4.11}$$

where we have used the relation that follows from the constraint $U^\alpha U_\alpha = c^2$:

$$U^\alpha\partial_\beta U_\alpha = 0. \tag{4.12}$$

From (4.11) and by the use of $(4.1)_2$ we get the balance equation for the internal energy density of a perfect fluid,

$$n\frac{de}{dt} + p\,\mathrm{div}\,\mathbf{v} + \frac{p}{\gamma}\frac{d\gamma}{dt} = 0. \tag{4.13}$$

An alternative form to write the balance equation for the internal energy density is by eliminating $d\gamma/dt$ from (4.13) through the use of the balance equation for the momentum density (4.8), that is

$$\frac{1}{\gamma}\frac{d\gamma}{dt} = \frac{\gamma^2}{c^2}\frac{d\mathbf{v}}{dt}\cdot\mathbf{v} \overset{(4.8)}{=} -\frac{1}{nh_E}\mathbf{v}\cdot\left[\vec{\nabla}p - \frac{\mathbf{v}}{c^2}\frac{\partial p}{\partial t}\right]. \tag{4.14}$$

Hence we have from (4.13) and (4.14),

$$n\frac{de}{dt} + p\,\mathrm{div}\,\mathbf{v} = \frac{p}{nh_E}\mathbf{v}\cdot\left[\vec{\nabla}p - \frac{\mathbf{v}}{c^2}\frac{\partial p}{\partial t}\right]. \tag{4.15}$$

The right-hand side of (4.15) vanishes in the non-relativistic limit and it reduces to the non-relativistic balance equation for the internal energy density of a perfect fluid.

If we transform the Gibbs equation (2.99) into an equation for the material time derivative and use the balance equations for the particle number density (4.1) and internal energy density (4.15) we get

$$\frac{ds_E}{dt} = \frac{1}{T}\left(\frac{de}{dt} - \frac{p}{n^2}\frac{dn}{dt}\right) = 0, \tag{4.16}$$

that is the flow of a perfect fluid is isentropic along each particle trajectory. Thus if the entropy per particle is the same at each point at $t = 0$, the flow will be globally isentropic. This is of course true only if the solutions are assumed to be continuous; if generalized, discontinuous solutions are permitted, then shock waves occur and, in general, the solution will not be isentropic everywhere.

Let us analyze a solution of the system of equations (4.1), (4.8) and (4.16) that corresponds to a problem of wave propagation. We consider that the particle number density is a function of the pressure p and of the entropy per particle s_E, i.e., $n = n(p, s_E)$. Further we search for a solution of the form

$$p(\mathbf{x},t) = p_0 + \tilde{p}(\mathbf{x},t), \qquad s_E(\mathbf{x},t) = s_0 + \tilde{s}(\mathbf{x},t), \qquad \mathbf{v}(\mathbf{x},t) = \tilde{\mathbf{v}}(\mathbf{x},t), \tag{4.17}$$

that corresponds to a non-perturbed state of constant pressure and entropy per particle and vanishing velocity superposed by a perturbed state characterized by $\tilde{p}$, $\tilde{s}$ and $\tilde{\mathbf{v}}$. By introducing (4.17) into the balance equations (4.1), (4.8) and (4.16) and by neglecting all products of the quantities with tildes and its derivatives, we get

$$\left(\frac{\partial n}{\partial p}\right)_s \frac{\partial \tilde{p}}{\partial t} + n_0\mathrm{div}\,\tilde{\mathbf{v}} = 0, \qquad \frac{n_0 h_0}{c^2}\frac{\partial \tilde{\mathbf{v}}}{\partial t} = -\vec{\nabla}\tilde{p}, \tag{4.18}$$

where in the first equation we have used the relation $\partial\tilde{s}/\partial t = 0$ and h_0 denotes the value of the enthalpy per particle in the non-perturbed state. By taking the time derivative of $(4.18)_1$ and the divergence of $(4.18)_2$ we can eliminate $\partial\mathrm{div}\,\tilde{\mathbf{v}}/\partial t$ from both equations and obtain the wave equation

$$\frac{1}{v_s^2}\frac{\partial^2\tilde{p}}{\partial t^2} = \nabla^2\tilde{p}, \tag{4.19}$$

where v_s, the adiabatic speed of sound, is given by

$$v_s = \sqrt{\frac{c^2}{h_0}\left(\frac{\partial p}{\partial n}\right)_s}. \tag{4.20}$$

In the non-relativistic limit where $h_0 \to mc^2$, the adiabatic speed of sound reduces to the well-known result $v_s = \sqrt{(\partial p/\partial \varrho)_s}$.

Problems

4.2.1 Obtain the balance equation for the internal energy density of a perfect fluid (4.13).

4.2.2 Show that equation (4.18) follow from the insertion of (4.17) into the balance equations (4.1), (4.8) and (4.16) by neglecting all products of quantities with tildes.

4.3 Eckart decomposition

In order to identify the physical meaning of the different quantities appearing in the balance equations, it is useful to introduce decompositions of these quantities with respect to the four-velocity U^α. The most usual decomposition is due to Eckart [1] and will be illustrated in this section. The alternative decomposition is due to Landau and Lifshitz [4], and will be discussed in the next section.

In the Eckart decomposition the particle four-flow N^α and the energy-momentum tensor $T^{\alpha\beta}$ for a viscous heat conducting fluid are written as

$$N^\alpha = nU^\alpha, \tag{4.21}$$

$$T^{\alpha\beta} = p^{\langle\alpha\beta\rangle} - (p+\varpi)\,\Delta^{\alpha\beta} + \frac{1}{c^2}\left(U^\alpha q^\beta + U^\beta q^\alpha\right) + \frac{en}{c^2}U^\alpha U^\beta. \tag{4.22}$$

We recall that $\Delta^{\alpha\beta}$ is the projector

$$\Delta^{\alpha\beta} = \eta^{\alpha\beta} - \frac{1}{c^2}U^\alpha U^\beta. \tag{4.23}$$

The above decompositions define the quantities n, $p^{\langle\alpha\beta\rangle}$, p, ϖ, q^α and ϵ, which are identified as:

$$n = \frac{1}{c^2}N^\alpha U_\alpha \quad \text{– particle number density,} \tag{4.24}$$

$$p^{\langle\alpha\beta\rangle} = \left(\Delta^\alpha_\gamma \Delta^\beta_\delta - \frac{1}{3}\Delta^{\alpha\beta}\Delta_{\gamma\delta}\right)T^{\gamma\delta} \quad \text{– pressure deviator,} \tag{4.25}$$

$$(p+\varpi) = -\frac{1}{3}\Delta_{\alpha\beta}T^{\alpha\beta} \quad \text{– hydrostatic pressure + dynamic pressure,} \tag{4.26}$$

$$q^\alpha = \Delta^\alpha_\gamma U_\beta T^{\beta\gamma} \quad \text{– heat flux,} \tag{4.27}$$

$$e = \frac{1}{nc^2} U_\alpha T^{\alpha\beta} U_\beta \quad \text{– internal energy per particle.} \tag{4.28}$$

The dynamic pressure is the non-equilibrium part of the trace of the energy-momentum tensor, since the pressure p and the energy per particle e refer to equilibrium quantities.

In the local Lorentz rest frame $(U^\alpha) = (c, \mathbf{0})$, and we have from (2.152) and (4.24) through (4.28):

a) the particle number density is given by

$$n = \frac{1}{c} N_R^0; \tag{4.29}$$

b) the non-vanishing components of the pressure deviator correspond to the spatial components of the energy-momentum tensor $T_R^{\langle ij \rangle}$,

$$p^{\langle 00 \rangle} = 0, \qquad p^{\langle 0i \rangle} = p^{\langle i0 \rangle} = 0, \qquad p^{\langle ij \rangle} = T_R^{\langle ij \rangle}, \tag{4.30}$$

c) the trace of the spatial components of the energy-momentum tensor is the sum of the hydrodynamic pressure and the dynamic pressure

$$p + \varpi = -\frac{1}{3} \eta_{ij} T_R^{ij}, \tag{4.31}$$

d) the non-zero components of the heat flux are only the spatial components

$$q^0 = 0, \qquad q^i = c T_R^{0i}, \tag{4.32}$$

e) the energy density is given by the temporal coordinates of the energy-momentum tensor

$$e = \frac{1}{n} T_R^{00}. \tag{4.33}$$

Another quantity that can be decomposed into parts parallel and perpendicular to the four-velocity is the entropy four-flow

$$S^\alpha = nsU^\alpha + \phi^\alpha. \tag{4.34}$$

The above decomposition defines the quantities

$$s = \frac{1}{nc^2} S^\alpha U_\alpha \quad \text{– entropy per particle,} \tag{4.35}$$

$$\phi^\alpha = \Delta^\alpha_\beta S^\beta \quad \text{– entropy flux.} \tag{4.36}$$

In a local Lorentz rest frame we obtain from (2.152), (4.35) and (4.36):

$$ns = \frac{1}{c} S_R^0, \qquad \phi^0 = 0, \qquad \phi^i = S_R^i, \tag{4.37}$$

that is the entropy density is the temporal component of the entropy four-flow while its spatial components represent the entropy flux.

We shall write now the balance equations for the particle four-flow and for the energy-momentum tensor in terms of the differential operators D and ∇^α defined in (2.157). First we insert the expression for the particle four-flow (4.21) into the balance equation $\partial_\alpha N^\alpha = 0$ and get, by the use of (2.157),

$$\frac{1}{c^2}U^\alpha D(nU_\alpha) + \nabla^\alpha(nU_\alpha) = 0, \qquad \text{or} \qquad Dn + n\nabla^\alpha U_\alpha = 0 \tag{4.38}$$

since $U^\alpha DU_\alpha = 0$ and $U_\alpha \nabla^\alpha = 0$. Equation $(4.38)_2$ has a form similar to the non-relativistic equation for the particle number density.

From the balance equation for the energy-momentum tensor $\partial_\beta T^{\alpha\beta} = 0$ we obtain by the use of (4.22) and (2.157)

$$\partial_\beta T^{\alpha\beta} = \left(\frac{1}{c^2}U_\beta D + \nabla_\beta\right)\Big[p^{\langle\alpha\beta\rangle} - (p+\varpi)\Delta^{\alpha\beta}$$
$$+\frac{1}{c^2}(U^\alpha q^\beta + U^\beta q^\alpha) + \frac{en}{c^2}U^\alpha U^\beta\Big] = 0, \tag{4.39}$$

or after some simplifications

$$\frac{1}{c^2}Dq^\alpha + \frac{en}{c^2}DU^\alpha + \frac{1}{c^2}U^\alpha(nDe + eDn) - \frac{1}{c^2}p^{\langle\alpha\beta\rangle}DU_\beta + \frac{1}{c^2}(p+\varpi)DU^\alpha$$
$$-\frac{1}{c^4}U^\alpha q^\beta DU_\beta + \nabla_\beta p^{\langle\alpha\beta\rangle} - \nabla^\alpha(p+\varpi) + \frac{1}{c^2}(p+\varpi)U^\alpha\nabla_\beta U^\beta$$
$$+\frac{1}{c^2}q^\alpha\nabla_\beta U^\beta + \frac{1}{c^2}U^\alpha\nabla_\beta q^\beta + \frac{1}{c^2}q^\beta\nabla_\beta U^\alpha + \frac{en}{c^2}U^\alpha\nabla_\beta U^\beta = 0. \tag{4.40}$$

If we take the scalar product of (4.40) by U_α it reduces to the balance equation for the energy density:

$$nDe = -(p+\varpi)\nabla_\alpha U^\alpha + p^{\langle\alpha\beta\rangle}\nabla_\beta U_\alpha - \nabla_\alpha q^\alpha + \frac{2}{c^2}q^\alpha DU_\alpha. \tag{4.41}$$

In order to derive the above equation we have used the balance of the particle number density (4.38) and the relationships

$$\begin{cases} p^{\langle\alpha\beta\rangle}U_\alpha = 0, \qquad q^\alpha U_\alpha = 0, \\ U_\alpha\nabla_\beta p^{\langle\alpha\beta\rangle} = -p^{\langle\alpha\beta\rangle}\nabla_\beta U_\alpha, \qquad U_\alpha Dq^\alpha = -q^\alpha DU_\alpha. \end{cases} \tag{4.42}$$

Without the last term on the right-hand side of (4.41) it has a form similar to the non-relativistic balance equation for the energy density.

On the other hand, if we multiply (4.40) by the projector Δ^β_α it follows that

$$\frac{nh_E}{c^2}DU^\alpha = \nabla^\alpha(p+\varpi) - \nabla_\beta p^{\langle\alpha\beta\rangle} + \frac{1}{c^2}\Big(p^{\langle\alpha\beta\rangle}DU_\beta - \varpi DU^\alpha$$

$$-Dq^\alpha - q^\alpha \nabla_\beta U^\beta - q^\beta \nabla_\beta U^\alpha - \frac{1}{c^2} U^\alpha q^\beta D U_\beta - U^\alpha p^{\langle\beta\gamma\rangle} \nabla_\beta U_\gamma \Big), \tag{4.43}$$

which is the balance equation for the momentum density. Apart from the last term on the right-hand side of (4.43) this equation has a form similar to the non-relativistic balance equation for the momentum density.

The balance equation for the entropy four-flow is

$$\partial_\alpha S^\alpha = \varsigma \geq 0, \tag{4.44}$$

where ς is the non-negative entropy production rate. In terms of the differential operators D and ∇_α the above equation can be written as

$$nDs + \nabla_\alpha \phi^\alpha + \frac{1}{c^2} U_\alpha D \phi^\alpha = \varsigma \geq 0. \tag{4.45}$$

Without the last term on the left-hand side of (4.45) this equation has also a form similar to the non-relativistic balance equation for the entropy density.

Problems

4.3.1 Obtain first (4.40) from (4.39) and show afterwards that the balance equations for the energy density and for the momentum density are given by (4.41) and (4.43), respectively.

4.3.2 Show that the balance equation for the entropy four-flow (4.44) written in terms of the operators D and ∇_α is given by (4.45).

4.4 Landau and Lifshitz decomposition

Instead of the decomposition for the particle four-flow (4.21) and for the energy-momentum tensor (4.22) due to Eckart, Landau and Lifshitz proposed an alternative decomposition given by

$$N^\alpha = n_L U_L^\alpha + \mathcal{J}^\alpha, \tag{4.46}$$

$$T^{\alpha\beta} = p_L^{\langle\alpha\beta\rangle} - (p_L + \varpi_L) \Delta_L^{\alpha\beta} + \frac{e_L n_L}{c^2} U_L^\alpha U_L^\beta, \tag{4.47}$$

where the index L refers to the quantities defined according to Landau and Lifshitz decomposition. The projector in this case is given by

$$\Delta_L^{\alpha\beta} = \eta^{\alpha\beta} - \frac{1}{c^2} U_L^\alpha U_L^\beta. \tag{4.48}$$

The four-vector $\mathcal{J}^\alpha$ denotes the non-equilibrium part of the particle four-flow which is perpendicular to the four-velocity, i.e., $\mathcal{J}_\alpha U_L^\alpha = 0$. We shall show that for processes close to equilibrium $\mathcal{J}^\alpha$ is related to the heat flux q^α of Eckart decomposition. In order to relate both decompositions let us write

$$U_L^\alpha = U^\alpha + \mathcal{U}^\alpha, \tag{4.49}$$

where $\mathcal{U}^\alpha$ represents a non-equilibrium quantity such that for processes close to equilibrium products $\mathcal{U}^\alpha\mathcal{U}_\alpha$ can be neglected. Hence for processes close to equilibrium we get that

$$U_L^\alpha U_{L\alpha} = c^2 = (U^\alpha + \mathcal{U}^\alpha)(U_\alpha + \mathcal{U}_\alpha) \approx c^2 + 2U^\alpha\mathcal{U}_\alpha. \tag{4.50}$$

This last equation implies that U^α and $\mathcal{U}^\alpha$ in this approximation are perpendicular to each other, that is $U^\alpha\mathcal{U}_\alpha = 0$. Further if we take the scalar product of (4.49) by U_α we get by the use of $U^\alpha\mathcal{U}_\alpha = 0$ that $U_L^\alpha U_\alpha = c^2$.

On the other hand, the particle four-flow can be written as

$$N^\alpha \stackrel{(4.21)}{=} nU^\alpha \stackrel{(4.46)}{=} n_L U_L^\alpha + \mathcal{J}^\alpha. \tag{4.51}$$

If we take the scalar product of (4.51) by $U_{L\alpha}$ and use the relationships $\mathcal{J}^\alpha U_{L\alpha} = 0$, $U^\alpha U_{L\alpha} = U_L^\alpha U_{L\alpha} = c^2$, we conclude that

$$n = n_L, \tag{4.52}$$

that is for processes close to equilibrium the particle number density does not depend on the description chosen. Further from the insertion of (4.49) into (4.51) and by considering (4.52) we are led to

$$\mathcal{U}^\alpha = -\frac{\mathcal{J}^\alpha}{n}. \tag{4.53}$$

By applying the same methodology for the energy-momentum tensor we have that

$$T^{\alpha\beta} \stackrel{(4.22)}{=} p^{\langle\alpha\beta\rangle} - (p + \varpi)\,\Delta^{\alpha\beta} + \frac{1}{c^2}\left(U^\alpha q^\beta + U^\beta q^\alpha\right) + \frac{en}{c^2}U^\alpha U^\beta$$

$$\stackrel{(4.47)}{=} p_L^{\langle\alpha\beta\rangle} - (p_L + \varpi_L)\,\Delta^{\alpha\beta} - \frac{h_E^L}{c^2}\left(U^\alpha \mathcal{J}^\beta + U^\beta \mathcal{J}^\alpha\right) + \frac{e_L n_L}{c^2}U^\alpha U^\beta, \tag{4.54}$$

by the use of (4.49), by neglecting all non-linear terms $\mathcal{J}^\alpha\mathcal{J}^\beta$, $\varpi_L\mathcal{J}^\alpha$ and by considering the definition of the enthalpy per particle $h_E^L = e_L + p_L/n_L$. From the above equation we conclude that

$$p_L = p, \quad e_L = e, \quad \varpi_L = \varpi, \quad p_L^{\langle\alpha\beta\rangle} = p^{\langle\alpha\beta\rangle}, \quad \mathcal{J}^\alpha = -\frac{1}{h_E}q^\alpha, \tag{4.55}$$

that is the hydrodynamic pressures, the energies per particle, the dynamic pressures and the pressure deviators are the same in both descriptions for processes close to equilibrium, while the non-equilibrium part of the particle four-flow $\mathcal{J}^\alpha$ in the Landau and Lifshitz decomposition is related to the heat flux q^α in the Eckart decomposition.

By considering the above results we can rewrite the particle four-flow and the energy-momentum tensor in the Landau and Lifshitz decomposition for processes close to equilibrium as:

$$N^\alpha = nU_L^\alpha - \frac{q^\alpha}{h_E}, \tag{4.56}$$

$$T^{\alpha\beta} = p^{\langle\alpha\beta\rangle} - (p+\varpi)\,\Delta_L^{\alpha\beta} + \frac{en}{c^2}U_L^\alpha U_L^\beta. \tag{4.57}$$

It is also interesting to write the relationship between the two four-velocities in the Eckart U^α and in the Landau and Lifshitz U_L^α descriptions and the heat flux q^α. By combining (4.49), (4.53) and $(4.55)_5$ we get that

$$U^\alpha = U_L^\alpha - \frac{1}{nh_E}q^\alpha. \tag{4.58}$$

Problems

4.4.1 Obtain the right-hand side of equation (4.54).

4.4.2 Show that if we consider up to second order terms in $\mathcal{J}^\alpha$, we have that the relationship between the particle number densities in the two decompositions is:

$$n_L = n\left(1 - \frac{1}{2n^2c^2}\mathcal{J}^\alpha\mathcal{J}_\alpha\right).$$

4.5 Thermodynamics of a single fluid

The objective of the thermodynamic theory of a single relativistic fluid is the determination of the fields of:

$$\begin{cases} N^\alpha(x^\beta) \text{ – particle four-flow,} \\ T(x^\beta) \text{ – temperature,} \end{cases} \tag{4.59}$$

in all events x^α.

The balance equations used to determine the fields (4.59) are the conservation laws for the particle four-flow and for the energy-momentum tensor which we reproduce below:

$$\partial_\alpha N^\alpha = 0, \qquad \partial_\beta T^{\alpha\beta} = 0. \tag{4.60}$$

The above equations are not field equations for the basic fields (4.59) since they depend also on the unknown quantities: the dynamic pressure ϖ, the pressure deviator $p^{\langle\alpha\beta\rangle}$ and the heat flux q^α, provided we know the thermal equation of state $p = p(n,T)$ and the energy equation of state $e = e(n,T)$. In the following we shall determine the constitutive equations for the above quantities in terms of the basic fields (4.59), based on the thermodynamic theory of irreversible processes and using the Eckart decomposition. The starting point of this theory is the Gibbs equation (2.99) which we write in a local Lorentz rest frame as

$$\frac{\partial s}{\partial t} = \frac{1}{T}\left(\frac{\partial e}{\partial t} - \frac{p}{n^2}\frac{\partial n}{\partial t}\right). \tag{4.61}$$

Since in the local Lorentz rest frame $\partial/\partial t \equiv U^0\partial_0$, in an arbitrary Lorentz frame we have $U^\alpha\partial_\alpha \equiv D$ and the Gibbs equation can be written in terms of the convective time derivative D as

$$Ds = \frac{1}{T}\left(De - \frac{p}{n^2}Dn\right). \tag{4.62}$$

If we eliminate the derivatives De and Dn by using the balance equation for the particle number density (4.38) and for the energy density (4.41) we get after some rearrangements

$$nDs + \nabla_\alpha\left(\frac{q^\alpha}{T}\right) + \frac{1}{c^2}U_\alpha D\left(\frac{q^\alpha}{T}\right)$$

$$= -\frac{\varpi}{T}\nabla_\alpha U^\alpha + \frac{p^{\langle\alpha\beta\rangle}}{T}\nabla_\alpha U_\beta - \frac{q^\alpha}{T^2}\left(\nabla_\alpha T - \frac{T}{c^2}DU_\alpha\right). \tag{4.63}$$

Note that we have dropped out the index E from the entropy per particle and transformed the Gibbs equation, which refers to equilibrium, into a non-equilibrium equation. This amounts to defining the non-equilibrium entropy density as a state variable related to the other variables by the same relation holding for equilibrium states. This is one of the shortcomings of the linear thermodynamics of irreversible processes. If we compare (4.45) with (4.63) we can identify the entropy flux ϕ^α and the entropy production rate ς as

$$\phi^\alpha = \frac{q^\alpha}{T}, \tag{4.64}$$

$$\varsigma = -\frac{\varpi}{T}\nabla_\alpha U^\alpha + \frac{p^{\langle\alpha\beta\rangle}}{T}\nabla_\alpha U_\beta - \frac{q^\alpha}{T^2}\left(\nabla_\alpha T - \frac{T}{c^2}DU_\alpha\right) \geq 0. \tag{4.65}$$

The entropy flux of a single relativistic fluid is just the heat flux divided by the temperature. This result is similar to that found in the non-relativistic theory of a single fluid.

In the thermodynamic theory of irreversible processes one calls each gradient a "(thermodynamic) force" and the factor multiplying it in the expression of the entropy production rate a "(thermodynamic) flux". Thus one identifies in (4.65) the quantities

$$\begin{cases} \text{forces: } \nabla_\alpha U^\alpha, \nabla^{\langle\alpha}U^{\beta\rangle}, \left[\nabla^\alpha T - (T/c^2)DU^\alpha\right]; \\ \text{fluxes: } \varpi, p^{\langle\alpha\beta\rangle}, q^\alpha. \end{cases} \tag{4.66}$$

Further it is assumed that a linear relationship between the thermodynamic fluxes and forces holds. In this case we have

$$\varpi = -\eta \nabla_\alpha U^\alpha, \tag{4.67}$$

$$p^{\langle\alpha\beta\rangle} = 2\mu \nabla^{\langle\alpha} U^{\beta\rangle} = 2\mu \left[\frac{1}{2}\left(\Delta^\alpha_\gamma \Delta^\beta_\delta + \Delta^\alpha_\delta \Delta^\beta_\gamma\right) - \frac{1}{3}\Delta^{\alpha\beta}\Delta_{\gamma\delta}\right] \nabla^\gamma U^\delta, \tag{4.68}$$

$$q^\alpha = \lambda \left(\nabla^\alpha T - \frac{T}{c^2} D U^\alpha\right), \tag{4.69}$$

that represent the constitutive equations for a single viscous and heat conducting relativistic fluid. Equations (4.67) and (4.68) are the constitutive equations of a Newtonian viscous fluid also known as Navier–Stokes law. Equation (4.69) is the Fourier law. The coefficients η, μ and λ are called bulk viscosity, shear viscosity and thermal conductivity, respectively. All coefficients are non-negative because this holds for the entropy production rate (Problem 4.5.2).

The relativistic Navier–Stokes law has an expression similar to the non-relativistic one, but the Fourier law has its peculiarities which we proceed to analyze. In a local Lorentz rest frame we get from (4.69)

$$q^0 = 0, \qquad \mathbf{q} = -\lambda \vec{\nabla} T. \tag{4.70}$$

The above equation has a form similar to the non-relativistic Fourier law. The term $-TDU^\alpha/c^2$ in (4.69), which has not a corresponding one in the non-relativistic case, represents an isothermal heat flux when matter is accelerated. This small term acts in a direction opposite to the acceleration and is due to the inertia of energy. We can also express (4.69) in another form by eliminating the time derivative of the four-velocity through (4.43) with $\varpi = 0$, $p^{\langle\alpha\beta\rangle} = 0$ and $q^\alpha = 0$, which characterizes the so-called Eulerian fluid. In this case (4.43) can be written as

$$\frac{nh_E}{c^2} D U^\alpha = \nabla^\alpha p, \tag{4.71}$$

that represents the balance equation for the momentum density of a non-viscous and non-conducting fluid. Hence we get from (4.69) and (4.71) that the constitutive equation for the heat flux can be written as

$$q^\alpha = \lambda \left(\nabla^\alpha T - \frac{T}{nh_E} \nabla^\alpha p\right), \tag{4.72}$$

i.e., even for isothermal processes there exists a heat flux due to a pressure gradient.

Problems

4.5.1 Show that the Gibbs equation (4.62) reduces to (4.63), when Dn and De are eliminated through the use of the balance equations for the particle number density (4.38) and for the energy density (4.41).

4.5.2. Prove that the coefficients of bulk viscosity, shear viscosity and thermal conductivity are non-negative if this holds for the entropy production rate (4.65).

4.5.3 Show that

$$p_{\langle\alpha\beta\rangle}\nabla^\alpha U^\beta = p_{\alpha\beta}\nabla^{\langle\alpha}U^{\beta\rangle} = p_{\langle\alpha\beta\rangle}\nabla^{\langle\alpha}U^{\beta\rangle}.$$

4.5.4 In the non-relativistic limiting case we have that the four-velocity reduces to $(U^\alpha) = (c, \mathbf{v})$ since $v \ll c$. Show that in this limiting case we have that

$$\Delta^{00} \to 0, \qquad \Delta^{ij} \to \eta^{ij}, \qquad \nabla^0 \to 0, \qquad \nabla^i \to -\vec{\nabla}^i.$$

4.5.5 Based on the above results show that in the non-relativistic limiting case the relativistic Navier–Stokes law (4.67), (4.68) and Fourier law (4.69) lead to

$$\varpi = -\eta\vec{\nabla}^i v^i = -\eta\,\mathrm{div}\,\mathbf{v}, \quad q^0 = 0, \quad q^i = -\lambda\vec{\nabla}^i T, \quad p^{\langle 00\rangle} = 0, \quad p^{\langle 0i\rangle} = 0,$$

$$p^{\langle ij\rangle} = -\mu\left(\vec{\nabla}^i v^j + \vec{\nabla}^j v^i - \frac{2}{3}\vec{\nabla}^k v^k \delta^{ij}\right) = -2\mu\vec{\nabla}^{\langle i}v^{i\rangle}.$$

References

[1] C. Eckart, The thermodynamics of irreversible processes, III. Relativistic theory of a simple fluid, *Phys. Rev.* **58**, 919–924 (1940).

[2] S. R. de Groot, W. A. van Leeuwen and Ch. G. van Weert, *Relativistic kinetic theory* (North-Holland, Amsterdam, 1980).

[3] G. M. Kremer, *Uma introdução à teoria cinética relativística* (Editora da UFPR, Curitiba, 1998).

[4] L. D. Landau and E. M. Lifshitz, *Fluid mechanics*, 2nd. ed. (Pergamon Press, Oxford, 1987).

[5] I. Müller and T. Ruggeri, *Rational extended thermodynamics* 2nd. ed., Springer Tracts in Natural Philosophy, vol. 37 (Springer, New York, 1998).

[6] J. M. Stewart, *Non-equilibrium relativistic kinetic theory*, Lecture Notes in Physics, vol. 10 (Springer, Heidelberg, 1971).

[7] S. Weinberg, *Gravitation and cosmology. Principles and applications of the theory of relativity* (Wiley, New York, 1972).

Chapter 5

Chapman–Enskog Method

5.1 Introduction

In this chapter we shall study a single non-degenerate relativistic gas in a non-equilibrium state, and determine from the Boltzmann equation the linear constitutive equations for the dynamic pressure ϖ, pressure deviator $p^{\langle\alpha\beta\rangle}$ and heat flux q^α that correspond to the laws of Navier–Stokes and Fourier of a single viscous and heat conducting relativistic fluid. The corresponding transport coefficients of bulk viscosity, shear viscosity and thermal conductivity will be determined in terms of the interaction law between the relativistic particles. For didactical purposes we shall present first a simplified version of the Chapman–Enskog method.

5.2 Simplified version

We begin by writing the Boltzmann equation (2.36) in the absence of external forces as

$$p^\alpha \frac{\partial f}{\partial x^\alpha} = \int (f_*' f' - f_* f) F \sigma d\Omega \frac{d^3 p_*}{p_{*0}}. \tag{5.1}$$

The basic idea of the method is to split the one-particle distribution function $f(\mathbf{x}, \mathbf{p}, t)$ into two additive terms. The first term corresponds to the Maxwell–Jüttner distribution function (3.27)

$$f^{(0)}(\mathbf{x}, \mathbf{p}, t) = \frac{n}{4\pi m^2 ckT K_2(\zeta)} e^{-\frac{U^\alpha p_\alpha}{kT}}, \tag{5.2}$$

which gives the local values of the particle number density n, temperature T, and four-velocity U^α. This term is known as the first approximation of the one-particle distribution function. The second approximation of the one-particle distribution function corresponds to a small deviation from the equilibrium described by the

Maxwell–Jüttner distribution function which leads to the linear constitutive equations for a gas in a non-equilibrium state that correspond to the laws of Navier–Stokes and Fourier. Hence we write

$$f(\mathbf{x},\mathbf{p},t) = f^{(0)}(\mathbf{x},\mathbf{p},t) + f^{(1)}(\mathbf{x},\mathbf{p},t) = f^{(0)}(\mathbf{x},\mathbf{p},t)[1+\phi(\mathbf{x},\mathbf{p},t)], \tag{5.3}$$

where $f^{(1)} \equiv f^{(0)}\phi$ is the deviation of the Maxwell–Jüttner distribution function. This does not define the rules for our approximation. In order to define the method in a better way, we must assume that the mean free path ℓ, i.e., the average length covered by a particle between two subsequent collisions, is small with respect to the typical length L over which particle number density, bulk velocity and temperature vary in a significant way. This is an appropriate assumption in order to obtain an approximate continuum theory from the Boltzmann equation. Then ϕ is assumed to be of the order of ℓ/L.

We remark that the concept of mean free path is a qualitative one, which can be made quantitative in several ways, i.e., by defining accurately how the average is performed. Different definitions lead to values that differ from each other by a numerical factor of order unity. Thus it is not necessary to specify ℓ in a more precise way when we discuss general aspects of the matter. The reader is warned, however, to check the definition of the mean free path, if any, used by authors who give numerical results in terms of this quantity. We also remark that ℓ is inversely proportional to the particle number density n and a typical value of the differential cross-section σ.

We recall that in Section 2.7 we have identified the five scalar constants of the one-particle distribution function in equilibrium through the definitions of the particle four-flow and energy per particle, i.e.,

$$N^\alpha = c\int p^\alpha f^{(0)}\frac{d^3p}{p_0}, \qquad nec^2 = U_\alpha U_\beta\left[c\int p^\alpha p^\beta f^{(0)}\frac{d^3p}{p_0}\right]. \tag{5.4}$$

The above equations define the variables: particle number density n, four-velocity U^α, and energy per particle e (or implicitly the temperature T). Since n, U^α, and e are variables their values are the same if we define them in terms of the equilibrium or in terms of the non-equilibrium distribution function. In this sense if we are using the Eckart decomposition (see Section 4.3) we have the following constraints for the one-particle distribution function:

$$N^\alpha = nU^\alpha = c\int p^\alpha f^{(0)}\frac{d^3p}{p_0} = c\int p^\alpha f\frac{d^3p}{p_0}, \tag{5.5}$$

$$nec^2 = U_\alpha U_\beta\left[c\int p^\alpha p^\beta f^{(0)}\frac{d^3p}{p_0}\right] = U_\alpha U_\beta\left[c\int p^\alpha p^\beta f\frac{d^3p}{p_0}\right]. \tag{5.6}$$

If we are using the Landau and Lifshitz decomposition the corresponding constraints read (see Section 4.4):

$$U_L^\alpha N_\alpha = n_L c^2 = U_L^\alpha\left[c\int p_\alpha f^{(0)}\frac{d^3p}{p_0}\right] = U_L^\alpha\left[c\int p_\alpha f\frac{d^3p}{p_0}\right], \tag{5.7}$$

$$n_L e_L c^2 = U_L^\alpha U_L^\beta \left[c \int p_\alpha p_\beta f^{(0)} \frac{d^3 p}{p_0} \right] = U_L^\alpha U_L^\beta \left[c \int p_\alpha p_\beta f \frac{d^3 p}{p_0} \right], \tag{5.8}$$

$$\Delta_L^{\alpha\beta} U_L^\gamma T_{\beta\gamma} = \Delta_L^{\alpha\beta} U_L^\gamma \left[c \int p_\beta p_\gamma f^{(0)} \frac{d^3 p}{p_0} \right] = \Delta_L^{\alpha\beta} U_L^\gamma \left[c \int p_\beta p_\gamma f \frac{d^3 p}{p_0} \right] = 0, \tag{5.9}$$

where (5.9) is an identity that follows from (4.47).

In the following we shall deal only with the Eckart decomposition. Hence the conditions (5.5) and (5.6) together with the representation of the one-particle distribution function (5.3) lead to the following constraints for the deviation $f^{(0)}\phi$ of the Maxwell–Jüttner distribution function

$$\int p^\alpha f^{(0)} \phi \frac{d^3 p}{p_0} = 0, \qquad U_\alpha U_\beta \int p^\alpha p^\beta f^{(0)} \phi \frac{d^3 p}{p_0} = 0. \tag{5.10}$$

In the Chapman–Enskog method the deviation $f^{(0)}\phi$ from the Maxwell–Jüttner distribution function is determined (at the lowest order in ℓ/L) from the Boltzmann integro-differential equation (5.1) in the following manner. We insert (5.3) into (5.1) and keep only the leading terms in both sides of the resulting equation, i.e., on the right-hand side we consider the linear terms in the deviation $f^{(0)}\phi$ and on the left-hand side we take into account only the derivatives of the Maxwell–Jüttner distribution function since these derivatives correspond to the thermodynamic forces that induce the appearance of the deviation $f^{(0)}\phi$. This is in agreement with the facts that the collision term produces a value inversely proportional to the mean free path, the left-hand side produces a value inversely proportional to the macroscopic length L, and we stated that we look for a ϕ proportional to ℓ/L. Hence we write

$$p^\alpha \frac{\partial f^{(0)}}{\partial x^\alpha} = \int f_*^{(0)} f^{(0)} \left[\phi_*' + \phi' - \phi_* - \phi\right] F \sigma d\Omega \frac{d^3 p_*}{p_{*0}} \tag{5.11}$$

where we have used the relationship $f_*'^{(0)} f'^{(0)} = f_*^{(0)} f^{(0)}$.

We can write (5.11) in a compact form as

$$\mathcal{D} f^{(0)} = \mathcal{I}[\phi], \tag{5.12}$$

where $\mathcal{D}$ and $\mathcal{I}$ are two operators defined by

$$\mathcal{D} \equiv p^\alpha \frac{\partial}{\partial x^\alpha}, \qquad \mathcal{I}[\phi] \equiv \int f_*^{(0)} f^{(0)} \left[\phi_*' + \phi' - \phi_* - \phi\right] F \sigma d\Omega \frac{d^3 p_*}{p_{*0}}. \tag{5.13}$$

Now we infer that (5.12) is no longer an integro-differential equation but an inhomogeneous integral equation for the deviation ϕ; this is due to the fact that we have treated the terms containing derivatives as a perturbation. The reader should remark that this procedure does not lead to a typical solution of the Boltzmann equation which should also contain terms rapidly changing on the scale of a mean free path; however, the solutions to be obtained are exactly those needed to build a bridge between kinetic theory and macroscopic fluid dynamics, since the latter is not concerned with changes on such a tiny scale.

The solution of the above integral equation for ϕ is one of the fundamental problems in the kinetic theory of gases, because it leads to the evaluation of transport coefficients, such as bulk and shear viscosities and thermal conductivity, in terms of interaction between particles.

Before we proceed with the analysis of (5.12) we shall state the following result: If $\chi(\mathbf{x}, \mathbf{p}, t)$ is an arbitrary function, then

$$\int \chi \mathcal{I}[\phi] \frac{d^3p}{p_0} = \int \phi \mathcal{I}[\chi] \frac{d^3p}{p_0}. \tag{5.14}$$

The proof of the above assertion is similar to that employed to derive the right-hand side of the transfer equation (2.53) in Section 2.3. An immediate consequence of (5.14) is that for a summational invariant $\chi \equiv \psi$, with $\psi = 1$ or $\psi = p^\alpha$, the above integral vanishes, i.e.,

$$\int \psi \mathcal{I}[\phi] \frac{d^3p}{p_0} = 0. \tag{5.15}$$

Hence we get from (5.12) that for a summational invariant ψ,

$$\int \psi \mathcal{D} f^{(0)} \frac{d^3p}{p_0} = 0, \tag{5.16}$$

which represents a constraint in the derivatives of the fields n, U^α and T that appear in the Maxwell–Jüttner distribution function.

If we perform the derivatives of $\mathcal{D}f^{(0)}$ by using (5.2), then

$$\mathcal{D}f^{(0)} = \frac{f^{(0)}}{n}\Bigg\{\frac{1}{c^2}U^\alpha p_\alpha Dn + p_\alpha \nabla^\alpha n + \frac{n}{kT^2}(p_\beta U^\beta - e)\left[\frac{1}{c^2}U^\alpha p_\alpha DT\right.$$
$$\left. + p_\alpha \nabla^\alpha T\right] - \frac{n}{kT}\left[\frac{1}{c^2}(p_\alpha U^\alpha) p_\beta DU^\beta + p_\alpha p_\beta \nabla^\alpha U^\beta\right]\Bigg\}. \tag{5.17}$$

In order to get the above equation we have employed the decomposition (2.158) which defines the operators D and ∇^α, as well as the decomposition of the momentum four-vector

$$p^\alpha = \Delta^\alpha_\beta p^\beta + \frac{1}{c^2} U^\alpha (p^\beta U_\beta), \tag{5.18}$$

and the relationship $U^\beta \partial_\alpha U_\beta = 0$.

If we insert (5.17) into (5.16), choose $\psi = 1$ and integrate the resulting equation over all values of d^3p/p_0, we have that

$$Dn + n\nabla^\alpha U_\alpha = 0, \tag{5.19}$$

by considering the integrals (5.225) and (5.226) of Section 5.6.

By using the same methodology above, but by choosing $\psi = p^\gamma$, we get that the projections with U_γ and ∇_γ^ϵ of the resulting equation lead respectively to the equations

$$nc_v DT + p\nabla_\alpha U^\alpha = 0, \tag{5.20}$$

$$\frac{nh_E}{c^2} DU^\epsilon = \nabla^\epsilon p, \tag{5.21}$$

again by taking into account the integrals (5.226) and (5.227) of Section 5.6. We recall that for a non-degenerate relativistic gas

$$c_v = k(\zeta^2 + 5G\zeta - G^2\zeta^2 - 1), \quad \text{and} \quad h_E = mc^2 G, \tag{5.22}$$

are the expressions for the heat capacity at constant volume and for the enthalpy per particle, respectively.

Equation (5.19) refers to the balance equation for the particle number density (see also (4.38)). If we compare (5.20) and (5.21) with the balance equations for the energy density (4.41) and for the momentum density (4.43), we conclude that the former are particular cases of the latter. Indeed, if we consider that the dynamic pressure ϖ the heat flux q^α and the pressure deviator $p^{\langle\alpha\beta\rangle}$ are all zero in (4.41) and (4.43) we get (5.20) and (5.21). As was pointed out in Section 4.5 a relativistic fluid whose constitutive equations are given by $\varpi = 0$, $q^\alpha = 0$ and $p^{\langle\alpha\beta\rangle} = 0$ is called a relativistic Eulerian fluid.

Now equations (5.19) through (5.21) are employed to eliminate the convective time derivatives Dn, DT and DU^ϵ from (5.17) yielding:

$$f^{(0)}\left\{-\frac{k^2T}{c^2c_v}\left[\frac{1}{3}\zeta^2\frac{c_v}{k} - (G^2\zeta^2 - 4G\zeta - \zeta^2)\left(\frac{U^\beta p_\beta}{kT}\right)\right.\right.$$
$$\left. -\frac{1}{3}(\zeta^2 + 5G\zeta - \zeta^2G^2 - 4)\left(\frac{U^\beta p_\beta}{kT}\right)^2\right]\nabla_\alpha U^\alpha - \frac{p_\alpha p_\beta}{kT}\nabla^{\langle\alpha}U^{\beta\rangle}$$
$$\left. + \frac{p_\alpha}{kT^2}(p_\beta U^\beta - h_E)\left(\nabla^\alpha T - \frac{T}{nh_E}\nabla^\alpha p\right)\right\} \overset{(5.12)}{=} \mathcal{I}[\phi]. \tag{5.23}$$

In the above equation we have used the relationship:

$$p_\alpha p_\beta \nabla^\alpha U^\beta = p_\alpha p_\beta\left(\nabla^{\langle\alpha}U^{\beta\rangle} + \nabla^{[\alpha}U^{\beta]} + \frac{1}{3}\nabla^\gamma U_\gamma \Delta^{\alpha\beta}\right)$$
$$= p_\alpha p_\beta \nabla^{\langle\alpha}U^{\beta\rangle} + \frac{1}{3}\left(m^2c^2 - \frac{(U^\beta p_\beta)^2}{c^2}\right)\nabla^\alpha U_\alpha. \tag{5.24}$$

By inspecting the inhomogeneous integral equation (5.23) we conclude that a good approximation to its solution can be written as

$$\phi = \left[a_0 + a_1 U^\alpha p_\alpha + a_2(U^\beta p_\beta)^2\right]\nabla_\alpha U^\alpha$$
$$+(a_3 + a_4 U^\beta p_\beta)p_\alpha\left[\nabla^\alpha T - \frac{T}{nh_E}\nabla^\alpha p\right] + a_5 p_\alpha p_\beta \nabla^{\langle\alpha}U^{\beta\rangle}, \tag{5.25}$$

where a_0 through a_5 are unknown quantities, which may depend only on n and T, and are constant with respect to p_α.

If we insert (5.25) into the constraints (5.10) and integrate the resulting equations by using the integrals of Section 5.6, we obtain that the unknowns a_0 and a_1 can be expressed in terms of a_2 and a_3 in terms of a_4 as follows:

$$a_0 = -\frac{\zeta(15G + 2\zeta - 6G^2\zeta + 5G\zeta^2 + \zeta^3 - \zeta^3 G^2)}{\zeta^2 + 5G\zeta - G^2\zeta^2 - 1}(kT)^2 a_2, \tag{5.26}$$

$$a_1 = -\frac{3\zeta(\zeta + 6G - G^2\zeta)}{\zeta^2 + 5G\zeta - G^2\zeta^2 - 1} kT a_2, \tag{5.27}$$

$$a_3 = -\zeta G kT a_4. \tag{5.28}$$

Hence we have only to determine the three unknowns a_2, a_4 and a_5, which can be obtained from

$$-f^{(0)}\frac{k^2T}{c^2c_v}\left[\frac{1}{3}\zeta^2\frac{c_v}{k} - (G^2\zeta^2 - 4G\zeta - \zeta^2)\left(\frac{U^\beta p_\beta}{kT}\right)\right.$$

$$\left.-\frac{1}{3}(\zeta^2 + 5G\zeta - \zeta^2G^2 - 4)\left(\frac{U^\beta p_\beta}{kT}\right)^2\right] = a_2 U_\alpha U_\beta \mathcal{I}[p^\alpha p^\beta], \tag{5.29}$$

$$f^{(0)}\Delta^\alpha_\beta \frac{p^\beta}{kT^2}\left[p_\gamma U^\gamma - h_E\right] = a_4 \Delta^\alpha_\beta U_\gamma \mathcal{I}[p^\beta p^\gamma], \tag{5.30}$$

$$-f^{(0)}\left[\Delta^\alpha_\gamma\Delta^\beta_\delta - \frac{1}{3}\Delta^{\alpha\beta}\Delta_{\gamma\delta}\right]\frac{p^\gamma p^\delta}{kT} = a_5\left[\Delta^\alpha_\gamma\Delta^\beta_\delta - \frac{1}{3}\Delta^{\alpha\beta}\Delta_{\gamma\delta}\right]\mathcal{I}[p^\gamma p^\delta]. \tag{5.31}$$

The above equations follow through the insertion of (5.25) into (5.23) and by equating the coefficients of $\nabla_\alpha U^\alpha$, $[\nabla^\alpha T - T/(nh_E)\nabla^\alpha p]$ and $\nabla^{\langle\alpha}U^{\beta\rangle}$ to zero. We note that $\mathcal{I}[\psi] = 0$ whenever ψ is a summational invariant.

The determination of the constants a_2, a_4 and a_5 from (5.29) through (5.31) proceeds as follows. First we multiply (5.29) by $U_\gamma p^\gamma U_\delta p^\delta$, integrate the resulting equation over all values of d^3p/p_0 and find, by using the integrals of Section 5.6, that a_2 is given by

$$a_2 = \frac{pkT}{cI_1}\frac{\zeta(20G + 3\zeta - 13G^2\zeta - 2G\zeta^2 + 2G^3\zeta^2)}{1 - 5G\zeta - \zeta^2 + G^2\zeta^2}. \tag{5.32}$$

Next we apply the same methodology and multiply (5.30) by $p_\alpha U^\delta p_\delta$ and (5.31) by $p_\alpha p_\beta$ and get respectively

$$a_4 = 3kpc\frac{\zeta(\zeta + 5G - G^2\zeta)}{I_1 - c^2 I_2}, \tag{5.33}$$

$$a_5 = -\frac{30mpc^3G}{2I_1 - 6c^2I_2 + 3c^4I_3}. \tag{5.34}$$

In (5.32) through (5.34) I_1, I_2 and I_3 represent the integrals

$$I_1 = U_\alpha U_\beta U_\gamma U_\delta \int p^\alpha p^\beta \mathcal{I}[p^\gamma p^\delta] \frac{d^3p}{p_0}, \tag{5.35}$$

$$I_2 = U_\alpha U_\gamma \int p^\alpha p^\beta \mathcal{I}[p^\gamma p_\beta] \frac{d^3p}{p_0}, \tag{5.36}$$

$$I_3 = \int p^\alpha p^\beta \mathcal{I}[p_\alpha p_\beta] \frac{d^3p}{p_0}. \tag{5.37}$$

Since we have found all coefficients a_0 through a_5, the deviation from the Maxwell–Jüttner distribution function (5.25) is known and it can be used to determine the constitutive equations for the dynamic pressure ϖ, heat flux q^α and pressure deviator $p^{\langle\alpha\beta\rangle}$. We begin with the determination of the dynamic pressure ϖ and recall that it is defined by the relationship (see (4.26))

$$p + \varpi = -\frac{1}{3}\Delta_{\alpha\beta}T^{\alpha\beta} = -\frac{1}{3}\Delta_{\alpha\beta}\left[c\int p^\alpha p^\beta f \frac{d^3p}{p_0}\right]. \tag{5.38}$$

Hence by inserting (5.3), together with (5.25) through (5.28) and (5.32) through (5.34), into (5.38) and integrating the resulting equation over all values of d^3p/p_0, we get the constitutive equation for the dynamic pressure:

$$\varpi = -\eta \nabla_\alpha U^\alpha. \tag{5.39}$$

To get the above equation we have used the integrals of Section 5.6. In (5.39) the coefficient of bulk viscosity reads

$$\eta = -\frac{p^2 kTm^2c^3}{I_1}\frac{(20G + 3\zeta - 13G^2\zeta - 2G\zeta^2 + 2G^3\zeta^2)^2}{(1 - 5G\zeta - \zeta^2 + G^2\zeta^2)^2}. \tag{5.40}$$

By following the same methodology we get from (see (4.27))

$$q^\alpha = \Delta^\alpha_\beta U_\gamma T^{\beta\gamma} = \Delta^\alpha_\beta U_\gamma \left[c\int p^\beta p^\gamma f \frac{d^3p}{p_0}\right], \tag{5.41}$$

the constitutive equation for the heat flux

$$q^\alpha = \lambda\left[\nabla^\alpha T - \frac{T}{nh_E}\nabla^\alpha p\right], \tag{5.42}$$

where the coefficient of thermal conductivity is given by

$$\lambda = -\frac{3kp^2m^2c^5(\zeta + 5G - G^2\zeta)^2}{I_1 - c^2 I_2}. \tag{5.43}$$

The constitutive equation for the pressure deviator is obtained from (see (4.25))

$$p^{\langle\alpha\beta\rangle} = \left(\Delta^\alpha_\gamma\Delta^\beta_\delta - \frac{1}{3}\Delta_{\alpha\beta}\Delta_{\gamma\delta}\right)T^{\gamma\delta}$$

$$= \left(\Delta^\alpha_\gamma\Delta^\beta_\delta - \frac{1}{3}\Delta_{\alpha\beta}\Delta_{\gamma\delta}\right)\left[c\int p^\gamma p^\delta f\frac{d^3p}{p_0}\right], \tag{5.44}$$

yielding

$$p^{\langle\alpha\beta\rangle} = 2\mu\nabla^{\langle\alpha}U^{\beta\rangle}. \tag{5.45}$$

The coefficient of shear viscosity reads

$$\mu = -\frac{30p^2kTm^2c^3G^2}{2I_1 - 6c^2I_2 + 3c^4I_3}. \tag{5.46}$$

In the next section we shall give the reduced forms of the integrals I_1, I_2 and I_3 defined by (5.35) through (5.37).

Problems

5.2.1 Show that the relationship (5.14) holds.

5.2.2 Obtain the balance equations for the particle number density (5.19), energy density (5.20) and momentum density (5.21) for an Eulerian relativistic gas from (5.16).

5.2.3 Show that the coefficients a_0, a_1 and a_3 can be expressed in terms of the coefficients a_2 and a_4 as stated in (5.26) through (5.28).

5.2.4 Obtain the expressions for the coefficients a_2, a_4 and a_5 given in (5.32) through (5.34) from (5.29) through (5.31).

5.2.5 Show that the coefficient of bulk viscosity, thermal conductivity and shear viscosity are given by (5.40), (5.43) and (5.46), respectively.

5.3 The integrals I_1, I_2 and I_3

For the determination of the integrals I_1, I_2 e I_3 we shall introduce the total momentum four-vector P^α and the relative momentum four-vector Q^α defined by

$$P^\alpha = p^\alpha + p^\alpha_*, \qquad P'^\alpha = p'^\alpha + p'^\alpha_*, \tag{5.47}$$

$$Q^\alpha = p^\alpha - p^\alpha_*, \qquad Q'^\alpha = p'^\alpha - p'^\alpha_*. \tag{5.48}$$

From the above equations together with the conservation law of momentum-energy $p^\alpha + p^\alpha_* = p'^\alpha + p'^\alpha_*$, we obtain the properties

$$P^\alpha = P'^\alpha, \qquad P^\alpha Q_\alpha = P'^\alpha Q'_\alpha = 0, \tag{5.49}$$

$$P^2 = 4m^2c^2 + Q^2, \tag{5.50}$$

where P^2 and Q^2 are defined by

$$P^\alpha P_\alpha = P^2, \qquad Q^\alpha Q_\alpha = -Q^2. \tag{5.51}$$

The inverse transformations of (5.47) and (5.48) are

$$p^\alpha = \frac{1}{2}\left(P^\alpha + Q^\alpha\right), \qquad p_*^\alpha = \frac{1}{2}\left(P^\alpha - Q^\alpha\right), \tag{5.52}$$

$$p'^\alpha = \frac{1}{2}\left(P^\alpha + Q'^\alpha\right), \qquad p_*'^\alpha = \frac{1}{2}\left(P^\alpha - Q'^\alpha\right). \tag{5.53}$$

By using (5.52) and (5.53) one can calculate the Jacobian of the transformation and show that (Problem 5.3.1)

$$d^3p\, d^3p_* = |J| d^3P d^3Q = \frac{1}{2^3} d^3P d^3Q. \tag{5.54}$$

According to (2.35) the invariant flux is defined by

$$F = \frac{p_0 p_{*0}}{c} g_\varnothing, \tag{5.55}$$

and we can write, based on (5.54) and (5.55), the expression

$$F \frac{d^3p}{p_0} \frac{d^3p_*}{p_{*0}} = \frac{1}{2^3} \frac{g_\varnothing}{c} d^3P d^3Q. \tag{5.56}$$

First we shall deal with the integral I_1, and for this purpose we substitute (5.52), (5.53) and (5.56) into (5.35) and get

$$I_1 = \frac{1}{2^6} U^\alpha U^\beta U^\gamma U^\delta \int f^{(0)} f_*^{(0)} (P_\alpha P_\beta + Q_\alpha P_\beta + P_\alpha Q_\beta$$

$$+ Q_\alpha Q_\beta) \left(Q'_\gamma Q'_\delta - Q_\gamma Q_\delta\right) \frac{g_\varnothing}{c} \sigma d\Omega d^3P d^3Q. \tag{5.57}$$

In order to evaluate the above integral, we choose the center-of-mass system where the spatial components of the total momentum four-vector vanish, i.e.,

$$(P^\alpha) = (P^0, \mathbf{0}), \qquad (Q^\alpha) = (0, \mathbf{Q}). \tag{5.58}$$

In this system the Møller relative speed $g_\varnothing$ has the representation

$$g_\varnothing = 2c\frac{Q}{P^0}. \tag{5.59}$$

Now we write the element of solid angle in (5.57) as $d\Omega = \sin\Theta d\Theta d\Phi$, where Θ and Φ are polar angles of Q'^α with respect to Q^α and such that Θ represents the scattering angle. Further we consider, without loss of generality, that Q^α is in the direction of the axis x^3, so that we can write Q^α and Q'^α as

$$(Q^\alpha) = Q\begin{pmatrix}0\\0\\0\\1\end{pmatrix}, \qquad (Q'^\alpha) = Q\begin{pmatrix}0\\ \sin\Theta\cos\Phi\\ \sin\Theta\sin\Phi\\ \cos\Theta\end{pmatrix}. \tag{5.60}$$

If we use the above representations it is easy to calculate the integral in the variable $0 \le \Phi \le 2\pi$, indeed

$$\int_0^{2\pi}\left(Q'^\alpha Q'^\beta - Q^\alpha Q^\beta\right)d\Phi$$

$$= \pi Q^2\sin^2\Theta\begin{pmatrix}0&0&0&0\\0&1&0&0\\0&0&1&0\\0&0&0&-2\end{pmatrix} = \pi Q^2\sin^2\Theta\left(\frac{P^\alpha P^\beta}{P^2} - \eta^{\alpha\beta} - 3\frac{Q^\alpha Q^\beta}{Q^2}\right). \tag{5.61}$$

As a first step we use (5.59) and (5.61) to write the integral I_1 in (5.57) as

$$I_1 = \frac{1}{2^5}U^\alpha U^\beta U^\gamma U^\delta\int f^{(0)}f_*^{(0)}\left(P_\alpha P_\beta + Q_\alpha P_\beta + P_\alpha Q_\beta + Q_\alpha Q_\beta\right)$$

$$\times\left(\frac{P_\gamma P_\delta}{P^2} - g_{\gamma\delta} - 3\frac{Q_\gamma Q_\delta}{Q^2}\right)\pi Q^3\sigma\sin^3\Theta\, d\Theta\, d^3Q\frac{d^3P}{P^0}. \tag{5.62}$$

We can now perform the integrations in the spherical angles of Q^α, denoted by θ and ϕ, i.e.,

$$(Q^\alpha) = Q\begin{pmatrix}0\\ \sin\theta\cos\phi\\ \sin\theta\sin\phi\\ \cos\theta\end{pmatrix}, \qquad d^3Q = Q^2\sin\theta\, d\theta\, d\phi\, dQ = Q^2 d\Omega^\star dQ, \tag{5.63}$$

with $d\Omega^\star = \sin\theta d\theta d\phi$ representing an element of solid angle. Indeed, one can obtain the following expression for the integrals:

$$\int d\Omega^\star = \int_0^{2\pi}\int_0^{\pi}\sin\theta\, d\theta\, d\phi = 4\pi, \tag{5.64}$$

$$\int Q^\alpha Q^\beta d\Omega^\star = \frac{4\pi}{3}Q^2\left(\frac{P^\alpha P^\beta}{P^2} - \eta^{\alpha\beta}\right), \tag{5.65}$$

$$\int Q^\alpha d\Omega^\star = 0, \qquad \int Q^\alpha Q^\beta Q^\sigma d\Omega^\star = 0, \tag{5.66}$$

$$\int Q^\alpha Q^\beta Q^\gamma Q^\delta d\Omega^\star = \frac{4\pi}{15}Q^4\left[3\frac{P^\alpha P^\beta P^\gamma P^\delta}{P^4} - \frac{1}{P^2}(\eta^{\alpha\beta}P^\gamma P^\delta\right.$$

$$+\eta^{\gamma\delta}P^\alpha P^\beta + \eta^{\alpha\gamma}P^\beta P^\delta + \eta^{\beta\delta}P^\alpha P^\gamma + \eta^{\alpha\delta}P^\beta P^\gamma + \eta^{\beta\gamma}P^\alpha P^\delta)$$

$$\left.+(\eta^{\alpha\beta}\eta^{\gamma\delta} + \eta^{\alpha\gamma}\eta^{\beta\delta} + \eta^{\alpha\delta}\eta^{\beta\gamma})\right]. \tag{5.67}$$

Now if we perform the integration of (5.62) with respect to the element of solid angle $d\Omega^\star$, I_1 reduces to

$$I_1 = -\frac{1}{480}\frac{n^2c^2}{[kTm^2K_2(\zeta)]^2}\int\left[Z^\star - \frac{2}{c^2}Z^\star_{\gamma\delta}\frac{U^\gamma U^\delta}{(Q^2+4m^2c^2)}\right.$$

$$\left.+\frac{1}{c^4}Z^\star_{\gamma\delta\epsilon\lambda}\frac{U^\gamma U^\delta U^\epsilon U^\lambda}{(Q^2+4m^2c^2)^2}\right]Q^7\sigma\sin^3\Theta\, d\Theta\, dQ \tag{5.68}$$

where $Z^\star$, $Z^\star_{\gamma\delta}$ and $Z^\star_{\gamma\delta\epsilon\lambda}$ denote the integrals

$$Z^\star = \int e^{-\frac{1}{kT}U_\alpha P^\alpha}\frac{d^3P}{P^0}, \qquad Z^\star_{\gamma\delta} = \int e^{-\frac{1}{kT}U_\alpha P^\alpha}P_\gamma P_\delta\frac{d^3P}{P^0}, \tag{5.69}$$

$$Z^\star_{\gamma\delta\epsilon\lambda} = \int e^{-\frac{1}{kT}U_\alpha P^\alpha}P_\gamma P_\delta P_\epsilon P_\lambda\frac{d^3P}{P^0}. \tag{5.70}$$

In order to obtain (5.68) we have used (5.50) and the expression for the product of the Maxwell–Jüttner distribution functions $f^{(0)}f^{(0)}_*$ written in terms of the total momentum four-vector P^α, i.e.,

$$f^{(0)}f^{(0)}_* = \frac{n^2}{[4\pi kTm^2cK_2(\zeta)]^2}e^{-\frac{1}{kT}U_\alpha P^\alpha}. \tag{5.71}$$

The reduced form of the integral I_1 is obtained through the integration in d^3P/P^0. The integrations on the modulus of the relative momentum four-vector Q and on the scattering angle Θ remain to be performed. These integrations can be done once we know the differential cross-section which in the general case is a function of both variables, i.e., $\sigma = \sigma(Q,\Theta)$. For the integration in d^3P/P^0 we choose a local Lorentz rest frame where $(U^\alpha) = (c,\mathbf{0})$. In this frame the integral $Z^\star$ above is written as

$$Z^\star = \int e^{-\frac{c}{kT}P^0}\frac{d^3P}{P^0} = 4\pi\int_0^\infty e^{-\frac{c}{kT}P^0}|\mathbf{P}|^2\frac{d|\mathbf{P}|}{P^0}. \tag{5.72}$$

In the last equality we have used spherical coordinates so that

$$d^3P = |\mathbf{P}|^2\sin\vartheta d\vartheta d\varphi d|\mathbf{P}|$$

and performed the integrations in the angles ϑ and φ. On the other hand, we have according to (5.50)

$$(P^0)^2 = |\mathbf{P}|^2 + Q^2 + 4m^2c^2. \tag{5.73}$$

If we introduce a new variable y defined by

$$y = \frac{P^0}{(Q^2 + 4m^2c^2)^{\frac{1}{2}}}, \tag{5.74}$$

we can express $|\mathbf{P}|^2$ and $d|\mathbf{P}|$ as

$$|\mathbf{P}|^2 = (Q^2 + 4m^2c^2)(y^2 - 1) \quad \text{and} \quad d|\mathbf{P}| = (Q^2 + 4m^2c^2)^{\frac{1}{2}} \frac{y\,dy}{(y^2 - 1)^{\frac{1}{2}}}. \tag{5.75}$$

In this case the integral (5.72) can be written as

$$Z^\star = 4\pi m^2c^2 Q^{\star 2} \int_1^\infty e^{-\zeta Q^\star y}(y^2 - 1)^{\frac{1}{2}}\, dy, \tag{5.76}$$

where we have introduced the abbreviation

$$Q^\star = \left(\frac{Q^2}{m^2c^2} + 4\right)^{\frac{1}{2}}. \tag{5.77}$$

The above integral is a modified Bessel function of the second kind and according to (3.19) it can be represented by

$$Z^\star = 4\pi m^2c^2 Q^{\star 2} \frac{K_1(\zeta Q^\star)}{\zeta Q^\star}, \tag{5.78}$$

By using the same methodology and the integrals (5.231) and (5.232) of Section 5.6, it is easy to show that

$$Z^\star - \frac{2}{c^2} Z^\star_{\gamma\delta} \frac{U^\gamma U^\delta}{(Q^2 + 4m^2c^2)} + \frac{1}{c^4} Z^\star_{\gamma\delta\epsilon\lambda} \frac{U^\gamma U^\delta U^\epsilon U^\lambda}{(Q^2 + 4m^2c^2)^2}$$

$$= 60\pi m^2c^2 Q^{\star 2} \frac{K_3(\zeta Q^\star)}{(\zeta Q^\star)^3}. \tag{5.79}$$

Now we are ready to obtain the final form of the integral I_1. We insert (5.79) into (5.68), introduce a new variable of integration $x = \zeta Q^\star$ – defined in the domain $2\zeta \leq x < \infty$ since $0 \leq Q < \infty$ – and get that the reduced form of the integral I_1 reads

$$I_1 = -\frac{\pi}{8} \frac{n^2m^2c^4(kT)^2}{K_2(\zeta)^2} \int_{2\zeta}^\infty \int_0^\pi \left(\frac{x^2}{\zeta^2} - 4\right)^3 K_3(x)\sigma \sin^3\Theta\, d\Theta\, dx. \tag{5.80}$$

As was pointed out previously, if we know the differential cross-section $\sigma = \sigma(Q, \Theta)$ we can perform the two remaining integrations.

The integrals I_2 and I_3, defined by (5.36) and (5.37), can be calculated by using the same methodology described above, and we have that

$$I_2 = \frac{-1}{192} \frac{n^2}{[kTm^2K_2(\zeta)]^2} \int \left[Z^\star - \frac{1}{c^2} Z^\star_{\delta\lambda} \frac{U^\delta U^\lambda}{(Q^2 + 4m^2c^2)} \right] Q^7 \sigma \sin^3 \Theta \, d\Theta \, dQ, \quad (5.81)$$

$$I_3 = -\frac{1}{64} \frac{n^2}{[kTm^2cK_2(\zeta)]^2} \int Z^\star Q^7 \sigma \sin^3 \Theta \, d\Theta \, dQ. \quad (5.82)$$

The reduced forms of I_2 and I_3 are

$$I_2 = \frac{\pi}{16} \frac{n^2 m^2 c^2 (kT)^2}{K_2(\zeta)^2} \int_{2\zeta}^{\infty} \int_0^{\pi} \left(\frac{x^2}{\zeta^2} - 4 \right)^3 x K_2(x) \sigma \sin^3 \Theta \, d\Theta \, dx, \quad (5.83)$$

$$I_3 = -\frac{\pi}{16} \frac{n^2 m^2 (kT)^2}{K_2(\zeta)^2} \int_{2\zeta}^{\infty} \int_0^{\pi} \left(\frac{x^2}{\zeta^2} - 4 \right)^3 x^2 K_1(x) \sigma \sin^3 \Theta \, d\Theta \, dx. \quad (5.84)$$

Problems

5.3.1 Show that the modulus of the Jacobian of the transformation between the elements of integration $d^3p d^3p_*$ and $d^3P d^3Q$ is $1/2^3$.

5.3.2 Show that the Møller relative speed in the center-of-mass system reduces to (5.59).

5.3.3 Get the expression (5.61) as a result of the corresponding integration.

5.3.4 Obtain the values of the integrals with respect to the element of solid angle $d\Omega^\star$ given by (5.65) through (5.67).

5.3.5 Obtain the expression (5.79).

5.3.6 Show that the reduced forms of the integrals I_2 and I_3 are given by (5.83) and (5.84), respectively.

5.4 Transport coefficients

For the complete determination of the integrals of the last section we shall need to know the differential cross-section as a function of the relative momentum four-vector and of the scattering angle $\sigma = \sigma(Q, \Theta)$. In this section we shall analyze two models of differential cross-section. The first is characterized by a constant differential cross-section, which might describe the collision between hadrons at very high energies and corresponds in the non-relativistic limiting case to a differential cross-section of hard-sphere particles. The second is the so-called Israel particles [7] which corresponds in the non-relativistic limiting case to a differential cross-section of Maxwellian particles [1, 2]. We shall see in the next chapter that this is not the unique choice that leads to a differential cross-section of Maxwellian particles in the non-relativistic limiting case.

5.4.1 Hard-sphere particles

A gas with hard-sphere particles is characterized by a constant differential cross-section. Under this simplifying assumption we can perform the two remaining integrations of (5.80), yielding

$$I_1 = -\frac{\pi}{6}\frac{n^2m^2c^4(kT)^2\sigma}{K_2(\zeta)^2}\int_{2\zeta}^{\infty}\left(\frac{x^2}{\zeta^2}-4\right)^3 K_3(x)dx$$

$$= -\frac{64\pi n^2(kT)^6\sigma}{m^2c^4K_2(\zeta)^2}\left[2K_2(2\zeta)+\zeta K_3(2\zeta)\right]. \tag{5.85}$$

Above we have used from [12] the relationship

$$\int_{2\zeta}^{\infty}\left(\frac{x^2}{4\zeta^2}-1\right)^3 K_n(x)x^{3-n}dx = K_{4-n}(2\zeta)+\frac{4}{\zeta}K_{5-n}(2\zeta). \tag{5.86}$$

In the same manner one can obtain respectively from (5.83) and (5.84):

$$I_2 = \frac{64\pi n^2(kT)^6\sigma}{m^2c^6K_2(\zeta)^2}\left[4\zeta K_3(2\zeta)+\zeta^2K_2(2\zeta)\right], \tag{5.87}$$

$$I_3 = -\frac{128\pi n^2(kT)^6\sigma}{m^2c^8K_2(\zeta)^2}\left[(12\zeta+\zeta^3)K_3(2\zeta)+4\zeta^2K_2(2\zeta)\right]. \tag{5.88}$$

Now we insert the integrals (5.85), (5.87) and (5.88) into the transport coefficients (5.40), (5.43) and (5.46) and obtain, respectively:

$$\eta = \frac{1}{64\pi}\frac{kT}{c\sigma}\frac{(20G+3\zeta-13G^2\zeta-2G\zeta^2+2G^3\zeta^2)^2\zeta^4K_2(\zeta)^2}{(1-5G\zeta-\zeta^2+G^2\zeta^2)^2(2K_2(2\zeta)+\zeta K_3(2\zeta))}, \tag{5.89}$$

$$\lambda = \frac{3}{64\pi}\frac{ck}{\sigma}\frac{(\zeta+5G-G^2\zeta)^2\zeta^4K_2(\zeta)^2}{(\zeta^2+2)K_2(2\zeta)+5\zeta K_3(2\zeta)}, \tag{5.90}$$

$$\mu = \frac{15}{64\pi}\frac{kT}{c\sigma}\frac{\zeta^4K_3(\zeta)^2}{(2+15\zeta^2)K_2(2\zeta)+(3\zeta^3+49\zeta)K_3(2\zeta)}. \tag{5.91}$$

The above expressions are the final form of the transport coefficients of bulk viscosity η, thermal conductivity λ and shear viscosity μ for a relativistic gas of hard-sphere particles.

In the limiting case of low temperatures ($\zeta \gg 1$) one can use the expansion (3.22) for the modified Bessel functions of second kind and get from (5.89) through (5.91) the non-relativistic limit:

$$\eta = \frac{25}{256\sigma}\left(\frac{mkT}{\pi}\right)^{\frac{1}{2}}\frac{1}{\zeta^2}\left(1 - \frac{183}{16}\frac{1}{\zeta} + \frac{41001}{512}\frac{1}{\zeta^2} - \frac{3635165}{8192}\frac{1}{\zeta^3} + \cdots\right), \tag{5.92}$$

$$\lambda = \frac{75}{256\sigma}\left(\frac{mkT}{\pi}\right)^{\frac{1}{2}}\frac{k}{m}\left(1 + \frac{13}{16}\frac{1}{\zeta} - \frac{1951}{512}\frac{1}{\zeta^2} + \frac{57335}{8192}\frac{1}{\zeta^3} + \cdots\right), \tag{5.93}$$

$$\mu = \frac{5}{64\sigma}\left(\frac{mkT}{\pi}\right)^{\frac{1}{2}}\left(1 + \frac{25}{16}\frac{1}{\zeta} - \frac{293}{1536}\frac{1}{\zeta^2} - \frac{20519}{24576}\frac{1}{\zeta^3} + \cdots\right). \tag{5.94}$$

The above expansions were first obtained by Stewart [12].

It is interesting to analyze the case where only the first term of (5.92) through (5.94) is taken into account and by considering that $\sigma = a^2/4$, which corresponds to a differential cross-section of a non-relativistic gas of hard-sphere particles of diameter a. In this case we have

$$\eta = \frac{25}{64a^2}\left(\frac{mkT}{\pi}\right)^{\frac{1}{2}}\frac{1}{\zeta^2}, \tag{5.95}$$

$$\lambda = \frac{75}{64a^2}\frac{k}{m}\left(\frac{mkT}{\pi}\right)^{\frac{1}{2}}, \tag{5.96}$$

$$\mu = \frac{5}{16a^2}\left(\frac{mkT}{\pi}\right)^{\frac{1}{2}}. \tag{5.97}$$

From the above equations we conclude that for low temperatures the coefficient of bulk viscosity vanishes since it is proportional to $1/\zeta^2$, while the coefficients of thermal conductivity and shear viscosity reduce to their expressions of a monatomic non-relativistic gas of hard-sphere particles [2]. Note that in the non-relativistic case the transport coefficients are proportional to $T^{\frac{1}{2}}$ and that the ratio between the thermal conductivity and the shear viscosity is given by $\lambda/\mu = 15k/(4m)$, which agree with the results obtained from the non-relativistic Boltzmann equation [2].

On the other hand, if we use the expansion (3.23) for the modified Bessel functions of second kind when the temperatures are very high or the rest mass is very small ($\zeta \ll 1$), we get from (5.89) through (5.91):

$$\eta = \frac{1}{288\pi}\frac{kT}{c\sigma}\zeta^4\left[1 + \left(\frac{49}{12} + 6\ln\left(\frac{\zeta}{2}\right) + 6\gamma\right)\zeta^2 + \cdots\right], \tag{5.98}$$

$$\lambda = \frac{1}{2\pi}\frac{ck}{\sigma}\left[1 - \frac{1}{4}\zeta^2 + \cdots\right], \tag{5.99}$$

$$\mu = \frac{3}{10\pi}\frac{kT}{c\sigma}\left[1 + \frac{1}{20}\zeta^2 + \cdots\right]. \tag{5.100}$$

These are the ultra-relativistic expressions of the transport coefficients. By considering only the first term in the above expansions we infer that: a) the bulk viscosity is also small in the ultra-relativistic case since it is proportional to ζ^4; b) the thermal conductivity is a constant; c) the shear viscosity is proportional to the temperature T; d) the ratio between the thermal conductivity and the shear viscosity depends on the temperature according to $\lambda/\mu = 5c^2/(3T)$.

Some values of the transport coefficients for intermediate values of ζ will be given in the chapter on the method of moments.

Problems

5.4.1.1 Obtain the expressions (5.87) and (5.88) for the integrals I_2 and I_3 of a relativistic gas of hard-sphere particles.

5.4.1.2 Obtain the expressions for the transport coefficients of a gas of hard-sphere particles in the non-relativistic limit (5.92) through (5.94) and in the ultra-relativistic limit (5.98) through (5.100). (Hint: Use any language for symbolic mathematical calculation like Maple or Mathematica.)

5.4.2 Israel particles

The differential cross-section for the Israel particles is given by [7]

$$\sigma = \frac{m}{Q}\frac{1}{\left(1+\frac{Q^2}{4m^2c^2}\right)}\mathcal{F}(\Theta), \tag{5.101}$$

where $\mathcal{F}(\Theta)$ is an arbitrary function of the scattering angle Θ.

We insert (5.101) into the expression (5.80) for I_1 and get by performing the integration of the resulting equation

$$I_1 = -\frac{15}{4}\pi^2\frac{p^2m^2c^3\mathcal{B}}{K_2(\zeta)^2\zeta^3e^{2\zeta}}. \tag{5.102}$$

Here we have used for the modified Bessel function of the second kind [4] the relationships

$$\int_1^\infty x^{-\frac{\nu}{2}}(x-1)^{\mu-1}K_\nu(a\sqrt{x})dx = \Gamma(\mu)\left(\frac{2}{a}\right)^\mu K_{\nu-\mu}(a), \quad \mu>0, \quad a>0, \tag{5.103}$$

$$K_\nu(x) = K_{-\nu}(x), \qquad K_{\frac{1}{2}}(2\zeta) = \sqrt{\frac{\pi}{4\zeta}}e^{-2\zeta}. \tag{5.104}$$

In (5.102) $\mathcal{B}$ denotes the integral over the scattering angle

$$\mathcal{B} = \int_0^\pi \mathcal{F}(\Theta)\sin^3\Theta\, d\Theta. \tag{5.105}$$

By following the same procedure we get from (5.83) and (5.84) that the expressions for the integrals I_2 and I_3 read

$$I_2 = \frac{15}{8}\pi^2 \frac{p^2 m^2 c(1+2\zeta)\mathcal{B}}{K_2(\zeta)^2 \zeta^3 e^{2\zeta}}, \tag{5.106}$$

$$I_3 = -\frac{15}{8}\pi^2 \frac{p^2 m^2 (3+6\zeta+4\zeta^2)\mathcal{B}}{cK_2(\zeta)^2 \zeta^3 e^{2\zeta}}. \tag{5.107}$$

The final form of the transport coefficients for Israel particles result from the insertion of (5.102), (5.106) and (5.107) into (5.40), (5.43), (5.46), yielding

$$\eta = \frac{4kT}{15\pi^2} \frac{K_2(\zeta)^2 \zeta^3 e^{2\zeta}}{\mathcal{B}} \frac{(20G + 3\zeta - 13G^2\zeta - 2G\zeta^2 + 2G^3\zeta^2)^2}{(1 - 5G\zeta - \zeta^2 + G^2\zeta^2)^2}, \tag{5.108}$$

$$\lambda = \frac{8kc^2}{5\pi^2} \frac{K_2(\zeta)^2 \zeta^3 e^{2\zeta}}{\mathcal{B}} \frac{(\zeta + 5G - G^2\zeta)^2}{(3+2\zeta)}, \tag{5.109}$$

$$\mu = \frac{16kT}{\pi^2} \frac{K_2(\zeta)^2 \zeta^3 e^{2\zeta}}{\mathcal{B}} \frac{G^2}{(19 + 30\zeta + 12\zeta^2)}. \tag{5.110}$$

In the limiting case of low temperatures ($\zeta \gg 1$), i.e., in the non-relativistic limit, the above expressions reduce to

$$\eta = \frac{5}{6}\frac{kT}{\pi\mathcal{B}}\frac{1}{\zeta^2}\left(1 - \frac{29}{4}\frac{1}{\zeta} + \frac{1149}{32}\frac{1}{\zeta^2} - \frac{19217}{128}\frac{1}{\zeta^3} + \cdots\right), \tag{5.111}$$

$$\lambda = \frac{5}{2}\frac{k^2 T}{\pi m\mathcal{B}}\left(1 + \frac{21}{4}\frac{1}{\zeta} + \frac{201}{32}\frac{1}{\zeta^2} - \frac{423}{128}\frac{1}{\zeta^3} + \cdots\right), \tag{5.112}$$

$$\mu = \frac{2}{3}\frac{kT}{\pi\mathcal{B}}\left(1 + \frac{25}{4}\frac{1}{\zeta} + \frac{1603}{96}\frac{1}{\zeta^2} + \frac{2525}{128}\frac{1}{\zeta^3} + \cdots\right), \tag{5.113}$$

by the use of the asymptotic expansion (3.22). If we consider only the first term in the above expansions we conclude that: a) the coefficient of bulk viscosity is small in the non-relativistic limit since it is proportional to $1/\zeta^2$; b) the thermal conductivity and the shear viscosity are proportional to T which agrees with the corresponding transport coefficients of a non-relativistic gas with Maxwellian particles [2]; c) the ratio between the thermal conductivity and the shear viscosity is the same as that found for hard-sphere particles, i.e., $\lambda/\mu = 15k/(4m)$.

The transport coefficients (5.108) through (5.110) reduce in the case of high temperatures ($\zeta \ll 1$) to:

$$\eta = \frac{16}{135}\frac{kT}{\pi^2\mathcal{B}}\zeta\left[1 + 2\zeta + \left(\frac{16}{3} + 6\ln\left(\frac{\zeta}{2}\right) + 6\gamma\right)\zeta^2 + \cdots\right], \tag{5.114}$$

$$\lambda = \frac{512}{15}\frac{kc^2}{\pi^2\mathcal{B}}\frac{1}{\zeta^3}\left[1 + \frac{4}{3}\zeta + \frac{13}{36}\zeta^2 + \cdots\right], \tag{5.115}$$

$$\mu = \frac{1024}{19}\frac{kT}{\pi^2\mathcal{B}}\frac{1}{\zeta^3}\left[1+\frac{8}{19}\zeta+\frac{655}{1444}\zeta^2+\cdots\right], \tag{5.116}$$

if we use the expansion (3.23). The above expressions correspond to the ultra-relativistic case. Here we also have that the coefficient of bulk viscosity is small since its leading term depends on ζ. However in this model the coefficients of thermal conductivity and viscosity are very large since their leading terms are proportional to $1/\zeta^3$. Further the ratio $\lambda/\mu = 19c^2/(30T)$ differs from the value of the hard-sphere model, but has the same dependence on the temperature.

Problems

5.4.2.1 Obtain the expressions (5.106) and (5.107) for the integrals I_2 and I_3 of a relativistic gas of Israel particles.

5.4.2.2 Obtain the expressions for the transport coefficients of a gas of Israel particles in the non-relativistic limit (5.111) through (5.113) and in the ultra-relativistic limit (5.114) through (5.116).

5.5 Formal version

5.5.1 Integral equations

Before discussing the formal version of the Chapman–Enskog method as a procedure to approximate systematically a solution in the presence of small gradients of the macroscopic quantities, we remark that one can consider several expansions of this kind. The first one was adopted by Hilbert [6] who considered solutions which can be expanded into a power series in the mean free path. The series for the one-particle distribution function reads as

$$f = f^{(0)} + \epsilon f^{(1)} + \epsilon^2 f^{(2)} + \cdots = \sum_{r=0}^{\infty}\epsilon^r f^{(r)}, \tag{5.117}$$

where ϵ is a small parameter. This parameter can be assumed to be the ratio between the mean free path ℓ and the macroscopic length L. In fact, we can make the space-time coordinates non-dimensional by dividing them by L and thus the parameter $\epsilon = \ell/L$ will occur automatically in the Boltzmann equation, if we take into account that the differential cross-section σ is inversely proportional to ℓ. Later ϵ can be set equal to unity, thus restoring dimensioned quantities.

If we write explicitly the parameter ϵ in the Boltzmann equation (5.1), the latter reads as

$$\epsilon p_\alpha \frac{\partial f}{\partial x_\alpha} \equiv \epsilon p_\alpha\left[\frac{1}{c^2}U^\alpha Df + \nabla^\alpha f\right] = Q(f,f), \tag{5.118}$$

where the collision term is written in terms of the bilinear quantity

$$Q(X,Y) = \frac{1}{2}\int (X'_*Y' + X'Y'_* - X_*Y - XY_*)F\sigma d\Omega \frac{d^3p_*}{p_{*0}}. \tag{5.119}$$

When the parameter ϵ is written explicitly in the left-hand side of the Boltzmann equation, it is clear that the derivatives of the distribution function are related to non-equilibrium processes in the gas, whereas the right-hand side contains the collision term which tends to keep the solution close to the equilibrium one. We call this form of the equation the *scaled Boltzmann equation.*

It is clear that we face a singular perturbation and thus, as remarked before, we shall not obtain the entire set of solutions of the Boltzmann equation, but only a subset of solutions, called *normal solutions.* The class obtained may depend on the rules that we are going to introduce. Hilbert proposed the simplest expansion in powers of ϵ (in the non-relativistic case): we just insert the power series (5.117) into the scaled Boltzmann equation (5.118) and equate the coefficients of each power of ϵ appearing in either side. In this way, however, we obtain a Maxwell–Jüttner distribution function at the lowest order, with parameters satisfying the relativistic Euler equations and corrections to this solution which are obtained by solving inhomogeneous linearized Euler equations (Problems 5.5.1.1-4). This result turns out to be inconvenient, because there are interesting solutions that cannot be expanded into a power series in ϵ. To recover these solutions and to investigate the relationship between the Boltzmann equation and the Navier–Stokes equations, Enskog [3] introduced, for the non-relativistic case, an expansion, usually called the Chapman–Enskog expansion [1, 2].

The idea behind this expansion is that the functional dependence of f upon the local particle number density, bulk velocity and energy per particle can be expanded into a power series. Although there are many formal similarities with the Hilbert expansion, the procedure is rather different. In particular the Maxwell–Jüttner distribution function occurring at the lower step is the local one in the Chapman–Enskog expansion, and hence no corrections to particle number density, bulk velocity, or energy per particle are introduced at the subsequent steps.

As a consequence, by inserting the series expansion (5.117) into the definitions of the dynamic pressure (5.38), heat flux (5.41) and pressure deviator (5.44) it follows that

$$\varpi = -\frac{1}{3}\Delta_{\alpha\beta}\left[c\int p^\alpha p^\beta \sum_{r=1}^{\infty}\epsilon^r f^{(r)}\frac{d^3p}{p_0}\right] = \sum_{r=1}^{\infty}\epsilon^r \varpi^{(r)}, \tag{5.120}$$

$$q^\alpha = \Delta^\alpha_\beta U_\gamma\left[c\int p^\beta p^\gamma \sum_{r=1}^{\infty}\epsilon^r f^{(r)}\frac{d^3p}{p_0}\right] = \sum_{r=1}^{\infty}\epsilon^r q^{\alpha(r)}, \tag{5.121}$$

$$p^{\langle\alpha\beta\rangle} = \left(\Delta^\alpha_\gamma\Delta^\beta_\delta - \frac{1}{3}\Delta^{\alpha\beta}\Delta_{\gamma\delta}\right)\left[c\int p^\gamma p^\delta \sum_{r=1}^{\infty}\epsilon^r f^{(r)}\frac{d^3p}{p_0}\right] = \sum_{r=1}^{\infty}\epsilon^r p^{\langle\alpha\beta\rangle(r)}, \tag{5.122}$$

i.e., they are described only by the non-equilibrium distribution functions $f^{(r)}$ with $r \geq 1$.

The next step in this method is to expand the convective time derivative D as

$$D = D_0 + \epsilon D_1 + \epsilon^2 D_2 + \cdots = \sum_{r=0}^{\infty} \epsilon^r D_r. \tag{5.123}$$

By using (5.120) through (5.123) we can decompose the balance of the particle number density (4.38) as

$$D_0 n + n \nabla^\alpha U_\alpha = 0, \tag{5.124}$$

$$D_r n = 0, \qquad r \geq 1. \tag{5.125}$$

Further the decomposition of the balance equation for the momentum density (4.43) reads

$$\frac{n h_E}{c^2} D_0 U^\alpha = \nabla^\alpha p, \tag{5.126}$$

$$\frac{n h_E}{c^2} D_r U^\alpha = \nabla^\alpha \varpi^{(r)} - \nabla_\beta p^{\langle\alpha\beta\rangle(r)} + \frac{1}{c^2}\left\{ \sum_{s=0}^{r-1} p^{\langle\alpha\beta\rangle(r-s)} D_s U_\beta \right.$$

$$- \sum_{s=0}^{r-1} \varpi^{(r-s)} D_s U^\alpha - \sum_{s=0}^{r-1} D_s q^{\alpha(r-s)} - q^{\alpha(r)} \nabla_\beta U^\beta - q^{\beta(r)} \nabla_\beta U^\alpha$$

$$\left. - \frac{1}{c^2} U^\alpha \sum_{s=0}^{r-1} q^{\beta(r-s)} D_s U_\beta - U^\alpha p^{\langle\beta\gamma\rangle(r)} \nabla_\beta U_\gamma \right\}, \qquad r \geq 1. \tag{5.127}$$

The decomposition of the balance equation for the energy density (4.41) is written as

$$n D_0 e = -p \nabla^\alpha U_\alpha, \tag{5.128}$$

$$n D_r e = -\varpi^{(r)} \nabla^\alpha U_\alpha + p^{\langle\alpha\beta\rangle(r)} \nabla_\alpha U_\beta - \nabla_\alpha q^{\alpha(r)} + \frac{2}{c^2} q^{\alpha(r)} D U_\alpha, \quad r \geq 1. \tag{5.129}$$

We insert now the expansions (5.117) and (5.123) into the Boltzmann equation (5.118) and equate the coefficients of equal power of ϵ, yielding

$$Q(f^{(0)}, f^{(0)}) = 0, \tag{5.130}$$

$$2Q(f^{(0)}, f^{(1)}) = p_\alpha \left[\frac{1}{c^2} U^\alpha D_0 f^{(0)} + \nabla^\alpha f^{(0)} \right], \tag{5.131}$$

$$2Q(f^{(0)}, f^{(2)}) + Q(f^{(1)}, f^{(1)}) = p_\alpha \left[\frac{1}{c^2} U^\alpha \left(D_0 f^{(1)} + D_1 f^{(0)} \right) + \nabla^\alpha f^{(1)} \right], \tag{5.132}$$

and so on. The above equations are the three first integral equations for $f^{(0)}$, $f^{(1)}$ and $f^{(2)}$.

The solutions $f^{(r)}$ must satisfy the constraints

$$\int p^\alpha f^{(r)} \frac{d^3p}{p_0} = 0, \qquad r \geq 1, \tag{5.133}$$

$$U_\alpha U_\beta \int p^\alpha p^\beta f^{(r)} \frac{d^3p}{p_0} = 0, \qquad r \geq 1, \tag{5.134}$$

which follow from the conditions (5.5) and (5.6).

As remarked by one of the authors [1] for the non-relativistic case, the Chapman–Enskog expansion seems to introduce spurious solutions, especially if one looks for steady states. This is essentially due to the fact that one really considers infinitely many time scales (of orders $\epsilon, \epsilon^2, \ldots, \epsilon^n, \ldots$). The same author introduced only two time scales (of orders ϵ and ϵ^2) and was able to recover the compressible Navier–Stokes equations. In order to explain the idea, we remark that the Navier–Stokes equations describe two kinds of processes, convection and diffusion, which act on two different time scales. If we consider only the first scale we obtain the compressible Euler equations; if we insist on the second one we can obtain the Navier–Stokes equations only at the price of losing compressibility. If we want both compressibility and diffusion, we have to keep both scales at the same time and think of f as

$$f(\mathbf{x}, \mathbf{p}, t) = f(\epsilon\mathbf{x}, \mathbf{p}, \epsilon t, \epsilon^2 t)\,. \tag{5.135}$$

This enables us to introduce two different time variables $t_1 = \epsilon t$, $t_2 = \epsilon^2 t$ and a new space variable $\mathbf{x}_1 = \epsilon\mathbf{x}$ such that $f = f(\mathbf{x}_1, \mathbf{p}, t_1, t_2)$. The fluid dynamical variables are functions of $\mathbf{x}_1, t_1, t_2$, and for both f and the fluid dynamical variables the time derivative is given by

$$\frac{\partial}{\partial t} = \epsilon \frac{\partial f}{\partial t_1} + \epsilon^2 \frac{\partial f}{\partial t_2}\,. \tag{5.136}$$

In particular, the Boltzmann equation can be rewritten as

$$\frac{p_0}{c}\left(\epsilon \frac{\partial f}{\partial t_1} + \epsilon^2 \frac{\partial f}{\partial t_2}\right) + \epsilon\mathbf{p} \cdot \frac{\partial f}{\partial \mathbf{x}} = Q(f, f)\,. \tag{5.137}$$

If we expand f formally into a power series in ϵ, we find that at the lowest order f is a Maxwell–Jüttner distribution function. The compatibility conditions at the first order give that the time derivatives of the fluid dynamic variables with respect to t_1 is determined by the Euler equations but the derivatives with respect to t_2 are determined only at the next level and are given by the terms of the compressible Navier–Stokes equations describing the effects of viscosity and heat conductivity. The two contributions are, of course, to be added as indicated in (5.136) to obtain the full time derivative and thus write the compressible Navier–Stokes equations.

The relativistic version of this procedure is easily obtained by replacing the time derivatives by the convective time derivatives D and the space derivatives by ∇^α.

Problems

5.5.1.1 Obtain the equations which determine the Hilbert expansion in powers of ϵ for the relativistic case, by inserting the power series (5.117) into the scaled Boltzmann equation (5.118) and equating the coefficients of each power of ϵ appearing in either side.

5.5.1.2 Solve the equations obtained in the previous problem, showing that one obtains a Maxwell–Jüttner distribution function at the lowest order, and that, in order to solve the integral equation at the next order, the source term must satisfy the relativistic Euler equations.

5.5.1.3 Remark that once one replaces the convective time derivative by the Euler equations, then the integral equation at the second approximation can be solved but the solution is determined up to a linear combination of collision invariants and that the parameters occurring in this combination satisfy inhomogeneous linearized Euler equations with a suitable source.

5.5.1.4 Prove that the situation met at the second level of the Hilbert expansion repeats itself at the next levels, the only difference being in the source term of the inhomogeneous linearized Euler equations.

5.5.2 Second approximation

The solution of the integral equation (5.130) was found in Section 2.7 and leads to the Maxwell–Jüttner distribution function (5.2).

For the second integral equation (5.131) we write the convective time derivative and the gradient as

$$D_0 f^{(0)} = \frac{\partial f^{(0)}}{\partial n} D_0 n + \frac{\partial f^{(0)}}{\partial U^\alpha} D_0 U^\alpha + \frac{\partial f^{(0)}}{\partial T} D_0 T, \tag{5.138}$$

$$\nabla^\alpha f^{(0)} = \frac{\partial f^{(0)}}{\partial n} \nabla^\alpha n + \frac{\partial f^{(0)}}{\partial U^\beta} \nabla^\alpha U^\beta + \frac{\partial f^{(0)}}{\partial T} \nabla^\alpha T, \tag{5.139}$$

since $f^{(0)}$ is only a function of n, U^α and T. First we eliminate from (5.138) the convective time derivatives $D_0 n$, $D_0 U^\alpha$ and $D_0 T$ by using (5.124), (5.126) and (5.128), respectively[1]. Next we use the transformed equation (5.138) and (5.139) to eliminate $D_0 f^{(0)}$ and $\nabla^\alpha f^{(0)}$ from (5.131), yielding

$$2Q(f^{(0)}, f^{(1)}) = f^{(0)} \Bigg\{ -\frac{k^2 T}{c^2 c_v} \left[\frac{1}{3}\zeta^2 \frac{c_v}{k} - (G^2\zeta^2 - 4G\zeta - \zeta^2) \left(\frac{U^\beta p_\beta}{kT} \right) \right.$$

$$\left. -\frac{1}{3}(\zeta^2 + 5G\zeta - \zeta^2 G^2 - 4) \left(\frac{U^\beta p_\beta}{kT} \right)^2 \right] \nabla_\alpha U^\alpha - \frac{p_\alpha p_\beta}{kT} \nabla^{\langle\alpha} U^{\beta\rangle}$$

[1]Recall that e is only a function of T so that $D_0 e = c_v D_0 T$.

$$+\frac{p_\alpha}{kT^2}\left(p_\beta U^\beta - h_E\right)\left(\nabla^\alpha T - \frac{T}{nh_E}\nabla^\alpha p\right)\Bigg\}. \tag{5.140}$$

Note that if we put $f^{(1)} = f^{(0)}\phi$, the above integral equation reduces to (5.23).

The general solution ϕ of the integral equation (5.140) can be written as

$$\phi = \phi_h + \phi_p. \tag{5.141}$$

ϕ_h is a solution of the homogeneous integral equation $Q(f^{(0)}, f^{(0)}\phi) = 0$, i.e., a summational invariant, and ϕ_p a particular solution. Further the particular solution must be proportional to the thermodynamic forces:

$$\nabla_\alpha U^\alpha, \quad \nabla^{\langle\alpha}U^{\beta\rangle}, \quad \nabla^\alpha T - \frac{T}{nh_E}\nabla^\alpha p. \tag{5.142}$$

Hence the general solution is written as

$$\phi = A + B_\alpha p^\alpha + C\nabla_\alpha U^\alpha + E_{\alpha\beta}\nabla^{\langle\alpha}U^{\beta\rangle} + F_\alpha\left(\nabla^\alpha T - \frac{T}{nh_E}\nabla^\alpha p\right), \tag{5.143}$$

where the scalar A and the four-vector B_α do not depend on the momentum four-vector p_α, while the scalar C, the four-vector F_α and the tensor $E_{\alpha\beta}$ may depend on n, T, U_α and p_α.

According to representation theorems of isotropic functions (see Pennisi [11]) one can conclude that:

i) C must be a function of

$$C = C(n, T, p_\alpha U^\alpha), \tag{5.144}$$

since $U_\alpha U^\alpha = c^2$ and $p_\alpha p^\alpha = m^2c^2$;

ii) F^α must be a linear combination of U^α and p^α but due to the relationship $U_\alpha \nabla^\alpha = 0$ it is reduced to

$$F^\alpha = F(n, T, p_\lambda U^\lambda)\Delta^\alpha_\beta p^\beta, \tag{5.145}$$

by using the decomposition (5.18);

iii) $E^{\alpha\beta}$ is a symmetric tensor which must be a linear combination of $\eta^{\alpha\beta}$, $U^\alpha U^\beta$, $(p^\alpha U^\beta + p^\beta U^\alpha)$ and $p^\alpha p^\beta$ but due to the relationships $\eta_{\alpha\beta}\nabla^{\langle\alpha}U^{\beta\rangle} = 0$, $(p_\alpha U_\beta + p_\beta U_\alpha)\nabla^{\langle\alpha}U^{\beta\rangle} = 0$, and $U_\alpha U_\beta \nabla^{\langle\alpha}U^{\beta\rangle} = 0$ it is reduced to

$$E^{\alpha\beta} = E(n, T, p_\lambda U^\lambda)\left(\Delta^\alpha_\gamma\Delta^\beta_\delta - \frac{1}{3}\Delta^{\alpha\beta}\Delta_{\gamma\delta}\right)p^\gamma p^\delta, \tag{5.146}$$

again by the use of the decomposition (5.18).

If we take into account the constraints (5.133) and (5.134) it follows that:

i) A is proportional to $\nabla^\alpha U_\alpha$;

ii) $B_\alpha p^\alpha \stackrel{(5.18)}{=} (B_\alpha U^\alpha) p^\beta U_\beta + \Delta^\alpha_\beta B^\beta p_\alpha$ has the projection $B_\alpha U^\alpha$ proportional to $\nabla^\alpha U_\alpha$ while the projection $\Delta^\alpha_\beta B^\beta$ is proportional to $\{\nabla^\alpha T - [T/(nh_E)]\nabla^\alpha p\}$.

We incorporate all terms proportional to $\nabla_\alpha U^\alpha$ in only one term as well as the terms proportional to $\{\nabla^\alpha T - [T/(nh_E)]\nabla^\alpha p\}$ and write (5.143) in the form

$$\phi = C^* \nabla_\alpha U^\alpha + E\left(\Delta^\gamma_\alpha \Delta^\delta_\beta - \frac{1}{3}\Delta_{\alpha\beta}\Delta^{\gamma\delta}\right) p_\gamma p_\delta \nabla^{\langle\alpha} U^{\beta\rangle}$$
$$+ F^* \Delta^\beta_\alpha p_\beta \left(\nabla^\alpha T - \frac{T}{nh_E}\nabla^\alpha p\right). \tag{5.147}$$

The coefficients C^* and F^* are subjected to the constraints that follow from (5.133) and (5.134):

$$\int (U_\alpha p^\alpha) C^* f^{(0)} \frac{d^3 p}{p_0} = 0, \qquad \int (U_\alpha p^\alpha)^2 C^* f^{(0)} \frac{d^3 p}{p_0} = 0, \tag{5.148}$$

$$\int \Delta_{\alpha\beta}\, p^\alpha p^\beta F^* f^{(0)} \frac{d^3 p}{p_0} = 0. \tag{5.149}$$

Problems

5.5.2.1 Obtain the right-hand side of (5.140).

5.5.2.2 Show that a) A and $B_\alpha U^\alpha$ are proportional to $\nabla^\alpha U_\alpha$; b) $\Delta^\alpha_\beta B^\beta$ is proportional to $\{\nabla^\alpha T - [T/(nh_E)]\nabla^\alpha p\}$. (Hint: Use the integrals (5.234) through (5.238) of Section 5.6.)

5.5.2.3 Show that C^* and F^* are subjected to the constraints (5.148) and (5.149).

5.5.3 Orthogonal polynomials

Before we go further in the analysis of the integral equation (5.140), we shall follow Kelly [8] and introduce relativistic orthogonal polynomials in the dimensionless variable $U^\alpha p_\alpha/(kT)$, i.e., the polynomials generated by the set

$$1, \frac{U^\alpha p_\alpha}{kT}, \left(\frac{U^\alpha p_\alpha}{kT}\right), \left(\frac{U^\alpha p_\alpha}{kT}\right)^2, \left(\frac{U^\alpha p_\alpha}{kT}\right)^3, \ldots. \tag{5.150}$$

The method used by Kelly to construct the relativistic orthogonal polynomials is based on the Schmidt orthogonalization process (see for example Margenau and Murphy [10]) which we proceed to describe here. First one introduces a weight function

$$dX_{l+\frac{1}{2}} = (-1)^l \omega_l e^{-\frac{U^\lambda p_\lambda}{kT}} (\Delta^{\alpha\beta} p_\alpha p_\beta)^l \frac{d^3 p}{p_0}, \tag{5.151}$$

such that the relativistic orthogonal polynomials $R^{(n)}_{l+\frac{1}{2}}$ of order n and index $l+\frac{1}{2}$ satisfy the orthogonality condition

$$\int R^{(n)}_{l+\frac{1}{2}} R^{(m)}_{l+\frac{1}{2}} dX_{l+\frac{1}{2}} = \delta^{(n,m)}, \tag{5.152}$$

where $\delta^{(n,m)}$ denotes the Kronecker symbol.

The relativistic orthogonal polynomials of zeroth order and index $l+\frac{1}{2}$ are set equal to one, i.e.,

$$R^{(0)}_{l+\frac{1}{2}} = 1. \tag{5.153}$$

In the following we shall proceed to find the first relativistic orthogonal polynomials of index 1/2, 3/2 and 5/2 and show that the non-relativistic limiting case of $R^{(n)}_{l+\frac{1}{2}}$ reduce to the Sonine polynomials $S^{(n)}_{l+\frac{1}{2}}$ (see for example Chapman and Cowling [2]). The relativistic orthogonal polynomials of index 1/2 we shall determine here do not correspond to those of Kelly [8], since his weight function for the index 1/2 differs from (5.151). Further the relativistic orthogonal polynomials of index 1/2 of Kelly do not reduce to the Sonine polynomials of index 1/2 in the non-relativistic limiting case.

$R^{(n)}_{\frac{1}{2}}$ polynomials

First we shall determine the constant of normalization ω_0 by using (5.151) through (5.153) and the integral (5.224) of Section 5.6

$$\int \left[R^{(0)}_{\frac{1}{2}}\right]^2 dX_{\frac{1}{2}} = 1 = \omega_0 \int e^{-\frac{U^\lambda p_\lambda}{kT}} \frac{d^3p}{p_0} = \omega_0 Z = \omega_0 4\pi mkT K_1(\zeta). \tag{5.154}$$

Hence it follows that

$$\omega_0 = \frac{1}{4\pi mkT K_1(\zeta)}. \tag{5.155}$$

Further we write the relativistic orthogonal polynomial of first order and index 1/2 as

$$R^{(1)}_{\frac{1}{2}} = \alpha_1 \left(\frac{U^\alpha p_\alpha}{kT}\right) + \alpha_{10}, \tag{5.156}$$

and determine the constants α_1 and α_{10} through the orthogonality condition (5.152)

$$\int R^{(0)}_{\frac{1}{2}} R^{(1)}_{\frac{1}{2}} dX_{\frac{1}{2}} = 0, \qquad \int \left[R^{(1)}_{\frac{1}{2}}\right]^2 dX_{\frac{1}{2}} = 1. \tag{5.157}$$

A straightforward calculation leads to the system of equations

$$\alpha_1 \frac{Z^\alpha U_\alpha}{kT} \omega_0 + \alpha_{10} = 0, \tag{5.158}$$

$$\omega_0\left[\alpha_1^2\frac{Z^{\alpha\beta}U_\alpha U_\beta}{(kT)^2}+2\alpha_{10}\alpha_1\frac{Z^\alpha U_\alpha}{kT}+\alpha_{10}^2 Z\right]=1, \tag{5.159}$$

which Z, Z^α and $Z^{\alpha\beta}$ are integrals defined in Section 5.6 by (5.224) through (5.226). If we solve the system of equations (5.158) and (5.159) for α_1 and α_{10} by using the integrals of Section 5.6, we get

$$R^{(1)}_{\frac{1}{2}}=\alpha_1\left[\frac{U^\alpha p_\alpha}{kT}-\frac{\zeta^2}{G\zeta-4}\right], \tag{5.160}$$

where the constant α_1 is given by

$$\alpha_1=\frac{G\zeta-4}{\zeta\sqrt{4+G^2\zeta^2-5G\zeta-\zeta^2}}. \tag{5.161}$$

If we consider a local Lorentz rest frame where $(U^\alpha)=(c,\mathbf{0})$, we get in the non-relativistic limiting case

$$\frac{U^\alpha p_\alpha}{kT}=\frac{cp_0}{kT}=\frac{c}{kT}\sqrt{m^2c^2+|\mathbf{p}|^2}\approx\frac{mc^2}{kT}\left(1+\frac{|\mathbf{p}|^2}{2m^2c^2}\right), \tag{5.162}$$

$$\alpha_1\approx\sqrt{\frac{2}{3}}+\mathcal{O}\left(\frac{1}{\zeta}\right),\qquad \frac{\zeta^2}{G\zeta-4}\approx\zeta+\frac{3}{2}+\mathcal{O}\left(\frac{1}{\zeta}\right). \tag{5.163}$$

By combining (5.160) through (5.163) we obtain

$$R^{(1)}_{\frac{1}{2}}\approx\sqrt{\frac{2}{3}}\left[\frac{|\mathbf{p}|^2}{2mkT}-\frac{3}{2}\right]=-\sqrt{\frac{2}{3}}S^{(1)}_{\frac{1}{2}}, \tag{5.164}$$

where $S^{(1)}_{\frac{1}{2}}$ denotes the Sonine polynomial of first order and index 1/2 in the variable $|\mathbf{p}|^2/(2mkT)$.

The relativistic orthogonal polynomial of second order and index 1/2 is written as

$$R^{(2)}_{\frac{1}{2}}=\alpha_2\left(\frac{U^\alpha p_\alpha}{kT}\right)^2+\alpha_{21}\left(\frac{U^\alpha p_\alpha}{kT}\right)+\alpha_{20}, \tag{5.165}$$

and the coefficients are again calculated by using the orthogonality condition (5.152) and the integrals of Section 5.6, yielding

$$R^{(2)}_{\frac{1}{2}}=\alpha_2\Bigg[\left(\frac{U^\alpha p_\alpha}{kT}\right)^2+\frac{3\zeta(\zeta+4G-G^2\zeta)}{4+G^2\zeta^2-5G\zeta-\zeta^2}\left(\frac{U^\alpha p_\alpha}{kT}\right)$$
$$+\frac{\zeta^2(\zeta^2+5G\zeta-1-G^2\zeta^2)}{4+G^2\zeta^2-5G\zeta-\zeta^2}\Bigg], \tag{5.166}$$

where the constant α_2 reads

$$\alpha_2 = \frac{1}{\zeta^2\sqrt{3}}\sqrt{\frac{\zeta(4+G^2\zeta^2-5G\zeta-\zeta^2)(G\zeta-4)}{20G+3\zeta-13G^2\zeta-2G\zeta^2+2G^3\zeta^2}}. \tag{5.167}$$

In the non-relativistic limiting case and in a local Lorentz rest frame (5.166) reduces to

$$R_{\frac{1}{2}}^{(2)} \approx \sqrt{\frac{2}{15}}\left[\left(\frac{|\mathbf{p}|^2}{2mkT}\right)^2 - 5\left(\frac{|\mathbf{p}|^2}{2mkT}\right) + \frac{15}{4}\right] = 2\sqrt{\frac{2}{15}}S_{\frac{1}{2}}^{(2)}, \tag{5.168}$$

where $S_{\frac{1}{2}}^{(2)}$ is the Sonine polynomial of second order and index 1/2.

$R_{\frac{3}{2}}^{(n)}$ polynomials

The constant of normalization ω_1 in this case is obtained from

$$\int \left[R_{\frac{3}{2}}^{(0)}\right]^2 dX_{\frac{3}{2}} = 1 = -\omega_1 \int \Delta_{\alpha\beta}p^\alpha p^\beta e^{-\frac{U^\lambda p_\lambda}{kT}}\frac{d^3p}{p_0}$$
$$= -\omega_1\Delta_{\alpha\beta}Z^{\alpha\beta} = \omega_1 12\pi(mkT)^2K_2(\zeta), \tag{5.169}$$

where we have used (5.151) through (5.153) and the integral (5.226) of Section 5.6. Hence the normalization constant ω_1 is given by

$$\omega_1 = \frac{1}{12\pi(mkT)^2K_2(\zeta)}. \tag{5.170}$$

If we apply the same methodology we have used to obtain the relativistic orthogonal polynomials of index 1/2, we get that the polynomials of first and second order and index 3/2 are

$$R_{\frac{3}{2}}^{(1)} = \beta_1\left[\frac{U^\alpha p_\alpha}{kT} - G\zeta\right], \tag{5.171}$$

$$R_{\frac{3}{2}}^{(2)} = \beta_2\Bigg[\left(\frac{U^\alpha p_\alpha}{kT}\right)^2 + \frac{5(\zeta+6G-G^2\zeta)}{G^2\zeta-\zeta-5G}\left(\frac{U^\alpha p_\alpha}{kT}\right)$$
$$+\frac{\zeta(\zeta^2+5G\zeta-5G^2-G^2\zeta^2)}{G^2\zeta-\zeta-5G}\Bigg]. \tag{5.172}$$

The constants β_1 and β_2 are given by

$$\beta_1 = \frac{1}{\zeta}\sqrt{\frac{\zeta}{5G+\zeta-G^2\zeta}}, \tag{5.173}$$

$$\beta_2 = \frac{1}{\zeta^2\sqrt{5}}\sqrt{\frac{\zeta^3(G^2\zeta - \zeta - 5G)}{2G^2\zeta^2 - 2\zeta^2 + 7G^3\zeta - 17G\zeta - 30G^2}}. \tag{5.174}$$

In the non-relativistic limiting case the polynomials (5.171) and (5.172) in a local Lorentz rest frame read

$$R^{(1)}_{\frac{3}{2}} \approx \sqrt{\frac{2}{5}}\left[\frac{|\mathbf{p}|^2}{2mkT} - \frac{5}{2}\right] = -\sqrt{\frac{2}{5}}S^{(1)}_{\frac{3}{2}}, \tag{5.175}$$

$$R^{(2)}_{\frac{3}{2}} \approx \sqrt{\frac{2}{35}}\left[\left(\frac{|\mathbf{p}|^2}{2mkT}\right)^2 - 7\left(\frac{|\mathbf{p}|^2}{2mkT}\right) + \frac{35}{4}\right] = 2\sqrt{\frac{2}{35}}S^{(2)}_{\frac{3}{2}}, \tag{5.176}$$

where $S^{(1)}_{\frac{3}{2}}$ and $S^{(2)}_{\frac{3}{2}}$ are the Sonine polynomial of first and second order and index 3/2, respectively.

$R^{(n)}_{\frac{5}{2}}$ polynomials

We begin with the determination of the normalization constant ω_2 by using the orthogonality condition (5.152)

$$\int \left[R^{(0)}_{\frac{5}{2}}\right]^2 dX_{\frac{5}{2}} = 1 = \omega_2 \int (\Delta_{\alpha\beta}p^\alpha p^\beta)(\Delta_{\gamma\delta}p^\gamma p^\delta)e^{-\frac{U^\lambda p_\lambda}{kT}}\frac{d^3p}{p_0}$$

$$= \omega_2 \Delta_{\alpha\beta}\Delta_{\gamma\delta}Z^{\alpha\beta\gamma\delta}, \tag{5.177}$$

where $Z^{\alpha\beta\gamma\delta}$ denotes the integral (5.228) of Section 5.6. From (5.228) and (5.177) it follows that

$$\omega_2 = \frac{1}{60\pi(mkT)^3K_3(\zeta)}. \tag{5.178}$$

The relativistic orthogonal polynomial of first order and index 5/2, obtained by the same procedure above, is

$$R^{(1)}_{\frac{5}{2}} = \gamma_1\left[\frac{U^\alpha p_\alpha}{kT} - \frac{6G + \zeta}{G}\right], \tag{5.179}$$

where the constant γ_1 is given by

$$\gamma_1 = \frac{G}{\sqrt{G^2\zeta^2 + 6G^2 - 5G\zeta - \zeta^2}}. \tag{5.180}$$

In the non-relativistic limiting case and in a local Lorentz rest frame, $R^{(1)}_{\frac{5}{2}}$ reduces to a Sonine polynomial of first order and index 5/2, i.e.,

$$R^{(1)}_{\frac{5}{2}} \approx \sqrt{\frac{2}{7}}\left[\frac{|\mathbf{p}|^2}{2mkT} - \frac{7}{2}\right] = -\sqrt{\frac{2}{7}}S^{(1)}_{\frac{5}{2}}. \tag{5.181}$$

The expression of the relativistic orthogonal polynomial of second order and index 5/2 is given in Section 5.6.

Problems

5.5.3.1 Show that the relativistic orthogonal polynomials of second order and index 1/2, of first order and index 3/2 and of first order and index 5/2 are given by (5.166), (5.171) and (5.179), respectively.

5.5.3.2 For the relativistic orthogonal polynomials of the previous problem, obtain the corresponding expressions in the non-relativistic limiting case.

5.5.4 Expansion in orthogonal polynomials

Now we expand the coefficients C^*, E and F^* which appear in the deviation of the Maxwell–Jüttner distribution function – given in (5.147) – into a series of relativistic orthogonal polynomials as follows:

$$C^* = \sum_{n=0}^{\infty} c_n R_{\frac{1}{2}}^{(n)}, \qquad F^* = \sum_{n=0}^{\infty} a_n R_{\frac{3}{2}}^{(n)}, \qquad E = \sum_{n=0}^{\infty} b_n R_{\frac{5}{2}}^{(n)}. \tag{5.182}$$

The constraints (5.148) lead to the following relationships between the first three coefficients c_0, c_1 and c_2 of C^*, e.g.,

$$\frac{1}{\alpha_1}c_1 + \frac{\zeta^2}{G\zeta - 4}c_0 = 0, \tag{5.183}$$

$$\frac{1}{\zeta^2\alpha_2}c_2 + \frac{1 + G^2\zeta^2 - 5G\zeta - \zeta^2}{4 + G^2\zeta^2 - 5G\zeta - \zeta^2}c_0 = 0, \tag{5.184}$$

if we use the orthogonality condition (5.152) and the expressions for the first three relativistic orthogonal polynomials of index 1/2 given by (5.153), (5.160) and (5.166). Further we have used (5.183) in order to obtain (5.184).

The constraint (5.149) ensures that

$$a_0 = 0, \qquad \text{so that} \qquad F^* = \sum_{n=1}^{\infty} a_n R_{\frac{3}{2}}^{(n)}. \tag{5.185}$$

If we insert the representation of ϕ, given by (5.147), together with the expansions (5.182) into the integral equation (5.140) and make use of the definitions of the orthogonal polynomials $R_{\frac{1}{2}}^{(2)}$, $R_{\frac{3}{2}}^{(1)}$ and $R_{\frac{5}{2}}^{(0)}$, we get the following set of integral equations:

$$f^{(0)}\frac{kT}{3c^2\alpha_2}\frac{\zeta^2 + 5G\zeta - \zeta^2G^2 - 4}{\zeta^2 + 5G\zeta - \zeta^2G^2 - 1}R_{\frac{1}{2}}^{(2)} = \sum_{n=2}^{\infty} c_n \mathcal{I}\left[R_{\frac{1}{2}}^{(n)}\right], \tag{5.186}$$

$$f^{(0)}\frac{1}{T\beta_1}\Delta^{\alpha}_{\beta}p^{\beta}R_{\frac{3}{2}}^{(1)} = \Delta^{\alpha}_{\beta}\sum_{n=1}^{\infty} a_n \mathcal{I}\left[R_{\frac{3}{2}}^{(n)}p^{\beta}\right], \tag{5.187}$$

$$-f^{(0)}\left[\Delta^\alpha_\gamma\Delta^\beta_\delta-\frac{1}{3}\Delta^{\alpha\beta}\Delta_{\gamma\delta}\right]\frac{p^\gamma p^\delta}{kT}R^{(0)}_{\frac{5}{2}}$$
$$=\left[\Delta^\alpha_\gamma\Delta^\beta_\delta-\frac{1}{3}\Delta^{\alpha\beta}\Delta_{\gamma\delta}\right]\sum_{n=0}^{\infty}b_n\mathcal{I}\left[R^{(n)}_{\frac{5}{2}}p^\gamma p^\delta\right]. \tag{5.188}$$

Note that the sum in (5.186) begins with $n=2$ while the sum in (5.187) begins with $n=1$. This is due to the fact that $\mathcal{I}[\phi]=0$ whenever ϕ is a summational invariant.

Constitutive equation for the dynamic pressure

The constitutive equation for the dynamic pressure is obtained from (5.120) with $\epsilon=1$, (5.147) and $(5.182)_1$, yielding

$$\varpi^{(1)}=-\frac{c}{3}\nabla^\alpha U_\alpha\int\left[m^2c^2-\left(\frac{U^\lambda p_\lambda}{c}\right)^2\right]f^{(0)}\sum_{n=0}^{\infty}c_nR^{(n)}_{\frac{1}{2}}\frac{d^3p}{p_0}=-\eta\nabla^\alpha U_\alpha. \tag{5.189}$$

Here the coefficient of bulk viscosity η is expressed by

$$\eta=\frac{nmc^2}{3}\left(G-\frac{4}{\zeta}\right)c_0, \tag{5.190}$$

if we use the orthogonality condition (5.152) and the constraint (5.148).

Hence the coefficient of bulk viscosity η is determined once we know the coefficient c_0 – or according to (5.184) – the coefficient c_2. In the following we shall proceed to determine from the integral equation (5.186) the coefficient c_2. To begin with we multiply (5.186) by $R^{(m)}_{\frac{1}{2}}$, integrate the resulting equation over all values of d^3p/p_0 and obtain

$$\frac{n(G\zeta-4)(\zeta^2+5G\zeta-\zeta^2G^2-4)}{3c\zeta^2\alpha_2(\zeta^2+5G\zeta-\zeta^2G^2-1)}\delta^{(m,2)}=\sum_{n=2}^{\infty}C^{(m,n)}c_n, \tag{5.191}$$

by the use of the orthogonality condition (5.152) and of the constant of normalization (5.155). Above $C^{(m,n)}$ represents the integral

$$C^{(m,n)}=\int R^{(m)}_{\frac{1}{2}}\mathcal{I}\left[R^{(n)}_{\frac{1}{2}}\right]\frac{d^3p}{p_0}. \tag{5.192}$$

If we use the constraints (5.14) and (5.15) we get that

$$C^{(m,n)}=C^{(n,m)},\qquad C^{(0,n)}=C^{(1,n)}=0. \tag{5.193}$$

Equation (5.191) represents an infinite set of algebraic equations for the coefficients c_n and these coefficients are determined by using a method of successive approximations. The two first approximations of the coefficient c_2 – denoted by $[c_2]_1$ and $[c_2]_2$ – read:

$$[c_2]_1 = \frac{n(G\zeta-4)(\zeta^2+5G\zeta-\zeta^2G^2-4)}{3c\zeta^2\alpha_2(\zeta^2+5G\zeta-\zeta^2G^2-1)}\frac{1}{C^{(2,2)}}, \tag{5.194}$$

$$[c_2]_2 = \frac{n(G\zeta-4)(\zeta^2+5G\zeta-\zeta^2G^2-4)}{3c\zeta^2\alpha_2(\zeta^2+5G\zeta-\zeta^2G^2-1)}\frac{C^{(3,3)}}{C^{(2,2)}C^{(3,3)}-(C^{(2,3)})^2}. \tag{5.195}$$

The element $C^{(2,2)}$ is given in terms of the integral I_1 defined in (5.35), i.e.,

$$C^{(2,2)} = \frac{\alpha_2^2}{2}\frac{I_1}{(kT)^4}, \tag{5.196}$$

so that the first approximation of the coefficient of bulk viscosity η reduces to

$$[\eta]_1 = -\frac{p^2kTm^2c^3}{I_1}\frac{(20G+3\zeta-13G^2\zeta-2G\zeta^2+2G^3\zeta^2)^2}{(1-5G\zeta-\zeta^2+G^2\zeta^2)^2}. \tag{5.197}$$

To obtain (5.197) we have used (5.184), (5.190) and (5.194). Note that the above expression is the same as the one given in (5.40).

We shall not give here the other approximations for the coefficient of bulk viscosity nor those for the other transport coefficients, namely thermal conductivity and shear viscosity.

Constitutive equation for the heat flux

The expression for the heat flux is obtained from (5.121) with $\epsilon = 1$ together with (5.147), (5.171) and $(5.182)_2$, yielding

$$q^{\alpha(1)} = \Delta^\alpha_\beta \frac{ckT}{\beta_1}\sum_{n=1}^\infty a_n\left[\int p^\beta p^\gamma R^{(1)}_{\frac{3}{2}}R^{(n)}_{\frac{3}{2}}f^{(0)}\frac{d^3p}{p_0}\right]\left(\nabla_\gamma T-\frac{T}{nh_E}\nabla_\gamma p\right). \tag{5.198}$$

The integral within brackets in (5.198) can be written as

$$\int p^\beta p^\delta R^{(1)}_{\frac{3}{2}}R^{(n)}_{\frac{3}{2}}f^{(0)}\frac{d^3p}{p_0} = J_1U^\beta U^\delta + J_2\Delta^{\beta\delta}, \tag{5.199}$$

where J_1 and J_2 represent the integrals

$$J_1 = \frac{1}{c^4}\int (p^\beta U_\beta)^2 R^{(1)}_{\frac{3}{2}}R^{(n)}_{\frac{3}{2}}f^{(0)}\frac{d^3p}{p_0}, \tag{5.200}$$

$$J_2 = \frac{1}{3}\int \Delta_{\beta\delta}p^\beta p^\delta R^{(1)}_{\frac{3}{2}}R^{(n)}_{\frac{3}{2}}f^{(0)}\frac{d^3p}{p_0} = -\frac{p}{c}\delta^{(1,n)}, \tag{5.201}$$

where the last equality in (5.201) follows from (5.152) and (5.170).

Hence the heat flux, thanks to (5.198) through (5.201), reads

$$q^{\alpha(1)} = \lambda\left(\nabla^\alpha T - \frac{T}{nh_E}\nabla^\alpha p\right). \tag{5.202}$$

In the above equation the coefficient of thermal conductivity is given in terms of the coefficient a_1 through

$$\lambda = -\frac{pkTa_1}{\beta_1}. \tag{5.203}$$

The coefficient a_1 can be determined from the set of algebraic equations

$$-\frac{3p}{c\beta_1 T}\delta^{(m,1)} = \sum_{n=1}^{\infty} A^{(m,n)} a_n, \tag{5.204}$$

which are obtained by multiplying the integral equation (5.187) by the factor $\Delta_\alpha^\gamma p_\gamma R_{\frac{3}{2}}^{(m)}$, integrating the resulting equation over all values of d^3p/p_0, by the use the orthogonality condition (5.152) and (5.170). $A^{(m,n)}$ represents the integral

$$A^{(m,n)} = \int \Delta_{\alpha\beta} R_{\frac{3}{2}}^{(m)} p^\alpha \mathcal{I}\left[R_{\frac{3}{2}}^{(n)} p^\beta\right]\frac{d^3p}{p_0}. \tag{5.205}$$

The constraints (5.14) and (5.15) imply that

$$A^{(m,n)} = A^{(n,m)}, \qquad A^{(0,n)} = 0. \tag{5.206}$$

The first approximation of the coefficient a_1 that follows from (5.204) reads

$$[a_1]_1 = -\frac{3p}{c\beta_1 T A^{(1,1)}}, \tag{5.207}$$

while (5.205) furnishes that $A^{(1,1)}$ is given by

$$A^{(1,1)} = -\frac{\beta_1^2}{2c^2}\frac{I_1 - c^2 I_2}{(kT)^2}. \tag{5.208}$$

The integral I_2 is defined in (5.36).

Now we can get from (5.203), (5.207) and (5.208) that the first approximation of the coefficient of thermal conductivity reduces to

$$[\lambda]_1 = -\frac{3kp^2m^2c^5(\zeta + 5G - G^2\zeta)^2}{I_1 - c^2 I_2}, \tag{5.209}$$

which is the same expression as the one given by (5.43).

Constitutive equation for the pressure deviator

We use (5.122) with $\epsilon = 1$, (5.147) and $(5.182)_3$ and write the pressure deviator as

$$p^{\langle\alpha\beta\rangle(1)} = \left(\Delta^\alpha_\gamma \Delta^\beta_\delta - \frac{1}{3}\Delta^{\alpha\beta}\Delta_{\gamma\delta}\right)$$

$$\times c\sum_{n=0}^{\infty} b_n \left[\int p^\gamma p^\delta p^\epsilon p^\theta R^{(0)}_{\frac{5}{2}} R^{(n)}_{\frac{5}{2}} f^{(0)} \frac{d^3p}{p_0}\right] \nabla_{\langle\epsilon} U_{\theta\rangle}. \tag{5.210}$$

The integral in brackets can be decomposed as

$$\int p^\gamma p^\delta p^\epsilon p^\theta R^{(0)}_{\frac{5}{2}} R^{(n)}_{\frac{5}{2}} f^{(0)} \frac{d^3p}{p_0} = L_1 U^\gamma U^\delta U^\epsilon U^\theta + L_2(\Delta^{\gamma\delta}U^\epsilon U^\theta + \Delta^{\gamma\epsilon}U^\delta U^\theta$$

$$+\Delta^{\gamma\theta}U^\epsilon U^\delta + \Delta^{\epsilon\delta}U^\gamma U^\theta + \Delta^{\epsilon\theta}U^\gamma U^\delta + \Delta^{\theta\delta}U^\epsilon U^\gamma)$$

$$+L_3(\Delta^{\gamma\delta}\Delta^{\epsilon\theta} + \Delta^{\gamma\epsilon}\Delta^{\delta\theta} + \Delta^{\gamma\theta}\Delta^{\epsilon\delta}), \tag{5.211}$$

where L_1, L_2 and L_3 are given by

$$L_1 = \frac{1}{c^8}\int (p_\gamma U^\gamma)^4 R^{(0)}_{\frac{5}{2}} R^{(n)}_{\frac{5}{2}} f^{(0)} \frac{d^3p}{p_0}, \tag{5.212}$$

$$L_2 = \frac{1}{3c^4}\int (p_\gamma U^\gamma)^2 \Delta_{\delta\epsilon} p^\delta p^\epsilon R^{(0)}_{\frac{5}{2}} R^{(n)}_{\frac{5}{2}} f^{(0)} \frac{d^3p}{p_0}, \tag{5.213}$$

$$L_3 = \frac{1}{15}\int (\Delta_{\delta\epsilon} p^\delta p^\epsilon)^2 R^{(0)}_{\frac{5}{2}} R^{(n)}_{\frac{5}{2}} f^{(0)} \frac{d^3p}{p_0} = \frac{pmkT}{c} G\delta^{(0,n)}. \tag{5.214}$$

The last equality in (5.214) was obtained by the use of (5.152) and (5.178).

Now we get from (5.210) through (5.214) that the pressure deviator reads

$$p^{\langle\alpha\beta\rangle(1)} = 2\mu\nabla^{\langle\alpha}U^{\beta\rangle}, \tag{5.215}$$

where the coefficient of shear viscosity μ depends only on the coefficient b_0, i.e.,

$$\mu = pmkTGb_0. \tag{5.216}$$

If we multiply the integral equation (5.188) by $\Delta_{\alpha\epsilon}\Delta_{\beta\theta}p^\theta p^\epsilon R^{(n)}_{\frac{5}{2}}$ and integrate the resulting equation over all values of d^3p/p_0, we obtain the following set of algebraic equations for b_n:

$$-10\frac{pm}{c}G\delta^{(0,m)} = \sum_{n=0}^{\infty} B^{(m,n)} b_n, \tag{5.217}$$

where $B^{(m,n)}$ represents the integral

$$B^{(m,n)} = \left(\Delta_{\gamma\epsilon}\Delta_{\delta\theta} - \frac{1}{3}\Delta_{\epsilon\theta}\Delta_{\gamma\delta}\right)\int R^{(m)}_{\frac{5}{2}} p^\epsilon p^\theta \mathcal{I}\left[R^{(n)}_{\frac{5}{2}} p^\gamma p^\delta\right]\frac{d^3p}{p_0}. \tag{5.218}$$

The first approximation of the coefficient of shear viscosity obtained from (5.216) through (5.218) is the same as the one given in (5.46), e.g.,

$$[\mu]_1 = -\frac{30p^2kTm^2c^3G^2}{2I_1 - 6c^2I_2 + 3c^4I_3}, \tag{5.219}$$

since b_0 and $B^{(0,0)}$ are given by

$$b_0 = -10\frac{pm}{B^{(0,0)}c}G, \qquad B^{(0,0)} = \frac{1}{6c^4}(2I_1 - 6c^2I_2 + 3c^4I_3). \tag{5.220}$$

The integral I_3 is defined in (5.37).

Problems

5.5.4.1 Show that the representation of the integral in (5.211) holds, where the coefficients are given by (5.212) through (5.214).

5.5.4.2 Obtain the expressions for the bulk viscosity (5.197), thermal conductivity (5.209) and shear viscosity (5.219) by using the method described in this section.

5.6 Appendix

We consider the integral

$$Z = \int e^{-\frac{1}{kT}U_\lambda p^\lambda}\frac{d^3p}{p_0}. \tag{5.221}$$

In a local Lorentz rest frame we have that $U_\lambda p^\lambda/(kT) = cp^0/(kT)$ and the above integral reduces to

$$Z = 4\pi\int_0^\infty e^{-\frac{c}{kT}p^0}|\mathbf{p}|^2\frac{d|\mathbf{p}|}{p^0}, \tag{5.222}$$

by introducing spherical coordinates $d^3p = |\mathbf{p}|^2\sin\theta d|\mathbf{p}|d\theta d\varphi$ and integrating in the angles $0 \le \theta \le \pi$ and $0 \le \varphi \le 2\pi$. If we use the relationship $p^0 = \sqrt{|\mathbf{p}|^2 + m^2c^2}$ and change the variable of integration through $y = p^0/(mc)$, we have that

$$|\mathbf{p}|^2 = m^2c^2(y^2 - 1), \qquad d|\mathbf{p}| = mc\frac{y\,dy}{(y^2-1)^{\frac{1}{2}}}, \tag{5.223}$$

and (5.222) reduces to

$$Z = 4\pi(mc)^2\int_1^\infty e^{-\zeta y}(y^2-1)^{\frac{1}{2}}dy = 4\pi mkTK_1(\zeta), \tag{5.224}$$

that is Z is given in terms of the modified Bessel function of second kind (see (3.19)).

The following integrals can be obtained through successive differentiation of (5.224) with respect to $U^\alpha/(kT)$:

$$Z^\alpha = \int p^\alpha e^{-\frac{1}{kT}U_\lambda p^\lambda} \frac{d^3p}{p_0} = 4\pi m^2 kT K_2(\zeta) U^\alpha, \tag{5.225}$$

$$Z^{\alpha\beta} = \int p^\alpha p^\beta e^{-\frac{1}{kT}U_\lambda p^\lambda} \frac{d^3p}{p_0} = -4\pi (mkT)^2 \Big[K_2(\zeta)\eta^{\alpha\beta} - \frac{m}{kT} K_3(\zeta) U^\alpha U^\beta\Big], \tag{5.226}$$

$$Z^{\alpha\beta\gamma} = \int p^\alpha p^\beta p^\gamma e^{-\frac{1}{kT}U_\lambda p^\lambda} \frac{d^3p}{p_0} = -4\pi m^3 (kT)^2 \Big[K_3(\zeta)(\eta^{\alpha\beta}U^\gamma + \eta^{\alpha\gamma}U^\beta + \eta^{\gamma\beta}U^\alpha) - \frac{m}{kT} K_4(\zeta) U^\alpha U^\beta U^\gamma\Big], \tag{5.227}$$

$$Z^{\alpha\beta\gamma\delta} = \int p^\alpha p^\beta p^\gamma p^\delta e^{-\frac{1}{kT}U_\lambda p^\lambda} \frac{d^3p}{p_0} = 4\pi (mkT)^3 \Big[K_3(\zeta)(\eta^{\alpha\beta}\eta^{\gamma\delta} + \eta^{\alpha\gamma}\eta^{\beta\delta} + \eta^{\gamma\beta}\eta^{\alpha\delta}) - \frac{m}{kT} K_4(\zeta)(\eta^{\alpha\beta}U^\gamma U^\delta + \eta^{\alpha\gamma}U^\beta U^\delta + \eta^{\gamma\beta}U^\alpha U^\delta + \eta^{\alpha\delta}U^\gamma U^\beta + \eta^{\delta\gamma}U^\beta U^\alpha + \eta^{\delta\beta}U^\alpha U^\gamma) + \left(\frac{m}{kT}\right)^2 K_5(\zeta) U^\alpha U^\beta U^\gamma U^\delta\Big], \tag{5.228}$$

$$\begin{aligned} Z^{\alpha\beta\gamma\delta\epsilon} = \int p^\alpha p^\beta p^\gamma p^\delta p^\epsilon e^{-\frac{1}{kT}U_\lambda p^\lambda} \frac{d^3p}{p_0} = 4\pi m^4 (kT)^3 \Big\{K_4(\zeta)[U^\epsilon(\eta^{\alpha\beta}\eta^{\gamma\delta} \\ +\eta^{\alpha\gamma}\eta^{\beta\delta} + \eta^{\gamma\beta}\eta^{\alpha\delta}) + U^\delta(\eta^{\alpha\beta}\eta^{\gamma\epsilon} + \eta^{\alpha\gamma}\eta^{\beta\epsilon} + \eta^{\gamma\beta}\eta^{\alpha\epsilon}) + U^\gamma(\eta^{\alpha\beta}\eta^{\epsilon\delta} + \eta^{\alpha\epsilon}\eta^{\beta\delta} \\ +\eta^{\epsilon\beta}\eta^{\alpha\delta}) + U^\beta(\eta^{\alpha\epsilon}\eta^{\gamma\delta} + \eta^{\alpha\gamma}\eta^{\epsilon\delta} + \eta^{\gamma\epsilon}\eta^{\alpha\delta}) + U^\alpha(\eta^{\epsilon\beta}\eta^{\gamma\delta} + \eta^{\epsilon\gamma}\eta^{\beta\delta} \\ +\eta^{\gamma\beta}\eta^{\epsilon\delta})] - \frac{m}{kT} K_5(\zeta)[\eta^{\alpha\beta}U^\gamma U^\delta U^\epsilon + \eta^{\alpha\gamma}U^\beta U^\delta U^\epsilon + \eta^{\alpha\delta}U^\beta U^\gamma U^\epsilon \\ +\eta^{\delta\gamma}U^\alpha U^\beta U^\epsilon + \eta^{\delta\beta}U^\alpha U^\gamma U^\epsilon + \eta^{\beta\gamma}U^\alpha U^\delta U^\epsilon + \eta^{\alpha\epsilon}U^\beta U^\gamma U^\delta + \eta^{\beta\epsilon}U^\alpha U^\gamma U^\delta \\ +\eta^{\gamma\epsilon}U^\alpha U^\beta U^\delta + \eta^{\delta\epsilon}U^\alpha U^\beta U^\gamma] + \left(\frac{m}{kT}\right)^2 K_6(\zeta) U^\alpha U^\beta U^\gamma U^\delta U^\epsilon\Big\}. \end{aligned} \tag{5.229}$$

The same methodology can be applied to obtain the integrals below, i.e., by successive differentiation of (5.78) with respect to $U_\lambda/(kT)$:

$$Z^\star_\alpha = \int e^{-\frac{1}{kT}U_\lambda P^\lambda} P_\alpha \frac{d^3P}{P^0} = 4\pi m^2 kT Q^{\star 2} K_2(\zeta Q^\star) U_\alpha, \tag{5.230}$$

$$Z^\star_{\alpha\beta} = \int e^{-\frac{1}{kT}U_\lambda P^\lambda} P_\alpha P_\beta \frac{d^3P}{P^0} = -4\pi (mkT)^2 Q^{\star 2} \Big[K_2(\zeta Q^\star)\eta_{\alpha\beta} - \frac{m}{kT} Q^\star K_3(\zeta Q^\star) U_\alpha U_\beta\Big], \tag{5.231}$$

$$Z^\star_{\alpha\beta\gamma} = \int e^{-\frac{1}{kT}U_\lambda P^\lambda} P_\alpha P_\beta P_\gamma \frac{d^3P}{P^0} = -4\pi m^3 (kT)^2 Q^{\star 3} \Big[K_3(\zeta Q^\star)(U_\gamma \eta_{\alpha\beta}$$

$$+U_\alpha\eta_{\gamma\beta}+U_\beta\eta_{\alpha\gamma})-\frac{m}{kT}Q^\star K_4(\zeta Q^\star)U_\alpha U_\beta U_\gamma\Big], \tag{5.232}$$

$$Z^\star_{\alpha\beta\gamma\delta}=\int e^{-\frac{1}{kT}U_\lambda P^\lambda}P_\alpha P_\beta P_\gamma P_\delta\frac{d^3P}{P^0}=4\pi(mkT)^3Q^{\star 3}\Big[K_3(\zeta Q^\star)(\eta_{\alpha\beta}\eta_{\gamma\delta}$$

$$+\eta_{\alpha\gamma}\eta_{\beta\delta}+\eta_{\gamma\beta}\eta_{\alpha\delta})-\frac{m}{kT}Q^\star K_4(\zeta Q^\star)(\eta_{\alpha\beta}U_\gamma U_\delta+\eta_{\alpha\gamma}U_\beta U_\delta+\eta_{\gamma\beta}U_\alpha U_\delta$$

$$+\eta_{\alpha\delta}U_\gamma U_\beta+\eta_{\delta\gamma}U_\beta U_\alpha+\eta_{\delta\beta}U_\alpha U_\gamma)+\left(\frac{m}{kT}\right)^2Q^{\star 2}K_5(\zeta Q^\star)U_\alpha U_\beta U_\gamma U_\delta\Big]. \tag{5.233}$$

Other integrals that are needed in this and subsequent chapters are given below. Let $\mathcal{Y}(p_\alpha U^\alpha)$ be a scalar which is a function of the momentum four-vector only through the scalar $p_\alpha U^\alpha$; then it is easy to show that the following decompositions hold:

$$\int p^\alpha\mathcal{Y}f^{(0)}\frac{d^3p}{p_0}=\frac{\mathcal{I}^{(1,0)}}{c^2}U^\alpha, \tag{5.234}$$

$$\int p^\alpha p^\beta\mathcal{Y}f^{(0)}\frac{d^3p}{p_0}=\frac{\mathcal{I}^{(2,0)}}{c^4}U^\alpha U^\beta+\frac{\mathcal{I}^{(0,1)}}{3}\Delta^{\beta\delta}, \tag{5.235}$$

$$\int p^\alpha p^\beta p^\gamma\mathcal{Y}f^{(0)}\frac{d^3p}{p_0}=\frac{\mathcal{I}^{(3,0)}}{c^6}U^\alpha U^\beta U^\gamma+\frac{\mathcal{I}^{(1,1)}}{3c^2}(\Delta^{\alpha\beta}U^\gamma+\Delta^{\alpha\gamma}U^\beta+\Delta^{\beta\gamma}U^\alpha), \tag{5.236}$$

$$\int p^\alpha p^\beta p^\gamma p^\delta\mathcal{Y}f^{(0)}\frac{d^3p}{p_0}=\frac{\mathcal{I}^{(4,0)}}{c^8}U^\alpha U^\beta U^\gamma U^\delta+\frac{\mathcal{I}^{(2,1)}}{3c^4}(\Delta^{\alpha\beta}U^\gamma U^\delta+\Delta^{\alpha\gamma}U^\beta U^\delta$$

$$+\Delta^{\alpha\delta}U^\beta U^\gamma+\Delta^{\beta\gamma}U^\alpha U^\delta+\Delta^{\beta\delta}U^\alpha U^\gamma+\Delta^{\gamma\delta}U^\alpha U^\beta)$$

$$+\frac{\mathcal{I}^{(0,2)}}{15}(\Delta^{\alpha\beta}\Delta^{\gamma\delta}+\Delta^{\alpha\gamma}\Delta^{\beta\delta}+\Delta^{\alpha\delta}\Delta^{\beta\gamma}), \tag{5.237}$$

where the integrals $\mathcal{I}^{(m,n)}$ are given by

$$\mathcal{I}^{(m,n)}=\int(p^\alpha U_\alpha)^m(\Delta_{\alpha\beta}p^\alpha p^\beta)^n\mathcal{Y}f^{(0)}\frac{d^3p}{p_0}. \tag{5.238}$$

The expression for the relativistic orthogonal polynomial of second order and index 5/2 is:

$$R^{(2)}_{\frac{5}{2}}=\gamma_2G^{\frac{1}{2}}\Bigg[\zeta(G^2\zeta^2+6G^2-5G\zeta-\zeta^2)\left(\frac{U^\alpha p_\alpha}{kT}\right)^2$$

$$+(\zeta^4+5G\zeta^3+7\zeta^2+84G\zeta-6G^2\zeta^2+252G^2-G^2\zeta^4)\left(\frac{U^\alpha p_\alpha}{kT}\right)$$

$$+\zeta^2(7\zeta^2-84G^2+28G\zeta-7G^2\zeta^2)\Bigg], \tag{5.239}$$

where the constant γ_2 reads

$$(\gamma_2)^{-2} = 123480G^2\zeta^3 + 740880G^3\zeta^2 + 125496G^3\zeta^4 + 10290G\zeta^4 - 784\zeta^6$$
$$-17234G^2\zeta^6 + 14256G^4\zeta^6 + 1260G^3\zeta^5 - 18032G\zeta^5 - 150528G^2\zeta^4 + 22G^2\zeta^8$$
$$+4626G^3\zeta^7 + 152G^4\zeta^8 - 3682G\zeta^7 - 1824G^5\zeta^7 + 700\zeta^7 + 343\zeta^5 - 14040G^5\zeta^5$$
$$+142128G^5\zeta^3 - 40G^5\zeta^9 - 2G^4\zeta^{10} + 1016064G^5\zeta + 12523G\zeta^6 - 224\zeta^8 + 2G\zeta^{10}$$
$$+2222640G^4\zeta - 65184G^4\zeta^5 + 10836G^5\zeta^4 - 13559G^3\zeta^6 + 672G\zeta^8 + 7548G^5\zeta^6$$
$$-2318G^4\zeta^7 - 968G^3\zeta^8 + 346G^5\zeta^8 + 2373G^2\zeta^7 - 508032G^3\zeta^3 + 146664G^4\zeta^4$$
$$+4G^2\zeta^{10} + 15G^4\zeta^9 - 4G^3\zeta^{10} + 2G^5\zeta^{10} + 2667168G^5 - 2\zeta^{10} - 50G^2\zeta^9 - 60G\zeta^9$$
$$+100G^3\zeta^9 + 35\zeta^9 + 73248G^2\zeta^5 - 120960G^4\zeta^3 - 208656G^5\zeta^2 - 338688G^4\zeta^2.$$

References

[1] C. Cercignani, *The Boltzmann equation and its applications* (Springer, New York, 1988).

[2] S. Chapman and T. G. Cowling, *The mathematical theory of nonuniform gases*, 3rd. ed. (Cambridge Univ. Press, Cambridge, 1970).

[3] D. Enskog, *Kinetische Theorie der Vorgänge in mässig verdünnten Gasen*, Diss. Uppsala (Almqvist & Wiksells, Uppsala, 1917).

[4] I. S. Gradshteyn and I. M. Ryzhik, *Tables of integrals, series and products* 5th ed. (Academic Press, San Diego, 1994).

[5] S. R. de Groot, W. A. van Leeuwen and Ch. G. van Weert, *Relativistic kinetic theory* (North-Holland, Amsterdam, 1980).

[6] D. Hilbert, *Grundzüge einer allgemeinen Theorie der linearen Integralgleichungen* (B. G. Teubner, Leipzig, 1912).

[7] W. Israel, Relativistic kinetic theory of a simple gas, *J. Math. Phys.* **4**, 1163–1181 (1963).

[8] D. C. Kelly, *The kinetic theory of a relativistic gas*, unpublished report (Miami University, Oxford, 1963).

[9] G. M. Kremer, *Uma introdução à teoria cinética relativística* (Editora da UFPR, Curitiba, 1998).

[10] H. Margenau and G. M. Murphy, *The mathematics of physics and chemistry*, 2nd. ed. (D. van Nostrand, Princeton, 1956).

[11] S. Pennisi, Some representation theorems in a 4-dimensional inner product space, *Suppl. B. U. M. I. Fisica Matematica* **5**, 191–214 (1986).

[12] J. M. Stewart, *Non-equilibrium relativistic kinetic theory*, Lecture Notes in Physics, vol. 10 (Springer, Heidelberg, 1971).

Chapter 6

Method of Moments

6.1 Introduction

In this chapter we shall deal only with a single non-degenerate relativistic gas, the main purpose being to describe the Grad method which is based on the moments of the one-particle distribution function. For a non-relativistic monatomic gas one often uses a thirteen moment description whose objective is the determination of the fields of particle number density, bulk velocity, temperature, pressure deviator and heat flux when we are not too far from a local equilibrium distribution. For a relativistic gas one has to include the field of the dynamic pressure – since as we have seen in the last chapter it is a non-vanishing quantity – and in this case we have fourteen basic fields. These fields are determined from the fourteen moment equations (2.73) through (2.75) which we reproduce below:

$$\partial_\alpha N^\alpha = 0, \qquad \partial_\beta T^{\alpha\beta} = 0, \qquad \partial_\gamma T^{\alpha\beta\gamma} = P^{\alpha\beta}. \tag{6.1}$$

We recall that N^α and $T^{\alpha\beta}$ are given in terms of the fourteen fields n, U^α, T, ϖ, $p^{\langle\alpha\beta\rangle}$ and q^α through the decompositions of Eckart (4.21) and (4.22). Hence we conclude that the system of partial differential equations (6.1) is not a system of field equations for the fourteen basic fields since till now we do not know how to express the third-order moment $T^{\alpha\beta\gamma}$ and the production term of its balance equation $P^{\alpha\beta}$ in term of the fourteen fields. In order to achieve this objective it is necessary to know the one-particle distribution function in terms of the fourteen basic fields. This will be the subject of the next section.

6.2 Grad distribution function

In order to obtain the non-equilibrium one-particle distribution function for a relativistic gas we shall maximize the entropy per particle s under fourteen constraints that follow from the definitions of the following moments, when the products of the same moments with U_α are assumed to be given: particle four-flow N^α, energy-momentum tensor $T^{\alpha\beta}$ and third-order moment $T^{\alpha\beta\gamma}$. The entropy per particle, obtained from (2.86) and (4.35), reads

$$s = -\frac{k}{nc}U^\alpha \int p_\alpha f \ln\left(\frac{fh^3}{eg_s}\right)\frac{d^3p}{p_0}, \tag{6.2}$$

while the fourteen constraints follow from the definitions N^α, $T^{\alpha\beta}$ and $T^{\alpha\beta\gamma}$ given by (2.69) through (2.71):

$$N^\alpha U_\alpha = cU_\alpha \int p^\alpha f \frac{d^3p}{p_0}, \tag{6.3}$$

$$T^{\alpha\beta}U_\alpha = cU_\alpha \int p^\alpha p^\beta f \frac{d^3p}{p_0}, \tag{6.4}$$

$$T^{\langle\gamma\beta\rangle\alpha}U_\alpha = cU_\alpha \int p^{\langle\gamma}p^{\beta\rangle}p^\alpha f \frac{d^3p}{p_0}. \tag{6.5}$$

At first, one might think that it would be better to choose the moments N^α, $T^{\alpha\beta}$, and indeed that would be more spontaneous, but then we should proceed in a much more unnatural and complicated way, arriving at an essentially equivalent result.

The problem of maximizing the entropy per particle (6.2) subjected to the constraints provided by (6.3) through (6.5) when their left-hand sides are assumed to be known, is equivalent, according to variational calculus, to the problem of maximizing, without constraints, the functional

$$F = -\frac{k}{nc}U_\alpha \int p^\alpha f \ln f \frac{d^3p}{p_0} - \lambda c U_\alpha \int p^\alpha f \frac{d^3p}{p_0} - \lambda_\beta c U_\alpha \int p^\beta p^\alpha f \frac{d^3p}{p_0}$$

$$-\lambda_{\langle\gamma\beta\rangle} c U_\alpha \int p^{\langle\gamma}p^{\beta\rangle}p^\alpha f \frac{d^3p}{p_0}, \tag{6.6}$$

where λ, λ_β and $\lambda_{\langle\gamma\beta\rangle}$ are Lagrange undetermined multipliers. The number of independent multipliers is 14 because $\Delta^{\gamma\beta}\lambda_{\langle\gamma\beta\rangle} = \eta^{\gamma\beta}\lambda_{\langle\gamma\beta\rangle} = 0$.

The Euler–Lagrange equation for this problem reduces to $\partial F/\partial f = 0$ and it follows from (6.6) that

$$f = \frac{g_s}{h^3}\exp\left[-\frac{nc^2}{k}\left(\lambda + \lambda_\beta p^\beta + \lambda_{\langle\gamma\beta\rangle}p^\gamma p^\beta\right)\right], \tag{6.7}$$

providing that $p_\alpha U^\alpha \neq 0$.

To identify the Lagrange multipliers one has to note that the distribution function (6.7) is such that in equilibrium it must reduce to the Maxwell–Jüttner distribution function (3.27). Further we shall decompose the Lagrange multipliers into two parts: one is related to an equilibrium state while the other refers to a non-equilibrium state. Hence we write

$$\lambda = \lambda^E + \lambda^{NE}, \qquad \lambda_\beta = \lambda_\beta^E + \lambda_\beta^{NE}, \qquad \lambda_{\langle\gamma\beta\rangle} = \lambda_{\langle\gamma\beta\rangle}^{NE}, \tag{6.8}$$

such that λ^{NE}, λ_β^{NE} and $\lambda_{\langle\gamma\beta\rangle}^{NE}$ are zero in a state of equilibrium. By using the above decompositions, the one-particle distribution function (6.7) can be written as

$$f = \frac{g_s}{h^3}\exp\left[-\frac{nc^2}{k}\left(\lambda^E + \lambda_\beta^E p^\beta\right)\right]$$

$$\times \exp\left[-\frac{nc^2}{k}\left(\lambda^{NE} + \lambda_\beta^{NE} p^\beta + \lambda_{\langle\gamma\beta\rangle}^{NE} p^\gamma p^\beta\right)\right]. \tag{6.9}$$

The first exponential function above is the Maxwell–Jüttner distribution function $f^{(0)}$. Since we are interested in processes close to local equilibrium we may treat the Lagrange multipliers λ^{NE}, λ_β^{NE} and $\lambda_{\langle\gamma\beta\rangle}^{NE}$ as small quantities, and by using the approximation $\exp(-x) \approx 1 - x$ – valid for all $x \ll 1$ – for the second exponential function we get

$$f \approx f^{(0)}\left\{1 - \frac{nc^2}{k}\left(\lambda^{NE} + \lambda_\beta^{NE} p^\beta + \lambda_{\langle\gamma\beta\rangle}^{NE} p^\gamma p^\beta\right)\right\}. \tag{6.10}$$

One can verify from the above expression that the non-equilibrium distribution function is determined once we know the Lagrange multipliers λ^{NE}, λ_β^{NE} and $\lambda_{\langle\gamma\beta\rangle}^{NE}$. For the determination of these Lagrange multipliers we make use of the following decomposition into two contributions orthogonal in space-time:

$$\lambda_\beta^{NE} = \tilde\lambda U_\beta + \tilde\lambda_\gamma \Delta_\beta^\gamma, \tag{6.11}$$

$$\lambda_{\langle\gamma\beta\rangle}^{NE} = \Lambda U_\gamma U_\beta + \frac{1}{2}\Lambda_\alpha\left(\Delta_\gamma^\alpha U_\beta + \Delta_\beta^\alpha U_\gamma\right) + \Lambda_{\alpha\delta}\left(\Delta_\gamma^\alpha \Delta_\beta^\delta - \frac{1}{3}\Delta^{\alpha\delta}\Delta_{\gamma\beta}\right), \tag{6.12}$$

and write (6.10) as

$$f = f^{(0)}\Bigg\{1 - \frac{nc^2}{k}\lambda^{NE} - \frac{nc^2}{k}\left(\tilde\lambda U_\alpha + \tilde\lambda_\gamma \Delta_\alpha^\gamma\right)p^\alpha - \frac{nc^2}{k}\Bigg[\Lambda U_\alpha U_\beta$$

$$+\frac{1}{2}\Lambda_\gamma\left(\Delta_\alpha^\gamma U_\beta + \Delta_\beta^\gamma U_\alpha\right) + \Lambda_{\gamma\delta}\left(\Delta_\alpha^\gamma \Delta_\beta^\delta - \frac{1}{3}\Delta^{\gamma\delta}\Delta_{\alpha\beta}\right)\Bigg]p^\alpha p^\beta\Bigg\}. \tag{6.13}$$

We insert the above expression for f into the definition of the particle four-flow

$$N^\alpha = c\int p^\alpha f \frac{d^3p}{p^0} = c\int p^\alpha f^{(0)} \frac{d^3p}{p^0}, \tag{6.14}$$

and integrate the resulting equation by using the integrals Z of the Appendix of Chapter 5 which are given in (5.224) through (5.229). If we take the product of the resulting equation by U_α and Δ^β_α, we get respectively

$$\lambda^{NE} + \tilde{\lambda} mc^2\left(G - \frac{1}{\zeta}\right) + \Lambda m^2c^4\left(1 + 3\frac{G}{\zeta}\right) = 0, \tag{6.15}$$

$$\tilde{\lambda}_\gamma \Delta^{\gamma\beta} + mc^2 G\Lambda_\gamma\Delta^{\gamma\beta} = 0. \tag{6.16}$$

On the other hand, the substitution of (6.13) into the definition of the energy-momentum tensor

$$T^{\alpha\beta} = c\int p^\alpha p^\beta f\frac{d^3p}{p^0}, \tag{6.17}$$

and by following the same procedure as above, we get an equation whose projections

$$\left(\Delta^\gamma_\alpha\Delta^\delta_\beta - \frac{1}{3}\Delta^{\gamma\delta}\Delta_{\alpha\beta}\right)T^{\alpha\beta} = p^{\langle\gamma\delta\rangle}, \quad \Delta_{\alpha\beta}T^{\alpha\beta} = -3(p+\varpi), \tag{6.18}$$

$$\Delta^\gamma_\alpha U_\beta T^{\alpha\beta} = q^\gamma, \qquad U_\alpha U_\beta T^{\alpha\beta} = nc^2 e, \tag{6.19}$$

lead to, respectively,

$$p^{\langle\gamma\delta\rangle} = -2n^2m^2c^4T\frac{G}{\zeta}\Lambda^{\langle\gamma\delta\rangle}, \tag{6.20}$$

$$\varpi = -n^2c^2T\left[\lambda^{NE} + Gmc^2\tilde{\lambda} + \left(1 + 5\frac{G}{\zeta}\right)m^2c^4\Lambda\right], \tag{6.21}$$

$$q^\gamma = n^2c^4T\left[Gm\Delta^{\delta\gamma}\tilde{\lambda}_\delta + \left(1 + 5\frac{G}{\zeta}\right)m^2c^2\Delta^{\delta\gamma}\Lambda_\delta\right], \tag{6.22}$$

$$\lambda^{NE}\left(G - \frac{1}{\zeta}\right) + \tilde{\lambda}mc^2\left(3\frac{G}{\zeta} + 1\right) + \Lambda m^2c^4\left(15\frac{G}{\zeta^2} + \frac{2}{\zeta} + G\right) = 0. \tag{6.23}$$

By solving the system of equations (6.15), (6.21) and (6.23) for $\lambda^{NE}, \tilde{\lambda}$ and Λ we get the scalar Lagrange multipliers

$$\lambda^{NE} = -\frac{(15G + 2\zeta - 6G^2\zeta + 5G\zeta^2 + \zeta^3 - G^2\zeta^3)}{20G + 3\zeta - 13G^2\zeta - 2G\zeta^2 + 2G^3\zeta^2}\frac{\varpi}{n^2c^2T}, \tag{6.24}$$

$$\tilde{\lambda} = -\frac{3\zeta(6G + \zeta - G^2\zeta)}{20G + 3\zeta - 13G^2\zeta - 2G\zeta^2 + 2G^3\zeta^2}\frac{\varpi}{mn^2c^4T}, \tag{6.25}$$

$$\Lambda = -\frac{\zeta(1 - 5G\zeta - \zeta^2 + G^2\zeta^2)}{20G + 3\zeta - 13G^2\zeta - 2G\zeta^2 + 2G^3\zeta^2}\frac{\varpi}{n^2m^2c^6T}. \tag{6.26}$$

From the system of equations for Λ_δ and $\tilde{\lambda}_\delta$, represented by (6.16) and (6.22), we get the vectorial Lagrange multipliers

$$\Delta^{\delta\gamma}\Lambda_\delta = \frac{\zeta}{\zeta + 5G - G^2\zeta}\frac{q^\gamma}{n^2m^2c^6T}, \tag{6.27}$$

$$\Delta^{\delta\gamma}\tilde{\lambda}_\delta = \frac{-G\zeta}{\zeta + 5G - G^2\zeta}\frac{q^\gamma}{mn^2c^4T}. \tag{6.28}$$

The tensorial Lagrange multiplier $\Lambda^{\langle\gamma\delta\rangle}$ is obtained from (6.20)

$$\Lambda^{\langle\gamma\delta\rangle} = -\frac{\zeta}{2G}\frac{p^{\langle\gamma\delta\rangle}}{n^2m^2c^4T}. \tag{6.29}$$

Since we have identified all Lagrange multipliers, the non-equilibrium distribution function in terms of the fourteen fields $n, U^\alpha, T, \varpi, p^{\langle\alpha\beta\rangle}$ and q^α is obtained by substituting (6.24) through (6.29) into (6.13), yielding

$$f = f^{(0)}\Bigg\{1 + \frac{\varpi}{p}\frac{1 - 5G\zeta - \zeta^2 + G^2\zeta^2}{20G + 3\zeta - 13G^2\zeta - 2G\zeta^2 + 2G^3\zeta^2}$$

$$\times\left[\frac{15G + 2\zeta - 6G^2\zeta + 5G\zeta^2 + \zeta^3 - G^2\zeta^3}{1 - 5G\zeta - \zeta^2 + G^2\zeta^2}\right.$$

$$\left. + \frac{3\zeta}{mc^2}\frac{6G + \zeta - G^2\zeta}{1 - 5G\zeta - \zeta^2 + G^2\zeta^2}U_\alpha p^\alpha + \frac{\zeta}{m^2c^4}U_\alpha U_\beta p^\alpha p^\beta\right]$$

$$+ \frac{q_\alpha}{p}\frac{\zeta}{\zeta + 5G - G^2\zeta}\left[\frac{G}{mc^2}p^\alpha - \frac{1}{m^2c^4}U_\beta p^\alpha p^\beta\right] + \frac{p_{\langle\alpha\beta\rangle}}{p}\frac{\zeta}{2G}\frac{1}{m^2c^2}p^\alpha p^\beta\Bigg\}. \tag{6.30}$$

The above equation represents the Grad distribution function for a relativistic gas.

Another way to obtain the Grad distribution function is by expanding the one-particle distribution function in a polynomial of the momentum four-vector p^α about the locally Maxwell–Jüttner distribution function, i.e.,

$$f = f^{(0)}\left(1 + a_\alpha p^\alpha + a_{\alpha\beta}p^\alpha p^\beta\right), \tag{6.31}$$

where the coefficients a_α and $a_{\alpha\beta}$ are calculated by using the definitions of the fourteen fields: particle four-flow N^α and energy-momentum tensor $T^{\alpha\beta}$.

Problems

6.2.1 Obtain (6.15) and (6.16) from the projections of the particle four-flow and (6.20) through (6.23) from the projections of the energy-momentum tensor.

6.2.2 By solving the system of equations (6.15), (6.21) and (6.23) obtain the expressions for $\lambda^{NE}, \tilde{\lambda}$ and Λ given in (6.24) through (6.26).

6.2.3 Solve (6.16) and (6.22) for Λ_δ and $\tilde{\lambda}_\delta$ and obtain (6.27) and (6.28).

6.2.4 Consider that the non-equilibrium distribution function has a representation given by (6.31). Obtain the coefficients a_α and $a_{\alpha\beta}$ by inserting (6.31) into the definition of the particle four-flow and of the momentum-energy tensor and show that this leads to Grad distribution function (6.30).

6.2.5 As in the problem 3.2.3.2 consider: a) a local Lorentz rest frame where $(U^\alpha) = (c, \mathbf{0})$; b) that the modulus of the bulk velocity $\mathbf{v} = c\mathbf{p}/p^0$ of the gas particles is small when compared to the speed of light, i.e., $v \ll c$ and c) the non-relativistic limiting condition $\zeta \gg 1$. By considering only the first term in the asymptotic expansion of G and by neglecting all terms of order $1/\zeta$ show that (6.15) is equivalent to (6.23) and infer from (6.21) that $\varpi = 0$. Next choose $\lambda^{NE} = \tilde{\lambda} = \Lambda = 0$ and show that the relativistic Grad distribution function (6.30) reduces to

$$f = \frac{n}{(2\pi kT)^{3/2}} e^{-\frac{|\mathbf{p}|^2}{2mkT}} \left\{ 1 + \frac{2\delta_{ij} q^i p^j}{5pkT} \left[\frac{|\mathbf{p}|^2}{2mkT} - \frac{5}{2} \right] + \frac{p_{\langle ij \rangle} p^i p^j}{2mpkT} \right\}.$$

This is the well-known Grad distribution function [6] for the thirteen moments: particle number density, bulk velocity, temperature, heat flux and pressure deviator. Note that the frame chosen here is that which is moving with the gas.

6.3 Constitutive equations for $T^{\alpha\beta\gamma}$ and $P^{\alpha\beta}$

In order to determine the constitutive equation for the third-order moment we insert Grad distribution function (6.30) into its definition

$$T^{\alpha\beta\gamma} = c \int p^\alpha p^\beta p^\gamma f \frac{d^3 p}{p_0}, \tag{6.32}$$

and integrate the resulting equation by using the integrals (5.227) through (5.229) of the Appendix of Chapter 5. Hence we get that the desired constitutive equation for $T^{\alpha\beta\gamma}$ in terms of the fourteen basic fields reads

$$T^{\alpha\beta\gamma} = (nC_1 + C_2\varpi) U^\alpha U^\beta U^\gamma + \frac{c^2}{6}(nm^2 - nC_1 - C_2\varpi)(\eta^{\alpha\beta} U^\gamma + \eta^{\alpha\gamma} U^\beta$$
$$+ \eta^{\beta\gamma} U^\alpha) + C_3(\eta^{\alpha\beta} q^\gamma + \eta^{\alpha\gamma} q^\beta + \eta^{\beta\gamma} q^\alpha) - \frac{6}{c^2} C_3 (U^\alpha U^\beta q^\gamma + U^\alpha U^\gamma q^\beta$$
$$+ U^\beta U^\gamma q^\alpha) + C_4 (p^{\langle\alpha\beta\rangle} U^\gamma + p^{\langle\alpha\gamma\rangle} U^\beta + p^{\langle\beta\gamma\rangle} U^\alpha). \tag{6.33}$$

The scalar coefficients C_1 through C_4 are given by

$$C_1 = \frac{m^2}{\zeta}(\zeta + 6G), \tag{6.34}$$

$$C_2 = -\frac{6m}{c^2\zeta} \frac{2\zeta^3 - 5\zeta + (19\zeta^2 - 30)G - (2\zeta^3 - 45\zeta)G^2 - 9\zeta^2 G^3}{20G + 3\zeta - 13G^2\zeta - 2\zeta^2 G + 2\zeta^2 G^3}, \tag{6.35}$$

$$C_3 = -\frac{m}{\zeta}\frac{\zeta + 6G - G^2\zeta}{\zeta + 5G - G^2\zeta}, \qquad C_4 = \frac{m}{G\zeta}(\zeta + 6G). \tag{6.36}$$

The determination of the linearized constitutive equation for the production term

$$P^{\alpha\beta} = \frac{c}{2}\int (p'^{\alpha}p'^{\beta} + p'^{\alpha}_{*}p'^{\beta}_{*} - p^{\alpha}p^{\beta} - p^{\alpha}_{*}p^{\beta}_{*}) f f_{*} F\sigma d\Omega \frac{d^3p_*}{p_{*0}}\frac{d^3p}{p_0}, \tag{6.37}$$

proceeds as follows. First we use (5.52), (5.53), (5.56) and (5.59) and write $P^{\alpha\beta}$ as

$$P^{\alpha\beta} = \frac{c}{16}\int (Q'^{\alpha}Q'^{\beta} - Q^{\alpha}Q^{\beta}) f f_{*} Q\sigma \sin\Theta d\Phi d\Theta d^3Q \frac{d^3P}{P^0}$$

$$\stackrel{(5.61)}{=} \frac{c\pi}{16}\int \left(\frac{P^{\alpha}P^{\beta}}{P^2} - \eta^{\alpha\beta} - 3\frac{Q^{\alpha}Q^{\beta}}{Q^2}\right) f f_{*} Q^3 \sigma \sin^3\Theta d\Theta d^3Q \frac{d^3P}{P^0}. \tag{6.38}$$

Next we build the product of the Grad distribution functions ff_* in terms of the variables P^α and Q^α and neglect all non-linear terms in ϖ, q^α and $p^{\langle\alpha\beta\rangle}$. Hence we obtain that (6.38) reduces to

$$P^{\alpha\beta} = \frac{c\pi n^2}{32(4\pi kTm^2cK_2)^2}\left\{\frac{\varpi}{p}\frac{\zeta(1 - 5G\zeta - \zeta^2 + G^2\zeta^2)}{20G + 3\zeta - 13G^2\zeta - 2G\zeta^2 + 2G^3\zeta^2}\frac{U_\gamma U_\delta}{m^2c^4}\right.$$

$$\left. - \frac{q_\gamma}{p}\frac{\zeta}{\zeta + 5G - G^2\zeta}\frac{U_\delta}{m^2c^4} + \frac{p_{\langle\gamma\delta\rangle}}{p}\frac{\zeta}{2G}\frac{1}{m^2c^2}\right\}$$

$$\times \int \left(\frac{P^{\alpha}P^{\beta}}{P^2} - \eta^{\alpha\beta} - 3\frac{Q^{\alpha}Q^{\beta}}{Q^2}\right) Q^\gamma Q^\delta Q^3 e^{-\frac{1}{kT}U^\lambda P_\lambda} \sigma \sin^3\Theta d\Theta d^3Q \frac{d^3P}{P^0}, \tag{6.39}$$

by noting that the terms that do not depend on Q^α or those that are odd in Q^α imply a vanishing integral when the integrations in the spherical angles of Q^α are performed (see (5.65) and (5.66)). If we integrate (6.39) in the spherical angles of Q^α and in d^3P/P^0 by using (5.231) through (5.233), we get the final form of the production term for a relativistic gas where the interaction potential between the relativistic particles is arbitrary:

$$P^{\alpha\beta} = B_1\left(\eta^{\alpha\beta} - \frac{4}{c^2}U^\alpha U^\beta\right)\varpi + \frac{B_2}{c^2}(U^\alpha q^\beta + U^\beta q^\alpha) + B_3 p^{\langle\alpha\beta\rangle}. \tag{6.40}$$

The scalar coefficients B_1 through B_3 are given by

$$B_1 = -\frac{1}{3m^2c^5}\frac{\zeta}{p}\frac{(1 - \zeta^2 - 5\zeta G + \zeta^2G^2)I_1}{(20G + 3\zeta - 13G^2\zeta - 2G\zeta^2 + 2G^3\zeta^2)}, \tag{6.41}$$

$$B_2 = \frac{1}{3m^2c^5}\frac{\zeta}{p}\frac{I_1 - c^2I_2}{\zeta + 5G - \zeta G^2}, \tag{6.42}$$

$$B_3 = \frac{1}{30m^2c^5}\frac{\zeta}{p}\frac{2I_1 - 6c^2I_2 + 3c^4I_3}{G}. \tag{6.43}$$

The integrals I_1, I_2 and I_3 above are given by (5.80), (5.83) and (5.84), respectively. For a relativistic gas of hard-sphere particles we combine (5.85), (5.87) and (5.88) together with (6.41) through (6.43), yielding

$$B_1 = \frac{64\pi p\sigma}{3cK_2(\zeta)^2\zeta^3}\frac{(1-\zeta^2-5\zeta G+\zeta^2G^2)\left[2K_2(2\zeta)+\zeta K_3(2\zeta)\right]}{(20G+3\zeta-13G^2\zeta-2G\zeta^2+2G^3\zeta^2)}, \tag{6.44}$$

$$B_2 = -\frac{64\pi p\sigma}{3cK_2(\zeta)^2\zeta^3}\frac{\left[(2+\zeta^2)K_2(2\zeta)+5\zeta K_3(2\zeta)\right]}{\zeta+5G-\zeta G^2}, \tag{6.45}$$

$$B_3 = -\frac{64\pi p\sigma}{15cK_2(\zeta)^2\zeta^3G}\left[(2+15\zeta^2)K_2(2\zeta)+(3\zeta^3+49\zeta)K_3(2\zeta)\right]. \tag{6.46}$$

Hence (6.33) and (6.40) are the representations of the third-order moment $T^{\alpha\beta\gamma}$ and of the production term $P^{\alpha\beta}$ as functions of the fourteen fields n, T, U^α, ϖ, $p^{\langle\alpha\beta\rangle}$ and q^α.

Problems

6.3.1 Show that the scalar coefficients of the third-order moment (6.33) are given by (6.34) through (6.36).

6.3.2 Let $\mathcal{X} = \mathcal{X}(Q)$ be a scalar function of the relative momentum four-vector Q^α. Show that

$$\int\left(\frac{P^\alpha P^\beta}{P^2} - \eta^{\alpha\beta} - 3\frac{Q^\alpha Q^\beta}{Q^2}\right)\mathcal{X}(Q)d^3Q = 0.$$

6.3.3 Show that the constitutive equation for the production term as function of the fourteen fields is (6.40) where the scalar coefficients are given by (6.41) through (6.43).

6.3.4 Show that the scalar coefficients of the production term for a relativistic gas of hard-sphere particles are given by (6.44) through (6.46).

6.4 Linearized field equations

We shall now obtain the linearized field equations for the fourteen fields n, T, U^α, ϖ, $p^{\langle\alpha\beta\rangle}$, q^α and we begin by writing the linearized forms of the balance equations for the particle number density (4.38), momentum density (4.43) and energy density (4.41) as

$$Dn + n\nabla^\alpha U_\alpha = 0, \tag{6.47}$$

$$\frac{nh_E}{c^2}DU^\alpha = \nabla^\alpha(p+\varpi) - \nabla_\beta p^{\langle\alpha\beta\rangle} - \frac{1}{c^2}Dq^\alpha, \tag{6.48}$$

$$nDe = -p\nabla_\beta U^\beta - \nabla_\beta q^\beta. \tag{6.49}$$

In the above equations we have neglected all products of ϖ, $p^{\langle\alpha\beta\rangle}$ and q^α with gradients ∇^α and with the convective time derivative D.

Next we insert the constitutive equations for $T^{\alpha\beta\gamma}$ and $P^{\alpha\beta}$, given by (6.33) and (6.40), into the balance equation for the third-order moment $(6.1)_3$. By neglecting all non-linear terms, we get that the product of the resulting equation with $U_\alpha U_\beta$, $\Delta^\gamma_\alpha U_\beta$ and $\Delta^\gamma_\alpha\Delta^\delta_\beta - \Delta_{\alpha\beta}\Delta^{\gamma\delta}/3$, yield

$$\frac{C_2}{2}D\varpi + \frac{1}{2}(m^2+C_1)Dn - \frac{\zeta}{2T}nC_1'DT - 5\frac{C_3}{c^2}\nabla^\alpha q_\alpha$$
$$+\frac{1}{6}(nm^2+5nC_1)\nabla^\alpha U_\alpha = \frac{1}{c^4}P^{\alpha\beta}U_\alpha U_\beta = -\frac{3}{c^2}B_1\varpi, \tag{6.50}$$

$$5C_3Dq^\alpha - \frac{c^4}{6}\left[(m^2-C_1)\nabla^\alpha n + \frac{\zeta}{T}nC_1'\nabla^\alpha T - C_2\nabla^\alpha\varpi\right] - c^2C_4\nabla_\beta p^{\langle\alpha\beta\rangle}$$
$$-\frac{c^2}{6}(nm^2+5nC_1)DU^\alpha = -P^{\beta\gamma}U_\beta\Delta^\alpha_\gamma = -B_2q^\alpha, \tag{6.51}$$

$$C_4Dp^{\langle\alpha\beta\rangle} + 2C_3\nabla^{\langle\alpha}q^{\beta\rangle} + \frac{c^2}{3}(nm^2-nC_1)\nabla^{\langle\alpha}U^{\beta\rangle}$$
$$= \left(\Delta^\alpha_\gamma\Delta^\beta_\delta - \frac{1}{3}\Delta^{\alpha\beta}\Delta_{\gamma\delta}\right)P^{\langle\gamma\delta\rangle} = B_3p^{\langle\alpha\beta\rangle}, \tag{6.52}$$

respectively. The prime now denotes the differentiation with respect to ζ.

Equations (6.47) through (6.52) are the linearized field equations for the fourteen fields of n, T, U^α, ϖ, $p^{\langle\alpha\beta\rangle}$, q^α and in Chapter 9 we shall find a solution of this system of equations that corresponds to the problem of propagation of harmonic waves of small amplitude.

Problem

6.4.1 Obtain (6.50) through (6.52) from the balance equation of the third-order moment written in terms of the fourteen fields by performing the projections $U_\alpha U_\beta$, $\Delta^\gamma_\alpha U_\beta$ and $\Delta^\gamma_\alpha\Delta^\delta_\beta - \Delta_{\alpha\beta}\Delta^{\gamma\delta}/3$.

6.5 Five-field theory

The five-field theory is characterized by the fields of particle number density n, four-velocity U^α and temperature T, and their balance equations are given by (4.38), (4.43) and (4.41), respectively. The corresponding field equations are obtained from the balance equations once the constitutive equations for the dynamic pressure ϖ, pressure deviator $p^{\langle\alpha\beta\rangle}$ and heat flux q^α are known functions of n, U^α and T. In order to get such constitutive equations we shall use the remaining equations of the fourteen-field theory – (6.50) through (6.52) – and a method akin to the Maxwellian iteration procedure used in the classical case [8].

6.5.1 Laws of Navier–Stokes and Fourier

First we eliminate the material time derivatives Dn, DU^α and DT from (6.50) and (6.51) by the use of (6.47) through (6.49) and get

$$\frac{C_2}{2}D\varpi - \frac{n}{2}(m^2+C_1)\nabla^\alpha U_\alpha + \frac{\zeta}{2kT}\frac{C_1'}{\zeta^2+5\zeta G-\zeta^2G^2-1}(\nabla^\alpha q_\alpha + p\nabla^\alpha U_\alpha)$$

$$-5\frac{C_3}{c^2}\nabla^\alpha q_\alpha + \frac{1}{6}(nm^2+5nC_1)\nabla^\alpha U_\alpha = -\frac{3}{c^2}B_1\varpi, \tag{6.53}$$

$$5C_3Dq^\alpha - \frac{c^4}{6}\left[(m^2-C_1)\nabla^\alpha n + \frac{\zeta}{T}nC_1'\nabla^\alpha T - C_2\nabla^\alpha\varpi\right] - c^2C_4\nabla_\beta p^{\langle\alpha\beta\rangle}$$

$$-\frac{c^4}{6nh_E}(nm^2+5nC_1)\left[\nabla^\alpha(p+\varpi) - \nabla_\beta p^{\langle\alpha\beta\rangle} - \frac{1}{c^2}Dq^\alpha\right] = -B_2q^\alpha, \tag{6.54}$$

by noting that $De = c_vDT$ where $c_v = k(\zeta^2+5G\zeta-\zeta^2G^2-1)$ is the heat capacity per particle at constant volume. Next we insert on the left-hand side of (6.52) through (6.54) the equilibrium values $\varpi^{[0]}=0$, $q^{\alpha[0]}=0$ and $p^{\langle\alpha\beta\rangle[0]}=0$ and then get on the right-hand side the first values of the iterates for the dynamic pressure $\varpi^{[1]}$, heat flux $q^{\alpha[1]}$ and pressure deviator $p^{\langle\alpha\beta\rangle[1]}$ that correspond to the constitutive equations for a viscous and heat conducting relativistic gas[1]:

$$\varpi^{[1]} = -\eta\nabla_\alpha U^\alpha, \quad p^{\langle\alpha\beta\rangle[1]} = 2\mu\nabla^{\langle\alpha}U^{\beta\rangle}, \quad q^{\alpha[1]} = \lambda\left(\nabla^\alpha T - \frac{T}{nh}\nabla^\alpha p\right). \tag{6.55}$$

The transport coefficients of bulk viscosity η, thermal conductivity λ and shear viscosity μ have the same values as those obtained by the use of the Chapman–Enskog method in a simplified form namely (5.40), (5.43) and (5.46), respectively.

Problems

6.5.1.1 Show that the elimination of the material time derivatives Dn, DU^α and DT from (6.50) and (6.51) by the use of (6.47) through (6.49) lead to (6.53) and (6.54).

6.5.1.2 Obtain the constitutive relations (6.55) and show that the transport coefficients are the same as those found by the use of the method of Chapman–Enskog (5.40), (5.43) and (5.46).

[1] Here the indexes [0], [1], [2], etc. denote the order of the iteration.

6.5.2 Linearized Burnett equations

The second iterate for the dynamic pressure is obtained by inserting the first iterates (6.55) on the left-hand side of (6.53), linearization and elimination of the convective time derivative DU^α by using (6.48), yielding

$$\varpi^{[2]} = -\eta\nabla_\alpha U^\alpha + \eta_1\nabla_\alpha\nabla^\alpha T - \eta_2\nabla_\alpha\nabla^\alpha p, \tag{6.56}$$

where the coefficients η_1 and η_2 are given by:

$$\eta_1 = -\frac{\eta\lambda}{p}\frac{(\zeta+6G-G^2\zeta)(5-10G\zeta+2G^2\zeta^2-2\zeta^2)}{\zeta(\zeta+5G-G^2\zeta)(20G+3\zeta-13\zeta G^2-2\zeta^2 G+2\zeta^2 G^3)}, \tag{6.57}$$

$$\eta_2 = \frac{\eta_1 T}{p\zeta G} - \frac{3\eta^2 T c_v}{p^2 m G\zeta}\frac{2\zeta^3-5\zeta+(19\zeta^2-30)G-(2\zeta^3-45\zeta)G^2-9\zeta^2G^3}{(20G+3\zeta-13\zeta G^2-2\zeta^2 G+2\zeta^2 G^3)^2}. \tag{6.58}$$

If we consider a differential cross-section of hard-sphere particles, we obtain that the coefficients η_1 and η_2 in the limiting case of low temperatures where $\zeta \gg 1$ can be written as

$$\eta_1 = \frac{375}{16384}\frac{k}{n\pi\sigma^2}\frac{1}{\zeta}\left[1-\frac{45}{8}\frac{1}{\zeta}+\frac{3069}{128}\frac{1}{\zeta^2}-\frac{100525}{1024}\frac{1}{\zeta^3}+\cdots\right], \tag{6.59}$$

$$\eta_2 = \frac{375}{32768}\frac{1}{n^2\pi\sigma^2}\frac{1}{\zeta^2}\left[1-\frac{55}{8}\frac{1}{\zeta}+\frac{5413}{128}\frac{1}{\zeta^2}-\frac{315491}{1024}\frac{1}{\zeta^3}+\cdots\right], \tag{6.60}$$

which correspond to their expressions in the non-relativistic limit.

The expressions for η_1 and η_2 in the case of high temperatures where $\zeta \ll 1$, which corresponds to the ultra-relativistic limiting case, read

$$\eta_1 = \frac{1}{96}\frac{k}{n\pi^2\sigma^2}\zeta^2\left[1+\left(\frac{15}{8}+3\gamma+3\ln\frac{\zeta}{2}\right)\zeta^2+\cdots\right], \tag{6.61}$$

$$\eta_2 = \frac{1}{384}\frac{1}{n^2\pi^2\sigma^2}\zeta^2\left[1+\left(\frac{3}{2}+3\gamma+3\ln\frac{\zeta}{2}\right)\zeta^2+\cdots\right]. \tag{6.62}$$

In Table 6.1 we give the values of the dimensionless coefficients of the dynamic pressure, namely

$$\eta^D = \frac{\eta c\sigma}{kT}, \qquad \eta_1^D = \frac{\eta_1\sigma^2 n}{k}, \qquad \eta_2^D = \eta_2 n^2\sigma^2. \tag{6.63}$$

By inspecting Table 6.1 we conclude that the three coefficients are very small in the two limiting cases of small and high temperatures, i.e., the dynamic pressure vanishes in these two limiting cases. Further we have that the coefficient η_1^D is larger than η^D and η_2^D. This means that the coefficient of the term with an inhomogeneous temperature field – expressed by the Laplacian of the temperature – is the leading one, i.e., it is larger than that related with the divergence of the

ζ	η^D	η_1^D	η_2^D
0.01	1.10309×10^{-11}	1.05438×10^{-7}	2.63586×10^{-8}
0.1	9.90485×10^{-8}	9.97525×10^{-6}	2.48483×10^{-6}
1	1.49562×10^{-4}	3.56097×10^{-4}	7.34398×10^{-5}
10	6.28594×10^{-4}	4.42699×10^{-4}	2.03935×10^{-5}
100	4.55297×10^{-5}	6.63103×10^{-5}	3.33169×10^{-7}

Table 6.1: η^D, η_1^D and η_2^D as functions of ζ

bulk velocity, which is the coefficient of bulk viscosity. This was first pointed out in [11].

To obtain the second iterate for the heat flux we insert the first iterates (6.55) on the left-hand side of (6.54), linearize the resulting equation and eliminate the convective time derivatives Dn and DT by using (6.47) and (6.49) respectively, yielding

$$q^{\alpha[2]} = \lambda\left(\nabla^\alpha T - \frac{T}{nh_E}\nabla^\alpha p\right) + \lambda_1 \nabla^\alpha \nabla_\beta U^\beta - \lambda_2 \nabla_\beta \nabla^{\langle\alpha} U^{\beta\rangle}. \tag{6.64}$$

The coefficients λ_1 and λ_2 read

$$\lambda_1 = \frac{\eta\lambda T}{p}\left[\frac{2\zeta^3 - 5\zeta + (19\zeta^2 - 30)G - (2\zeta^3 - 45\zeta)G^2 - 9\zeta^2 G^3}{\zeta(\zeta + 5G - G^2\zeta)(20G + 3\zeta - 13G^2\zeta - 2G\zeta^2 + 2G^3\zeta^2)}\right.$$

$$\left. + \frac{(\zeta + 5G)}{\zeta(\zeta + 5G - G^2\zeta)G}\right] + \frac{m\lambda^2 T}{pk\zeta^2 G^2}\frac{(\zeta^2 + 5G\zeta - G^2\zeta^2 - 5G^2)(\zeta + 4G - G^2\zeta)}{(\zeta + 5G - G^2\zeta)^2(\zeta^2 + 5G\zeta - G^2\zeta^2 - 1)}, \tag{6.65}$$

$$\lambda_2 = \frac{2\mu\lambda T}{p}\frac{1}{\zeta(\zeta + 5G - G^2\zeta)}. \tag{6.66}$$

For hard-sphere particles at low temperatures we have that $\zeta \gg 1$ and the coefficients λ_1 and λ_2 in the non-relativistic limiting case reduce to

$$\lambda_1 = \frac{375}{16384}\frac{kT}{n\pi\sigma^2}\left[1 + \frac{5}{8}\frac{1}{\zeta} - \frac{1723}{128}\frac{1}{\zeta^2} + \frac{51777}{1024}\frac{1}{\zeta^3} + \cdots\right], \tag{6.67}$$

$$\lambda_2 = \frac{75}{4096}\frac{kT}{n\pi\sigma^2}\left[1 + \frac{7}{8}\frac{1}{\zeta} - \frac{689}{384}\frac{1}{\zeta^2} + \frac{9301}{3072}\frac{1}{\zeta^3} + \cdots\right]. \tag{6.68}$$

Their expressions for high temperatures, i.e., in the ultra-relativistic limiting case where $\zeta \ll 1$ are

$$\lambda_1 = \frac{7}{192}\frac{kT}{n\pi^2\sigma^2}\zeta^2\left[1 + \left(\frac{53}{84} + \frac{11}{7}\gamma + \frac{11}{7}\ln\frac{\zeta}{2}\right)\zeta^2 + \cdots\right], \tag{6.69}$$

ζ	λ^D	λ_1^D	λ_2^D
0.01	1.59150×10^{-1}	3.69187×10^{-7}	7.59903×10^{-3}
0.1	1.58768×10^{-1}	3.57532×10^{-5}	7.59358×10^{-3}
1	1.38658×10^{-1}	1.83716×10^{-3}	7.32279×10^{-3}
10	5.48613×10^{-2}	7.05420×10^{-3}	6.24937×10^{-3}
100	1.66572×10^{-2}	7.31853×10^{-3}	5.87841×10^{-3}

Table 6.2: λ^D, λ_1^D and λ_2^D as functions of ζ

$$\lambda_2 = \frac{3}{40}\frac{kT}{n\pi^2\sigma^2}\left[1-\frac{3}{40}\zeta^2+\cdots\right]. \tag{6.70}$$

The values of the dimensionless coefficients of the heat flux

$$\lambda^D = \frac{\lambda\sigma}{ck}, \qquad \lambda_1^D = \frac{\lambda_1\sigma^2 n}{kT}, \qquad \lambda_2^D = \frac{\lambda_2 n\sigma^2}{kT}, \tag{6.71}$$

are given in Table 6.2. We note that the dimensionless coefficients λ_1^D and λ_2^D are smaller than the dimensionless coefficient of thermal conductivity λ^D in all cases, i.e., in the non-relativistic, relativistic and ultra-relativistic cases.

The same methodology above is applied to obtain the second iterate for the pressure deviator, i.e., we insert the first iterates (6.55) into the left-hand side of (6.52) and get by considering only linear terms and by eliminating the convective time derivative DU^α by the use of (6.48):

$$p^{\langle\alpha\beta\rangle[2]} = 2\mu\nabla^{\langle\alpha}U^{\beta\rangle} + \mu_1\nabla^{\langle\alpha}\nabla^{\beta\rangle}T - \mu_2\nabla^{\langle\alpha}\nabla^{\beta\rangle}p. \tag{6.72}$$

The coefficients μ_1 and μ_2 are given by

$$\mu_1 = \frac{2\mu\lambda}{p}\frac{\zeta+6G-G^2\zeta}{\zeta G(\zeta+5G-G^2\zeta)}, \tag{6.73}$$

$$\mu_2 = \frac{2T}{p^2}\left[\lambda\mu\frac{\zeta+6G-G^2\zeta}{\zeta^2G^2(\zeta+5G-G^2\zeta)} + \frac{k}{m}\mu^2\frac{(\zeta+6G)}{\zeta G^3}\right]. \tag{6.74}$$

The expressions of the coefficients μ_1 and μ_2 for hard-sphere particles at low temperatures reduce to

$$\mu_1 = \frac{75}{4096}\frac{k}{n\pi\sigma^2}\left[1+\frac{27}{8}\frac{1}{\zeta}-\frac{809}{384}\frac{1}{\zeta^2}-\frac{7893}{1024}\frac{1}{\zeta^3}+\cdots\right], \tag{6.75}$$

$$\mu_2 = \frac{25}{2048}\frac{1}{n^2\pi\sigma^2}\left[1+\frac{25}{8}\frac{1}{\zeta}+\frac{215}{384}\frac{1}{\zeta^2}-\frac{40321}{3072}\frac{1}{\zeta^3}+\cdots\right], \tag{6.76}$$

which correspond to the non-relativistic limiting case where $\zeta \gg 1$.

Their expressions for high temperatures, i.e., in the ultra-relativistic limiting case where $\zeta \ll 1$, are

$$\mu_1 = \frac{3}{20}\frac{k}{n\pi^2\sigma^2}\left[1-\frac{1}{5}\zeta^2+\cdots\right], \tag{6.77}$$

ζ	μ^D	μ_1^D	μ_2^D
0.01	9.54934×10^{-2}	1.51978×10^{-2}	1.06385×10^{-2}
0.1	9.55402×10^{-2}	1.51686×10^{-2}	1.06193×10^{-2}
1	9.90606×10^{-2}	1.36084×10^{-2}	9.51384×10^{-3}
10	1.60806×10^{-1}	7.64197×10^{-3}	5.07907×10^{-3}
100	4.47651×10^{-1}	6.02388×10^{-3}	4.00721×10^{-3}

Table 6.3: μ^D, μ_1^D and μ_2^D as functions of ζ

$$\mu_2 = \frac{21}{200}\frac{1}{n^2\pi^2\sigma^2}\left[1 - \frac{13}{70}\zeta^2 + \cdots\right]. \tag{6.78}$$

In Table 6.3 we give the values of the dimensionless coefficients of the pressure deviator:

$$\mu^D = \frac{\mu c\sigma}{kT}, \qquad \mu_1^D = \frac{\mu_1\sigma^2 n}{k}, \qquad \mu_2^D = \mu_2 n^2\sigma^2. \tag{6.79}$$

We conclude from this table that the dimensionless coefficients μ_1^D and μ_2^D are of the same order but are smaller than the dimensionless coefficient of shear viscosity μ^D in the three cases namely non-relativistic, relativistic and ultra-relativistic limiting cases.

Equations (6.56), (6.64) and (6.72) are the linearized relativistic Burnett equations and their expressions in the non-relativistic limiting case read:

$$\varpi^{[2]} = 0, \tag{6.80}$$

$$q^{i[2]} = -\lambda\delta^{ij}\frac{\partial T}{\partial x^j} - \frac{15\mu^2}{4nm}\delta^{ij}\frac{\partial^2 v^k}{\partial x^j x^k} + \frac{3\mu^2}{mn}\frac{\partial D^{\langle ij\rangle}}{\partial x^j}, \tag{6.81}$$

$$p^{\langle ij\rangle[2]} = -2\mu D^{\langle ij\rangle} - \frac{2\mu^2}{pmn}\frac{\partial^2 p}{\partial x_{\langle i}\partial x_{j\rangle}} + \frac{3\mu^2}{Tmn}\frac{\partial^2 T}{\partial x_{\langle i}\partial x_{j\rangle}}, \tag{6.82}$$

if we neglect all terms of order up to $1/\zeta$. Above we have introduced the deformation rate (or stretching) tensor

$$D^{ij} = \frac{1}{2}\left(\frac{\partial v^i}{\partial x_j} + \frac{\partial v^j}{\partial x_i}\right), \qquad \text{with} \qquad D^{\langle ij\rangle} = D^{ij} - \frac{1}{3}\delta_{rs}D^{rs}\delta^{ij}. \tag{6.83}$$

Equations (6.81) and (6.82) correspond to the linearized Burnett equations of Wang-Chang and Uhlenbeck [16].

Problems

6.5.2.1 Obtain the linearized relativistic Burnett equations given by (6.56), (6.64) and (6.72).

6.5.2.2 Obtain the non-relativistic and the ultra-relativistic expressions for the coefficients $\eta_1, \eta_2, \lambda_1, \lambda_2, \mu_1$ and μ_2.

6.5.2.3 Show that (6.56), (6.64) and (6.72) reduce to (6.80) through (6.82) in the non-relativistic limiting case. (Hint: Use the results of problem 4.5.4.)

6.6 Maxwellian particles

In 1867 Maxwell [12] presented a model in which the interparticle force of a non-relativistic gas was inversely proportional to the fifth power of the relative distance between the particles. The particles that obey this kind of interparticle force are known in the literature as Maxwellian particles [4, 3]. The advantage of this potential function is that the knowledge of the distribution function is not required to proceed with the integration of the production term. According to Maxwell: ... *It will be shewn that we have reason from experiments on the viscosity of gases to believe that* $n = 5$. *In this case* V *will disappear from the expressions of the form (3) and they will be capable of immediate integration with respect to* dN_1 *and* dN_2. ... *If, however,* n *is not equal to 5, so that* V *does not disappear, we should require to know the form of the function* f_2 *before we could proceed further with the integration.* ...[2]

We are going to determine the relativistic Maxwellian cross-section by analyzing the production term $P^{\alpha\beta}$ of the third-order moment $T^{\alpha\beta\gamma}$. The production term $P^{\alpha\beta}$, given by (6.38), could be written thanks to (5.55), (5.56) and (5.59) as

$$\begin{aligned} P^{\alpha\beta} &= \frac{c\pi}{4}\int\left(\frac{P^\alpha P^\beta}{P^2} - \eta^{\alpha\beta} - 3\frac{Q^\alpha Q^\beta}{Q^2}\right) Q^2 f f_* F\sigma \sin^3\Theta d\Theta d\Phi \frac{d^3p_*}{p_{*0}}\frac{d^3p}{p_0} \\ &= c\pi\int\frac{1}{4m^2c^2+Q^2}\Big[p^\gamma p_{*\gamma}(2p_*^\alpha p^\beta + 2p^\alpha p_*^\beta - p_*^\alpha p_*^\beta - p^\alpha p^\beta) \\ &\quad + m^2c^2(p_*^\alpha p^\beta + p^\alpha p_*^\beta - 2p_*^\alpha p_*^\beta - 2p^\alpha p^\beta) \\ &\quad + \eta^{\alpha\beta}(m^4c^4 - p^\gamma p^\delta p_{*\gamma} p_{*\delta})\Big] f_* f F\sigma\sin^3\Theta d\Theta\frac{d^3p_*}{p_{*0}}\frac{d^3p}{p_0}. \end{aligned} \tag{6.84}$$

To obtain the last equality we have used (5.47) and (5.48).

We note from (6.84) that the choice [10]

$$4c\sigma F = (4m^2c^2 + Q^2)\mathcal{F}(\Theta), \tag{6.85}$$

where $\mathcal{F}(\Theta)$ is an arbitrary function of the scattering angle Θ, leads to a complete determination of the production term as a function of the moments without the

[2] V is the relative velocity, the equation (3) corresponds to the collision term and f_2 is the one-particle distribution function.

knowledge of the form of the one-particle distribution function. Indeed we obtain from (6.84) by adopting (6.85)

$$P^{\alpha\beta} = \frac{\pi\mathcal{B}}{4c^2}\Bigg[2m^2c^2N^\alpha N^\beta + \eta^{\alpha\beta}(T^\gamma_\gamma T^\delta_\delta - T^{\gamma\delta}T_{\gamma\delta})$$

$$-2T^{\alpha\beta\gamma}N_\gamma + 4T^\alpha_\gamma T^{\gamma\beta} - 4T^{\alpha\beta}T^\gamma_\gamma\Bigg], \tag{6.86}$$

where $\mathcal{B}$ is defined in (5.105).

Since $F = PQ/2$ one can write (6.85) as

$$\sigma = \frac{m}{Q}\left(1 + \frac{Q^2}{4m^2c^2}\right)^{\frac{1}{2}}\mathcal{F}(\Theta). \tag{6.87}$$

The above equation differs from the differential cross-section of the Israel particles (5.101) and from that of Polak et al. [13] which is given by

$$\sigma = \frac{m}{Q}\mathcal{F}(\Theta), \tag{6.88}$$

although all of them – i.e., (5.101), (6.85) and (6.88) – tend to the differential cross-section of Maxwellian particles in the non-relativistic limiting case, which is inversely proportional to the relative velocity.

Now we shall determine the transport coefficients for gases of relativistic particles whose differential cross-section is given by (6.85). Although we have evaluated the collision term appearing in the moment equations in terms of moments without specifying the distribution function as in the non-relativistic case, we cannot, at variance with the latter case, consider our work completed, because one of these moments is of third order, i.e., of higher order with respect to those that we set out to determine. This is a weak point of the concept of the Maxwellian cross-section in the case of a relativistic gas. To determine the moments of second order, we need to express the third-order tensor $T^{\alpha\beta\gamma}$ in terms of lower order tensors. For that purpose we shall use the fourteen balance equations (6.1) and close the fourteen equations by expressing the third-order moment $T^{\alpha\beta\gamma}$ in terms of the fourteen fields $n, T, U^\alpha, \varpi, p^{\langle\alpha\beta\rangle}$ and q^α as stated in (6.33). In this case the coefficients B_1 through B_3 of the linearized production term (6.40) read

$$B_1 = \frac{p\pi\mathcal{B}}{c^2}\frac{(\zeta + 5G - \zeta G^2)(1 - \zeta^2 - 5\zeta G + \zeta^2G^2)}{(3\zeta + 20G - 2\zeta^2 G - 13\zeta G^2 + 2\zeta^2 G^3)}, \tag{6.89}$$

$$B_2 = -\frac{p\pi\mathcal{B}}{2c^2}\frac{\zeta + 10G - \zeta G^2}{\zeta + 5G - \zeta G^2}, \tag{6.90}$$

$$B_3 = -\frac{p\pi\mathcal{B}}{2c^2}\frac{2G + \zeta + 2\zeta G^2}{G}. \tag{6.91}$$

The transport coefficients are obtained by applying the same methodology we have described in Section 6.5.1, i.e., by using the Maxwellian iteration procedure. Here their expressions are

$$\eta = \frac{kT}{3\pi\mathcal{B}} \frac{\zeta(20G + 3\zeta - 13G^2\zeta - 2G\zeta^2 + 2G^3\zeta^2)^2}{(1 - 5G\zeta - \zeta^2 + G^2\zeta^2)^2(\zeta + 5G - \zeta G^2)}, \tag{6.92}$$

$$\mu = \frac{2kT}{\pi\mathcal{B}} \frac{G^2\zeta}{2G + \zeta + 2\zeta G^2}, \tag{6.93}$$

$$\lambda = \frac{2kc^2}{\pi\mathcal{B}} \frac{\zeta(\zeta + 5G - G^2\zeta)^2}{\zeta + 10G - \zeta G^2}, \tag{6.94}$$

which represent the transport coefficients of bulk viscosity, shear viscosity and thermal conductivity for a relativistic gas whose particles have differential cross-section given by (6.85).

In the non-relativistic limiting case where $\zeta \gg 1$ they reduce to

$$\eta = \frac{5kT}{6\pi\mathcal{B}} \frac{1}{\zeta^2} \left[1 - \frac{25}{2}\frac{1}{\zeta} + \frac{93}{\zeta^2} + \cdots\right], \tag{6.95}$$

$$\mu = \frac{2kT}{3\pi\mathcal{B}} \left[1 + \frac{1}{\zeta} - \frac{7}{3}\frac{1}{\zeta^2} + \cdots\right], \tag{6.96}$$

$$\lambda = \frac{5k^2T}{2m\pi\mathcal{B}} \left[1 - \frac{39}{8}\frac{1}{\zeta^2} + \cdots\right]. \tag{6.97}$$

We note that if we consider only the first term in the above expansions we could infer that: a) the coefficient of bulk viscosity is small since it is proportional to $1/\zeta^2$; b) the coefficients of shear viscosity and thermal conductivity, as in the case of the Israel particles, are proportional to the temperature; c) the ratio $\lambda/\mu = 15k/(4m)$ is the same one as in all models of cross-section discussed previously.

In the ultra-relativistic limiting case where $\zeta \ll 1$ the expressions for the transport coefficients are

$$\eta = \frac{kT}{108\pi\mathcal{B}}\zeta^4 \left[1 + \left(\frac{95}{24} + 6\gamma + 6\ln\frac{\zeta}{2}\right)\zeta^2 + \cdots\right], \tag{6.98}$$

$$\mu = \frac{4kT}{5\pi\mathcal{B}} \left[1 + \frac{1}{320}\left(1 + 4\gamma + 4\ln\frac{\zeta}{2}\right)\zeta^4 + \cdots\right], \tag{6.99}$$

$$\lambda = \frac{4kc^2}{3\pi\mathcal{B}} \left[1 - \frac{1}{3}\zeta^2 + \ldots\right]. \tag{6.100}$$

Again by considering only the first term in the expansions we conclude that: a) the coefficient of bulk viscosity is small since it depends on ζ^4; b) unlike the case of the Israel particles the leading term of the shear viscosity and of the thermal conductivity do not depend on ζ; c) the ratio $\lambda/\mu = 5c^2/(3T)$ agrees with that found for hard-sphere particles.

Problems

6.6.1 For the cross-section (6.85) obtain that the production term is given by (6.86).

6.6.2 Show that $F = PQ/2$.

6.6.3 Obtain the expressions (6.92) through (6.94) for the transport coefficients, and show that their limiting values are given by (6.95) through (6.100).

6.7 Combined method of Chapman–Enskog and Grad

We shall now determine the constitutive equations for the dynamic pressure ϖ, pressure deviator $p^{\langle\alpha\beta\rangle}$ and heat flux q^α by using a method that combines the features of the methods of Chapman–Enskog and Grad, but neither a solution of the integral equation is needed – as in the method of Chapman–Enskog – nor the field equations for the moments are used – as in the method of Grad. This is the so-called combined Chapman–Enskog and Grad method [2, 10].

To begin with we note that Grad distribution function (6.30) can be written as

$$f = f^{(0)}(1+\phi), \tag{6.101}$$

which have a form similar to the second approximation of the one-particle distribution function in the Chapman–Enskog method. In (6.101) ϕ depends on the non-equilibrium quantities: dynamic pressure ϖ, pressure deviator $p^{\langle\alpha\beta\rangle}$ and heat flux q^α.

Further if we apply the same methodology described in Section 5.2, we obtain equation (5.23), which for the Grad distribution function reads

$$f^{(0)}\Bigg\{-\frac{k^2T}{c^2c_v}\left[\frac{1}{3}\zeta^2\frac{c_v}{k}-(G^2\zeta^2-4G\zeta-\zeta^2)\left(\frac{U^\beta p_\beta}{kT}\right)\right.$$

$$\left.-\frac{1}{3}(\zeta^2+5G\zeta-\zeta^2G^2-4)\left(\frac{U^\beta p_\beta}{kT}\right)^2\right]\nabla_\alpha U^\alpha-\frac{p_\alpha p_\beta}{kT}\nabla^{\langle\alpha}U^{\beta\rangle}$$

$$+\frac{p_\alpha}{kT^2}(p_\beta U^\beta-h_E)\left(\nabla^\alpha T-\frac{T}{nh_E}\nabla^\alpha p\right)\Bigg\}=\mathcal{I}[\phi]$$

$$=\int f^{(0)}f_*^{(0)}\left[p_*'^\alpha p_*'^\beta+p'^\alpha p'^\beta-p_*^\alpha p_*^\beta-p^\alpha p^\beta\right]F\sigma d\Omega\frac{d^3p_*}{p_{*0}}$$

$$\times\Bigg\{\frac{\varpi}{p}\frac{(1-5G\zeta-\zeta^2+G^2\zeta^2)\zeta}{20G+3\zeta-13G^2\zeta-2G\zeta^2+2G^3\zeta^2}\frac{1}{m^2c^4}U_\alpha U_\beta$$

$$-\frac{q_\alpha}{p}\frac{\zeta}{\zeta+5G-G^2\zeta}\frac{U_\beta}{m^2c^4}+\frac{p_{\langle\alpha\beta\rangle}}{p}\frac{\zeta}{2G}\frac{1}{m^2c^2}\Bigg\}. \tag{6.102}$$

We stress here that (6.102) is not an integral equation for ϕ since ϕ is a known function of dynamic pressure ϖ, pressure deviator $p^{\langle\alpha\beta\rangle}$ and heat flux q^α. However, (6.102) is not an exact equation in momentum space, because we assumed a fixed dependence of f upon p^α and there is no reason to expect that there is such a solution for the integral equation (5.23) in ϕ that we obtained when we did not specify this dependence. Thus we can only satisfy a few moment equations obtained from (6.102). But one can expand the one-particle distribution function in an infinite sum of moments [2] and obtain the same results as those that follows by using the Chapman–Enskog method.

Next we multiply (6.102) by $p_\epsilon p_\lambda$ and integrate the resulting equation over all values of d^3p/p_0 and get by using the integrals (5.226) through (5.228) of the appendix of Chapter 5:

$$\frac{pkT}{c^3}\Bigg\{\frac{20G+3\zeta-13G^2\zeta-2G\zeta^2+2G^3\zeta^2}{3(5G\zeta-G^2\zeta^2+\zeta^2-1)}\left(\zeta\eta_{\epsilon\lambda}-\frac{4m}{kT}U_\epsilon U_\lambda\right)\nabla_\alpha U^\alpha$$

$$-2\zeta G\nabla_{\langle\epsilon}U_{\lambda\rangle}+\zeta(\zeta G^2-5G-\zeta)(\eta_{\alpha\epsilon}U_\lambda+\eta_{\alpha\lambda}U_\epsilon)\left(\nabla^\alpha T-\frac{T}{nh_E}\nabla^\alpha p\right)\Bigg\}$$

$$=\int\mathcal{I}[\phi]p_\epsilon p_\lambda\frac{d^3p}{p_0}$$

$$=I_{\epsilon\lambda\gamma\delta}\Bigg\{\frac{\varpi}{p}\frac{(1-5G\zeta-\zeta^2+G^2\zeta^2)\zeta}{20G+3\zeta-13G^2\zeta-2G\zeta^2+2G^3\zeta^2}\frac{1}{m^2c^4}U^\gamma U^\delta$$

$$-\frac{q^\delta}{p}\frac{\zeta}{\zeta+5G-G^2\zeta}\frac{U^\gamma}{m^2c^4}+\frac{p^{\langle\delta\gamma\rangle}}{p}\frac{\zeta}{2G}\frac{1}{m^2c^2}\Bigg\}, \tag{6.103}$$

where $I_{\epsilon\lambda\gamma\delta}$ denotes the integral

$$I_{\epsilon\lambda\gamma\delta}=\int p_\epsilon p_\lambda\mathcal{I}[p_\gamma p_\delta]\frac{d^3p}{p_0}. \tag{6.104}$$

We can decompose the integral (6.104) as:

$$I_{\epsilon\lambda\gamma\delta}=A_1U_\epsilon U_\lambda U_\gamma U_\delta+A_2(\Delta_{\epsilon\lambda}U_\gamma U_\delta+\Delta_{\gamma\delta}U_\epsilon U_\lambda)+A_3(\Delta_{\epsilon\gamma}U_\lambda U_\delta$$

$$+\Delta_{\epsilon\delta}U_\lambda U_\gamma+\Delta_{\lambda\gamma}U_\epsilon U_\delta+\Delta_{\lambda\delta}U_\gamma U_\epsilon)+A_4(\Delta_{\lambda\delta}\Delta_{\gamma\epsilon}+\Delta_{\lambda\gamma}\Delta_{\epsilon\delta})+A_5\Delta_{\lambda\epsilon}\Delta_{\gamma\delta}, \tag{6.105}$$

where the coefficients A_1 through A_5 can be determined as functions of the integrals I_1, I_2 and I_3 which are given by (5.35), (5.36) and (5.37), respectively. Indeed if we contract (6.105) we get that $I^{\epsilon}_{\epsilon\gamma\delta}=0$ and the projections of this last equation with $U^\gamma U^\delta$ and $\Delta^{\gamma\delta}$ lead respectively to

$$A_1c^6+3A_2c^4=0, \tag{6.106}$$

$$3c^2A_2+6A_4+9A_5=0. \tag{6.107}$$

Further the projections $U^\epsilon U^\lambda U^\gamma U^\delta$, $\eta^{\epsilon\gamma}U^\lambda U^\delta$ and $\eta^{\epsilon\gamma}\eta^{\lambda\delta}$ of $I_{\epsilon\lambda\gamma\delta}$ imply the equations

$$c^8A_1=I_1, \tag{6.108}$$

$$c^6A_1+3c^4A_3=I_2, \tag{6.109}$$

$$c^4A_1+6c^2A_3+12A_4+3A_5=I_3. \tag{6.110}$$

We solve the system of equations (6.106) through (6.110) for A_1 through A_5 and get that the integral (6.105) can be written as

$$I_{\epsilon\lambda\gamma\delta}=\frac{1}{c^8}I_1U_\epsilon U_\lambda U_\gamma U_\delta+\frac{1}{30}\left[\frac{2}{c^4}I_1-\frac{6}{c^2}I_2+3I_3\right](\Delta_{\lambda\delta}\Delta_{\gamma\epsilon}+\Delta_{\lambda\gamma}\Delta_{\epsilon\delta})$$
$$+\frac{1}{3c^4}\left[I_2-\frac{1}{c^2}I_1\right](\Delta_{\epsilon\gamma}U_\lambda U_\delta+\Delta_{\epsilon\delta}U_\lambda U_\gamma+\Delta_{\lambda\gamma}U_\epsilon U_\delta+\Delta_{\lambda\delta}U_\gamma U_\epsilon)$$
$$-\frac{1}{3c^6}I_1(\Delta_{\epsilon\lambda}U_\gamma U_\delta+\Delta_{\gamma\delta}U_\epsilon U_\lambda)+\frac{1}{15}\left[\frac{1}{c^4}I_1+\frac{2}{c^2}I_2-I_3\right]\Delta_{\lambda\epsilon}\Delta_{\gamma\delta}. \tag{6.111}$$

We use now (6.111) and write the right-hand side of (6.103) in the form

$$\int\mathcal{I}[\phi]p_\epsilon p_\lambda\frac{d^3p}{p_0}=-\frac{1}{p}\frac{\zeta}{\zeta+5G-G^2\zeta}\frac{1}{3m^2c^6}\left[I_2-\frac{1}{c^2}I_1\right](q_\epsilon U_\lambda+q^\lambda U_\epsilon)$$
$$+\frac{\varpi}{p}\frac{(1-5G\zeta-\zeta^2+G^2\zeta^2)\zeta}{20G+3\zeta-13G^2\zeta-2G\zeta^2+2G^3\zeta^2}\frac{I_1}{m^2c^4}\left[\frac{1}{c^4}U_\epsilon U_\lambda-\frac{1}{3c^2}\Delta_{\epsilon\lambda}\right]$$
$$+\frac{p_{\langle\epsilon\lambda\rangle}}{p}\frac{\zeta}{G}\frac{1}{30m^2c^2}\left[\frac{2}{c^4}I_1-\frac{6}{c^2}I_2+3I_3\right]. \tag{6.112}$$

Now we are ready to determine the constitutive equations for the dynamic pressure ϖ, pressure deviator $p^{\langle\alpha\beta\rangle}$ and heat flux q^α. First we multiply (6.103) whose right-hand side is given by (6.112) by $U^\epsilon U^\lambda$ (or $\Delta^{\epsilon\lambda}$) and get

$$\varpi=-\eta\nabla_\alpha U^\alpha. \tag{6.113}$$

Further by multiplying (6.103) by $\Delta^\epsilon_\theta U^\lambda$ it follows that

$$q^\alpha=\lambda\left(\nabla^\alpha T-\frac{T}{nh}\nabla^\alpha p\right). \tag{6.114}$$

Finally we multiply (6.103) by $\Delta_\theta^{(\epsilon}\Delta_\phi^{\lambda)} - \frac{1}{3}\Delta^{\epsilon\lambda}\Delta_{\theta\phi}$ yielding

$$p^{\langle\alpha\beta\rangle} = 2\mu\nabla^{\langle\alpha}U^{\beta\rangle}. \tag{6.115}$$

The transport coefficients which appear in the constitutive equations for the dynamic pressure (6.113), heat flux (6.114) and pressure deviator (6.115) are given by (5.40), (5.43) and (5.46) and correspond to the coefficients of bulk viscosity, thermal conductivity and shear viscosity, respectively.

Problems

6.7.1 Obtain (6.103) by multiplying (6.102) by $p_\epsilon p_\lambda$ and integrating the resulting equation over all values of d^3p/p_0.

6.7.2 Show that the integral $I_{\epsilon\lambda\gamma\delta}$ has the representation which is given in (6.111).

6.7.3 Obtain the laws of Navier–Stokes and Fourier from (6.103) together with (6.112) through the projections $U^\epsilon U^\lambda$ (or $\Delta^{\epsilon\lambda}$); $\Delta_\theta^\epsilon U^\lambda$ and $\Delta_\theta^{(\epsilon}\Delta_\phi^{\lambda)} - \frac{1}{3}\Delta^{\epsilon\lambda}\Delta_{\theta\phi}$.

References

[1] J. L. Anderson, Relativistic Boltzmann theory and Grad method of moments, in *Relativity* pp. 109–124, eds. M. Carmeli, S. I. Fickler and L. Witten (Plenum, New York, 1970).

[2] A. G. Bezerra Jr., S. Reinecke and G. M. Kremer, A combined Chapman–Enskog and Grad method. I. Monatomic gases and mixtures, *Continuum Mech. Thermodyn.* **6**, 149–160 (1994).

[3] C. Cercignani, *The Boltzmann equation and its applications* (Springer, New York, 1988).

[4] S. Chapman and T. G. Cowling, *The mathematical theory of nonuniform gases*, 3rd. ed. (Cambridge Univ. Press, Cambridge, 1970).

[5] W. Dreyer, Maximisation of entropy for relativistic and degenerate gases in non-equilibrium, in *Kinetic theory and extended thermodynamics* pp. 107–123, eds. I. Müller and T. Ruggeri (Pitagora, Bologna, 1987).

[6] H. Grad, On the kinetic theory of rarefied gases, *Commun. Pure Appl. Math.* **2**, 331–407 (1949).

[7] S. R. de Groot, W. A. van Leeuwen and Ch. G. van Weert, *Relativistic kinetic theory* (North-Holland, Amsterdam, 1980).

[8] E. Ikenberry and C. Truesdell, On the pressures and the flux of energy in a gas according to Maxwell's kinetic theory, I.,*J. Rational Mech. Anal.* **5**, 1–54 (1956).

[9] G. M. Kremer, *Uma introdução à teoria cinética relativística* (Editora da UFPR, Curitiba, 1998).

[10] G. M. Kremer, On the kinetic theory of relativistic gases, *Continuum Mech. Thermodyn.* **9**, 13–21 (1997).

[11] G. M. Kremer and I. Müller, Dynamic pressure in relativistic thermodynamics, *Ann. Inst. Henri Poincaré* **67**, 111–121 (1997).

[12] J. C. Maxwell, On the dynamical theory of gases, *Phil. Trans. R. Soc. London* **157**, 49–88 (1867).

[13] P. H. Polak, W. A. van Leeuwen and S. R. de Groot, On relativistic kinetic gas theory. X. Transport coefficients in the intermediate relativistic regime. Values for special models, *Physica* **66**, 455–473 (1973).

[14] L. L. Samojeden and G. M. Kremer, The Burnett equations for a relativistic gas, *Continuum Mech. Thermodyn.* **12**, 387–401 (2000).

[15] J. M. Stewart, *Non-equilibrium relativistic kinetic theory*, Lecture Notes in Physics, vol. 10 (Springer, Heidelberg, 1971).

[16] C. S. Wang-Chang and G. E. Uhlenbeck, On the transport phenomena in monatomic gases, in *Studies in Statistical Mechanics* pp. 17–26, eds. J. de Boer and G. E. Uhlenbeck, Vol. V (North-Holland, Amsterdam, 1970).

Chapter 7

Chemically Reacting Gas Mixtures

7.1 Introduction

In this chapter we analyze mixtures of relativistic gases in which chemical or nuclear reactions occur. First we introduce the Boltzmann equation for chemically reacting gas mixtures and derive the transfer equations for the constituents and for the mixture. The equilibrium properties are discussed on the basis of the Maxwell–Jüttner distribution function of the constituents. The non-equilibrium processes in chemically reacting gas mixtures are analyzed first within the framework of a thermodynamic theory of irreversible processes. Next we apply the Chapman and Enskog method and derive the constitutive equations for the dynamic pressure, pressure deviator and heat flux of the mixture, for the diffusion flux of each constituent of the mixture and for the reaction rate density. The Onsager reciprocity relations which relate the coefficients of cross-effects are proved as well as the relationship which shows that the matrix of the diffusion coefficients is symmetric.

7.2 Boltzmann and transfer equations

We consider a relativistic gas mixture of four constituents of generic chemical symbols A, B, C and D, that reacts according to a single bimolecular reaction of the type

$$A + B \rightleftharpoons C + D. \tag{7.1}$$

There exist two types of collisions between the particles of the constituents, i.e., the elastic collisions that refer to non-reactive processes and the inelastic collisions which are associated with the reactions. For the inelastic collisions the energy-momentum conservation law (1.103) is valid so that we can write

$$p_A^\alpha + p_B^\alpha = p_C'^\alpha + p_D'^\alpha. \tag{7.2}$$

However for reactive processes the masses of the particles are not conserved since one has to consider that the sum

$$\sum_{a=A}^{D} \nu_a m_a = m_A + m_B - m_C - m_D = \Delta m, \tag{7.3}$$

is the mass defect Δm. Here m_a denotes the rest mass of a particle of the constituent a, ν_a its stoichiometric coefficient. For the reaction of the type (7.1) we have that $\nu_A = \nu_B = -\nu_C = -\nu_D = 1$.

This applies to chemical as well as to nuclear reactions; the main difference is that the mass defect is negligible in (ordinary) chemical reactions. On the other hand, if we convert the mass defect into a reaction energy, multiplying by c^2, we obtain significant energies for chemical reactions and enormous energies for nuclear reactions.

The mass defect is the difference between the atomic mass and the sum of the protons, neutrons and electrons that make up the atom. As an example consider the mass of the helium atom ^{4}He:

$$m_{^4He} = 4.002\,603\, m_u, \tag{7.4}$$

where m_u=1.660 540×10^{-27} kg is the unified atomic mass unity. The mass of its constituent parts is the mass of two protons ($m_p = 1.007\,276\, m_u$) plus the mass of two neutrons ($m_n = 1.008\,664\, m_u$) plus the mass of two electrons ($m_e = 5.485\,799 \times 10^{-4}\, m_u$):

$$2m_p + 2m_n + 2m_e = 4.032\,977\, m_u. \tag{7.5}$$

The mass defect is then

$$\Delta m = 4.032\,977\, m_u - 4.002\,603\, m_u = 0.030\,374\, m_u. \tag{7.6}$$

Hence the mass of an atom is less than the mass of its constituent parts, the difference Δm being the mass defect. If we make use of Einstein mass-energy equivalence and multiply the mass defect by c^2 we get the binding energy of the nucleus $\Delta E = \Delta m c^2$ which is the energy that must be applied to the nucleus in order to break it apart. In this case we have

$$\Delta E = 4.54 \times 10^{-12}\,\mathrm{J} = 28.3\,\mathrm{MeV}, \tag{7.7}$$

since 1 eV= $1.602\,177 \times 10^{-19}$ J.

A state of the mixture is characterized by the set of one-particle distribution functions

$$f(\mathbf{x}, \mathbf{p}_a, t) \equiv f_a, \qquad a = A, B, C, D \tag{7.8}$$

such that $f(\mathbf{x}, \mathbf{p}_a, t) d^3x d^3p_a$ gives at time t the number of particles of constituent a in the volume element d^3x about $\mathbf{x}$ and with momenta in the range d^3p_a about

$\mathbf{p}_a$. The one-particle distribution function of constituent a satisfies a Boltzmann equation of the form

$$p_a^\alpha \frac{\partial f_a}{\partial x^\alpha} = \sum_{b=A}^{D} \int (f'_a f'_b - f_a f_b) F_{ba} \sigma_{ab}\, d\Omega_{ba} \frac{d^3 p_b}{p_{b0}} + R_a, \tag{7.9}$$

which is a generalization of (2.45) to chemically reacting gas mixtures when external forces are not taken into account. The first term on the right-hand side of (7.9) refers to the elastic collisions with F_{ba}, σ_{ab} and $d\Omega_{ba}$ denoting the invariant flux, the invariant differential elastic cross-section and the element of solid angle that characterizes a binary collision between the particles of constituent a with those of constituent b, respectively. We use (2.35) in order to write the invariant flux as

$$F_{ba} = \frac{p_a^0 p_b^0}{c} g_{ba} = \frac{p_a^0 p_b^0}{c} \sqrt{(\mathbf{v}_a - \mathbf{v}_b)^2 - \frac{1}{c^2}(\mathbf{v}_a \times \mathbf{v}_b)^2}. \tag{7.10}$$

In addition we used the abbreviations

$$f'_a \equiv f(\mathbf{x}, \mathbf{p}'_a, t), \quad f'_b \equiv f(\mathbf{x}, \mathbf{p}'_b, t), \quad f_a \equiv f(\mathbf{x}, \mathbf{p}_a, t), \quad f_b \equiv f(\mathbf{x}, \mathbf{p}_b, t), \tag{7.11}$$

where the primes denote post-collisional momenta. R_a is the collision term due to the reactions, which is written as

$$R_A = \int (f'_C f'_D - f_A f_B) F_{BA} \sigma^\star_{AB} d\Omega \frac{d^3 p_B}{p_{B0}}, \tag{7.12}$$

$$R_C = \int (f_A f_B - f'_C f'_D) F'_{DC} \sigma^\star_{CD} d\Omega' \frac{d^3 p'_D}{p'_{D0}}, \tag{7.13}$$

with similar expressions for R_B and R_D. The differential reactive cross-section for the forward reaction $\sigma^\star_{AB}$ is connected to the one for the backward reaction $\sigma^\star_{CD}$ through (see (2.32))

$$\int_\Omega g_{BA} \sigma^\star_{AB} d\Omega d^3 p_A d^3 p_B = \int_{\Omega'} g'_{DC} \sigma^\star_{CD} d\Omega' d^3 p'_C d^3 p'_D. \tag{7.14}$$

The general equation of transfer for the constituent a of the mixture is obtained through the multiplication of the Boltzmann equation (7.9) by an arbitrary function $\psi_a \equiv \psi(\mathbf{x}, \mathbf{p}_a, t)$ and integration of the resulting equation over all values of $d^3 p_a / p_{a0}$, yielding

$$\frac{\partial}{\partial x^\alpha} \int \psi_a p_a^\alpha f_a \frac{d^3 p_a}{p_{a0}} - \int p_a^\alpha \frac{\partial \psi_a}{\partial x^\alpha} f_a \frac{d^3 p_a}{p_{a0}}$$

$$= \sum_{b=A}^{D} \int \psi_a (f'_a f'_b - f_a f_b) F_{ba} \sigma_{ab} d\Omega_{ba} \frac{d^3 p_b}{p_{b0}} \frac{d^3 p_a}{p_{a0}} + \int \psi_a R_a \frac{d^3 p_a}{p_{a0}}$$

$$= \sum_{b=A}^{D} \int (\psi_a' - \psi_a) f_a f_b F_{ba} \sigma_{ab} d\Omega_{ba} \frac{d^3 p_b}{p_{b0}} \frac{d^3 p_a}{p_{a0}} + \int \psi_a R_a \frac{d^3 p_a}{p_{a0}}. \tag{7.15}$$

In order to deduce the right-hand side of the above equation we have used the same reasoning as in Section 2.3.

If we sum (7.15) over all constituents we get a general equation of transfer for the mixture that reads

$$\frac{\partial}{\partial x^\alpha} \sum_{a=A}^{D} \int \psi_a p_a^\alpha f_a \frac{d^3 p_a}{p_{a0}} - \sum_{a=A}^{D} \int p_a^\alpha \frac{\partial \psi_a}{\partial x^\alpha} f_a \frac{d^3 p_a}{p_{a0}} = \sum_{a=A}^{D} \int \psi_a R_a \frac{d^3 p_a}{p_{a0}}$$

$$+\frac{1}{2} \sum_{a,b=A}^{D} \int (\psi_a' + \psi_b' - \psi_a - \psi_b) f_a f_b F_{ba} \sigma_{ab} d\Omega_{ba} \frac{d^3 p_b}{p_{b0}} \frac{d^3 p_a}{p_{a0}}. \tag{7.16}$$

The last term on the right-hand side of (7.16) was obtained by using the symmetry properties of the collision term and by interchanging the dummy indexes a and b in the sums.

The moments of the distribution function we are interested in are the partial particle four-flow N_a^α and the partial energy-momentum tensor $T_a^{\alpha\beta}$ which are defined through

$$N_a^\alpha = c \int p_a^\alpha f_a \frac{d^3 p_a}{p_{a0}}, \qquad T_a^{\alpha\beta} = c \int p_a^\alpha p_a^\beta f_a \frac{d^3 p_a}{p_{a0}}. \tag{7.17}$$

The particle four-flow of the mixture N^α and the energy-momentum tensor of the mixture $T^{\alpha\beta}$ are obtained by summing the partial quantities, i.e.,

$$N^\alpha = \sum_{a=A}^{D} N_a^\alpha, \qquad T^{\alpha\beta} = \sum_{a=A}^{D} T_a^{\alpha\beta}. \tag{7.18}$$

The balance equation for the partial particle four-flow is obtained by choosing $\psi_a = c$ in (7.15), yielding

$$\partial_\alpha N_a^\alpha = \nu_a \ell, \tag{7.19}$$

where, in this chapter, ℓ denotes the reaction rate density which is defined by

$$\ell = c \int (f_C' f_D' - f_A f_B) F_{BA} \sigma_{AB}^\star d\Omega \frac{d^3 p_B}{p_{B0}} \frac{d^3 p_A}{p_{A0}}. \tag{7.20}$$

If we sum (7.19) over all constituents of the mixture we get the balance equation for the particle four-flow of the mixture, which reads

$$\partial_\alpha N^\alpha = 0, \qquad \text{since} \qquad \sum_{a=A}^{D} \nu_a = 0. \tag{7.21}$$

The particle four-flow N_a^α of constituent a can be decomposed in the Eckart description into a term proportional to the four-velocity plus a term perpendicular to it, viz.,

$$N_a^\alpha = n_a U^\alpha + J_a^\alpha, \qquad \text{such that} \qquad J_a^\alpha U_\alpha = 0, \tag{7.22}$$

where n_a is the particle number density and J_a^α the diffusion flux of constituent a. If we sum $(7.22)_1$ over all constituents, using (7.18) and the decomposition for the particle four-flow of the mixture $N^\alpha = nU^\alpha$, we find that

$$n = \sum_{a=A}^{D} n_a, \qquad \sum_{a=A}^{D} J_a^\alpha = 0. \tag{7.23}$$

Hence the particle number density of the mixture n is the sum of the partial particle number densities n_a, and due to the constraint $(7.23)_2$ there exist in the case of four constituents, only three linearly independent diffusion fluxes.

The balance equation for the energy-momentum tensor of the mixture is obtained by choosing $\psi_a = cp_a^\beta$ in the general equation of transfer (7.16), yielding

$$\partial_\beta T^{\alpha\beta} = 0, \tag{7.24}$$

since the energy-momentum conservation law is valid for elastic as well as for inelastic collisions.

Problems

7.2.1 Explain why it is not possible to change ψ_a into ψ_b in (7.15).

7.2.2 Obtain the right-hand sides of (7.15) and (7.16).

7.3 Maxwell–Jüttner distribution function

In chemically reacting mixtures it is usual to write the first approximation of the one-particle distribution function of constituent a as

$$f_a^{(0)} = \frac{1}{h^3} e^{\frac{\mu_a}{kT} - \frac{U_\alpha p_a^\alpha}{kT}}, \tag{7.25}$$

which is based on the Maxwell–Jüttner distribution function of a single gas (2.123) with T and U^α denoting respectively the temperature and the four-velocity of the mixture, while μ_a denotes the chemical potential of the constituent a of the mixture.

If we insert the Maxwell–Jüttner distribution function (7.25) into the collision terms, it is easy to verify that the elastic collision term vanishes while the inelastic collision term does not. The reason is that the chemical potential in the Maxwell–Jüttner distribution function (7.25) is not the equilibrium one. For the equilibrium chemical potential μ_a^E the so-called equation of chemical equilibrium holds, viz.,

$$\sum_{a=A}^{D} \nu_a \mu_a^E = 0, \qquad \text{or} \qquad \mu_A^E + \mu_B^E = \mu_C^E + \mu_D^E. \tag{7.26}$$

For a non-equilibrium process the sum above defines the affinity $\mathcal{A}$ for the forward reaction

$$\mathcal{A} = \sum_{a=A}^{D} \nu_a \mu_a = \mu_A + \mu_B - \mu_C - \mu_D. \tag{7.27}$$

The affinity is a quantity that vanishes in equilibrium, due to the constraint (7.26).

Thanks to the definition of the affinity (7.27) one may express the product of the two distribution functions $f_C'^{(0)} f_D'^{(0)}$ as

$$f_C'^{(0)} f_D'^{(0)} = e^{-\frac{\mathcal{A}}{kT}} f_A^{(0)} f_B^{(0)}. \tag{7.28}$$

One can relate the chemical potential μ_a to the particle number density n_a of constituent a. Indeed, by inserting (7.25) into the definition of the particle four-flow $(7.17)_1$ of constituent a, by integrating the resulting equation over all values of d^3p_a/p_{a0} and by performing the projection with the four-velocity U_α, it follows that

$$\mu_a = kT \ln \left\{ \frac{n_a h^3}{4\pi c m_a^2 kT K_2(\zeta_a)} \right\}, \quad \text{where} \quad \zeta_a = \frac{m_a c^2}{kT}. \tag{7.29}$$

An important relationship which will be used later is the gradient of the chemical potential (7.29) that reads

$$\partial_\alpha \left(\frac{\mu_a}{T} \right) = \frac{k}{n_a} \partial_\alpha n_a + \frac{k}{T}(1 - \zeta_a G_a) \partial_\alpha T = \frac{1}{n_a T} \partial_\alpha p_a - \frac{h_E^a}{T^2} \partial_\alpha T, \tag{7.30}$$

where $G_a = K_3(\zeta_a)/K_2(\zeta_a)$, $p_a = n_a kT$ is the pressure and $h_E^a = m_a c^2 G_a$ the enthalpy per particle of constituent a.

If we sum (7.30) over all constituents and introduce the particle fraction $x_a = n_a/n$ of constituent a, we get

$$\sum_{a=A}^{D} x_a \partial_\alpha \left(\frac{\mu_a}{T} \right) = -\frac{h_E}{T^2} \left[\partial_\alpha T - \frac{T}{n h_E} \partial_\alpha p \right], \tag{7.31}$$

where the enthalpy per particle h_E and the pressure p of the mixture are defined through

$$n h_E = \sum_{a=A}^{D} n_a h_E^a, \qquad p = \sum_{a=A}^{D} p_a. \tag{7.32}$$

The above relationship will also be used later in this chapter.

Problems

7.3.1 Show that the chemical potential of the constituent a is given by (7.29).

7.3.2 Obtain the two equations (7.30) and (7.31) for the gradient of the chemical potential.

7.4 Thermodynamics of mixtures

The objective of the thermodynamic theory of mixtures of r chemically reacting fluids is the determination of the fields:

$$\begin{cases} N_a^\alpha(x^\beta) \text{ – partial particle four-flow}, & (a = 1, 2, \dots, r), \\ T(x^\beta) \text{ – temperature}, \end{cases} \tag{7.33}$$

at all events x^β. We shall restrict ourselves to the case where all constituents are at the same temperature, which is the temperature of the mixture T. Further we assume that there occurs only a single reaction of the generic type

$$\nu_1 A_1 + \cdots + \nu_b A_b \rightleftharpoons \nu_{b+1} A_{b+1} + \cdots + \nu_r A_r, \tag{7.34}$$

where ν_a denotes the stoichiometric coefficient and A_a the chemical symbol of the constituent a.

The balance equations for the partial particle four-flow N_a^α and for the energy-momentum tensor of the mixture $T^{\alpha\beta}$ are given by (7.19) and (7.24) which we reproduce here:

$$\partial_\alpha N_a^\alpha = \nu_a \ell, \qquad \partial_\beta T^{\alpha\beta} = 0. \tag{7.35}$$

We note that the above equations, in principle, determine the fields (7.33).

If we introduce the decompositions (2.157) and (7.22) into $(7.35)_1$ it follows that the balance equation for the partial particle four-flow – written in terms of the differential operators $D = U^\alpha \partial_\alpha$ and $\nabla^\alpha = \Delta^{\alpha\beta}\partial_\beta$ – reads

$$Dn_a + n_a \nabla_\alpha U^\alpha + \nabla_\alpha J_a^\alpha - \frac{1}{c^2} J_a^\alpha DU_\alpha = \nu_a \ell. \tag{7.36}$$

By summing the above equation over all constituents and making use of $(7.23)_2$ we get the balance equation for the particle four-flow of the mixture

$$Dn + n\nabla_\alpha U^\alpha = \sum_{a=1}^{r} \nu_a \ell. \tag{7.37}$$

Note that for a reaction of the type (7.34) the sum of the stoichiometric coefficients does not vanish. From (7.36) and (7.37) it is easy to derive the balance equation for the particle fractions $x_a = n_a/n$:

$$nDx_a + x_a \sum_{b=1}^{r} \nu_b \ell + \nabla_\alpha J_a^\alpha - \frac{1}{c^2} J_a^\alpha DU_\alpha = \nu_a \ell. \tag{7.38}$$

On the other hand, the introduction of Eckart decomposition (4.22) into the balance equation for the energy-momentum tensor of the mixture $(7.35)_2$ leads to the following balance equation for the energy density of the mixture:

$$nDe = -(p+\varpi)\nabla_\alpha U^\alpha + p^{\langle\alpha\beta\rangle}\nabla_\beta U_\alpha - \nabla_\alpha q^\alpha + \frac{2}{c^2}q^\alpha DU_\alpha - e\sum_{a=1}^{r}\nu_a \ell, \tag{7.39}$$

while the balance equation for the momentum density (4.43) remains unchanged. The last term on the right-hand side of (7.39) has appeared since we have used (7.37) in order to eliminate Dn from the balance equation for the energy density of the mixture.

The balance equation for the entropy density of the mixture has also one term that depends on the reaction rate density. Indeed instead of (4.45) we have

$$nDs + \nabla_\alpha \phi^\alpha + \frac{1}{c^2}U_\alpha D\phi^\alpha + s\sum_{a=1}^{r}\nu_a \ell = \varsigma \geq 0. \tag{7.40}$$

The balance equations (7.36), (7.39) and (4.43) are not field equations for the fields (7.33) since they depend on the following unknown quantities: reaction rate density ℓ, diffusion fluxes J_a^α, dynamic pressure ϖ, pressure deviator $p^{\langle\alpha\beta\rangle}$ and heat flux q^α. We proceed to derive the constitutive equations for these quantities on the basis of the thermodynamic theory of irreversible processes. We start with the Gibbs equation for mixtures written in terms of the differential operator D:

$$Ds = \frac{1}{T}\left[De - \frac{p}{n^2}Dn - \sum_{a=1}^{r}\mu_a Dx_a\right]. \tag{7.41}$$

The quantity μ_a is the chemical potential of the constituent a in the mixture and for mixtures of ideal gases it is given by

$$\mu_a = e_a - Ts_a + \frac{p_a}{n_a}, \tag{7.42}$$

where e_a, s_a and p_a are the energy per particle, the entropy per particle and the pressure of constituent a, respectively. The relationships between the partial quantities and the corresponding one for the mixture are

$$ne = \sum_{a=1}^{r} n_a e_a, \qquad ns = \sum_{a=1}^{r} n_a s_a, \qquad p = \sum_{a=1}^{r} p_a. \tag{7.43}$$

Hence if we multiply (7.42) by n_a and sum the resulting equation over all constituents we get

$$\sum_{a=1}^{r} x_a \mu_a = e - Ts + \frac{p}{n}. \tag{7.44}$$

We eliminate from (7.41) the differentials Dn, Dx_a and De by using (7.37) through (7.39) and get, after some rearrangements,

$$\begin{aligned}
nDs + \nabla_\alpha\left(\frac{q^\alpha}{T} - \sum_{a=1}^{r}\frac{\mu_a J_a^\alpha}{T}\right) + \frac{1}{c^2}U_\alpha D\left(\frac{q^\alpha}{T} - \sum_{a=1}^{r}\frac{\mu_a J_a^\alpha}{T}\right) + s\sum_{a=1}^{r}\nu_a \ell \\
= -\frac{\varpi}{T}\nabla_\alpha U^\alpha + \frac{p^{\langle\alpha\beta\rangle}}{T}\nabla_\alpha U_\beta - \frac{q^\alpha}{T^2}\left(\nabla_\alpha T - \frac{T}{c^2}DU_\alpha\right) \\
-\sum_{a=1}^{r-1} J_a^\alpha \nabla_\alpha\left(\frac{\mu_a - \mu_r}{T}\right) - \mathcal{A}\ell.
\end{aligned} \tag{7.45}$$

In the above equation we have also used the relationship

$$\begin{aligned}
\sum_{a=1}^{r} J_a^\alpha \nabla_\alpha\left(\frac{\mu_a}{T}\right) &= \sum_{a=1}^{r-1} J_a^\alpha \nabla_\alpha\left(\frac{\mu_a}{T}\right) + J_r^\alpha \nabla_\alpha\left(\frac{\mu_r}{T}\right) \\
&\overset{(7.23)_2}{=} \sum_{a=1}^{r-1} J_a^\alpha \nabla_\alpha\left(\frac{\mu_a - \mu_r}{T}\right).
\end{aligned} \tag{7.46}$$

If we compare (7.45) with (7.40) we can identify the entropy flux ϕ^α and the entropy production rate ς as:

$$\phi^\alpha = \left(\frac{q^\alpha}{T} - \sum_{a=1}^{r}\frac{\mu_a J_a^\alpha}{T}\right), \tag{7.47}$$

$$\begin{aligned}
\varsigma = -\frac{\varpi}{T}\nabla_\alpha U^\alpha + \frac{p^{\langle\alpha\beta\rangle}}{T}\nabla_\alpha U_\beta - \frac{q^\alpha}{T^2}\left(\nabla_\alpha T - \frac{T}{c^2}DU_\alpha\right) \\
-\sum_{a=1}^{r-1} J_a^\alpha \nabla_\alpha\left(\frac{\mu_a - \mu_r}{T}\right) - \mathcal{A}\ell \geq 0.
\end{aligned} \tag{7.48}$$

From the equation for the entropy production rate (7.48) one identifies the following quantities as well:

$$\begin{cases} \text{forces: } \nabla_\alpha U^\alpha, \mathcal{A}, \nabla^{\langle\alpha}U^{\beta\rangle}, [\nabla^\alpha T - (T/c^2)DU^\alpha], \nabla^\alpha[(\mu_a - \mu_r)/T)]; \\ \text{fluxes: } \varpi, \ell, p^{\langle\alpha\beta\rangle}, q^\alpha, J_a^\alpha. \end{cases} \tag{7.49}$$

According to the thermodynamic theory of irreversible processes we assume linear relationships between thermodynamic forces and fluxes and write the constitutive equations as:

$$\varpi = -\eta\nabla_\alpha U^\alpha - L\mathcal{A}, \tag{7.50}$$

$$\ell = -L^\star\nabla_\alpha U^\alpha - \mathcal{L}\mathcal{A}, \tag{7.51}$$

$$p^{\langle\alpha\beta\rangle} = 2\mu\nabla^{\langle\alpha}U^{\beta\rangle}, \tag{7.52}$$

$$q^\alpha = \lambda^\star \left(\nabla^\alpha T - \frac{T}{c^2} DU^\alpha\right) + T^2 \sum_{a=1}^{r-1} L_a \nabla^\alpha \left(\frac{\mu_a - \mu_r}{T}\right), \tag{7.53}$$

$$J_a^\alpha = L_a^\star \left(\nabla^\alpha T - \frac{T}{c^2} DU^\alpha\right) + \sum_{b=1}^{r-1} L_{ab} \nabla^\alpha \left(\frac{\mu_b - \mu_r}{T}\right). \tag{7.54}$$

The scalar coefficients in (7.50) to (7.54) have the following meaning: η is the bulk viscosity, L is the chemical viscosity, μ is the shear viscosity, $\lambda^\star$ is related to the thermal conductivity, L_{ab} is related to the matrix of the diffusion coefficients and $L_a^\star$, L_a are related to cross-effects of thermal-diffusion and diffusion-thermo, that is the flow of particles caused by a temperature gradient and the flow of heat caused by a gradient of particle number density, respectively. The coefficients $L^\star$ and $\mathcal{L}$ do not have generally accepted names.

In the thermodynamic theory of irreversible processes the following relationships, known as Onsager reciprocity relations, are postulated:

$$L_{ab} = L_{ba}, \qquad L_a = L_a^\star, \qquad \text{and} \qquad L = -L^\star, \tag{7.55}$$

that is the matrix of the diffusion coefficients is symmetric, the coefficient of the thermal-diffusion effect is equal to the coefficient of the diffusion-thermo effect and, apart from the minus sign, the coefficients of cross-effects in the constitutive equations for the dynamic pressure ϖ and for the reaction rate density ℓ are equal. The minus sign is due to the fact that the thermodynamic force $\nabla_\alpha U^\alpha$ changes its sign under time reversal while the thermodynamic force $\mathcal{A}$ does not change. In Section 7.6 we shall prove the Onsager reciprocity relations (7.55) on the basis of the Boltzmann equation.

Problems

7.4.1 Obtain the balance for the particle fractions (7.38).

7.4.2 Show that (7.39) represents the balance equation for the energy density of the mixture.

7.4.3 Obtain the balance equation for the entropy density of the mixture (7.45) from the Gibbs equation (7.41).

7.3.4 Give a physical reason why the thermodynamic force represented by $\nabla_\alpha U^\alpha$ changes its sign under time reversal while the thermodynamic force represented by $\mathcal{A}$ does not change.

7.5 Transport coefficients

We shall refer to Hermens [3, 4] and Hermens et al. [5] in order to determine the transport coefficients of chemically reacting gas mixtures by using the method of

Chapman and Enskog. Here we shall consider the bimolecular of the type (7.1) and assume that all constituents of the mixture are at the same temperature, which is the temperature of the mixture.

We start by writing the one-particle distribution function of constituent a as

$$f_a = f_a^{(0)}(1+\phi_a), \tag{7.56}$$

where $f_a^{(0)}\phi_a$ represents the deviation of the Maxwell–Jüttner distribution function (7.25).

Since N^α, $N_a^\alpha U_\alpha$ and $T^{\alpha\beta}U_\alpha U_\beta$ represent equilibrium quantities in the Eckart decomposition, we obtain the following constraints for the deviation $f_a^{(0)}\phi_a$ of the Maxwell–Jüttner distribution function:

$$\sum_{a=A}^{D}\int p_a^\alpha f_a^{(0)}\phi_a \frac{d^3p_a}{p_{a0}} = 0, \tag{7.57}$$

$$U_\alpha \int p_a^\alpha f_a^{(0)}\phi_a \frac{d^3p_a}{p_{a0}} = 0, \tag{7.58}$$

$$U_\alpha U_\beta \sum_{a=A}^{D}\int p_a^\alpha p_a^\beta f_a^{(0)}\phi_a \frac{d^3p_a}{p_{a0}} = 0. \tag{7.59}$$

In many chemical reactions the number of reactive collisions is smaller than the number of non-reactive collisions. In this case the relaxation time of the elastic collisions is smaller than the relaxation time of the inelastic collisions and it is usual to consider[1] the reactive collision term R_a of the same order as the streaming term $p_a^\alpha\partial_\alpha f_a$. Hence if we insert the distribution function (7.56) into the Boltzmann equation (7.9), we get by using the Chapman and Enskog methodology:

$$p_a^\alpha \frac{\partial f_a^{(0)}}{\partial x^\alpha} - R_a^{(0)} = \sum_{b=A}^{D}\int f_a^{(0)} f_b^{(0)}(\phi_a' + \phi_b' - \phi_a - \phi_b)F_{ba}\sigma_{ab}\, d\Omega_{ba}\frac{d^3p_b}{p_{b0}}. \tag{7.60}$$

Let us transform the left-hand side of (7.60) by inserting the Maxwell–Jüttner distribution function (7.25) into the streaming term $p_a^\alpha\partial_\alpha f_a$ and into the reactive collision term R_a. First we note that, thanks to the relationship (7.28), the reactive collision term can be represented as

$$R_a^{(0)} = -(1-e^{-\frac{\mathcal{A}}{kT}})F_a^{(0)}, \tag{7.61}$$

where we have introduced the term $F_a^{(0)}$ given by

$$F_a^{(0)} = \int f_A^{(0)} f_B^{(0)}\left[F_{BA}\sigma_{AB}^\star\left(\delta_{aA}\frac{d^3p_B}{p_{B0}} + \delta_{aB}\frac{d^3p_A}{p_{A0}}\right)d\Omega\right.$$

[1]To the best of our knowledge this hypothesis was first formulated by Prigogine and Xhrouet [6].

$$-F'_{DC}\sigma^{\star}_{CD}\left(\delta_{aC}\frac{d^3p'_D}{p'_{D0}}+\delta_{aD}\frac{d^3p'_C}{p'_{C0}}\right)d\Omega'\Bigg]. \tag{7.62}$$

Further, the streaming term can be written in terms of the convective time derivative $D = U^\alpha\partial_\alpha$ and of the gradient $\nabla^\alpha = \Delta^{\alpha\beta}\partial_\beta$ as

$$p^\alpha_a\frac{\partial f^{(0)}_a}{\partial x^\alpha} = f^{(0)}_a\left\{\frac{p^\beta_a}{k}\left[\nabla_\beta\left(\frac{\mu_a}{T}\right)+\frac{U_\beta}{c^2}D\left(\frac{\mu_a}{T}\right)\right]-\frac{\Delta_{\alpha\beta}p^\alpha_a p^\beta_a}{3kT}\nabla_\gamma U^\gamma\right.$$

$$\left.-\frac{p^\alpha_a p^\beta_a}{kT}\nabla_{\langle\alpha}U_{\beta\rangle}+\frac{U_\alpha p^\alpha_a p^\beta_a}{kT^2}\left[\nabla_\beta T+\frac{U_\beta}{c^2}DT-\frac{T}{c^2}DU_\beta\right]\right\}. \tag{7.63}$$

Hence by taking into account (7.62) and (7.63) it follows that equation (7.60) reduces to

$$f^{(0)}_a\left\{\frac{p^\beta_a}{k}\left[\nabla_\beta\left(\frac{\mu_a}{T}\right)+\frac{U_\beta}{c^2}D\left(\frac{\mu_a}{T}\right)\right]-\frac{\Delta_{\alpha\beta}p^\alpha_a p^\beta_a}{3kT}\nabla_\gamma U^\gamma-\frac{p^\alpha_a p^\beta_a}{kT}\nabla_{\langle\alpha}U_{\beta\rangle}\right.$$

$$\left.+\frac{U_\alpha p^\alpha_a p^\beta_a}{kT^2}\left[\nabla_\beta T+\frac{U_\beta}{c^2}DT-\frac{T}{c^2}DU_\beta\right]\right\}+(1-e^{-\frac{\mathcal{A}}{kT}})F^{(0)}_a$$

$$=\sum_{b=A}^{D}\int f^{(0)}_a f^{(0)}_b(\phi'_a+\phi'_b-\phi_a-\phi_b)F_{ba}\sigma_{ab}\,d\Omega_{ba}\frac{d^3p_b}{p_{b0}}. \tag{7.64}$$

If we integrate the above equation over all values of d^3p_a/p_{a0}, we obtain the balance equation for the particle number density of constituent a, with a vanishing diffusion flux ($J^\alpha_a = 0$) when compared with (7.36). This equation reads

$$\frac{n_a}{k}D\left(\frac{\mu_a}{T}\right)-\frac{n_a}{T}(1-\zeta_a G_a)DT+n_a\nabla^\alpha U_\alpha \stackrel{(7.30)}{=} Dn_a+n_a\nabla^\alpha U_\alpha=\nu_a\ell^{(0)}, \tag{7.65}$$

where $\ell^{(0)}$ denotes the first approximation to the reaction rate density

$$\ell^{(0)} = c\int(f'^{(0)}_C f'^{(0)}_D - f^{(0)}_A f^{(0)}_B)F_{BA}\sigma^{\star}_{AB}d\Omega\frac{d^3p_B}{p_{B0}}\frac{d^3p_A}{p_{A0}}$$

$$\stackrel{(7.28)}{=} -(1-e^{-\frac{\mathcal{A}}{kT}})ck^{(0)}_f n_A n_B. \tag{7.66}$$

In (7.66) $k^{(0)}_f$ is the first approximation to the rate constant of the forward reaction which is defined by

$$k^{(0)}_f = \frac{1}{n_A n_B}\int f^{(0)}_A f^{(0)}_B F_{BA}\sigma^{\star}_{AB}d\Omega\frac{d^3p_B}{p_{B0}}\frac{d^3p_A}{p_{A0}}. \tag{7.67}$$

In a relativistic Eulerian fluid mixture the fields of dynamic pressure, pressure deviator and heat flux of the mixture vanish as well as the diffusion fluxes of the constituents. Hence the balance equations for the energy density and momentum density of the mixture reduce to the balance equations for a single Eulerian fluid that read (see (5.20) and (5.21))

$$nc_v DT + p\nabla_\alpha U^\alpha = 0, \tag{7.68}$$

$$\frac{nh_E}{c^2} DU^\alpha = \nabla^\alpha p, \tag{7.69}$$

where c_v is the heat capacity per particle at constant volume of the mixture.

We eliminate now the convective time derivatives $D(\mu_a/T)$, DT and DU^α from (7.64) by using the balance equations for an Eulerian fluid mixture (7.65), (7.68) and (7.69) and obtain the following integral equation for the deviation of the Maxwell–Jüttner distribution function ϕ_a:

$$f_a^{(0)}\Bigg\{\frac{p_a^\beta}{k}\nabla_\beta\left(\frac{\mu_a}{T}\right) + \frac{U_\alpha p_a^\alpha p_a^\beta}{kT^2}\left[\nabla_\beta T - \frac{T}{nh_E}\nabla_\beta p\right] - \frac{\Delta_{\alpha\beta}p_a^\alpha p_a^\beta}{3kT}\nabla_\gamma U^\gamma$$
$$-\frac{p_a^\alpha p_a^\beta}{kT}\nabla_{\langle\alpha}U_{\beta\rangle} - \frac{U_\beta p_a^\beta}{c^2}\left[\left(1+\frac{k}{c_v}(1-\zeta_a G_a)\right)\nabla_\alpha U^\alpha - \frac{\nu_a}{n_a}\ell^{(0)}\right]$$
$$-\frac{1}{c_v T c^2}p_a^\alpha p_a^\beta U_\alpha U_\beta \nabla_\gamma U^\gamma\Bigg\} + (1-e^{-\frac{A}{kT}})F_a^{(0)}$$
$$= \sum_{b=A}^{D}\int f_a^{(0)} f_b^{(0)}(\phi_a' + \phi_b' - \phi_a - \phi_b)F_{ba}\sigma_{ab}\,d\Omega_{ba}\frac{d^3 p_b}{p_{b0}}. \tag{7.70}$$

In (7.70) we have to take into account the constraint (7.31) which means that there exist only three linearly independent gradients of the chemical potentials. Hence the two first terms of (7.70) can be represented as

$$\frac{p_a^\beta}{k}\nabla_\beta\left(\frac{\mu_a}{T}\right) + \frac{U_\alpha p_a^\alpha p_a^\beta}{kT^2}\left[\nabla_\beta T - \frac{T}{nh_E}\nabla_\beta p\right]$$
$$= -\frac{p_a^\beta}{k}\sum_{b=A}^{C}(x_b - \delta_{ab})\nabla_\beta\left(\frac{\mu_b-\mu_D}{T}\right) + \frac{p_a^\beta}{kT^2}(U_\alpha p_a^\alpha - h_E)\left[\nabla_\beta T - \frac{T}{nh_E}\nabla_\beta p\right], \tag{7.71}$$

so that we can write the integral equation (7.70) thanks to the relationships (7.66) and (7.71) as

$$f_a^{(0)}\Bigg\{-\frac{p_a^\beta}{k}\sum_{b=A}^{C}(x_b-\delta_{ab})\nabla_\beta\Big(\frac{\mu_b-\mu_D}{T}\Big) + \frac{p_a^\beta}{kT^2}(U_\alpha p_a^\alpha - h_E)\Big[\nabla_\beta T - \frac{T}{nh_E}\nabla_\beta p\Big]$$
$$-\frac{p_a^\alpha p_a^\beta}{kT}\nabla_{\langle\alpha}U_{\beta\rangle} - \Big[\frac{U_\beta p_a^\beta}{c^2}\Big(1+\frac{k}{c_v}(1-\zeta_a G_a)\Big) + \frac{1}{c_v T c^2}p_a^\alpha p_a^\beta U_\alpha U_\beta$$

$$
+\frac{\Delta_{\alpha\beta}p_a^\alpha p_a^\beta}{3kT}\Big]\nabla_\gamma U^\gamma+(1-e^{-\frac{A}{kT}})\Big[\frac{F_a^{(0)}}{f_a^{(0)}}-\frac{\nu_a}{cn_a}U_\beta p_a^\beta k_f^{(0)}n_A n_B\Big]\Big\}
$$

$$
=\sum_{b=A}^{D}\int f_a^{(0)}f_b^{(0)}(\phi_a'+\phi_b'-\phi_a-\phi_b)F_{ba}\sigma_{ab}\,d\Omega_{ba}\frac{d^3p_b}{p_{b0}}. \tag{7.72}
$$

The integral equation (7.72) has a solution of the form

$$
\phi_a=A_a p_a^\beta\Big[\nabla_\beta T-\frac{T}{nh_E}\nabla_\beta p\Big]+\sum_{b=A}^{C}B_{ab}p_a^\beta\nabla_\beta\Big(\frac{\mu_b-\mu_D}{T}\Big)+D_a\nabla_\alpha U^\alpha
$$

$$
+C_a p_a^\alpha p_a^\beta\nabla_{\langle\alpha}U_{\beta\rangle}+E_a(1-e^{-\frac{A}{kT}}). \tag{7.73}
$$

The scalars A_a, B_{ab}, C_a, D_a and E_a, which are functions of n_a, T, U^α and p_a^α, are determined from the integral equations that are obtained by inserting (7.73) into (7.72), viz.,

$$
-\frac{p_a^\beta}{k}\Delta_{\alpha\beta}(x_d-\delta_{ad})=\sum_{b=A}^{D}\int f_b^{(0)}\Delta_{\alpha\beta}\big[B_{ad}'p_a'^\beta+B_{bd}'p_b'^\beta
$$

$$
-B_{ad}p_a^\beta-B_{bd}p_b^\beta\big]F_{ba}\sigma_{ab}\,d\Omega_{ba}\frac{d^3p_b}{p_{b0}},\quad (d=A,B,C), \tag{7.74}
$$

$$
\frac{p_a^\beta}{kT^2}\Delta_{\alpha\beta}(U_\gamma p_a^\gamma-h_E)=\sum_{b=A}^{D}\int f_b^{(0)}\Delta_{\alpha\beta}\big[A_a'p_a'^\beta+A_b'p_b'^\beta
$$

$$
-A_a p_a^\beta-A_b p_b^\beta\big]F_{ba}\sigma_{ab}\,d\Omega_{ba}\frac{d^3p_b}{p_{b0}}, \tag{7.75}
$$

$$
-\frac{p_a^\gamma p_a^\delta}{kT}\Delta_\gamma^{\langle\alpha}\Delta_\delta^{\beta\rangle}=\sum_{b=A}^{D}\int f_b^{(0)}\Delta_\gamma^{\langle\alpha}\Delta_\delta^{\beta\rangle}\big[C_a'p_a'^\gamma p_a'^\delta+C_b'p_b'^\gamma p_b'^\delta
$$

$$
-C_a p_a^\gamma p_a^\delta-C_b p_b^\gamma p_b^\delta\big]F_{ba}\sigma_{ab}\,d\Omega_{ba}\frac{d^3p_b}{p_{b0}}, \tag{7.76}
$$

$$
-\Big\{\frac{U_\beta p_a^\beta}{c^2}\Big[1+\frac{k}{c_v}(1-\zeta_a G_a)\Big]+\frac{p_a^\alpha p_a^\beta U_\alpha U_\beta}{c_v T c^2}+\frac{\Delta_{\alpha\beta}p_a^\alpha p_a^\beta}{3kT}\Big\}
$$

$$
=\sum_{b=A}^{D}\int f_b^{(0)}\big[D_a'+D_b'-D_a-D_b\big]F_{ba}\sigma_{ab}\,d\Omega_{ba}\frac{d^3p_b}{p_{b0}}, \tag{7.77}
$$

$$
\frac{F_a^{(0)}}{f_a^{(0)}}-\frac{\nu_a}{cn_a}U_\beta p_a^\beta k_f^{(0)}n_A n_B
$$

$$= \sum_{b=A}^{D} \int f_b^{(0)} \left[E_a' + E_b' - E_a - E_b\right] F_{ba}\sigma_{ab}\, d\Omega_{ba} \frac{d^3 p_b}{p_{b0}}. \tag{7.78}$$

Since we are interested only in general expressions for the transport coefficients we shall not solve the integral equations (7.74) through (7.78). The transport coefficients shall be expressed in terms of the coefficients A_a, B_{ab}, C_a, D_a and E_a, which can be obtained from the above mentioned integral equations.

Let us calculate the constitutive terms in the second Chapman and Enskog approximation of a chemically reacting mixture of relativistic gases. These constitutive terms are: the reaction rate density ℓ, the dynamic pressure of the mixture ϖ, the pressure deviator of the mixture $p^{\langle\alpha\beta\rangle}$, the heat flux of the mixture q^α and the diffusion flux $\underset{a}{J}^\alpha$ of the constituent a.

We begin with the determination of the constitutive equation for the reaction rate density, which can be obtained through the insertion of (7.56) together with (7.73) into its definition (7.20), yielding

$$\ell = -L^\star \nabla_\alpha U^\alpha - \mathcal{L}\mathcal{A}. \tag{7.79}$$

In order to get (7.79) we have considered that $\mathcal{A}/(kT) \ll 1$ and approximated $[1 - e^{-\mathcal{A}/(kT)}] \approx \mathcal{A}/(kT)$. This condition is valid in the last stage of a chemical reaction where the processes are close to chemical equilibrium. Further we have neglected all non-linear terms in the scalar thermodynamic forces $\nabla_\alpha U^\alpha$ and $\mathcal{A}$. The coefficients $L^\star$ and $\mathcal{L}$ are given by

$$L^\star = -c \int f_A^{(0)} f_B^{(0)} \left[D_C' + D_D' - D_A - D_B\right] F_{BA}\sigma_{AB}^\star d\Omega \frac{d^3 p_B}{p_{B0}} \frac{d^3 p_A}{p_{A0}}, \tag{7.80}$$

$$\mathcal{L} = \frac{c k_f^{(0)} n_A n_B}{kT} - \frac{c}{kT} \int f_A^{(0)} f_B^{(0)} \left[E_C' + E_D'\right.$$

$$\left. - E_A - E_B\right] F_{BA}\sigma_{AB}^\star d\Omega \frac{d^3 p_B}{p_{B0}} \frac{d^3 p_A}{p_{A0}}. \tag{7.81}$$

We insert now (7.56) together with (7.73) into the definition of the energy-momentum tensor of the mixture, which is given by $(7.17)_2$ and $(7.18)_2$, and by performing the projection $\Delta_{\alpha\beta}T^{\alpha\beta} = -3(p + \varpi)$ we get the constitutive equation for the dynamic pressure of the mixture that reads

$$\varpi = -\eta \nabla_\alpha U^\alpha - L\mathcal{A}. \tag{7.82}$$

In (7.82) the coefficients of bulk viscosity η and of chemical viscosity L are given by

$$\eta = \frac{c}{3} \sum_{a=A}^{D} \int \Delta_{\alpha\beta} p_a^\alpha p_a^\beta f_a^{(0)} D_a \frac{d^3 p_a}{p_{a0}}, \tag{7.83}$$

$$L = \frac{c}{3kT}\sum_{a=A}^{D}\int \Delta_{\alpha\beta}p_a^\alpha p_a^\beta f_a^{(0)} E_a \frac{d^3p_a}{p_{a0}}. \tag{7.84}$$

If we perform the projection $\Delta_\gamma^{\langle\alpha}\Delta_\delta^{\beta\rangle}T^{\gamma\delta} = p^{\langle\alpha\beta\rangle}$ it follows the constitutive equation for the pressure deviator for the mixture, viz.,

$$p^{\langle\alpha\beta\rangle} = 2\mu\nabla^{\langle\alpha}U^{\beta\rangle}, \tag{7.85}$$

where the expression for the coefficient of shear viscosity of the mixture μ becomes

$$\mu = \frac{c}{15}\sum_{a=A}^{D}\int (\Delta_{\alpha\beta}p_a^\alpha p_a^\beta)^2 f_a^{(0)} C_a \frac{d^3p_a}{p_{a0}}. \tag{7.86}$$

Further, the heat flux of the mixture can be obtained from the projection $\Delta^{\alpha\beta}U^\gamma T_{\beta\gamma} = q^\alpha$, yielding

$$q^\alpha = \lambda^\star\left(\nabla^\alpha T - \frac{T}{nh_E}\nabla^\alpha p\right) + T^2\sum_{b=A}^{C} L_b\nabla^\alpha\left(\frac{\mu_b-\mu_D}{T}\right). \tag{7.87}$$

The expressions for the coefficients related to the thermal conductivity of the mixture $\lambda^\star$ and to the diffusion-thermo cross-effect L_b read

$$\lambda^\star = \frac{c}{3}\sum_{a=A}^{D}\int \Delta_{\alpha\beta}U_\gamma p_a^\alpha p_a^\beta p_a^\gamma f_a^{(0)} A_a \frac{d^3p_a}{p_{a0}}, \tag{7.88}$$

$$L_b = \frac{c}{3T^2}\sum_{a=A}^{D}\int \Delta_{\alpha\beta}U_\gamma p_a^\alpha p_a^\beta p_a^\gamma f_a^{(0)} B_{ab} \frac{d^3p_a}{p_{a0}}. \tag{7.89}$$

The last constitutive equation to be determined is the one for the partial diffusion flux J_a^α, which is obtained by inserting (7.56) together with (7.73) into the definition of the partial particle four-flow $(7.17)_1$ and by using its decomposition $(7.22)_1$:

$$J_a^\alpha = L_a^\star\left(\nabla^\alpha T - \frac{T}{nh_E}\nabla^\alpha p\right) + \sum_{b=A}^{C} L_{ab}\nabla^\alpha\left(\frac{\mu_b-\mu_D}{T}\right). \tag{7.90}$$

In (7.90) the coefficients $L_a^\star$ and L_{ab} are related to the thermal-diffusion cross-effect and to the matrix of the diffusion coefficients, respectively. Their expressions are given by

$$L_a^\star = \frac{c}{3}\int \Delta_{\alpha\beta}p_a^\alpha p_a^\beta f_a^{(0)} A_a \frac{d^3p_a}{p_{a0}}, \tag{7.91}$$

$$L_{ab} = \frac{c}{3}\int \Delta_{\alpha\beta}p_a^\alpha p_a^\beta f_a^{(0)} B_{ab} \frac{d^3p_a}{p_{a0}}. \tag{7.92}$$

In the following we shall analyze the relationships between the transport coefficients of a chemically reacting mixture of relativistic gases which are known as the Onsager reciprocity relations.

Problems

7.5.1 Check that (7.63) holds.

7.5.2 Obtain the balance equation for the partial particle number density (7.65) by integrating (7.64) over all values of d^3p_a/p_{a0}.

7.5.3 Check that (7.71) holds.

7.5.4 Obtain the constitutive equation and the corresponding transport coefficients for: a) the reaction rate density (7.79); b) the dynamic pressure of the mixture (7.82); c) the pressure deviator of the mixture (7.85); d) the heat flux of the mixture (7.87) and e) the partial diffusion flux (7.90). (Hint: Make use of the integrals (5.234) through (5.238) of the Appendix to Chapter 5.)

7.6 Onsager reciprocity relations

In order to determine the Onsager reciprocity relations we shall need the following relationship which is valid for arbitrary functions ϕ_a and ψ_a:

$$\sum_{a,b=A}^{D}\int f_a^{(0)}f_b^{(0)}\psi_a(\phi_a'+\phi_b'-\phi_a-\phi_b)F_{ba}\sigma_{ab}\,d\Omega_{ba}\frac{d^3p_b}{p_{b0}}\frac{d^3p_a}{p_{a0}}$$

$$=\sum_{a,b=A}^{D}\int f_a^{(0)}f_b^{(0)}\phi_a(\psi_a'+\psi_b'-\psi_a-\psi_b)F_{ba}\sigma_{ab}\,d\Omega_{ba}\frac{d^3p_b}{p_{b0}}\frac{d^3p_a}{p_{a0}}. \tag{7.93}$$

The above equation can be proved by using the symmetry properties of the elastic collision term and by interchanging the dummy indices a and b in the sums.

Let us analyze first the coefficient of cross-effect $L^\star$ of the reaction rate density ℓ. According to (7.62) we can write the expression (7.80) for $L^\star$ as

$$L^\star = c\sum_{a=A}^{D}\int \mathcal{D}_a F_a^{(0)}\frac{d^3p_a}{p_{a0}}. \tag{7.94}$$

Further if we multiply the integral equation (7.78) by $cf_a^{(0)}\mathcal{D}_a$ and integrate the resulting equation over all values of d^3p_a/p_{a0}, then after summing over all constituents $a=A,B,C,D$ we get that

$$c\sum_{a=A}^{D}\int\left[F_a^{(0)}-\frac{\nu_a}{cn_a}U_\beta p_a^\beta k_f^{(0)}n_An_Bf_a^{(0)}\right]\mathcal{D}_a\frac{d^3p_a}{p_{a0}}$$

$$=c\sum_{a,b=A}^{D}\int f_a^{(0)}f_b^{(0)}\mathcal{D}_a\big[\mathcal{E}_a'+\mathcal{E}_b'-\mathcal{E}_a-\mathcal{E}_b\big]F_{ba}\sigma_{ab}\,d\Omega_{ba}\frac{d^3p_b}{p_{b0}}\frac{d^3p_a}{p_{a0}}=L^\star. \tag{7.95}$$

The last equality was written thanks to the expression (7.94) of $L^\star$ and to the constraint (7.58).

The coefficient of cross-effect L of the dynamic pressure ϖ of the mixture can be obtained from the integral equation (7.77). Indeed, the multiplication of (7.77) by $-cf_a^{(0)}E_a$, the integration of the resulting equation over all values of d^3p_a/p_{a0} and subsequent sum over all constituents $a = A, B, C, D$ leads to

$$c\sum_{a=A}^{D}\int\left\{\frac{U_\beta p_a^\beta}{c^2}\left[(1+\frac{k}{c_v}(1-\zeta_a G_a)\right]+\frac{p_a^\alpha p_a^\beta U_\alpha U_\beta}{c_v T c^2}+\frac{\Delta_{\alpha\beta}p_a^\alpha p_a^\beta}{3kT}\right\}f_a^{(0)}E_a\frac{d^3p_a}{p_{a0}}$$

$$=-c\sum_{a,b=A}^{D}\int f_a^{(0)}f_b^{(0)}E_a\left[D_a'+D_b'-D_a-D_b\right]F_{ba}\sigma_{ab}\,d\Omega_{ba}\frac{d^3p_b}{p_{b0}}\frac{d^3p_a}{p_{a0}}=L. \quad (7.96)$$

In order to obtain the last equality in (7.96) we have used the constraints (7.58), (7.59) and the definition (7.84) of L.

We compare now (7.95) with (7.96) and by using the relationship (7.93) we find the first Onsager reciprocity relation between the coefficients of cross-effects which occur in the constitutive equations for the reaction rate density and for the dynamic pressure, i.e., $L = -L^\star$. We recall here that the minus sign which has appeared in this relationship is due to the fact that under time reversal the affinity $\mathcal{A}$ does not change its sign while the divergence of the four-velocity does change.

For the determination of the next Onsager reciprocity relations we note first that due to the constraint that there exist only three linearly independent diffusion fluxes since $\sum_{a=A}^{D}J_a^\alpha=0$ we must have according to (7.90) through (7.92):

$$\sum_{a=A}^{D}L_a^\star=\frac{c}{3}\sum_{a=A}^{D}\int\Delta_{\alpha\beta}p_a^\alpha p_a^\beta f_a^{(0)}A_a\frac{d^3p_a}{p_{a0}}=0, \quad (7.97)$$

$$\sum_{a=A}^{D}L_{ab}=\frac{c}{3}\sum_{a=A}^{D}\int\Delta_{\alpha\beta}p_a^\alpha p_a^\beta f_a^{(0)}B_{ab}\frac{d^3p_a}{p_{a0}}=0, \quad (b=A,B,C). \quad (7.98)$$

We observe the above constraints and the constraint between the particle fractions $\sum_{A=1}^{D}x_a=1$ to write the two coefficients $L_a^\star$ and L_{ab} as

$$L_a^\star=-\frac{c}{3}\sum_{b=A}^{D}\int\Delta_{\alpha\beta}p_b^\alpha p_b^\beta f_b^{(0)}A_b(x_a-\delta_{ab})\frac{d^3p_b}{p_{b0}}, \quad (7.99)$$

$$L_{ab}=-\frac{c}{3}\sum_{d=A}^{D}\int\Delta_{\alpha\beta}p_d^\alpha p_d^\beta f_d^{(0)}B_{db}(x_a-\delta_{ad})\frac{d^3p_d}{p_{d0}}. \quad (7.100)$$

Next we multiply (7.74) by $ckp_a^\alpha f_a^{(0)} B_{ac}/3$, integrate the resulting equation over all values of d^3p_a/p_{a0} and after that we sum over all constituents, yielding

$$-\frac{c}{3}\sum_{a=A}^{D}\int \Delta_{\alpha\beta}p_a^\alpha p_a^\beta f_a^{(0)} B_{ac}(x_d-\delta_{ad})\frac{d^3p_a}{p_{a0}} \stackrel{(7.100)}{=} L_{dc}$$

$$=\frac{ck}{3}\sum_{a,b=A}^{D}\int f_a^{(0)} f_b^{(0)}\Delta_{\alpha\beta}p_a^\alpha B_{ac}\big[B'_{ad}p_a'^\beta + B'_{bd}p_b'^\beta - B_{ad}p_a^\beta$$

$$-B_{bd}p_b^\beta\big]F_{ba}\sigma_{ab}\,d\Omega_{ba}\frac{d^3p_b}{p_{b0}}\frac{d^3p_a}{p_{a0}} \stackrel{(7.93)}{=} \frac{ck}{3}\sum_{a,b=A}^{D}\int f_a^{(0)} f_b^{(0)}\Delta_{\alpha\beta}p_a^\alpha B_{ad}\big[B'_{ac}p_a'^\beta$$

$$+B'_{bc}p_b'^\beta - B_{ac}p_a^\beta - B_{bc}p_b^\beta\big]F_{ba}\sigma_{ab}\,d\Omega_{ba}\frac{d^3p_b}{p_{b0}}\frac{d^3p_a}{p_{a0}} = L_{cd}. \tag{7.101}$$

Hence we have proved the second Onsager reciprocity relation which shows that the matrix of the diffusion coefficients is symmetric, $L_{cd}=L_{dc}$.

The expression for the coefficient of the thermal-diffusion effect $L_d^\star$ is obtained through the multiplication of (7.74) by $ckp_a^\alpha f_a^{(0)} A_a/3$, integration of the resulting equation over all values of d^3p_a/p_{a0} and subsequent summing over all constituents:

$$-\frac{c}{3}\sum_{a=A}^{D}\int \Delta_{\alpha\beta}p_a^\alpha p_a^\beta (x_d-\delta_{ad}) f_a^{(0)} A_a\frac{d^3p_a}{p_{a0}} = L_d^\star$$

$$=\frac{ck}{3}\sum_{a,b=A}^{D}\int f_a^{(0)} f_b^{(0)}\Delta_{\alpha\beta}p_a^\alpha A_a\big[B'_{ad}p_a'^\beta + B'_{bd}p_b'^\beta$$

$$-B_{ad}p_a^\beta - B_{bd}p_b^\beta\big]F_{ba}\sigma_{ab}\,d\Omega_{ba}\frac{d^3p_b}{p_{b0}}\frac{d^3p_a}{p_{a0}}, \tag{7.102}$$

where in the first equality we have used the relationship (7.99).

In the same way we obtain the expression for the coefficient of the diffusion-thermo effect L_d. Indeed the multiplication of equation (7.75) by $ckp_a^\alpha f_a^{(0)} B_{ad}/3$, integration of the resulting equation over all values of d^3p_a/p_{a0} and the subsequent summing over all constituents leads to:

$$\frac{c}{3T^2}\sum_{a=A}^{D}\int \Delta_{\alpha\beta}p_a^\alpha p_a^\beta (U_\gamma p_a^\gamma - h_E) f_a^{(0)} B_{ad}\frac{d^3p_a}{p_{a0}} = L_d$$

$$=\frac{ck}{3}\sum_{a,b=A}^{D}\int f_a^{(0)} f_b^{(0)}\Delta_{\alpha\beta}p_a^\alpha B_{ad}\big[A'_a p_a'^\beta + A'_b p_b'^\beta$$

$$-A_a p_a^\beta - A_b p_b^\beta\big]F_{ba}\sigma_{ab}\,d\Omega_{ba}\frac{d^3p_b}{p_{b0}}\frac{d^3p_a}{p_{a0}}. \tag{7.103}$$

The first equality above can be written thanks to the constraint (7.59) and to the definition of the thermal-diffusion coefficient (7.89).

We compare now the expressions (7.102) and (7.103) and by using the property (7.93) we can infer the last Onsager reciprocity relation that relates the coefficients of the thermal-diffusion effect and of the diffusion-thermo effect $L_d^\star = L_d$.

Problems

7.6.1 Show that the coefficient of cross-effect $L^\star$ of the reaction rate density can be written as (7.94).

7.6.2 Check that (7.99) and (7.100) hold.

References

[1] J. L. Anderson, Variational principles for calculation of transport coefficients of relativistic multi-component systems, *Physica* **87 A**, 287–309 (1976).

[2] S. R. de Groot, W. A. van Leeuwen and Ch. G. van Weert, *Relativistic kinetic theory* (North-Holland, Amsterdam, 1980).

[3] W. T. Hermens, Relativistic kinetic theory of chemical reactions and other transport phenomena. Part I: General discussion, *Proc. Kon. Ned. Akad. Wetensch.* **74 B**, 376–397 (1971).

[4] W. T. Hermens, Relativistic kinetic theory of chemical reactions and other transport phenomena. Part II: Normally reactive mixture, *Proc. Kon. Ned. Akad. Wetensch.* **74 B**, 461–477 (1971).

[5] W. T. Hermens, W. A. van Leeuwen, C. G. van Weert and S. R. de Groot, On relativistic kinetic gas theory. VIII. Reciprocal relations for a reactive mixture, *Physica* **60**, 472–487 (1972).

[6] I. Prigogine and E. Xhrouet, On the perturbation of Maxwell distribution function by chemical reactions in gases, *Physica* **15**, 913–932 (1949).

[7] J. Ross and P. Mazur, Some deductions from a formal statistical mechanical theory of chemical kinetics, *J. Chem. Phys.* **35**, 19–28 (1961).

[8] J. M. Stewart, *Non-equilibrium relativistic kinetic theory*, Lecture Notes in Physics, vol. 10 (Springer, Heidelberg, 1971).

Chapter 8

Model Equations

8.1 Introduction

As we have seen one of the difficulties in dealing with the Boltzmann equation is related with the expression for its collision term since it depends on the product of distribution functions. In order to simplify the structure of the collision term while maintaining its basic properties, simpler expressions – known as collision models – have been proposed. In a non-relativistic theory the most widely known model of the Boltzmann equation is the so-called BGK model which was formulated independently by Bhatnagar, Gross and Krook [4] and Welander [22]. Other model equations for non-relativistic gases were proposed later and for more details we refer to the book by Cercignani [7].

The properties that the collision model $J(f)$ must fulfill are the same as those of the true collision term $Q(f,f)$. For a single non-degenerate relativistic gas the properties are:

1) For all summational invariants ψ, $Q(f,f)$ and $J(f)$ must satisfy

$$\int \psi Q(f,f)\frac{d^3p}{p_0} = 0, \qquad \text{hence} \qquad \int \psi J(f)\frac{d^3p}{p_0} = 0. \tag{8.1}$$

2) The tendency of the one-particle distribution function to the equilibrium distribution function – or equivalently the $\mathcal{H}$-theorem – must hold, i.e.,

$$\int Q(f,f)\ln f\frac{d^3p}{p_0} \le 0, \qquad \text{hence} \qquad \int J(f)\ln f\frac{d^3p}{p_0} \le 0. \tag{8.2}$$

The first model equation was proposed by Marle [18, 19, 20]; it was an extension of the non-relativistic BGK model to the relativistic case. Its expression is given by

$$J(f) = -\frac{m}{\tau}(f - f^{(0)}), \tag{8.3}$$

where τ represents a characteristic time of order of the time between collisions and m is the rest mass of a particle of the relativistic gas.

If we insert (8.3) into (8.1) it follows that we must have for the summational invariants $\psi = c$ and cp^α,

$$c\int f\frac{d^3p}{p_0} = A = c\int f^{(0)}\frac{d^3p}{p_0} = A_E \overset{(3.27)}{=} \frac{n}{m}\frac{K_1(\zeta)}{K_2(\zeta)}, \tag{8.4}$$

$$c\int p^\alpha f\frac{d^3p}{p_0} = N^\alpha = c\int p^\alpha f^{(0)}\frac{d^3p}{p_0} = N^\alpha_E, \tag{8.5}$$

respectively, so that the moments of zeroth-order A and of first order N^α represent only equilibrium quantities. Since N^α is the particle four-flow and (8.5) is a condition which is valid in the decomposition of Eckart (see (4.21)), such description is adopted when one is dealing with the model of Marle.

The second property (8.2) can be verified as follows. First we build the relationship

$$\int J(f)\ln f\frac{d^3p}{p_0} = -\frac{m}{\tau}\int (f - f^{(0)})\ln f\frac{d^3p}{p_0}. \tag{8.6}$$

Next we observe that due to the fact that $\ln f^{(0)}$ is a summational invariant we have that

$$\int f\ln f^{(0)}\frac{d^3p}{p_0} = \int f^{(0)}\ln f^{(0)}\frac{d^3p}{p_0}, \tag{8.7}$$

and (8.6) can be written as

$$\int J(f)\ln f\frac{d^3p}{p_0} = -\frac{m}{\tau}\int f^{(0)}\left(\frac{f}{f^{(0)}} - 1\right)\ln\left(\frac{f}{f^{(0)}}\right)\frac{d^3p}{p_0} \le 0, \tag{8.8}$$

thanks to the inequality $(x-1)\ln x \ge 0$ which is valid for all $x > 0$.

Let us analyze the kinetic model of Marle whose Boltzmann equation for a single non-degenerate relativistic gas without external forces can be written as

$$p^\alpha\frac{\partial f}{\partial x^\alpha} = -\frac{m}{\tau}(f - f^{(0)}). \tag{8.9}$$

If the one-particle distribution function does not depend on the spatial coordinates, (8.9) reduces to

$$\frac{E}{c^2}\frac{df}{dt} = -\frac{m}{\tau}(f - f^{(0)}), \tag{8.10}$$

where $E = cp^0$ is the energy of a particle of the gas. If we integrate the above equation we get that

$$f(t) = \left[f(0) + \frac{1}{\tau^\star}\int_0^t e^{\frac{t'}{\tau^\star}} f^{(0)}(t')dt'\right]e^{-\frac{t}{\tau^\star}}. \tag{8.11}$$

In (8.11), $\tau^\star = E\tau/mc^2$ is the relaxation time of the distribution function which depends on the energy of a gas particle. In the non-relativistic limiting case the particle energy reduces to $E \approx mc^2$ so that $\tau^\star = \tau$ and we recover the expression of the non-relativistic BGK model. However for the case of particles with zero rest mass the particle energy is given by $E = c|\mathbf{p}|$ and the relaxation time of the distribution function $\tau^\star$ tends to infinity.

The above arguments are due to Anderson and Witting [2] who proposed the model equation

$$J(f) = -\frac{U_L^\alpha p_\alpha}{c^2\tau}(f - f^{(0)}), \tag{8.12}$$

where U_L^α is the four-velocity in the Landau and Lifshitz description.

For the summational invariants $\psi = c$ and cp^α we obtain from (8.1) and (8.12), respectively,

$$c\int p^\alpha U_{L\alpha} f \frac{d^3p}{p_0} = N^\alpha U_{L\alpha} = c\int p^\alpha U_{L\alpha} f^{(0)} \frac{d^3p}{p_0} = N_E^\alpha U_{L\alpha}, \tag{8.13}$$

$$c\int p^\alpha p^\beta U_{L\beta} f \frac{d^3p}{p_0} = T^{\alpha\beta} U_{L\beta} = c\int p^\alpha p^\beta U_{L\beta} f^{(0)} \frac{d^3p}{p_0} = T_E^{\alpha\beta} U_{L\beta}. \tag{8.14}$$

The two above equations are verified for the Landau and Lifshitz decomposition of the particle four-flow and energy-momentum tensor (see (4.46) and (4.47)).

For the verification of the second property (8.2) we note that

$$\int p^\alpha U_{L\alpha} f \ln f^{(0)} \frac{d^3p}{p_0} = \int p^\alpha U_{L\alpha} f^{(0)} \ln f^{(0)} \frac{d^3p}{p_0}, \tag{8.15}$$

thanks to the fact that $\ln f^{(0)}$ is a summational invariant and to the relationships (8.13) and (8.14). Hence we can write

$$\int J(f) \ln f \frac{d^3p}{p_0} = -\frac{1}{c^2\tau}\int p^\alpha U_{L\alpha} f^{(0)} \left(\frac{f}{f^{(0)}} - 1\right) \ln \frac{f}{f^{(0)}} \frac{d^3p}{p_0} \leq 0. \tag{8.16}$$

The Boltzmann equation without external forces for the kinetic model of Anderson and Witting is

$$p^\alpha \frac{\partial f}{\partial x^\alpha} = -\frac{U_L^\alpha p_\alpha}{c^2\tau}(f - f^{(0)}). \tag{8.17}$$

If we consider that the distribution function does not depend on the spatial coordinates and choose a local Lorentz rest frame where $(U_L^\alpha) = (c, \mathbf{0})$, we can write (8.17) as

$$\frac{p^0}{c}\frac{df}{dt} = -\frac{cp^0}{c^2\tau}(f - f^{(0)}). \tag{8.18}$$

The solution of the above equation is given by

$$f(t) = \left[f(0) + \frac{1}{\tau}\int_0^t e^{\frac{t'}{\tau}} f^{(0)}(t')dt'\right] e^{-\frac{t}{\tau}}, \tag{8.19}$$

which has the same expression as that for the non-relativistic BGK model.

Problems

8.1.1 Check that (8.7) and (8.15) hold.

8.1.2 Show that the solution of the differential equation (8.10) is given by (8.11).

8.2 The characteristic time

In this section we shall show how to estimate the characteristic time τ which appears in the model equations.

In elementary kinetic theory of non-relativistic gases the time between collisions of hard-sphere particles of diameter a is given by

$$\tau_c = \frac{1}{n\pi a^2 \langle g \rangle}, \tag{8.20}$$

where $g = |\mathbf{v} - \mathbf{v}_*|$ denotes the relative velocity of two particles of velocities $\mathbf{v}$ and $\mathbf{v}_*$ and $\langle g \rangle$ is its mean value. In terms of the temperature of the gas T and of the rest mass m of a particle of the gas, the mean value of the relative velocity reads

$$\langle g \rangle = 4\sqrt{\frac{kT}{\pi m}}. \tag{8.21}$$

The characteristic time τ usually adopted in the model equations is proportional to the time between collisions τ_c. Since the mean relative velocity is of order of the mean velocity $\langle v \rangle$ and of the adiabatic sound speed v_s, i.e.,

$$\langle v \rangle = \sqrt{\frac{8kT}{\pi m}}, \qquad v_s = \sqrt{\frac{5kT}{3m}}, \qquad \langle g \rangle = \sqrt{2}\langle v \rangle = \sqrt{\frac{48}{5\pi}} v_s, \tag{8.22}$$

we can write the time between collisions in terms of a mean velocity $\mathcal{V}$ which is proportional to $\langle v \rangle$, or $\langle g \rangle$, or v_s:

$$\tau = \frac{1}{n\pi a^2 \mathcal{V}}. \tag{8.23}$$

Let us estimate the mean velocity $\langle v \rangle$ of a non-degenerate relativistic gas by calculating the mean value of the modulus of the velocity $\mathbf{v} = c\mathbf{p}/p_0$ with the Maxwell–Jüttner equilibrium distribution function (3.27), that is

$$\langle v \rangle = \int c\frac{|\mathbf{p}|}{p_0} f^{(0)} \frac{d^3p}{p_0} \Bigg/ \left(\int f^{(0)} \frac{d^3p}{p_0} \right) \overset{(8.4)}{=} \frac{mc^2 K_2}{nK_1} \int_0^\infty 4\pi |\mathbf{p}|^3 f^{(0)} \frac{d|\mathbf{p}|}{(p_0)^2}$$

$$\overset{(5.223)}{=} \frac{\zeta c}{K_1} \int_1^\infty e^{-\zeta y} \left(y - \frac{1}{y} \right) dy = \frac{\zeta c}{K_1} \left[e^{-\zeta} \frac{(1+\zeta)}{\zeta^2} - \int_1^\infty \frac{e^{-\zeta y}}{y} dy \right]. \tag{8.24}$$

The above integral was calculated in a local Lorentz rest frame by following the same methodology described in the Appendix of Chapter 5.

In the two limiting cases we have

a) non-relativistic limit ($\zeta \gg 1$): $\langle v \rangle \to \sqrt{\frac{8kT}{\pi m}}$,

b) ultra-relativistic limit ($\zeta \ll 1$): $\langle v \rangle \to c$,

i.e., in the non-relativistic limit the mean velocity tends to $(8.22)_1$ while in the ultra-relativistic limit it tends to the speed of light.

For degenerate gases the mean velocity $\langle v \rangle$ is calculated by using the equilibrium distribution function

$$f^{(0)} = \frac{g_s/h^3}{e^{-\frac{\mu_E}{kT}+\frac{U^\alpha p_\alpha}{kT}} \pm 1}. \tag{8.25}$$

The calculation is performed in the same manner as in Section 2.7, i.e., by considering a local Lorentz rest frame, yielding

$$\langle v \rangle = c\frac{J_{3,-1}}{J_{20}}, \tag{8.26}$$

where $J_{nm}(\zeta, \mu_E)$ are integrals defined in Chapter 2 (see also (8.206)).

We can also estimate the mean value of the Møller relative speed

$$g_ø = \sqrt{(\mathbf{v}-\mathbf{v}_*)^2 - \frac{1}{c^2}(\mathbf{v}\times\mathbf{v}_*)^2}, \tag{8.27}$$

by integrating $g_ø$ over all values of the velocities $\mathbf{v} = c\mathbf{p}/p_0$ and $\mathbf{v}_* = c\mathbf{p}_*/p_{*0}$ and by using the Maxwell–Jüttner equilibrium distribution function, i.e.,

$$\langle g_ø \rangle = \int g_ø f^{(0)} f_*^{(0)} \frac{d^3p}{p_0}\frac{d^3p_*}{p_{*0}} \Big/ \left(\int f^{(0)}\frac{d^3p}{p_0}\right)^2. \tag{8.28}$$

If we introduce the total momentum four-vector P^α, the relative momentum four-vector Q^α and choose the center-of-mass system where the spatial components of P^α vanish (see Section 5.3), we can write (8.28) as

$$\langle g_ø \rangle = \frac{c}{K_1^2}\int_{2\zeta}^\infty x\left(\frac{x^2}{\zeta^2}-4\right)dx \int_1^\infty \sqrt{y^2-1}\,e^{-xy}\frac{dy}{y^2}. \tag{8.29}$$

In the above equation we have introduced the new variables of integration x, y through the relationships

$$|\mathbf{P}| = \frac{mcx}{\zeta}\sqrt{y^2-1}, \qquad Q = mc\sqrt{\frac{x^2}{\zeta^2}-4}, \qquad P^0 = \frac{mcxy}{\zeta}, \tag{8.30}$$

so that differentials of $|\mathbf{P}|$ and Q can be written as

$$d|\mathbf{P}| = \frac{mcx}{\zeta}\frac{ydy}{\sqrt{y^2-1}}, \qquad dQ = \frac{mcx}{\zeta^2}\frac{dx}{\sqrt{\frac{x^2}{\zeta^2}-4}}. \tag{8.31}$$

By performing the integral in the variable x we get

$$\langle g_\varnothing \rangle = \frac{2c}{K_1^2}\int_1^\infty \frac{\sqrt{y^2-1}}{y^6\zeta^2}(3+6\zeta y+4\zeta^2 y^2)e^{-2\zeta y}dy. \tag{8.32}$$

If we change again the variable of integration by introducing $y = \cosh t$ it follows that

$$\langle g_\varnothing \rangle = \frac{2c}{\zeta^2 K_1^2}[4\zeta^2\mathrm{Ki}_2(2\zeta)+6\zeta\mathrm{Ki}_3(2\zeta)+(3-4\zeta^2)\mathrm{Ki}_4(2\zeta)$$
$$-6\zeta\mathrm{Ki}_5(2\zeta)-3\mathrm{Ki}_6(2\zeta)]. \tag{8.33}$$

In (8.33) Ki_1 denotes the integral for the modified Bessel functions (see Abramowitz and Stegun [1] p. 483):

$$\mathrm{Ki}_n(\zeta) = \int_\zeta^\infty \mathrm{Ki}_{n-1}(t)dt = \int_0^\infty \frac{e^{-\zeta\cosh t}}{\cosh^n t}dt, \tag{8.34}$$

where the relationships hold

$$\mathrm{Ki}_0(\zeta) = K_0(\zeta), \qquad \mathrm{Ki}_{-n} = (-1)^n \frac{d^n K_0(\zeta)}{d\zeta^n}. \tag{8.35}$$

In the limiting case where $\zeta \to 0$ the integrals reduce to

$$\mathrm{Ki}_{2n}(0) = \frac{\Gamma(n)\Gamma(3/2)}{\Gamma(n+1/2)}, \qquad \mathrm{Ki}_{2n+1}(0) = \frac{\pi}{2}\frac{\Gamma(n+1/2)}{\Gamma(1/2)\Gamma(n+1)}, \tag{8.36}$$

and one can prove that the asymptotic formula in the limiting case where $\zeta \gg 1$ is

$$\mathrm{Ki}_n(\zeta) = e^{-\zeta}\sqrt{\frac{\pi}{2\zeta}}\left[1-\frac{4n+1}{8\zeta}+\frac{3(16n^2+24n+3)}{128\zeta^2}\right.$$
$$\left.-\frac{5(64n^3+240n^2+212n+15)}{1024\zeta^3}+\cdots\right]. \tag{8.37}$$

The two limiting cases of the mean value of the Møller relative speed are

a) non-relativistic limit ($\zeta \gg 1$): $\langle g_\varnothing \rangle \to 4\sqrt{\frac{kT}{\pi m}}$,
b) ultra-relativistic limit ($\zeta \ll 1$): $\langle g_\varnothing \rangle \to \frac{4}{5}c$,

i.e., in the non-relativistic limit the Møller relative speed tends to the relative velocity (8.21) while in the ultra-relativistic limit it tends to 4/5 of the speed of light.

The adiabatic sound speed of a non-degenerate relativistic gas is calculated in Section 9.3.2 and its value reads (see (9.78))

$$v_s = \sqrt{\frac{kTc_pc^2}{h_E c_v}} = \sqrt{\frac{\zeta^2+5G\zeta-G^2\zeta^2}{G(\zeta^2+5G\zeta-G^2\zeta^2-1)}\frac{kT}{m}}. \tag{8.38}$$

For a degenerate relativistic gas the expression of the adiabatic sound speed is given by (see Problem 9.3.2.1)

$$v_s = \sqrt{\zeta \frac{[J^\bullet_{23}(J^\bullet_{40})^2 - 2J^\bullet_{22}J^\bullet_{41}J^\bullet_{40} + J^\bullet_{21}(J^\bullet_{41})^2]}{3[J^\bullet_{21}J^\bullet_{23} - (J^\bullet_{22})^2]J^\bullet_{41}} \frac{kT}{m}}. \tag{8.39}$$

In (8.39) $J^\bullet_{nm}(\zeta, \mu_E)$ denotes the partial derivative of $J_{nm}(\zeta, \mu_E)$ with respect to $\mu_E/(kT)$. The relationships between the integrals $J_{nm}(\zeta, \mu_E)$ and the integrals $I_n(\zeta, \mu_E)$ – defined in Chapter 3 (see also (8.206) and (8.207)) – for some values of m and n are given in the Appendix of this chapter.

We use the expression (8.23) of the characteristic time for the non-relativistic model equations and write the corresponding characteristic time for the relativistic model equations as

$$\tau = \frac{1}{4n\pi\sigma\mathcal{V}}, \tag{8.40}$$

where $\sigma = a^2/4$ is the differential cross-section of a relativistic gas of hard-sphere particles (see Section 5.4.1) and $\mathcal{V}$ is a mean velocity which is proportional to the mean velocities (8.24), (8.26), or to the Møller relative speed (8.33) or to the adiabatic sound speeds (8.38) and (8.39).

Problems

8.2.1 Obtain from the expressions for the mean velocity (8.24) and for the mean Møller relative speed (8.33) their corresponding values in the non-relativistic and ultra-relativistic limiting cases.

8.2.2 Show that the mean velocity of the particles of a degenerate gas is given by (8.26).

8.2.3 Obtain the asymptotic formula (8.37) for $\mathrm{Ki}_n(\zeta)$ in the limiting case where $\zeta \gg 1$.

8.3 Single non-degenerate gas

8.3.1 The model of Marle

Let us apply the model of Marle to the determination of the transport coefficients of a single non-degenerate relativistic gas. To achieve this objective we first derive a transfer equation which is obtained through the multiplication of the Boltzmann equation (8.9) by an arbitrary function $\psi(x^\beta, p^\beta)$ and integration of the resulting equation over all values of d^3p/p_0, yielding

$$\frac{\partial}{\partial x^\alpha}\int \psi p^\alpha f \frac{d^3p}{p_0} - \int p^\alpha \frac{\partial \psi}{\partial x^\alpha} f \frac{d^3p}{p_0} = -\frac{m}{\tau}\left[\int \psi f \frac{d^3p}{p_0} - \int \psi f^{(0)} \frac{d^3p}{p_0}\right]. \tag{8.41}$$

By choosing $\psi = c$ and $\psi = cp^\beta$ in the above equation and by using the relationships (8.4) and (8.5) one obtains the conservation laws for the particle four-flow $\partial_\alpha N^\alpha = 0$ and for the energy-momentum tensor $\partial_\alpha T^{\alpha\beta} = 0$, respectively. Further if we choose $\psi = cp^\beta p^\gamma$ the equation for the third-order moment reduces to

$$\partial_\alpha T^{\alpha\beta\gamma} = -\frac{m}{\tau}(T^{\beta\gamma} - T_E^{\beta\gamma}). \tag{8.42}$$

In order to obtain constitutive equations for the dynamic pressure, pressure deviator and heat flux we shall use a method akin to the Maxwellian iteration procedure which was described in Section 6.5. First we write from (6.33) through (6.36) the equilibrium value of the third-order moment as

$$T_E^{\alpha\beta\gamma} = nm^2\left[\left(1+6\frac{G_E}{\zeta_E}\right)U^\alpha U^\beta U^\gamma - c^2\frac{G_E}{\zeta_E}(\eta^{\alpha\beta}U^\gamma + \eta^{\alpha\gamma}U^\beta + \eta^{\gamma\beta}U^\alpha)\right], \tag{8.43}$$

since in equilibrium $\varpi = 0$, $q^\alpha = 0$ and $p^{\langle\alpha\beta\rangle} = 0$. Next we consider on the left-hand side of (8.42) only the equilibrium value of the third-order moment so that the non-equilibrium part of the energy-momentum tensor is given by

$$T^{\beta\gamma} - T_E^{\beta\gamma} = -\frac{\tau}{m}\partial_\alpha T_E^{\alpha\beta\gamma}. \tag{8.44}$$

If we insert (8.43) into (8.44) and multiply the resulting equation by $U_\beta U_\gamma$, it follows that the non-equilibrium part of the energy per particle reads

$$e - e_E = -\tau mc^2\left[2\frac{G_E}{\zeta_E}\nabla^\alpha U_\alpha + 3D\left(\frac{G_E}{\zeta_E}\right)\right], \tag{8.45}$$

since $T^{\beta\gamma}U_\beta U_\gamma = nec^2$. The convective time derivative of (8.45) can be eliminated by using the balance equation of the energy density (4.41) for an Eulerian gas ($p^{\langle\alpha\beta\rangle} = 0, \varpi = 0, q^\alpha = 0$), i.e.,

$$c_v D\zeta_E = k\zeta_E\nabla^\alpha U_\alpha, \qquad \text{where} \qquad c_v = k(\zeta_E^2 + 5G_E\zeta_E - G_E^2\zeta_E^2 - 1). \tag{8.46}$$

Hence we can write[1]

$$D\left(\frac{G_E}{\zeta_E}\right) = \frac{G_E^2\zeta_E - 6G_E - \zeta_E}{\zeta_E(\zeta_E^2 + 5G_E\zeta_E - G_E^2\zeta_E^2 - 1)}\nabla^\alpha U_\alpha, \tag{8.47}$$

and (8.45) reduces to

$$e - e_E = \tau mc^2\frac{20G_E + 3\zeta_E - 13G_E^2\zeta_E - 2G_E\zeta_E^2 + 2G_E^3\zeta_E^2}{\zeta_E(\zeta_E^2 + 5G_E\zeta_E - G_E^2\zeta_E^2 - 1)}\nabla^\alpha U_\alpha. \tag{8.48}$$

The non-equilibrium part of the pressure can be obtained from the expression of the non-equilibrium part of the energy per particle (see (3.28))

$$e - e_E = mc^2\left(G(\zeta) - \frac{1}{\zeta}\right) - mc^2\left(G(\zeta_E) - \frac{1}{\zeta_E}\right), \tag{8.49}$$

[1] Recall that the derivative of G with respect to ζ is $G' = G^2 - 5G/\zeta - 1$.

by expanding $G(\zeta)$ about ζ_E and by neglecting terms up to the second order in $(\zeta - \zeta_E)$:

$$G(\zeta) \approx G(\zeta_E)+G'(\zeta_E)(\zeta-\zeta_E) = G(\zeta_E)+\left[G(\zeta_E)^2 - 5\frac{G(\zeta_E)}{\zeta_E} - 1\right](\zeta-\zeta_E). \quad (8.50)$$

Insertion of (8.50) into (8.49) and by considering that

$$p - p_E = nmc^2\left(\frac{1}{\zeta} - \frac{1}{\zeta_E}\right), \quad (8.51)$$

we find the desired expression for the non-equilibrium part of the pressure

$$p - p_E = \frac{-n(e-e_E)}{1 - 5G_E\zeta_E - \zeta_E^2 + G_E^2\zeta_E^2}$$

$$\overset{(8.48)}{=} \tau p_E \frac{20G_E + 3\zeta_E - 13G_E^2\zeta_E - 2G_E\zeta_E^2 + 2G_E^3\zeta_E^2}{(1 - 5G_E\zeta_E - \zeta_E^2 + G_E^2\zeta_E^2)^2}\nabla^\alpha U_\alpha. \quad (8.52)$$

We are now ready to obtain the constitutive equation for the dynamic pressure. Indeed the projection $\Delta_{\beta\gamma}$ of (8.44) leads to

$$\varpi = -(p - p_E) - \tau p_E\left[D\left(\frac{G_E}{\zeta_E}\right) + \frac{G_E}{\zeta_E}\left(\frac{Dn}{n} + \frac{5}{3}\nabla^\alpha U_\alpha\right)\right] = -\eta\nabla^\alpha U_\alpha. \quad (8.53)$$

In order to derive (8.53) we have used the relationship $\Delta_{\beta\gamma}T^{\beta\gamma} = -3(p+\varpi)$, the balance equation for the particle number density $Dn + n\nabla^\alpha U_\alpha = 0$, and equations (8.47) and (8.52). The coefficient of bulk viscosity for the model equation of Marle is given by

$$\eta = \frac{\tau p_E}{3}\frac{20G_E + 3\zeta_E - 13G_E^2\zeta_E - 2G_E\zeta_E^2 + 2G_E^3\zeta_E^2}{1 - 5G_E\zeta_E - \zeta_E^2 + G_E^2\zeta_E^2}$$

$$\times\frac{4 - \zeta_E^2 - 5G_E\zeta_E + G_E^2\zeta_E^2}{1 - 5G_E\zeta_E - \zeta_E^2 + G_E^2\zeta_E^2}. \quad (8.54)$$

The projection $\Delta^\delta_\beta U_\gamma$ of (8.44) leads to the constitutive equations for the heat flux $q^\delta = \Delta^\delta_\beta U_\gamma T^{\beta\gamma}$ that reads

$$q^\delta = nm^2c^4\left[\left(1 + 5\frac{G_E}{\zeta_E}\right)\frac{DU^\delta}{c^2} - \nabla^\delta\left(\frac{G_E}{\zeta_E}\right) - \frac{G_E}{\zeta_E}\frac{\nabla^\delta n}{n}\right]$$

$$= \lambda\left(\nabla^\delta T_E - \frac{T_E}{nh_E}\nabla^\delta p_E\right). \quad (8.55)$$

We have used the balance of the momentum density for an Eulerian gas i.e., $(nh_E/c^2)DU^\alpha = \nabla^\alpha p_E$ in order to eliminate the convective time derivative of the four-velocity from the constitutive equation for the heat flux. The coefficient of thermal conductivity λ for the model equation of Marle is given by

$$\lambda = \tau p_E \frac{k}{m} \zeta_E(\zeta_E + 5G_E - G_E^2 \zeta_E). \tag{8.56}$$

Finally the projection $\Delta_\beta^{(\delta}\Delta_\gamma^{\epsilon)} - \Delta_{\beta\gamma}\Delta^{\delta\epsilon}/3$ of (8.44) gives the constitutive equation for the pressure deviator $p^{\langle\delta\epsilon\rangle} = (\Delta_\beta^{(\delta}\Delta_\gamma^{\epsilon)} - \Delta_{\beta\gamma}\Delta^{\delta\epsilon}/3)T^{\beta\gamma}$:

$$p^{\langle\delta\epsilon\rangle} = 2\mu\nabla^{\langle\delta}U^{\epsilon\rangle}, \quad \text{where} \quad \mu = \tau p_E G_E. \tag{8.57}$$

Equation $(8.57)_2$ represents the coefficient of shear viscosity μ for the model of Marle.

In the limiting case of low temperatures where $\zeta_E \gg 1$ – which corresponds to a non-relativistic gas – the transport coefficients (8.54), (8.56) and $(8.57)_2$ reduce to

$$\eta = \frac{5}{6}\frac{p_E \tau}{\zeta_E^2}\left[1 - \frac{21}{2\zeta_E} + \cdots\right], \tag{8.58}$$

$$\lambda = \frac{5k}{2m} p_E \tau \left[1 + \frac{3}{2\zeta_E} + \cdots\right], \tag{8.59}$$

$$\mu = p_E \tau \left[1 + \frac{5}{2\zeta_E} + \cdots\right], \tag{8.60}$$

by using the asymptotic expansion (3.22). It is interesting to note that for this model the leading term of the ratio between the thermal conductivity (8.60) and the shear viscosity (8.59) in the non-relativistic limiting case tends to $\lambda/\mu \approx 5k/2m$, while this ratio for the full Boltzmann equation is $\lambda/\mu \approx 15k/4m$ (see Section 5.4). This is a shortcoming of all kinetic models of BGK type.

For high temperatures, i.e., in the ultra-relativistic limiting case where $\zeta_E \ll 1$ the transport coefficients can be written as

$$\eta = \frac{p_E \tau}{54}\zeta_E^3\left[1 + \left(\frac{31}{12} + \frac{9}{2}\ln\left(\frac{\zeta_E}{2}\right) + \frac{9}{2}\gamma\right)\zeta_E^2 + \cdots\right], \tag{8.61}$$

$$\lambda = \frac{4c^2}{\zeta_E T} p_E \tau \left[1 - \frac{1}{8}\zeta_E^2 + \cdots\right], \tag{8.62}$$

$$\mu = \frac{4}{\zeta_E} p_E \tau \left[1 + \frac{1}{8}\zeta_E^2 + \cdots\right]. \tag{8.63}$$

thanks to the expression (3.23). The leading term of the ratio between the coefficients of thermal conductivity and shear viscosity in the ultra-relativistic limit is $\lambda/\mu = c^2/T$ which differs from those found for hard-sphere particles and Israel particles (see Section 5.4).

Problems

8.3.1.1 Show that the non-equilibrium part of the energy per particle is given by (8.45).

8.3.1.2 Check that the equation (8.47) holds.

8.3.1.3 Show that the non-equilibrium part of the pressure is given by (8.52). (Hint: use the fact that the difference between ζ and ζ_E is small.)

8.3.1.4 For the model equation of Marle obtain the constitutive equation and the corresponding expression of the transport coefficient for: a) the dynamic pressure (8.53); b) the heat flux $(8.55)_1$ and c) the pressure deviator $(8.55)_2$.

8.3.2 The model of Anderson and Witting

The kinetic model of Anderson and Witting provides the following transfer equation for an arbitrary function $\psi(x^\beta, p^\beta)$:

$$\frac{\partial}{\partial x^\alpha}\int \psi p^\alpha f \frac{d^3p}{p_0} - \int p^\alpha \frac{\partial \psi}{\partial x^\alpha} f \frac{d^3p}{p_0}$$
$$= -\frac{U_L^\alpha}{c^2\tau}\left[\int \psi p_\alpha f \frac{d^3p}{p_0} - \int \psi p_\alpha f^{(0)} \frac{d^3p}{p_0}\right]. \tag{8.64}$$

If we choose in (8.64) $\psi = c$ and $\psi = cp^\beta$ and use the relationships (8.13) and (8.14), we get the conservation laws of the particle four-flow and of the energy-momentum tensor, respectively. Furthermore the choice $\psi = cp^\beta p^\gamma$ leads to the equation for the third-order moment

$$\partial_\alpha T^{\alpha\beta\gamma} = -\frac{U_{L\alpha}}{c^2\tau}(T^{\alpha\beta\gamma} - T_E^{\alpha\beta\gamma}). \tag{8.65}$$

We shall use (8.65) to determine the transport coefficients of a viscous heat conducting relativistic gas. For this purpose we proceed as follows. First we express the third-order moment $T^{\alpha\beta\gamma}$ in terms of the fourteen fields $n, T, U_L^\alpha, \varpi, p^{\langle\alpha\beta\rangle}$ and q^α:

$$T^{\alpha\beta\gamma} = (nC_1 + C_2\varpi)U_L^\alpha U_L^\beta U_L^\gamma + \frac{c^2}{6}(nm^2 - nC_1 - C_2\varpi)(\eta^{\alpha\beta}U_L^\gamma + \eta^{\alpha\gamma}U_L^\beta$$
$$+\eta^{\beta\gamma}U_L^\alpha) + \tilde{C}_3(\eta^{\alpha\beta}q^\gamma + \eta^{\alpha\gamma}q^\beta + \eta^{\beta\gamma}q^\alpha) - \left(\frac{m^2}{h_E} + \frac{6}{c^2}\tilde{C}_3\right)(U_L^\alpha U_L^\beta q^\gamma$$
$$+U_L^\alpha U_L^\gamma q^\beta + U_L^\beta U_L^\gamma q^\alpha) + C_4(p^{\langle\alpha\beta\rangle}U_L^\gamma + p^{\langle\alpha\gamma\rangle}U_L^\beta + p^{\langle\beta\gamma\rangle}U_L^\alpha). \tag{8.66}$$

The scalar coefficients C_1, C_2 and C_4 are given by (6.34), (6.35) and $(6.36)_2$, respectively, while

$$\tilde{C}_3 = -\frac{m}{\zeta}\frac{G}{\zeta + 5G - G^2\zeta}. \tag{8.67}$$

One can obtain (8.66) from (6.33) by replacing the four-velocity in the Eckart description $U^\alpha = U_L^\alpha - q^\alpha/(nh_E)$ (see Section 4.4) and by neglecting all non-linear terms. Further we use the Maxwellian iteration method and write (8.65) as

$$(T^{\alpha\beta\gamma} - T_E^{\alpha\beta\gamma})U_{L\alpha} = -c^2\tau\partial_\alpha T_E^{\alpha\beta\gamma}. \tag{8.68}$$

If we insert (8.66) into (8.68) and take the projections $U_{L\beta}U_{L\gamma}$ (or $\Delta_{L\beta\gamma}$), $\Delta_{L\beta}^{\delta}U_{L\gamma}$ and $\Delta_{L\beta}^{(\delta}\Delta_{L\gamma}^{\epsilon)} - \Delta_{L\beta\gamma}\Delta_L^{\delta\epsilon}/3$, we get respectively

$$\varpi = -\eta\nabla_\alpha U_L^\alpha, \quad q^\delta = \lambda\left(\nabla^\delta T - \frac{T}{nh}\nabla^\delta p\right), \quad p^{\langle\delta\epsilon\rangle} = 2\mu\nabla^{\langle\delta}U_L^{\epsilon\rangle}, \tag{8.69}$$

which express the Navier–Stokes and Fourier laws. Here, as in the model of Marle, we have eliminated the convective time derivatives DT and DU^α through the balance equations of the energy per particle and of the momentum density of an Eulerian relativistic gas, respectively. We note also that in the linearized case $D = U^\alpha\partial_\alpha \approx U_L^\alpha\partial_\alpha$ and $\nabla^\alpha = \Delta^{\alpha\beta}\partial_\beta \approx \Delta_L^{\alpha\beta}\partial_\beta$. The transport coefficients of bulk viscosity η, thermal conductivity λ and shear viscosity μ – for the model equation of Anderson and Witting by using the equation for the third-order moment (8.65) – are given by

$$\eta = \frac{\tau p}{3}\frac{\zeta(20G + 3\zeta - 13G^2\zeta - 2G\zeta^2 + 2G^3\zeta^2)^2}{(1 - 5G\zeta - \zeta^2 + G^2\zeta^2)}$$

$$\times\frac{1}{(30G + 5\zeta - 45G^2\zeta - 19G\zeta^2 + 9G^3\zeta^2 - 2\zeta^3 + 2G^2\zeta^3)}, \tag{8.70}$$

$$\lambda = \tau p\frac{k}{m}\frac{\zeta^2G(\zeta + 5G - G^2\zeta)^2}{G^2\zeta^2 + 5G^2 - 5G\zeta - \zeta^2}, \tag{8.71}$$

$$\mu = \tau p\frac{\zeta G^2}{\zeta + 6G}. \tag{8.72}$$

For low temperatures where $\zeta \gg 1$ the transport coefficients (8.70) through (8.72) reduce to

$$\eta = \frac{5}{6}\frac{p\tau}{\zeta^2}\left[1 - \frac{16}{\zeta} + \cdots\right], \tag{8.73}$$

$$\lambda = \frac{5k}{2m}p\tau\left[1 - \frac{3}{\zeta} + \cdots\right], \tag{8.74}$$

$$\mu = p\tau\left[1 - \frac{1}{\zeta} + \cdots\right], \tag{8.75}$$

which are the expressions in the limiting case of a non-relativistic gas. Note that the leading terms of (8.74) and (8.75) imply that $\lambda/\mu = 5k/(2m)$ which is the same result found by using the model of Marle.

In the ultra-relativistic limiting case where $\zeta \ll 1$ we obtain from (8.70) through (8.72)

$$\eta = \frac{p\tau}{216}\zeta^4\left[1+\left(\frac{25}{6}+6\ln\left(\frac{\zeta}{2}\right)+6\gamma\right)\zeta^2+\cdots\right], \tag{8.76}$$

$$\lambda = \frac{4c^2}{5T}p\tau\left[1-\frac{13}{40}\zeta^2+\cdots\right], \tag{8.77}$$

$$\mu = \frac{2}{3}p\tau\left[1+\frac{1}{12}\zeta^2+\cdots\right]. \tag{8.78}$$

The method described above is based on the equation for the third-order moment (8.65) and on the Maxwellian iteration procedure. The method used by Anderson and Witting was the Chapman and Enskog procedure. In the following we shall show that the latter leads to different expressions for the transport coefficients.

We begin by writing the one-particle distribution function as

$$f = f^{(0)}(1+\phi), \tag{8.79}$$

where $f^{(0)}\phi$ is the deviation of the Maxwell–Jüttner equilibrium distribution function. If we insert (8.79) into the Boltzmann equation (8.17) it follows by using the Chapman and Enskog methodology (see Section 5.2) that

$$p^\alpha\frac{\partial f^{(0)}}{\partial x^\alpha} = -\frac{U_L^\alpha p_\alpha}{c^2\tau}f^{(0)}\phi. \tag{8.80}$$

The elimination of the convective time derivatives Dn, DT and DU^α of (8.80) – by using the balance equations for the particle number density, energy density and momentum density of an Eulerian relativistic gas – leads to

$$\phi = -\frac{c^2\tau}{U_L^\alpha p_\alpha}\left\{-\frac{k^2T}{c^2c_v}\left[\frac{1}{3}\zeta^2\frac{c_v}{k}-(G^2\zeta^2-4G\zeta-\zeta^2)\left(\frac{U_L^\beta p_\beta}{kT}\right)\right.\right.$$
$$\left.-\frac{1}{3}(\zeta^2+5G\zeta-\zeta^2G^2-4)\left(\frac{U_L^\beta p_\beta}{kT}\right)^2\right]\nabla_\alpha U_L^\alpha-\frac{p_\alpha p_\beta}{kT}\nabla^{\langle\alpha}U_L^{\beta\rangle}$$
$$\left.+\frac{p_\alpha}{kT^2}(p_\beta U_L^\beta-h_E)\left(\nabla^\alpha T-\frac{T}{nh_E}\nabla^\alpha p\right)\right\}. \tag{8.81}$$

We note that the only difference between the expression in the above parenthesis and the corresponding one in (5.23) is that the four-velocity in the Eckart description U^α in the latter is replaced by the four-velocity in the description of Landau and Lifshitz, U_L^α.

We infer from (8.81) that the deviation of the Maxwell–Jüttner distribution function $f^{(0)}\phi$ is a known function of the thermodynamic forces $\nabla_\alpha U_L^\alpha$, $\nabla^{\langle\alpha}U_L^{\beta\rangle}$

and $[\nabla^\alpha T - T/(nh_E)\nabla^\alpha p]$. We insert now (8.79) together with (8.81) into the definition of the particle four-flow

$$N^\alpha = c\int p^\alpha f \frac{d^3p}{p_0}, \tag{8.82}$$

integrate the resulting equation over all values of d^3p/p_0 in a local Lorentz rest frame, perform the projection $\Delta^\alpha_{L\beta}N^\beta = \mathcal{J}^\alpha = -q^\alpha/h_E$ and get the constitutive equation for the heat flux $(8.69)_2$. The coefficient of thermal conductivity, for the model of Anderson and Witting by using the Chapman and Enskog method, reads

$$\lambda = \frac{\tau k p}{3m}\zeta^4 G\left[G\left(\frac{1}{\zeta} - \frac{K_1}{K_2} + \frac{\mathrm{Ki}_1}{K_2}\right) - \frac{3}{\zeta^2}\right]. \tag{8.83}$$

The constitutive equations for the dynamic pressure $(8.69)_1$ and for the pressure deviator $(8.69)_3$ are obtained in the same manner as above, i.e., we insert (8.79) together with (8.81) into the definition of the energy-momentum tensor

$$T^{\alpha\beta} = c\int p^\alpha p^\beta f \frac{d^3p}{p_0}, \tag{8.84}$$

integrate the resulting equation over all values of d^3p/p_0 in a local Lorentz rest frame and perform respectively the projections

$$\Delta_{L\alpha\beta}T^{\alpha\beta} = -3(p+\varpi) \qquad \text{and} \qquad (\Delta^{(\gamma}_{L\alpha}\Delta^{\delta)}_{L\beta} - \Delta^{\gamma\delta}_L\Delta_{L\alpha\beta}/3)T^{\alpha\beta} = p^{\langle\gamma\delta\rangle}.$$

The resulting coefficients of bulk and shear viscosity, for the model of Anderson and Witting by using the Chapman and Enskog method, are

$$\eta = \frac{\tau p}{3}\zeta\left[\frac{3(G^2\zeta - 5G - \zeta)}{\zeta^2 + 5G\zeta - G^2\zeta^2 - 1} + \frac{\zeta^2}{3}\left(\frac{3G}{\zeta^2} - \frac{1}{\zeta} + \frac{K_1}{K_2} - \frac{\mathrm{Ki}_1}{K_2}\right)\right], \tag{8.85}$$

$$\mu = \frac{\tau p}{15}\zeta^3\left[\frac{3G}{\zeta^2} - \frac{1}{\zeta} + \frac{K_1}{K_2} - \frac{\mathrm{Ki}_1}{K_2}\right]. \tag{8.86}$$

Note that (8.83), (8.85) and (8.86) differ from (8.71), (8.70) and (8.72), respectively.

The expressions for the transport coefficients in the non-relativistic limit ($\zeta \gg 1$) are the same as those found by using the equation for the third-order moment (8.65), i.e., (8.73) through (8.75). In the ultra-relativistic limiting case ($\zeta \ll 1$) they read

$$\eta = \frac{p\tau}{54}\zeta^4\left[1 - \frac{3\pi}{2}\zeta + \cdots\right], \tag{8.87}$$

$$\lambda = \frac{4c^2}{3T}p\tau\left[1 - \frac{11}{8}\zeta^2 + \cdots\right], \tag{8.88}$$

$$\mu = \frac{4}{5}p\tau\left[1 + \frac{1}{24}\zeta^2 + \cdots\right]. \tag{8.89}$$

We infer from (8.77) and (8.78) that the leading term for the ratio between the coefficients of thermal conductivity and shear viscosity is $\lambda/\mu = 6c^2/(5T)$ whereas from (8.88) and (8.89) the same ratio is $\lambda/\mu = 5c^2/(3T)$. The latter has the same value as those found from the full Boltzmann equation for hard-sphere particles and for Maxwellian particles (see Sections 5.4 and 6.6).

Problems

8.3.2.1 Obtain the expression for the third-order moment (8.66) from (6.33) by replacing $U^\alpha = U_L^\alpha - q^\alpha/(nh_E)$ and by neglecting all non-linear terms.

8.3.2.2 For the model equation of Anderson and Witting obtain the constitutive equations for the dynamic pressure, heat flux and pressure deviator and the corresponding transport coefficients by using: a) the equation for the third-order moment (8.68) and b) the deviation of the Maxwell–Jüttner distribution function (8.81).

8.3.2.3 Show that the expressions for the transport coefficients, obtained from the model equation of Anderson and Witting by using the equation for the third-order moment and the method of Chapman and Enskog, coincide in the non-relativistic limiting case.

8.3.3 Comparison of the models

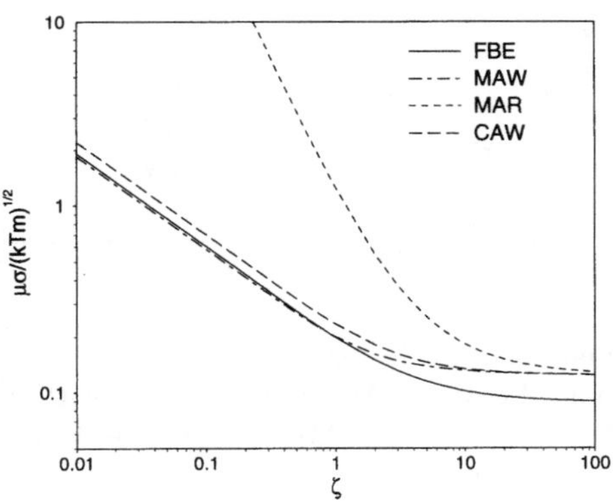

Figure 8.1: Shear viscosity as a function of ζ: FBE – full Boltzmann equation; MAW – Anderson–Witting with moment equation; MAR – Marle model; CAW – Anderson–Witting with Chapman–Enskog method.

We begin by noting that the transport coefficients obtained from the Marle model and from the Anderson and Witting model by using the equation for the third-order moment or by using the Chapman and Enskog method have the same expressions in the non-relativistic limiting case, if one considers only the leading term of each coefficient in their expansions (compare (8.58) through (8.60) with (8.73) through (8.75)).

In the ultra-relativistic limiting case we have from (8.61) through (8.63), (8.76) through (8.78) and (8.87) through (8.89):

$$\frac{\eta_{MAW}}{\eta_{CAW}} = \frac{1}{4}, \qquad \frac{\eta_{MAW}}{\eta_{MAR}} = \frac{\zeta}{4}, \tag{8.90}$$

$$\frac{\lambda_{MAW}}{\lambda_{CAW}} = \frac{3}{5}, \qquad \frac{\lambda_{MAW}}{\lambda_{MAR}} = \frac{\zeta}{5}, \tag{8.91}$$

$$\frac{\mu_{MAW}}{\mu_{CAW}} = \frac{5}{6}, \qquad \frac{\mu_{MAW}}{\mu_{MAR}} = \frac{\zeta}{6}, \tag{8.92}$$

where the index MAR refers to the transport coefficients obtained from the model of Marle, MAW from the model of Anderson and Witting by using the equation for the third-order moment and CAW from the model of Anderson and Witting by using the Chapman and Enskog method. Further we shall denote by FBE the transport coefficients (5.89) through (5.91) of a relativistic gas of hard-sphere particles obtained from the full Boltzmann equation.

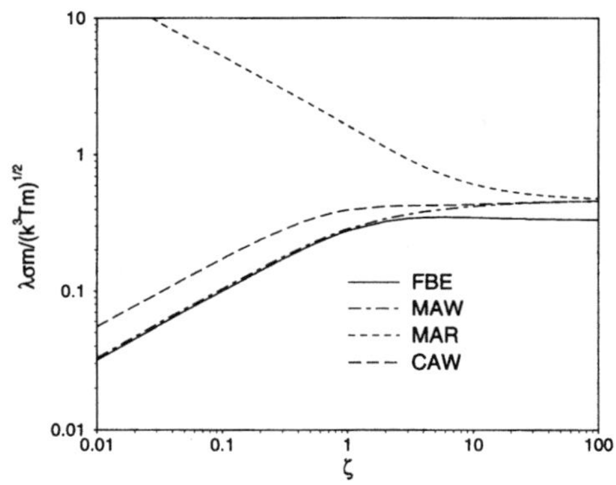

Figure 8.2: Thermal conductivity as a function of ζ: FBE – full Boltzmann equation; MAW – Anderson–Witting with moment equation; MAR – Marle model; CAW – Anderson–Witting with Chapman–Enskog method.

In Figure 8.1 the behavior of the coefficient of shear viscosity $\mu\sigma/(kTm)^{\frac{1}{2}}$ is shown as a function of the parameter ζ. The mean velocity $\mathcal{V}$ chosen for the characteristic time (8.40) was the adiabatic sound speed (8.38) without any correction factor. One observes that the MAW coefficient agrees with the FBE coefficient in the ultra-relativistic and in the relativistic regime up to $\zeta = 1$. The CAW coefficient is parallel to the MAW coefficient in the region $\zeta \leq 1$, and one can adjust the CAW coefficient with the FBE coefficient in this region, if we multiply it by the factor 5/6 of the relationship $(8.92)_1$. The MAR coefficient increases with the decrease of ζ and there exists no parameter by which one can fit it with the FBE coefficient of shear viscosity in the ultra-relativistic limiting case.

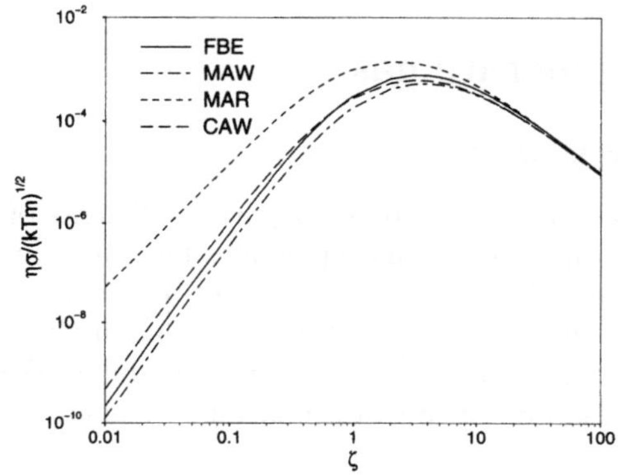

Figure 8.3: Bulk viscosity as a function of ζ: FBE – full Boltzmann equation; MAW – Anderson–Witting with moment equation; MAR – Marle model; CAW – Anderson–Witting with Chapman–Enskog method.

The coefficient of thermal conductivity $\lambda\sigma m/(k^3Tm)^{\frac{1}{2}}$ is plotted in Figure 8.2 as a function of the parameter ζ. As was pointed out in Section 8.3.1 in the non-relativistic limiting case the ratio between the coefficients of thermal conductivity and shear viscosity for the Marle and the Anderson and Witting models tends to $\lambda/\mu \approx 5k/2m$, while the same ratio for the full Boltzmann equation is $\lambda/\mu \approx 15k/4m$. Hence we cannot adjust the coefficient of shear viscosity *and* the coefficient of thermal conductivity with only one parameter. In other words if we adjust the coefficient of shear viscosity as was done in Figure 8.1 we have to use the same characteristic time multiplied by the factor 3/2 which gives the correct value of the ratio λ/μ. This was done in Figure 8.2 and one observes that the MAW coefficient coincides with the FBE coefficient in the ultra-relativistic and in the relativistic regime up to $\zeta = 2$. The CAW coefficient is parallel to the MAW coefficient in the region $\zeta < 0.6$ so that the MAW coefficient fits better the FBE coefficient than the CAW coefficient does. The MAR coefficient increases as ζ decreases and there exists no parameter in which one can adjust it to fit with the FBE coefficient of thermal conductivity in the ultra-relativistic limiting case.

The behavior of the coefficient of bulk viscosity $\eta\sigma/(kTm)^{\frac{1}{2}}$ as a function of the parameter ζ is showed in Figure 8.3. The same characteristic time was used as that for the coefficient of shear viscosity. One may infer that all coefficients obtained from the model equations do not fit the FBE coefficient in the ultra-relativistic and relativistic regime, but all tend to almost the same value in the non-relativistic limiting case.

The same above conclusions on the behavior of the transport coefficients of the model equations could be obtained if we choose the mean velocity $\mathcal{V}$ as the mean Møller relative speed $\langle g_\varnothing \rangle$ or the mean velocity $\langle v \rangle$ and appropriate factors to adjust them to the FBE transport coefficients.

8.4 Single degenerate gas

8.4.1 Non-zero rest mass

In this section we shall determine the transport coefficients of a single degenerate relativistic gas with non-zero rest mass particles by using the model of Anderson and Witting and the equation for the third-order moment (8.68). To that end we need first to know the distribution function in terms of the 14 fields $n, T, U_L^\alpha, q^\alpha$ and $p^{\langle\alpha\beta\rangle}$ in the decomposition of Landau and Lifshitz. We shall use here the same methodology described in Section 6.2, i.e., we shall maximize the entropy per particle

$$s = \frac{1}{nc^2} S^\alpha U_{L\alpha} = -\frac{k}{nc} U_{L\alpha} \int p^\alpha f \left[\ln\left(\frac{fh^3}{g_s}\right) - \left(1 + \frac{g_s}{\varepsilon f h^3}\right) \ln\left(1 + \frac{\varepsilon f h^3}{g_s}\right)\right] \frac{d^3p}{p_0}, \tag{8.93}$$

which follows from (2.81), subjected to the 14 constraints

$$N^\alpha U_{L\alpha} = cU_{L\alpha} \int p^\alpha f \frac{d^3p}{p_0}, \tag{8.94}$$

$$T^{\alpha\beta} U_{L\alpha} = cU_{L\alpha} \int p^\alpha p^\beta f \frac{d^3p}{p_0}, \tag{8.95}$$

$$T^{\langle\gamma\beta\rangle\alpha} U_{L\alpha} = cU_{L\alpha} \int p^{\langle\gamma} p^{\beta\rangle} p^\alpha f \frac{d^3p}{p_0}. \tag{8.96}$$

We construct first a functional by introducing the Lagrange multipliers λ, λ_β and $\lambda_{\langle\gamma\beta\rangle}$ associated with $N^\alpha U_{L\alpha}$, $T^{\alpha\beta} U_{L\alpha}$ and $T^{\langle\gamma\beta\rangle\alpha} U_{L\alpha}$, respectively, then we maximize the functional without constraints and obtain from the Euler–Lagrange equation

$$f = \frac{g_s/h^3}{\exp\left[\frac{nc^2}{k}(\lambda + \lambda_\alpha p^\alpha + \lambda_{\langle\alpha\beta\rangle} p^\alpha p^\beta)\right] \pm 1}. \tag{8.97}$$

As usual the plus sign in (8.97) refers to Fermi–Dirac statistics while the minus sign to Bose–Einstein statistics. Further we use the decomposition (6.8) to write the Lagrange multipliers as a sum of equilibrium and non-equilibrium parts and use the approximation

$$\frac{1}{\exp(x)\exp(y)\pm 1} \approx \frac{1}{\exp(x)\pm 1}\left[1-\frac{y\exp(x)}{\exp(x)\pm 1}\right], \tag{8.98}$$

valid for all $|y| \ll 1$, to write (8.97) as

$$f = f^{(0)}\left\{1-\frac{e^{-\frac{\mu_E}{kT}+\frac{U_L^\alpha p_\alpha}{kT}}}{e^{-\frac{\mu_E}{kT}+\frac{U_L^\alpha p_\alpha}{kT}}\pm 1}\frac{nc^2}{k}\left[\lambda^{NE}+\lambda_\alpha^{NE}p^\alpha+\lambda_{\langle\alpha\beta\rangle}^{NE}p^\alpha p^\beta\right]\right\}. \tag{8.99}$$

In the above equation we have identified, according to Section 2.7, the Lagrange multipliers in equilibrium with the equilibrium chemical potential μ_E, four-velocity U_L^α and temperature T, so that we get the equilibrium distribution function

$$f^{(0)} = \frac{g_s/h^3}{e^{-\frac{\mu_E}{kT}+\frac{U_L^\alpha p_\alpha}{kT}}\pm 1}. \tag{8.100}$$

If we insert the distribution function (8.99) into the definitions of the particle four-flow N^α and of the energy-momentum tensor $T^{\alpha\beta}$ and follow the same methodology stated in Section 6.2, it is possible to identify the Lagrange multipliers $\lambda^{NE}, \lambda_\alpha^{NE}$ and $\lambda_{\langle\alpha\beta\rangle}^{NE}$ by performing the integrations in a local Lorentz rest frame. The final expression for the distribution function in terms of the 14 fields n, T, U_L^α, ϖ, q^α and $p^{\langle\alpha\beta\rangle}$ is

$$f = f^{(0)}\Bigg\{1-\frac{e^{-\frac{\mu_E}{kT}+\frac{U_L^\alpha p_\alpha}{kT}}}{e^{-\frac{\mu_E}{kT}+\frac{U_L^\alpha p_\alpha}{kT}}\pm 1}\frac{3h^3}{4\pi m^4c^5 g_s}\Bigg[\frac{-\varpi}{\mathcal{D}_4^0}\Bigg([J_{22}^\bullet J_{24}^\bullet-(J_{23}^\bullet)^2]$$
$$+\frac{U_{L\alpha}p^\alpha}{mc^2}[J_{22}^\bullet J_{23}^\bullet - J_{21}^\bullet J_{24}^\bullet]+\left(\frac{U_{L\alpha}p^\alpha}{mc^2}\right)^2[J_{21}^\bullet J_{23}^\bullet-(J_{22}^\bullet)^2]\Bigg)$$
$$-\frac{q_\alpha p^\alpha}{[J_{40}^\bullet J_{42}^\bullet-(J_{41}^\bullet)^2]h_E}\left(J_{42}^\bullet - J_{41}^\bullet\frac{U_{L\alpha}p^\alpha}{mc^2}\right)-\frac{5p^{\langle\alpha\beta\rangle}p_\alpha p_\beta}{2J_{60}^\bullet (mc)^2}\Bigg]\Bigg\}. \tag{8.101}$$

This is Grad's distribution function for a degenerate relativistic gas written in terms of the decomposition of Landau and Lifshitz. We recall that $J_{nm}^\bullet(\zeta,\mu_E)$ denotes the partial derivative of the integral $J_{nm}(\zeta,\mu_E)$ – given in (8.206) – with respect to $\mu_E/(kT)$. The relationships between the integrals $J_{nm}^\bullet(\zeta,\mu_E)$ and the integrals $I_n(\zeta,\mu_E)$ – defined in (8.207) – for some values of m and n are given in the Appendix of this chapter. Further $\mathcal{D}_4^0$ denotes the determinant

$$\mathcal{D}_4^0 = \begin{vmatrix} J_{21}^\bullet & J_{22}^\bullet & J_{23}^\bullet \\ J_{22}^\bullet & J_{23}^\bullet & J_{24}^\bullet \\ J_{40}^\bullet & J_{41}^\bullet & J_{42}^\bullet \end{vmatrix}. \tag{8.102}$$

In order to express the third-order moment $T^{\alpha\beta\gamma}$ as a function of the 14 basic fields we insert Grad's distribution function (8.101) into its definition (6.32) and obtain by performing the integration over all values of d^3p/p_0 in a local Lorentz rest frame:

$$T^{\alpha\beta\gamma} = (nC_1^Q + C_2^Q\varpi)U_L^\alpha U_L^\beta U_L^\gamma + \frac{c^2}{6}(nm^2 - nC_1^Q - C_2^Q\varpi)(\eta^{\alpha\beta}U_L^\gamma + \eta^{\alpha\gamma}U_L^\beta$$

$$+\eta^{\beta\gamma}U_L^\alpha) + \tilde{C}_3^Q(\eta^{\alpha\beta}q^\gamma + \eta^{\alpha\gamma}q^\beta + \eta^{\beta\gamma}q^\alpha) - \left(\frac{m^2}{h_E} + \frac{6}{c^2}\tilde{C}_3^Q\right)(U_L^\alpha U_L^\beta q^\gamma$$

$$+U_L^\alpha U_L^\gamma q^\beta + U_L^\beta U_L^\gamma q^\alpha) + C_4^Q(p^{\langle\alpha\beta\rangle}U_L^\gamma + p^{\langle\alpha\gamma\rangle}U_L^\beta + p^{\langle\beta\gamma\rangle}U_L^\alpha). \tag{8.103}$$

The scalar coefficients C_1^Q through C_4^Q of the third-order moment are given by

$$C_1^Q = m^2\left(1 + 2\frac{J_{41}}{J_{21}}\right), \qquad C_2^Q = \frac{6m}{c^2}\frac{\mathcal{D}_4^1}{\mathcal{D}_4^0}, \tag{8.104}$$

$$\tilde{C}_3^Q = \frac{m^2c^2}{5h_E}\frac{J_{42}^\bullet J_{60}^\bullet - J_{41}^\bullet J_{61}^\bullet}{J_{40}^\bullet J_{42}^\bullet - (J_{41}^\bullet)^2}, \qquad C_4^Q = m\frac{J_{61}^\bullet}{J_{60}^\bullet}. \tag{8.105}$$

In $(8.104)_2$ $\mathcal{D}_4^1$ is the determinant

$$\mathcal{D}_4^1 = \begin{vmatrix} J_{21}^\bullet & J_{22}^\bullet & J_{23}^\bullet \\ J_{22}^\bullet & J_{23}^\bullet & J_{24}^\bullet \\ J_{41}^\bullet & J_{42}^\bullet & J_{43}^\bullet \end{vmatrix}. \tag{8.106}$$

Before we substitute the expression for the third-order moment (8.103) into the moment equation (8.68) we shall derive some useful relationships relating the derivatives of the equilibrium chemical potential μ_E. We begin by taking the gradient of

$$\frac{\mu_E}{kT} = \frac{1}{kT}\left(e - Ts_E + \frac{p}{n}\right), \tag{8.107}$$

which by the use of the Gibbs equation

$$\partial_\alpha s_E = \frac{1}{T}\left(\partial_\alpha e - \frac{p}{n^2}\partial_\alpha n\right), \tag{8.108}$$

and of the definition of the enthalpy per particle $h_E = e + p/n$, reads

$$\nabla^\alpha\left(\frac{\mu_E}{kT}\right) = -\frac{h_E}{kT^2}\left(\nabla^\alpha T - \frac{T}{nh_E}\nabla^\alpha p\right). \tag{8.109}$$

Next we build the convective time derivative of the particle number density and of the energy per particle (see (3.4) and (3.5)) expressed in terms of the integral $J_{nm}(\zeta, \mu_E)$, i.e.,

$$n = 4\pi(mc)^3\frac{g_s}{h^3}J_{21}, \qquad e = mc^2\frac{J_{22}}{J_{21}}, \tag{8.110}$$

yielding

$$\frac{Dn}{n} = \frac{J^\bullet_{21}}{J_{21}} D\left(\frac{\mu_E}{kT}\right) + \frac{J^\bullet_{22}}{J_{21}} \frac{\zeta}{T} DT, \tag{8.111}$$

$$De = \frac{mc^2}{J^2_{21}} \left[(J_{21}J^\bullet_{22} - J_{22}J^\bullet_{21}) D\left(\frac{\mu_E}{kT}\right) + \frac{\zeta}{T}(J_{21}J^\bullet_{23} - J_{22}J^\bullet_{22}) DT \right]. \tag{8.112}$$

In (8.111) and (8.112) it has been used the relationship $J'_{nm} = -J^\bullet_{n,m+1}$ between the derivatives with respect to ζ and $\mu_E/(kT)$.

If we use the balance equation of the particle number density and of the energy density for an Eulerian relativistic gas, i.e.,

$$Dn + n\nabla_\alpha U^\alpha_L = 0, \qquad nDe + p\nabla_\alpha U^\alpha_L = 0, \tag{8.113}$$

we get from (8.111) through (8.113)

$$D\left(\frac{\mu_E}{kT}\right) = -\frac{3J_{21}J^\bullet_{23} - \zeta J^\bullet_{22}J^\bullet_{41}}{3[J^\bullet_{23}J^\bullet_{21} - (J^\bullet_{22})^2]} \nabla_\alpha U^\alpha_L, \tag{8.114}$$

$$\frac{\zeta}{T} DT = \frac{3J_{21}J^\bullet_{22} - \zeta J^\bullet_{21}J^\bullet_{41}}{3[J^\bullet_{23}J^\bullet_{21} - (J^\bullet_{22})^2]} \nabla_\alpha U^\alpha_L. \tag{8.115}$$

In the above equations we have used the equations (3.4) and (3.6) to express the ratio p/n in terms of the integral $J_{nm}(\zeta, \mu_E)$:

$$\frac{p}{n} = \frac{mc^2 J_{40}}{3J_{21}} = \frac{mc^2(\zeta J^\bullet_{41} - 3J_{22})}{3J_{21}}. \tag{8.116}$$

Now we insert the representation of the third-order moment (8.103) into the moment equation (8.68) and – as in Section 8.3.2 – perform the projections $U_{L\beta}U_{L\gamma}$ (or $\Delta_{L\beta\gamma}$), $\Delta^\delta_{L\beta}U_{L\gamma}$ and $\Delta^{(\delta}_{L\beta}\Delta^{\epsilon)}_{L\gamma} - \Delta_{L\beta\gamma}\Delta^{\delta\epsilon}_L/3$ of the resulting equation. If we eliminate the convective time derivatives $D(\mu_E/kT)$, DT and DU^α – by using (8.114), (8.115) and $DU^\alpha = [c^2/(nh_E)]\nabla^\alpha p$ – from the projected equations, we get the constitutive equations for a viscous heat conducting relativistic gas (8.69) in which the transport coefficients read

$$\eta = \frac{\tau p}{3} \frac{\mathcal{D}^0_4}{\mathcal{D}^1_4} \left[5\frac{J_{41}}{J_{40}} + \frac{3[J^\bullet_{23}J^\bullet_{41} - J^\bullet_{22}J^\bullet_{42}]J_{21} + \zeta[J^\bullet_{21}J^\bullet_{42} - J^\bullet_{22}J^\bullet_{41}]J^\bullet_{41}}{[(J^\bullet_{22})^2 - J^\bullet_{23}J^\bullet_{21}]J_{40}} \right], \tag{8.117}$$

$$\lambda = \frac{\tau p}{9} \frac{k}{m} \frac{\zeta^3 J^\bullet_{41}}{J_{40}(J_{21})^2} \frac{[3J^\bullet_{42}J_{21} - \zeta(J^\bullet_{41})^2][J^\bullet_{40}J^\bullet_{42} - (J^\bullet_{41})^2]}{J^\bullet_{41}J^\bullet_{61} + (J^\bullet_{41})^2 - J^\bullet_{42}J^\bullet_{60} - J^\bullet_{40}J^\bullet_{42}}, \tag{8.118}$$

$$\mu = \tau p \frac{J_{41}J^\bullet_{60}}{J_{40}J^\bullet_{61}}. \tag{8.119}$$

The above expressions for the transport coefficients were obtained from the equation for the third-order moment (8.68) and they differ from those found by Anderson and Witting [3], since the authors have used the Chapman and Enskog method. Equations (8.117) through (8.119) are general expressions for the transport coefficients of a Bose or a Fermi relativistic gas and in the following we shall analyze the two cases separately. In both cases we have chosen the adiabatic sound speed (8.39) as the mean velocity $\mathcal{V}$ in the expression for the characteristic time (8.40).

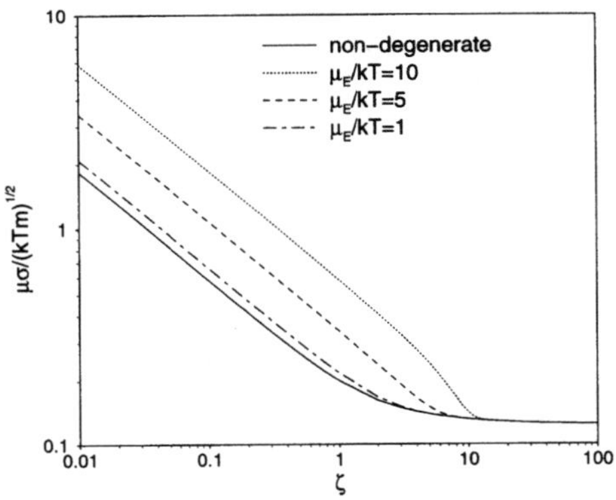

Figure 8.4: Shear viscosity of a non-degenerate gas and of a Fermi gas as a function of μ_E/kT and ζ.

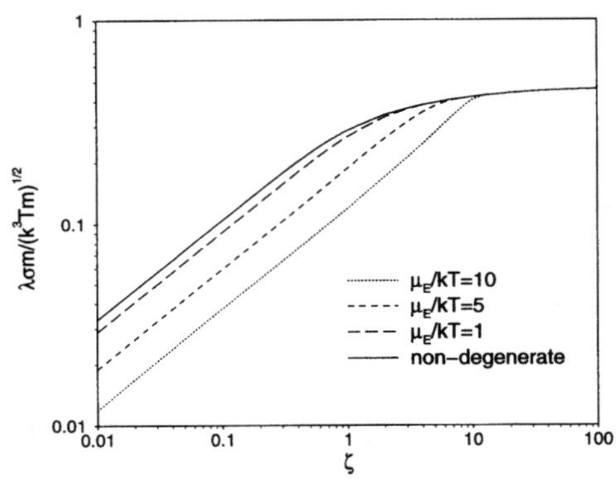

Figure 8.5: Thermal conductivity of a non-degenerate gas and of a Fermi gas as a function of μ_E/kT and ζ.

The transport coefficients for a Fermi gas are plotted in Figures 8.4 through 8.6 as functions of the parameter ζ. One may infer that the coefficient of shear viscosity $\mu\sigma/(kTm)^{\frac{1}{2}}$ increases when μ_E/kT increases and tends to the non-degenerate MAW coefficient when μ_E/kT becomes small and assumes negative values. The coefficient of thermal conductivity $\lambda\sigma m/(k^3Tm)^{\frac{1}{2}}$ decreases with increasing values

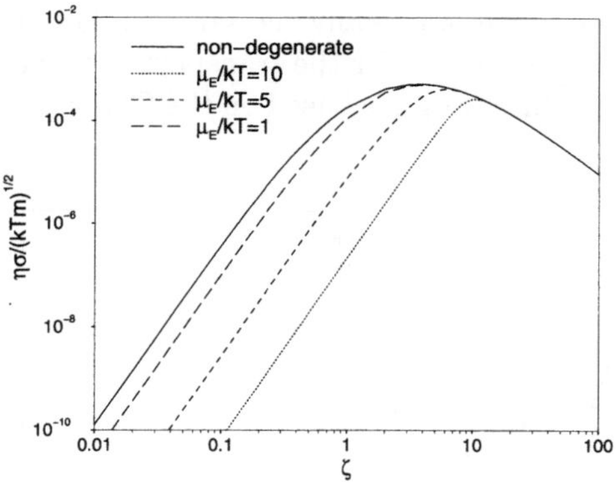

Figure 8.6: Bulk viscosity of a non-degenerate gas and of a Fermi gas as a function of μ_E/kT and ζ.

of μ_E/kT and also tends to the corresponding non-degenerate MAW coefficient for small and negative values of μ_E/kT. The coefficient of bulk viscosity $\eta\sigma/(kTm)^{\frac{1}{2}}$ has a similar behavior to the coefficient of thermal conductivity, i.e., it decreases with increasing values of μ_E/kT.

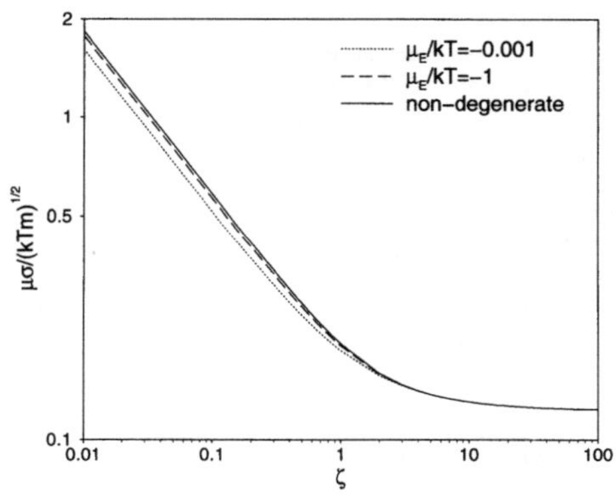

Figure 8.7: Shear viscosity of a non-degenerate gas and of a Bose gas as a function of μ_E/kT and ζ.

In Figures 8.7 through 8.9 the transport coefficients of a Bose gas are plotted as functions of the parameter ζ. For a Bose gas, μ_E/kT is always negative and one may infer that the coefficient of shear viscosity $\mu\sigma/(kTm)^{\frac{1}{2}}$ increases while the coefficients of thermal conductivity $\lambda\sigma m/(k^3Tm)^{\frac{1}{2}}$ and bulk viscosity $\eta\sigma/(kTm)^{\frac{1}{2}}$ decrease by increasing $|\mu_E/kT|$. All coefficients tend to the corresponding non-degenerate MAW coefficient when $|\mu_E/kT|$ is very large.

The above conclusions are practically the same as those of Anderson and Witting [3], although their expressions for the transport coefficients are different from (8.117) through (8.119) since they are based on the Chapman and Enskog method.

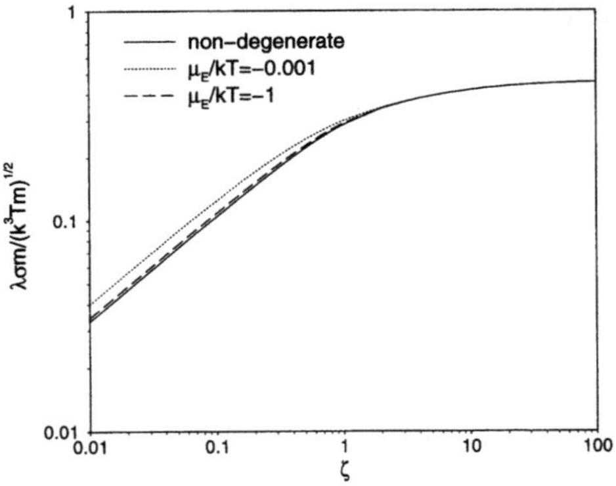

Figure 8.8: Thermal conductivity of a non-degenerate gas and of a Bose gas as a function of μ_E/kT and ζ.

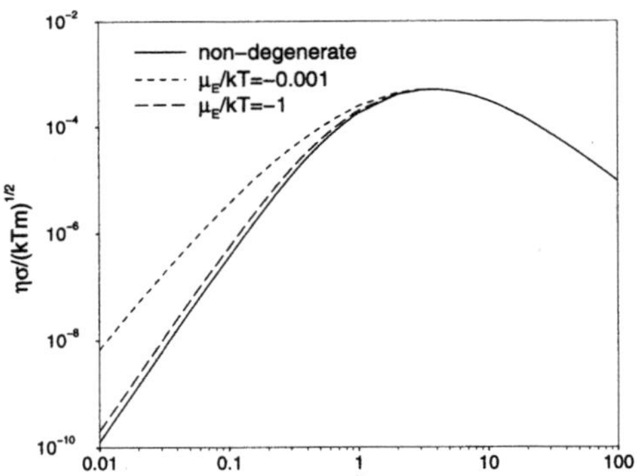

Figure 8.9: Bulk viscosity of a non-degenerate gas and of a Bose gas as a function of μ_E/kT and ζ.

Problems

8.4.1.1 Obtain Grad's distribution function (8.101) for a degenerate relativistic gas in the Landau and Lifshitz description.

8.4.1.2 Obtain the expression for the third-order moment in terms of the 14 fields (8.103) and show that the scalar coefficients are given by (8.104) and (8.105).

8.4.1.3 Show that in the limiting case of a non-degenerate relativistic gas the scalar coefficients $C_1^Q, C_2^Q, \tilde{C}_3^Q$ and C_4^Q ((8.104) and (8.105)) reduce to $C_1, C_2, \tilde{C}_3$ and C_4 ((6.34), (6.35), $(6.36)_2$ and (8.67)), respectively. (Hint: Use the relationships between the integrals $J^{\bullet}_{nm}(\zeta, \mu_E)$ and $I_n(\zeta, \mu_E)$ given in the Appendix to this chapter, and the relationship between the latter and the modified Bessel function of second kind given in Section 3.2.1.)

8.4.1.4 Check that (8.114) and (8.115) hold.

8.4.1.5 Obtain the expressions (8.117) through (8.119) for the transport coefficients of a Bose or a Fermi relativistic gas.

8.4.2 Zero rest mass

For a relativistic gas of particles of vanishing rest mass the dynamic pressure vanishes thanks to the relationships $p^\alpha p_\alpha = 0$, $3p = ne$ and

$$\eta_{\alpha\beta}T^{\alpha\beta} = 0 = -3(p + \varpi) + ne. \tag{8.120}$$

Hence instead of having 14 basic fields we have only 13 fields that are: the particle number density n, the four-velocity U_L^α, the temperature T, the pressure deviator $p^{\langle\alpha\beta\rangle}$ and the heat flux q^α.

The determination of Grad's distribution function for a relativistic gas with particles of zero rest mass proceeds in the same manner as stated in the last section, yielding

$$f = f^{(0)}\left\{1 + \frac{e^{-\frac{\mu_E}{kT} + \frac{U_L^\alpha p_\alpha}{kT}}}{e^{-\frac{\mu_E}{kT} + \frac{U_L^\alpha p_\alpha}{kT}} \pm 1} \frac{3h^3c^3}{4\pi(kT)^4 g_s}\left[\frac{1}{2i_4}\left(\frac{c}{kT}\right)^2 p^{\langle\alpha\beta\rangle}p_\alpha p_\beta \right.\right.$$

$$\left.\left. - \frac{q^\alpha p_\alpha}{[16i_3^2 - 15i_4 i_2]h_E}\left(5i_4 - 4i_3\frac{U_{L\alpha}p^\alpha}{kT}\right)\right]\right\}. \tag{8.121}$$

In (8.121) $f^{(0)}$ denotes the equilibrium distribution function (8.100), and $i_n(\mu_E)$ the integrals (see (3.8))

$$i_n(\mu_E) = \int_0^\infty \frac{x^n}{e^{-\frac{\mu_E}{kT}+x} \pm 1}, \qquad \text{with} \qquad i_n^\bullet(\mu_E) = n i_{n-1}(\mu_E). \tag{8.122}$$

In the above equation $i_n^\bullet(\mu_E)$ represents the differentiation of $i_n(\mu_E)$ with respect to μ_E/kT.

In order to obtain the constitutive equation for the third-order moment $T^{\alpha\beta\gamma}$ in terms of the 13 fields, we insert Grad's distribution function (8.121) into its definition (6.32) and get, by integrating the resulting equation over all values of d^3p/p_0 in a local Lorentz rest frame:

$$T^{\alpha\beta\gamma} = nC_1^Z U_L^\alpha U_L^\beta U_L^\gamma - \frac{nc^2}{6} C_1^Z (\eta^{\alpha\beta} U_L^\gamma + \eta^{\alpha\gamma} U_L^\beta + \eta^{\beta\gamma} U_L^\alpha)$$

$$+\tilde{C}_3^Z (\eta^{\alpha\beta} q^\gamma + \eta^{\alpha\gamma} q^\beta + \eta^{\beta\gamma} q^\alpha) - \frac{6}{c^2} \tilde{C}_3^Z (U_L^\alpha U_L^\beta q^\gamma + U_L^\alpha U_L^\gamma q^\beta$$

$$+U_L^\beta U_L^\gamma q^\alpha) + C_4^Z (p^{\langle\alpha\beta\rangle} U_L^\gamma + p^{\langle\alpha\gamma\rangle} U_L^\beta + p^{\langle\beta\gamma\rangle} U_L^\alpha). \tag{8.123}$$

The scalar coefficients C_1^Z through C_4^Z are given by

$$C_1^Z = \frac{2(kT)^2 i_4}{c^4 i_2}, \qquad C_4^Z = \frac{6kT i_5}{5c^2 i_4}, \tag{8.124}$$

$$\tilde{C}_3^Z = -\frac{(kT)^2 [25 i_4^2 - 24 i_3 i_5]}{5c^2 h_E [16 i_3^2 - 15 i_2 i_4]}, \tag{8.125}$$

where the enthalpy per particle reads $h_E = e + p/n = 4kT i_3/(3i_2)$ thanks to $e = kT i_3/i_2$ and $p/n = e/3$ (see (3.10) and (3.11)).

The gradient of μ_E/kT is given by (8.109), but if we build the convective time derivative of the particle number density n and of the energy per particle e (see (3.9) and (3.10))

$$n = 4\pi \left(\frac{kT}{c}\right)^3 \frac{g_s}{h^3} i_2, \qquad e = kT \frac{i_3}{i_2}, \tag{8.126}$$

we get that

$$Dn = 4\pi \left(\frac{k}{c}\right)^3 \frac{g_s}{h^3} \left[3T^2 i_2 DT + 2T^3 i_1 D\left(\frac{\mu_E}{kT}\right)\right] \overset{(8.113)_1}{=} -n\nabla_\alpha U_L^\alpha, \tag{8.127}$$

$$De = k\left[\frac{i_3}{i_2} DT + T\frac{3i_2^2 - 2i_3 i_1}{i_2^2} D\left(\frac{\mu_E}{kT}\right)\right] \overset{(8.113)_2}{=} -\frac{p}{n}\nabla_\alpha U_L^\alpha. \tag{8.128}$$

One concludes from the above system of equations that

$$D\left(\frac{\mu_E}{kT}\right) = 0, \qquad \text{and} \qquad \frac{3}{T} DT = -\nabla_\alpha U_L^\alpha. \tag{8.129}$$

The constitutive equations for the pressure deviator $p^{\langle\alpha\beta\rangle}$ and for the heat flux q^α are obtained by inserting the representation of the third-order moment (8.123) into the moment equation (8.68). Indeed, if we perform the projections $\Delta_{L\beta}^\delta U_{L\gamma}$ and $\Delta_{L\beta}^{(\delta} \Delta_{L\gamma}^{\epsilon)} - \Delta_{L\beta\gamma} \Delta_L^{\delta\epsilon}/3$ of the resulting equation, eliminate the convective time derivatives $D(\mu_E/kT)$, DT and DU^α from the projected equations by using (8.129) and $(nh_E/c^2)DU^\alpha = \nabla^\alpha p$, we get the Navier–Stokes and Fourier laws:

$$q^\alpha = \lambda\left(\nabla^\alpha T - \frac{T}{nh_E}\nabla^\alpha p\right), \qquad \lambda = \frac{4c^2}{9T}\tau p \frac{[16 i_3^2 - 15 i_2 i_4]^2}{i_2^2 [24 i_3 i_5 - 25 i_4^2]}, \tag{8.130}$$

$$p^{\langle\alpha\beta\rangle} = 2\mu\nabla^{\langle\alpha}U^{\beta\rangle}, \qquad \mu = \frac{5}{6}\tau p\frac{i_4^2}{i_3 i_5}. \tag{8.131}$$

The above coefficients of shear viscosity μ and thermal conductivity λ differ from those found by Anderson and Witting [3] since we have used here the equation for the third-order moment instead of the Chapman and Enskog method.

Ultra-relativistic Bose gas can be modeled as a relativistic gas with particles with zero rest mass. The equilibrium chemical potential in this case must be taken equal to zero, since it is impossible to fix the number of particles in the zeroth energy level. The justification of this assumption is that one can create a particle of rest mass m by spending an energy equal to mc^2. However for particles of vanishing rest mass one can create arbitrarily many particles in the zeroth energy level without any cost of energy. The particles at the zeroth energy level do not contribute to the thermodynamic and transport properties of the gas, since the energy of these particles is zero.

The integrals $i_n(\mu_E)$ for a ultra-relativistic Bose gas in the limiting case where $\mu_E = 0$ reduce to

$$i_n(0) = \Gamma(n+1)\zeta_{\mathcal{R}}(n+1), \tag{8.132}$$

where $\Gamma(n)$ is the gamma function and $\zeta_{\mathcal{R}}(n)$ the Riemann zeta function.

Since the ultra-relativistic limit of the adiabatic sound speeds (8.38) and (8.39) become $v_s = c/\sqrt{3}$, the characteristic time (8.40) reads

$$\tau = \frac{\sqrt{3}}{4n\pi\sigma c}, \tag{8.133}$$

and the transport coefficients $(8.131)_2$ and $(8.130)_2$ for an ultra-relativistic Bose gas are expressed by

$$\mu \approx 0.08973\frac{kT}{c\sigma}, \qquad \lambda \approx 0.14833\frac{kc}{\sigma}, \tag{8.134}$$

thanks to the values of the Riemann zeta function given in the Appendix. It is interesting to note that the ratio $\lambda/\mu \approx 5c^2/(3T)$ obtained here agrees with those found for hard-sphere particles (Section 5.4.1) and for Maxwellian particles (Section 6.6).

Problems

8.4.2.1 Obtain Grad's distribution function (8.121) for a relativistic gas with vanishing rest mass in the Landau and Lifshitz description.

8.4.2.2 Obtain the expression for the third-order moment in terms of the 13 fields (8.123) and show that the scalar coefficients are given by (8.124) and (8.125).

8.4.2.3 Obtain the expressions (8.130) and (8.131) for the transport coefficients of a relativistic gas with vanishing rest mass.

8.4.2.4 Show that for the case of a relativistic gas with vanishing rest mass the adiabatic sound speeds (8.38) and (8.39) reduce to $v_s = c/\sqrt{3}$.

8.5 Relativistic ionized gases

In this section we shall use the Anderson and Witting model in order to determine Ohm's law for binary mixtures of electrons and ions and of electrons and photons subjected to external electromagnetic fields. These two systems are important in astrophysics since they could describe magnetic white dwarfs or cosmological fluids in the plasma period and in the radiation dominated period.

8.5.1 Boltzmann and balance equations

We recall – see Section 7.2 – that a mixture of r constituents is characterized by the set of one-particle distribution functions

$$f(\mathbf{x}, \mathbf{p}_a, t) \equiv f_a, \qquad a = 1, 2, \ldots, r \tag{8.135}$$

that satisfies a Boltzmann equation of the type

$$p_a^\alpha \frac{\partial f_a}{\partial x^\alpha} + m_a \frac{\partial f_a K_a^\alpha}{\partial p_a^\alpha} = \sum_{b=1}^{r} \int (f_a' f_b' - f_a f_b) F_{ba} \sigma_{ab} \, d\Omega_{ba} \frac{d^3 p_b}{p_{b0}}. \tag{8.136}$$

The particles of the constituent a with rest mass m_a are supposed to have an electric charge q_a. The external force that acts on the charged particles of constituent a is of electromagnetic nature and given by (1.154)

$$K_a^\alpha = \frac{q_a}{c} F^{\alpha\beta} \frac{p_{a\beta}}{m_a}, \tag{8.137}$$

where $F^{\alpha\beta}$ is the electromagnetic field tensor.

In order to get the general equation of transfer for the constituent a of the mixture we multiply the Boltzmann equation (8.136) by an arbitrary function $\psi_a \equiv \psi(\mathbf{x}, \mathbf{p}_a, t)$ integrate the resulting equation over all values of $d^3 p_a / p_{a0}$ and use the same reasoning of Section 2.3, yielding

$$\frac{\partial}{\partial x^\alpha} \int \psi_a p_a^\alpha f_a \frac{d^3 p_a}{p_{a0}} - \int \left[p_a^\alpha \frac{\partial \psi_a}{\partial x^\alpha} + \frac{q_a}{c} F^{\alpha\beta} p_{a\beta} \frac{\partial \psi_a}{\partial p_a^\alpha} \right] f_a \frac{d^3 p_a}{p_{a0}}$$

$$= \sum_{b=1}^{r} \int \psi_a (f_a' f_b' - f_a f_b) F_{ba} \sigma_{ab} d\Omega_{ba} \frac{d^3 p_b}{p_{b0}} \frac{d^3 p_a}{p_{a0}}$$

$$= \sum_{b=1}^{r} \int (\psi_a' - \psi_a) f_a f_b F_{ba} \sigma_{ab} d\Omega_{ba} \frac{d^3 p_b}{p_{b0}} \frac{d^3 p_a}{p_{a0}}. \tag{8.138}$$

By summing (8.138) over all constituents of the mixture we get the general equation of transfer for the mixture, which reads

$$\frac{\partial}{\partial x^\alpha} \sum_{a=1}^{r} \int \psi_a p_a^\alpha f_a \frac{d^3 p_a}{p_{a0}} - \sum_{a=1}^{r} \int \left[p_a^\alpha \frac{\partial \psi_a}{\partial x^\alpha} + \frac{q_a}{c} F^{\alpha\beta} p_{a\beta} \frac{\partial \psi_a}{\partial p_a^\alpha} \right] f_a \frac{d^3 p_a}{p_{a0}}$$

$$= \frac{1}{2}\sum_{a,b=1}^{r}\int(\psi_a' + \psi_b' - \psi_a - \psi_b) f_a f_b F_{ba}\sigma_{ab} d\Omega_{ba} \frac{d^3 p_b}{p_{b0}} \frac{d^3 p_a}{p_{a0}}. \tag{8.139}$$

The right-hand sides of (8.138) and (8.139) were obtained by using the symmetry properties of the collision term; for more details one is referred to Chapter 7.

As in Section 7.2 we introduce the partial particle four-flow N_a^α and the partial energy-momentum tensor $T_a^{\alpha\beta}$ which are defined through

$$N_a^\alpha = c\int p_a^\alpha f_a \frac{d^3 p_a}{p_{a0}}, \qquad T_a^{\alpha\beta} = c\int p_a^\alpha p_a^\beta f_a \frac{d^3 p_a}{p_{a0}}, \tag{8.140}$$

and the corresponding quantities for the mixture

$$N^\alpha = \sum_{a=1}^{r} N_a^\alpha, \qquad T^{\alpha\beta} = \sum_{a=1}^{r} T_a^{\alpha\beta}. \tag{8.141}$$

Another quantity that is important in the analysis of ionized gases is the electric charge four-vector J^α, which is defined in terms of the partial particle four-flows N_a^α and of the partial electric charges q_a

$$J^\alpha = \sum_{a=1}^{r} q_a N_a^\alpha. \tag{8.142}$$

The balance equations for the particle four-flow of constituent a and of the mixture are obtained by choosing $\psi_a = c$ in (8.138) and (8.139), yielding

$$\partial_\alpha N_a^\alpha = 0, \qquad \partial_\alpha N^\alpha = 0, \tag{8.143}$$

respectively.

By choosing $\psi_a = c p_a^\beta$ in the general equation of transfer (8.139) the balance equation for the energy-momentum tensor of the mixture is

$$\partial_\beta T^{\alpha\beta} = \frac{1}{c} F^{\alpha\beta} \sum_{a=1}^{r} q_a N_{a\beta} \overset{(8.142)}{=} \frac{1}{c} F^{\alpha\beta} J_\beta. \tag{8.144}$$

We note that the above equation when compared with (1.161) has an opposite sign on its right-hand side. If we denote the energy-momentum tensor of (8.144) by an index pt that refers to the particles, we get the conservation law (see Landau and Lifshitz [15])

$$\partial_\alpha (T_{\text{pt}}^{\alpha\beta} + T_{\text{em}}^{\alpha\beta}) = 0, \tag{8.145}$$

i.e., the sum of the energy-momentum tensors of the particles and of the electromagnetic field satisfies a conservation equation.

Problems

8.5.1.1 Check that (8.138) holds.

8.5.1.2 Obtain the balance equation for the energy-momentum tensor of the mixture (8.144).

8.5.2 Decomposition with respect to the four-velocity

We recall that in Eckart's description the partial particle four-flow is decomposed according to

$$N_a^\alpha = n_a U^\alpha + J_a^\alpha, \quad \text{such that} \quad J_a^\alpha U_\alpha = 0, \tag{8.146}$$

where n_a is the particle number density and J_a^α the diffusion flux of the constituent a in the mixture. From the definitions $(8.140)_1$ and (8.146) we can represent the diffusion flux as

$$J_a^\alpha = \Delta_\beta^\alpha c \int p_a^\beta f_a \frac{d^3 p_a}{p_{a0}}. \tag{8.147}$$

Further by summing of (8.146) over all constituents it follows that

$$n = \sum_{a=1}^{r} n_a, \qquad \sum_{a=1}^{r} J_a^\alpha = 0, \tag{8.148}$$

which gives the particle number density of the mixture as the sum of the partial particle number densities and shows – due to the constraint $(8.148)_2$ – that there exist only $(r-1)$ partial diffusion fluxes that are linearly independent for a mixture of r constituents.

The electric current four-vector I^α is defined in terms of the partial diffusion fluxes J_a^α and of the partial electric charges q_a:

$$I^\alpha = \sum_{a=1}^{r} q_a J_a^\alpha. \tag{8.149}$$

One can relate the electric current four-vector I^α with the electric charge four-vector J^α since

$$I^\alpha = J^\alpha - \sum_{a=1}^{r} n_a q_a U^\alpha, \tag{8.150}$$

thanks to (8.142), (8.146) and (8.149).

We refer to Stueckelberg and Wanders [21] and to Hebenstreit [13] to write the decomposition of the partial energy-momentum tensor in the Eckart description as

$$T_a^{\alpha\beta} = p_a^{\langle\alpha\beta\rangle} - (p_a + \varpi_a)\Delta^{\alpha\beta} + \frac{1}{c^2} U^\alpha \left(q_a^\beta + h_E^a J_a^\beta\right)$$
$$+ \frac{1}{c^2} U^\beta \left(q_a^\alpha + h_E^a J_a^\alpha\right) + \frac{e_a n_a}{c^2} U^\alpha U^\beta, \tag{8.151}$$

where $p_a^{\langle\alpha\beta\rangle}$ is the partial pressure deviator, p_a the partial pressure, ϖ_a the partial dynamic pressure, q_a^α the partial heat flux, h_E^a the partial enthalpy per particle and e_a the partial energy per particle. If we sum the partial energy-momentum tensor over all constituents of the mixture and compare the resulting equation with the energy-momentum tensor of the mixture

$$T^{\alpha\beta} = p^{\langle\alpha\beta\rangle} - (p+\varpi)\Delta^{\alpha\beta} + \frac{1}{c^2}\left(U^\alpha q^\beta + U^\beta q^\alpha\right) + \frac{en}{c^2}U^\alpha U^\beta, \tag{8.152}$$

we can identify the quantities of the mixture

$$p^{\langle\alpha\beta\rangle} = \sum_{a=1}^{r} p_a^{\langle\alpha\beta\rangle}, \quad p = \sum_{a=1}^{r} p_a, \quad \varpi = \sum_{a=1}^{r} \varpi_a, \quad ne = \sum_{a=1}^{r} n_a e_a, \tag{8.153}$$

$$q^\alpha = \sum_{a=1}^{r} \left(q_a^\alpha + h_E^a J_a^\alpha\right). \tag{8.154}$$

Note that according to (8.154) the heat flux of the mixture is a sum of the partial heat fluxes and a term which represents the transport of heat due to diffusion.

According to de Groot and Suttorp [11] and van Erkelens and van Leeuwen [9] we can also decompose the electromagnetic field tensor $F^{\alpha\beta}$ into one part which is parallel to the four-velocity U^α and another which is perpendicular to it, i.e.,

$$F^{\alpha\beta} = \frac{1}{c^2}\left(F^{\alpha\gamma}U_\gamma U^\beta - F^{\beta\gamma}U_\gamma U^\alpha\right) + \Delta_\gamma^\alpha F^{\gamma\delta}\Delta_\delta^\beta. \tag{8.155}$$

Further we introduce the tensors E^α and $B^{\alpha\beta}$ defined by

$$E^\alpha = \frac{1}{c}F^{\alpha\beta}U_\beta, \qquad B^{\alpha\beta} = -\Delta_\gamma^\alpha F^{\gamma\delta}\Delta_\delta^\beta, \tag{8.156}$$

such that the electromagnetic field tensor can be written as

$$F^{\alpha\beta} = \frac{1}{c}\left(E^\alpha U^\beta - E^\beta U^\alpha\right) - B^{\alpha\beta}. \tag{8.157}$$

In a local Lorentz rest frame where $(U^\alpha) = (c, \mathbf{0})$ equations (8.156) and (1.143) imply

$$(E^\alpha) = (0, \mathbf{E}), \qquad B^{0\alpha} = B^{\alpha 0} = 0, \qquad B^{ij} = -c\epsilon^{ijk}B_k, \tag{8.158}$$

so that we can identify E^α as the electric field and $B^{\alpha\beta}$ as the magnetic flux induction tensor.

Since $F^{\alpha\beta}$ is an antisymmetric tensor we have that $F^{\alpha\beta}U_\alpha U_\beta = 0$ and one can get from (8.156) and (8.157) the relationships

$$E^\alpha U_\alpha = 0, \qquad B^{\alpha\beta}U_\beta = 0, \qquad \text{and} \qquad B^{\alpha\beta} = -B^{\beta\alpha}. \tag{8.159}$$

A similar decomposition for the electromagnetic field tensor (8.157) can be written in the Landau and Lifshitz description, since it is enough to replace U^α by U_L^α. However if we want to express the partial particle four-flow N_a^α and the partial energy-momentum tensor $T_a^{\alpha\beta}$ in terms of the quantities of the Eckart description $n_a, p_a, \varpi_a, q_a^\alpha, p_a^{\langle\alpha\beta\rangle}$ and of the four-velocity of the Landau and Lifshitz description U_L^α, we have to insert (see (4.58))

$$U^\alpha = U_L^\alpha - \frac{1}{nh_E}q^\alpha \tag{8.160}$$

into (8.146) and (8.151) and get by neglecting all non-linear terms

$$N_a^\alpha = n_a U_L^\alpha + J_a^\alpha - \frac{n_a q^\alpha}{nh_E}, \tag{8.161}$$

$$T_a^{\alpha\beta} = p_a^{\langle\alpha\beta\rangle} - (p_a+\varpi_a)\Delta_L^{\alpha\beta} + \frac{1}{c^2}U_L^\alpha\left(q_a^\beta + h_E^a J_a^\beta - \frac{n_a h_E^a}{nh_E}q^\beta\right)$$
$$+\frac{1}{c^2}U_L^\beta\left(q_a^\alpha + h_E^a J_a^\alpha - \frac{n_a h_E^a}{nh_E}q^\alpha\right) + \frac{e_a n_a}{c^2}U_L^\alpha U_L^\beta. \tag{8.162}$$

If we sum (8.161) and (8.162) over all constituents of the mixture and make use of equations (8.141), (8.148), (8.153), (8.154) as well as the relationship $\sum_{a=1}^r n_a h_E^a = nh_E$, we get

$$N^\alpha = nU_L^\alpha - \frac{q^\alpha}{h_E}, \tag{8.163}$$

$$T^{\alpha\beta} = p^{\langle\alpha\beta\rangle} - (p+\varpi)\Delta_L^{\alpha\beta} + \frac{en}{c^2}U_L^\alpha U_L^\beta, \tag{8.164}$$

that is, they reduce to the equations (4.56) and (4.57), respectively.

Problems

8.5.2.1 Check that in a local Lorentz rest frame (8.158) holds.

8.5.2.2 Obtain the representations for the particle number density and for the energy-momentum tensor of constituent a which are given by (8.161) and (8.162), respectively.

8.5.3 Ohm's law

As was stated previously the aim of this section on relativistic ionized gases is to derive Ohm's law for a binary mixture of electrons and ions and of electrons and photons. For that end we note that:

i) for a binary mixture of electrons ($a = e$) and ions ($a = i$), the electric current four-vector (8.149) can be written as

$$I^\alpha = -(Z+1)\mathrm{e}J_e^\alpha, \tag{8.165}$$

since the electric charges are given by $q_e = -\mathrm{e}$, $q_i = Z\mathrm{e}$ and the relationship between the diffusion fluxes is $J_e^\alpha = -J_i^\alpha$. In (8.165) Z denotes the atomic number and $\mathrm{e} = 1.602\,177 \times 10^{-19}$ C the elementary charge. Further we shall analyze the so-called Lorentzian plasma [17] where the collisions between the electrons can be neglected in comparison to the collisions between the electrons and ions. A Lorentzian plasma must fulfill two conditions: a) the mass of one constituent (ion) is much larger than the mass of the other constituent (electron) since $m_p/m_e \approx 1836$, where m_p denotes the mass of a proton; b) the modulus of the electric charge of the heavier constituent (ion) is much larger than that of the lighter one (electron), i.e., $Z \gg 1$;

ii) for a binary mixture of electrons ($a = e$) and photons ($a = \gamma$), the electric current four-vector (8.149) reduces to

$$I^\alpha = -\mathrm{e}J_e^\alpha, \tag{8.166}$$

because the electric charge of the photons is zero ($q_\gamma = 0$). Besides the collisions between electrons can also be neglected in comparison to the collisions between electrons and photons, which is the Compton scattering.

The Boltzmann equation for the electrons – (8.136) together with (8.137) – by considering the Anderson and Witting model (8.12) and the above arguments, is

$$p_e^\alpha \frac{\partial f_e}{\partial x^\alpha} + \frac{q_e}{c} F^{\alpha\beta} p_{e\beta} \frac{\partial f_e}{\partial p_e^\alpha} = -\frac{U_L^\alpha p_{e\alpha}}{c^2 \tau_{eb}} (f_e - f_e^{(0)}), \tag{8.167}$$

where τ_{eb} with $b = i$ or $b = \gamma$ is the time between collisions of electrons-ions or electrons-photons, respectively.

If we multiply (8.167) by c and integrate the resulting equation over all values of d^3p_e/p_{e0}, we get the balance equation for the particle four-flow of the electrons, which reads

$$\partial_\alpha N_e^\alpha = -\frac{U_{L\alpha}}{c^2 \tau_{eb}} (N_e^\alpha - N_e^\alpha|_E) \overset{(8.161)}{=} 0. \tag{8.168}$$

The balance equation for the energy-momentum tensor of the electrons is obtained through the multiplication of (8.167) by cp_e^γ and integration of the resulting equation over all values of d^3p_e/p_{e0}, yielding

$$\partial_\alpha T_e^{\alpha\gamma} - \frac{q_e}{c} F^{\gamma\beta} N_{e\beta} = -\frac{U_{L\alpha}}{c^2 \tau_{eb}} (T_e^{\alpha\gamma} - T_e^{\alpha\gamma}|_E)$$

$$\overset{(8.162)}{=} -\frac{1}{c^2 \tau_{eb}} \left(q_e^\gamma + h_E^e J_e^\gamma - \frac{n_e h_E^e}{n h_E} q^\gamma \right). \tag{8.169}$$

Since we are interested in deriving only Ohm's law, we can neglect the partial heat fluxes q_e^α and q_b^α so that the heat flux of the mixture (8.154) reduces to the transport of heat due to diffusion, i.e.,

$$q^\alpha = h_E^e J_e^\alpha + h_E^b J_b^\alpha = (h_E^e - h_E^b) J_e^\alpha. \tag{8.170}$$

In the right-hand side of (8.170) we have used the constraint $J_b^\alpha = -J_e^\alpha$. Further the particle four-flow of the electrons (8.161), by considering $q_e^\alpha = q_b^\alpha = 0$ and (8.170), reads

$$N_e^\alpha = n_e U_L^\alpha + \frac{h_E^b}{h_E} J_e^\alpha. \tag{8.171}$$

Now we use the same methodology of the previous sections and insert the equilibrium value of the energy-momentum tensor of the electrons, i.e.,

$$T_e^{\alpha\gamma}|_E = \frac{n_e h_E^e}{c^2} U_L^\alpha U_L^\gamma - p_e \eta^{\alpha\gamma}, \tag{8.172}$$

into the gradient term of (8.169) and get by performing the projection $\Delta_{L\gamma}^\delta$ of the resulting equation,

$$\frac{n_e h_E^e}{c^2} DU^\delta - \nabla^\delta p_e - n_e q_e E^\delta + \frac{q_e h_E^b}{c h_E} B^{\delta\beta} J_{e\beta} = -\frac{h_E^b h_E^e}{\tau_{eb} h_E c^2} J_e^\delta. \tag{8.173}$$

This is the balance equation for the momentum density of the electrons. In order to derive it we have used (8.170), (8.171) and the definitions of E^α and $B^{\alpha\beta}$ given by (8.156).

From the balance equation for the energy-momentum tensor of the mixture (8.144) one can obtain the balance equation for the momentum density of the mixture

$$\frac{n h_E}{c^2} DU^\delta = \nabla^\delta p + E^\delta (n_e q_e + n_b q_b) - \frac{1}{c} B^{\delta\gamma} I_\gamma. \tag{8.174}$$

The above equation is used to eliminate from (8.173) the convective time derivative DU^δ which leads to an equation that can be written as

$$a\mathcal{E}^\alpha = (b\eta^{\alpha\beta} + dB^{\alpha\beta}) I_\beta. \tag{8.175}$$

In (8.175) the constants a, b and d and the four-vector $\mathcal{E}^\alpha$ are defined by

$$a = \frac{q_e}{h_E^e} - \frac{q_b}{h_E^b}, \tag{8.176}$$

$$b = \frac{n}{\tau_{eb} c^2 n_e n_b (q_e - q_b)}, \tag{8.177}$$

$$d = \frac{1}{c(q_e - q_b)} \left[\frac{q_e n}{n_e n_b h_E^e} - \frac{q_e - q_b}{n_b h_E^b} \right], \tag{8.178}$$

$$\mathcal{E}^\alpha = E^\alpha + \left(\frac{q_e}{h_E^e} - \frac{q_b}{h_E^b}\right)^{-1}\left[\frac{1}{n_e h_E^e}\nabla^\alpha p_e - \frac{1}{n_b h_E^b}\nabla^\alpha p_b\right]. \tag{8.179}$$

The four-vector $\mathcal{E}^\alpha$ is the total electric field which is a combination of the external electric field and of the difference between the pressure gradients.

If we write (8.175) in terms of its components and solve the system of equations for the components of the electric current four-vector I^α, Ohm's law follows:

$$I^\alpha = \sigma^{\alpha\beta}\mathcal{E}_\beta, \tag{8.180}$$

which relates the electric current four-vector to the total electric field. The tensor $\sigma^{\alpha\beta}$ is identified with the electrical conductivity tensor and its expression, which is a function of the constants a, b, d and of the magnetic flux induction tensor $B^{\alpha\beta}$, is given by

$$\sigma^{\alpha\beta} = \frac{a}{b\left(1+\frac{d^2}{2b^2}B^{\gamma\delta}B_{\gamma\delta}\right)}\left[\left(1+\frac{d^2}{2b^2}B^{\gamma\delta}B_{\gamma\delta}\right)\eta^{\alpha\beta} - \frac{d}{b}B^{\alpha\beta} + \frac{d^2}{b^2}B^{\alpha\gamma}B_\gamma^{\ \beta}\right]. \tag{8.181}$$

In order to get a better physical interpretation of the components of the electrical conductivity tensor, it is usual in the theory of ionized gases to decompose the total electric field into parts parallel, perpendicular and transverse to the magnetic flux induction. To this end we follow van Erkelens and van Leeuwen [9] and introduce the dual $\tilde{B}^{\alpha\beta}$ of the magnetic flux induction tensor $B^{\alpha\beta}$ defined by

$$\tilde{B}^{\alpha\beta} = \frac{1}{2}\epsilon^{\alpha\beta\gamma\delta}B_{\gamma\delta}. \tag{8.182}$$

It is easy to verify from (8.182) and (8.158) that in a local Lorentz rest frame the only non-zero components of $\tilde{B}^{\alpha\beta}$ are $\tilde{B}^{0i} = cB^i$ since $\tilde{B}^{ij} = 0$ and $\tilde{B}^{00} = 0$.

The decomposition of the total electric field into parallel $\mathcal{E}_\parallel^\alpha$, perpendicular $\mathcal{E}_\perp^\alpha$ and transverse $\mathcal{E}_t^\alpha$ parts read

$$\mathcal{E}_\parallel^\alpha = \frac{1}{\left(\frac{1}{2}B^{\gamma\delta}B_{\gamma\delta}\right)}\tilde{B}^{\alpha\beta}\tilde{B}_{\beta\gamma}\mathcal{E}^\gamma, \tag{8.183}$$

$$\mathcal{E}_\perp^\alpha = \frac{-1}{\left(\frac{1}{2}B^{\gamma\delta}B_{\gamma\delta}\right)}B^{\alpha\beta}B_{\beta\gamma}\mathcal{E}^\gamma, \tag{8.184}$$

$$\mathcal{E}_t^\alpha = \frac{1}{\left(\frac{1}{2}B^{\gamma\delta}B_{\gamma\delta}\right)^{\frac{1}{2}}}B^{\alpha\beta}\mathcal{E}_\beta. \tag{8.185}$$

In a local Lorentz rest frame, (8.183) through (8.185) reduce to

$$\mathcal{E}_\parallel^0 = \mathcal{E}_\perp^0 = \mathcal{E}_t^0 = 0, \qquad \boldsymbol{\mathcal{E}}_\parallel = \frac{1}{B^2}(\mathbf{B}\cdot\boldsymbol{\mathcal{E}})\mathbf{B}, \tag{8.186}$$

$$\boldsymbol{\mathcal{E}}_\perp = \frac{1}{B^2}[(\mathbf{B}\cdot\boldsymbol{\mathcal{E}})\mathbf{B} - (\mathbf{B}\cdot\mathbf{B})\boldsymbol{\mathcal{E}}], \qquad \boldsymbol{\mathcal{E}}_t = \frac{1}{B}(\boldsymbol{\mathcal{E}}\times\mathbf{B}), \tag{8.187}$$

since $(B^{\gamma\delta}B_{\gamma\delta}/2)^{\frac{1}{2}} = c(\mathbf{B}\cdot\mathbf{B})^{\frac{1}{2}} = cB$. From (8.186) and (8.187) it is easy to verify that $\boldsymbol{\mathcal{E}}_{\|}$ is parallel to the magnetic flux induction $\mathbf{B}$, $\boldsymbol{\mathcal{E}}_{\perp}$ perpendicular to it while $\boldsymbol{\mathcal{E}}_t$ is perpendicular to both $\boldsymbol{\mathcal{E}}_{\|}$ and $\boldsymbol{\mathcal{E}}_{\perp}$.

We can now rewrite Ohm's law in terms of $\mathcal{E}_{\|}^{\alpha}$, $\mathcal{E}_{\perp}^{\alpha}$ and $\mathcal{E}_t^{\alpha}$ by using the relationship

$$\frac{1}{2}B^{\gamma\delta}B_{\gamma\delta}\eta^{\alpha\beta} = \tilde{B}^{\alpha\gamma}\tilde{B}_{\gamma}^{\ \beta} - B^{\alpha\gamma}B_{\gamma}^{\ \beta}. \tag{8.188}$$

Indeed, if we substitute (8.188) into Ohm's law (8.180) together with the expression for the electrical conductivity tensor (8.181) and make use of the definitions (8.183) through (8.185), we get that the electric current four-vector reads

$$I^{\alpha} = \sigma_{\|}\mathcal{E}_{\|}^{\alpha} + \sigma_{\perp}\mathcal{E}_{\perp}^{\alpha} + \sigma_t\mathcal{E}_t^{\alpha}. \tag{8.189}$$

The scalars $\sigma_{\|}$, $\sigma_{\perp}$ and σ_t are called the parallel, perpendicular and transverse components of the electrical conductivity tensor. Their expressions in terms of the constants a, b, d and of the modulus of the magnetic flux induction B are

$$\sigma_{\|} = \frac{a}{b}, \qquad \sigma_{\perp} = \frac{\sigma_{\|}}{1+\left(\frac{dBc}{b}\right)^2}, \qquad \sigma_t = -\sigma_{\perp}\frac{dBc}{b}. \tag{8.190}$$

We shall use the above results to obtain the electrical conductivities for binary mixtures of electrons and ions and of electrons and photons.

Electron-ion mixtures

For this system we shall consider the condition of quasi-neutrality which states that the particle number density of the electrons with respect to the ions is $n_e \approx Zn_i$. In this case the constants a, b and d, given by (8.176) through (8.178), reduce to

$$a = -\frac{enh_E}{n_i h_E^e h_E^i}, \qquad b = -\frac{1}{c^2\tau_{ei}en_e}, \qquad d = \frac{n_i h_E^i - n_e h_E^e}{cn_e n_i h_E^e h_E^i}, \tag{8.191}$$

and the electrical conductivities (8.190) read

$$\sigma_{\|} = \frac{\tau_{ei}e^2 n_e}{m_e}\left(1+\frac{Zh_E^e}{h_E^i}\right)\left(\frac{m_e c^2}{h_E^e}\right), \tag{8.192}$$

$$\sigma_{\perp} = \frac{\sigma_{\|}}{1+\left(1-\frac{Zh_E^e}{h_E^i}\right)^2\left(\frac{m_e c^2}{h_E^e}\right)^2(\omega_e\tau_{ei})^2}, \tag{8.193}$$

$$\sigma_t = \sigma_{\perp}\left(1-\frac{Zh_E^e}{h_E^i}\right)\left(\frac{m_e c^2}{h_E^e}\right)(\omega_e\tau_{ei}). \tag{8.194}$$

In the above equations we have introduced the electron cyclotron frequency $\omega_e = eB/m_e$.

In the limit of a non-degenerate and non-relativistic ionized gas we have that

$$h_E^e \to m_e c^2, \quad h_E^i \to m_i c^2, \quad \text{so that} \quad 1 \pm \frac{Z h_E^e}{h_E^i} = 1 \pm \frac{Z m_e}{m_i} \approx 1. \tag{8.195}$$

Note that we can use the approximation $Z \approx M/2$, where M is the relative atomic mass so that $Z m_e/m_i \approx m_e/(2m_i) \ll 1$. Hence (8.192) through (8.194) become

$$\sigma_\parallel = \frac{\tau_{ei} \mathrm{e}^2 n_e}{m_e}, \qquad \sigma_\perp = \frac{\sigma_\parallel}{1 + (\omega_e \tau_{ei})^2}, \qquad \sigma_t = \frac{\sigma_\parallel \omega_e \tau_{ei}}{1 + (\omega_e \tau_{ei})^2}, \tag{8.196}$$

which are well-known values for the electrical conductivities in the theory of non-degenerate and non-relativistic ionized gases (see for example Cap [5]).

Further in the limit of a non-degenerate ultra-relativistic electron gas and non-relativistic ion gas we have that

$$h_E^e \to 4kT, \qquad h_E^i \to m_i c^2, \qquad \zeta_e \ll 1, \qquad \zeta_i \gg 1, \tag{8.197}$$

so that in this limit the electrical conductivities (8.196) reduce to

$$\sigma_\parallel = \sigma_\perp = \frac{\tau_{ei} \mathrm{e}^2 n_e c^2}{4kT}, \qquad \sigma_t \to 0. \tag{8.198}$$

Hence in this limit the parallel and the perpendicular electrical conductivities are equal to each other while the transverse electrical conductivity vanishes.

Electron-photon mixtures

The values for the constants a, b, d – given by (8.176) through (8.178) – for a mixture of electrons and photons read

$$a = -\frac{\mathrm{e}}{h_E^e}, \qquad b = -\frac{1}{c^2 \tau_{e\gamma} \mathrm{e} n x_e x_\gamma}, \qquad d = \frac{1}{c n h_E^e x_e x_\gamma} \left[1 - \frac{x_e \zeta_e}{4} \frac{h_E^e}{m_e c^2}\right], \tag{8.199}$$

since $q_\gamma = 0$ and $h_E^\gamma = 4kT$. In (8.199) we have introduced the particle fractions $x_e = n_e/n$ and $x_\gamma = n_\gamma/n$. The electrical conductivities (8.190) for the electron-photon mixture are

$$\sigma_\parallel = \frac{\tau_{e\gamma} \mathrm{e}^2 n x_e x_\gamma c^2}{h_E^e}, \tag{8.200}$$

$$\sigma_\perp = \frac{\sigma_\parallel}{1 + \left[1 - \frac{x_e \zeta_e}{4} \frac{h_E^e}{m_e c^2}\right]^2 \left(\frac{m_e c^2}{h_E^e}\right)^2 (\omega_e \tau_{e\gamma})^2}, \tag{8.201}$$

$$\sigma_t = \sigma_\perp \left[1 - \frac{x_e \zeta_e}{4} \frac{h_E^e}{m_e c^2}\right] \left(\frac{m_e c^2}{h_E^e}\right) (\omega_e \tau_{e\gamma}). \tag{8.202}$$

The electrical conductivities (8.200) through (8.202) for the case of a non-degenerate and non-relativistic electron gas reduce to

$$\sigma_{\|} = \frac{\tau_{e\gamma} e^2 n x_e x_\gamma kT}{m_e}, \qquad \sigma_{\perp} \to 0, \qquad \sigma_t \to 0, \tag{8.203}$$

since $\zeta_e \gg 1$ and $h_E^e \to m_e c^2$. In this limit only the parallel electrical conductivity has a non-vanishing value.

In the limiting case of a non-degenerate and ultra-relativistic electron gas we have that $\zeta_e \ll 1$ and $h_E^e \to 4kT$ so that the electrical conductivities (8.200) through (8.202) become

$$\sigma_{\|} = \sigma_{\perp} = \frac{\tau_{e\gamma} e^2 n x_e x_\gamma c^2}{4kT}, \qquad \sigma_t \to 0. \tag{8.204}$$

Hence the parallel and the perpendicular electrical conductivity coincide while the transverse electrical conductivity vanishes.

If we compare the parallel electrical conductivity (8.200) with that found by de Groot et al. [12] which is based on the full Boltzmann equation, we find that they have similar expressions. Indeed if we use equation (XII.21) of [12] and: i) neglect the thermal-diffusion coefficient since here we have not considered the partial heat fluxes; ii) use the expressions for the diffusion coefficient D in (XII.24) as a function of the mean free path of the electrons $l_{e\gamma}$ in (XII.8), i.e., $D = cl_{e\gamma}n_\gamma/4n$; iii) introduce the time between successive electron-photon collisions $\tau_{e\gamma} = cl_{e\gamma}$, it follows that (XII.21) reduces to

$$\sigma_{\|} = \frac{\tau_{e\gamma} e^2 n_\gamma x_e x_\gamma c^2}{h_E}, \tag{8.205}$$

which has a similar expression when compared to (8.200).

Problems

8.5.3.1 Obtain the balance equation for the energy-momentum tensor of the electrons (8.169).

8.5.3.2 Obtain the balance equation for the momentum density of the electrons (8.173) and of the mixture (8.174).

8.5.3.3 Obtain Ohm's law (8.180) and show that the electrical conductivity tensor is given by (8.181).

8.5.3.4 Show that in a local Lorentz rest frame (8.186) and (8.187) hold.

8.5.3.5 Check that the relationship (8.188) holds.

8.5.3.6 Obtain the decomposition (8.189) for the electric current four-vector and show that the parallel, perpendicular and transverse components of the electrical conductivity tensor are given by (8.190).

8.5.3.7 Show that the electrical conductivities for electron-ion mixtures are given by (8.192) through (8.194) and obtain their limiting values (8.196) and (8.198) which correspond to the non-relativistic and ultra-relativistic electron gas, respectively.

8.5.3.8 Check that for electron-photon mixtures the electrical conductivities are given by (8.200) through (8.202) and obtain also their limiting values (8.203) and (8.204) which correspond to the non-relativistic and to the ultra-relativistic electron gas, respectively.

8.6 Appendix

The relationships between the integrals

$$J_{nm}(\zeta,\mu_E)=\int_0^\infty \frac{\sinh^n\vartheta\cosh^m\vartheta}{e^{\zeta\cosh\vartheta-\mu_E/kT}\pm 1}d\vartheta, \tag{8.206}$$

$$I_n(\zeta,\mu_E)=\int_0^\infty \frac{\cosh(n\vartheta)}{e^{\zeta\cosh\vartheta-\mu_E/kT}\pm 1}d\vartheta, \tag{8.207}$$

are obtained by the use of elementary relationships between hyperbolic functions, but unfortunately there exists no general recurrence formula relating these two integrals. Here we give only the relations for the integrals that appear in this chapter:[2]

$$J_{21}=\frac{1}{4}(I_3-I_1),\qquad J_{22}=\frac{1}{8}(I_4-I_0),\qquad J_{40}=\frac{1}{8}(I_4-4I_2+3I_0), \tag{8.208}$$

$$J_{41}=\frac{1}{16}(I_5-3I_3+2I_1),\qquad J_{21}^\bullet=\frac{1}{\zeta}I_2,\qquad J_{22}^\bullet=\frac{1}{4\zeta}(3I_3+I_1), \tag{8.209}$$

$$J_{23}^\bullet=\frac{1}{2\zeta}(I_4+I_2),\qquad J_{24}^\bullet=\frac{1}{16\zeta}(5I_5+9I_3+2I_1),\qquad J_{40}^\bullet=\frac{3}{4\zeta}(I_3-I_1), \tag{8.210}$$

$$J_{41}^\bullet=\frac{1}{2\zeta}(I_4-I_2),\qquad J_{42}^\bullet=\frac{1}{16\zeta}(5I_5-3I_3-2I_1),\qquad J_{43}^\bullet=\frac{3}{16\zeta}(I_6-I_2), \tag{8.211}$$

$$J_{60}^\bullet=\frac{5}{16\zeta}(I_5-3I_3+2I_1),\qquad J_{61}^\bullet=\frac{1}{16\zeta}(3I_6-8I_4+5I_2). \tag{8.212}$$

Some values of the Riemann zeta function are given below:

$$\begin{cases}\zeta_{\mathcal{R}}(3)\approx 1.20206, & \zeta_{\mathcal{R}}(4)\approx 1.08232,\\ \zeta_{\mathcal{R}}(5)\approx 1.03693, & \zeta_{\mathcal{R}}(6)\approx 1.01734.\end{cases} \tag{8.213}$$

[2]This list is the same as that given by Dreyer [8].

References

[1] M. Abramowitz and I. A. Stegun, *Handbook of mathematical functions* (Dover, New York, 1968).

[2] J. L. Anderson and H. R. Witting, A relativistic relaxation-time model for the Boltzmann equation, *Physica* **74**, 466–488 (1974).

[3] J. L. Anderson and H. R. Witting, Relativistic quantum transport coefficients, *Physica* **74**, 489–495 (1974).

[4] P. L. Bhatnagar, E. P. Gross, M. Krook, A model for collision processes in gases. Small amplitude processes in charged and neutral one-component systems, *Phys. Rev.* **94**, 511–525 (1954).

[5] F. Cap, *Lehrbuch der Plasmaphysik und Magnetohydrodynamik* (Springer, Wien, 1994).

[6] C. Cercignani and G. M. Kremer, Moment closure of the relativistic Anderson and Witting model equation, *Physica A* **290**, 192–202 (2001).

[7] C. Cercignani, *The Boltzmann equation and its applications* (Springer, New York, 1988).

[8] W. Dreyer, Maximisation of entropy for relativistic and degenerate gases in non-equilibrium, in *Kinetic theory and extended thermodynamics* pp. 107–123, eds. I. Müller and T. Ruggeri (Pitagora, Bologna, 1987).

[9] H. van Erkelens and W. A. van Leeuwen, Relativistic Boltzmann theory for a plasma, I. The entropy production, *Physica* **89 A**, 113–126 (1977).

[10] H. van Erkelens and W. A. van Leeuwen, Relativistic Boltzmann theory for a plasma, II. Reciprocal relations, *Physica* **89 A**, 225–244 (1977).

[11] S. R. de Groot and L. G. Suttorp, *Foundations of electrodynamics* (North-Holland, Amsterdam, 1972).

[12] S. R. de Groot, W. A. van Leeuwen and Ch. G. van Weert, *Relativistic kinetic theory* (North-Holland, Amsterdam, 1980).

[13] H. Hebenstreit, Balance equations for a relativistic plasma, I. Differential term, *Physica* **117 A**, 631–652 (1983).

[14] G. M. Kremer, *Uma introdução à teoria cinética relativística* (Editora da UFPR, Curitiba, 1998).

[15] L. D. Landau and E. M. Lifshitz, *The classical theory of fields*, 4th ed. (Pergamon Press, Oxford, 1980).

[16] L. Landau, E. M. Lifshitz, *Fluid mechanics* (Pergamon, Oxford, 1987).

[17] E. M. Lifshitz and L. P. Pitaevskii, *Physical Kinetics* (Pergamon, Oxford, 1981).

[18] C. Marle, Modèle cinétique pour l'établissement des lois de la conduction de la chaleur et de la viscosité en théorie de la relativité, *C. R. Acad. Sc. Paris* **260**, 6539–6541 (1965).

[19] C. Marle, Sur l'établissement des équations de l'hydrodynamique des fluides relativistes dissipatifs, I. L'équation de Boltzmann relativiste, *Ann. Inst. Henri Poincaré* **10**, 67–126 (1969).

[20] C. Marle, Sur l'établissement des équations de l' hydrodynamique des fluides relativistes dissipatifs, II. Méthodes de résolution approchée de l'équation de Boltzmann relativiste, *Ann. Inst. Henri Poincaré* **10**, 127–194 (1969).

[21] E. C. G. Stueckelberg and G. Wanders, Thermodynamique en relativité générale, *Helv. Phys. Acta* **26**, 417–420 (1953).

[22] P. Welander, On the temperature jump in a rarefied gas, *Arkiv för Fysik* **7**, 507–553 (1954).

Chapter 9

Wave Phenomena in a Relativistic Gas

9.1 Introduction

As is the case for other media, the propagation of disturbances plays an important role in a gas. When the gas is sufficiently rarefied, the variety of regimes and the unusual form of the basic equation make their study rather different from the corresponding one in continuum mechanics. Yet, the classification of these phenomena is not so different, and the term *wave* is applied indifferently to completely different situations. The common feature seems to be the propagation of a peculiar aspect, such as a sharp change or an oscillating behavior, which travels between different parts of the medium. The phase speed and the possible attenuation coefficient of the waves are typical objects of study, since they are general features which can help us in understanding many qualitative features of more complicated situations.

We can classify the propagation phenomena in a rarefied gas (classical or relativistic) into three classes:

1. The propagation of discontinuities in the distribution function or its derivatives. This phenomenon does not occur spontaneously in the kinetic model, at variance with other models, such as the Euler equations for an ideal fluid, but can be a consequence of discontinuous initial data or be induced by the presence of boundaries.

2. The propagation of small oscillations (such as sound waves or shear waves) against an equilibrium background. This phenomenon has the interesting feature of changing smoothly with the frequency or wavelength. Also, one can distinguish between free waves, in which the wave number is fixed, and forced waves, in which the frequency is fixed.

3. Traveling waves, typically shock waves. As mentioned before, the shock waves cannot be classified among the discontinuities, as is the case for an Euler fluid. In kinetic theory the situation is, in a certain sense, similar to that occurring for the Navier–Stokes equations. The similarity is related to the circumstance that the equations are linear in the higher order derivatives. The fact, however, that these

derivatives are the only ones occurring in the Boltzmann equation produces significant differences with respect to the case of Navier–Stokes equations. In both cases, however, we find that the solution is smooth but changes very rapidly through a thin layer. The name of traveling waves is due to the fact that we deal with solutions which travel at a constant speed without changing their shape.

In this classification we should have mentioned a fourth type of wave, intermediate between type 2 and type 3. There might be waves which have an oscillatory character, but cannot be strictly classified under 2, either because the background is not an equilibrium state or because their amplitude is large. Virtually nothing is known on this aspect of the solutions of the Boltzmann equation (even in the non-relativistic case). In the same category we should also put solutions, if any, which have the aspect of a solitary wave, i.e., solutions dominated by dispersion and nonlinearity, rather than by dissipation. Since dissipation and nonlinearity are very deeply tied with each other in the Boltzmann equation, it seems reasonable to expect that these phenomena might occur only for very small mean free paths, when a sort of decoupling between the left- and right-hand sides of the Boltzmann equation occurs.

In the next few sections we shall proceed to a study of the propagation phenomena, in the order suggested by the above classification.

9.2 Propagation of discontinuities

In this section we shall deal with solutions of the Boltzmann equation exhibiting discontinuities in the distribution function f or its derivatives with respect to space-time coordinates.

Let us start from the problem of studying solutions with discontinuities in the distribution function itself. We shall assume, for simplicity, that, although discontinuous, f remains finite and the collision term also remains finite.

Since the left-hand side of the Boltzmann equation contains the derivatives of f with respect to space-time coordinates, a solution discontinuous in these variables cannot be an ordinary solution of the Boltzmann equation, but must be what is called a *weak* solution. Given the simple structure of the left-hand side of the equation, we do not need to enter into mathematical subtleties, but we need only remark that that side of the equation is a directional derivative, a derivative of f taken along the trajectories that the particles would have in the absence of collisions (to be called simply the *trajectories* henceforth). This directional derivative must be finite since it equals the collision term which has been assumed to have this property. Hence the distribution function cannot have discontinuities along the trajectories. Thus the latter are the only possible *loci* of discontinuities, or to adopt a standard terminology, are possible *wave fronts*.

We have reached the conclusion that the possible wave fronts are given by the equation

$$\mathbf{x} = \mathbf{x}_0 + \boldsymbol{\xi} t. \tag{9.1}$$

where $\boldsymbol{\xi} = \mathbf{p}/p_0$ and $\mathbf{x}_0$ may depend on $\boldsymbol{\xi}$. We have adopted a three-dimensional notation, because it makes the matter transparent, but there is no problem in giving a (more cumbersome) description in space-time.

Taken literally, the above fronts reduce to a point at any given instant of time. This is satisfactory if the solution depends on just one space coordinate. The fronts expected in a two-dimensional case are typically curves. Thus in the plane case we can expect an arbitrary curve at time $t = 0$; the points of this curve can propagate with an arbitrary speed $|\boldsymbol{\xi}|$ in an arbitrary direction and produce a moving curve. Thus if there is a parabola of discontinuity $y_0 = ax_0^2$ at $t = 0$, it propagates as the parabola $y = a(x - \xi_1 t)^2 + \xi_2 t$ at subsequent times. We remark that the curve at any given time instant can depend on $\boldsymbol{\xi}$. In the three-dimensional case, the wave fronts are usually surfaces (a line would usually not be very significant as a *locus* of discontinuity, just as a point in the plane case); again if we assign the form of the surface at $t = 0$, equation (9.1) gives us a rule to construct the motion of the points of this surface. Another way of obtaining possible discontinuity curves in the two-dimensional case and discontinuity surfaces in the three-dimensional case is to keep $\mathbf{x}_0$ fixed and let $\boldsymbol{\xi}$ vary under a given geometric condition (such as tangency to a given boundary). In the three-dimensional case the surfaces cannot be arbitrarily shaped, but must be ruled surfaces, such as cylinders, cones, one-sheet hyperboloids, etc. or parts of these surfaces, which contain families of straight lines of the form (9.1) As in the plane case these wave fronts can propagate with an arbitrary speed in an arbitrary direction. Then we have a weaker singularity, because for each $\boldsymbol{\xi}$ we have just a point for any fixed value of t.

It is also interesting to discuss discontinuous solutions in the steady case, i.e., the case when a reference frame exists where the solution is time-independent. In this case, it is even more important to distinguish the different situations according to the number of space coordinates occurring in the solution. In the case that f does not depend on y and z, then the only possible discontinuity occurs at $\xi_1 = 0$; this occurs usually at the boundary of a problem in a slab or a half-space. It disappears at interior points, except in the case of free-molecular flows (no collisions), as we shall see later, when discussing the attenuation of the discontinuities. In the two-dimensional case, the discontinuities may be located on straight lines of equation

$$\xi_2 x - \xi_1 y = C(\boldsymbol{\xi}). \tag{9.2}$$

A limiting case is $\boldsymbol{\xi} = 0$. Similarly, in three dimensions, we have the relation

$$\boldsymbol{\xi} \times \mathbf{x} = \mathbf{C}(\boldsymbol{\xi}), \tag{9.3}$$

where $\mathbf{C}(\boldsymbol{\xi})$ is an arbitrary vector orthogonal to $\boldsymbol{\xi}$. As above, equations (9.3) essentially give the equation of a straight line in a three-dimensional space, whereas we expect a discontinuity surface. If we rewrite it (assuming $\boldsymbol{\xi} \neq 0$) in the equivalent form

$$\mathbf{x} = \mathbf{x}_0 + s\boldsymbol{\xi}, \tag{9.4}$$

where s is a real-valued parameter, we can proceed as above. We can assign a surface in space for $s = 0$ and compute a possible discontinuity surface by elimination, or take $\mathbf{x}_0$ fixed and let $\boldsymbol{\xi}$ vary.

Let us underline two aspects of the discontinuous solutions that we are investigating. The propagation takes place at a definite speed in a definite direction, if we fix a value of $\boldsymbol{\xi}$, but all kinds of speeds and directions arise if we consider the full distribution function.

This circumstance shows up in the fact that the moments of the distribution function are usually continuous, even when the distribution function is not. An exception occurs when the point, line or surface is independent of $\boldsymbol{\xi}$ for some value t, or when the range of integration of some variable is restricted by the geometry of the problem. In order to show this, let us first consider the case in which the solution depends on just one space coordinate x and time t. At time $t = 0$ we have just one discontinuity point. Without loss of generality, we can assume it to be the origin. We can disregard the components of $\boldsymbol{\xi}$ different from ξ_1 (which we simply denote by ξ) and consider, as a typical moment, the particle number density. The discontinuity line in a space-time diagram is $x = \xi t$. We have, for $t > 0$,

$$n(x,t) = \int f(x,\xi,t)d\xi = \int_{-c}^{x/t} f(x,\xi,t)d\xi + \int_{x/t}^{c} f(x,\xi,t)d\xi. \tag{9.5}$$

Since the function f is continuous at each point of the x, t plane except those of the straight line $x = \xi t$, the above formula shows that the discontinuity disappears by integration. To give an explicit example, let us assume that $f = A\exp(-\xi)H(\xi)$ for $x < \xi t$, and $f = B\exp(\xi)H(-\xi)$ for $x > \xi t$ ($A \neq B$, and H denotes the Heaviside step function). Then

$$n(x,t) = B\int_{-c}^{\min(x/t,0)} \exp(\xi)d\xi + A\int_{\max(x/t,0)}^{c} \exp(-\xi)d\xi, \tag{9.6}$$

and the discontinuity is smoothed out.

The occurrence of discontinuities in (non-relativistic) steady problems has been known for a long time in free-molecular flow; when the mean free path is finite, the discontinuities are damped out by collisions (see below), but still exist. When the mean free path is very small, they are only felt in the Knudsen layers, small layers a few mean free paths thick near the walls. In the classical case this fact was discovered by Sone [26], who worked with the BGK model equation. The relativistic case has been discussed by one of the authors [10].

Let us consider now the case of a discontinuity in the derivatives, when we assume the function to be continuous. Without any loss of generality we can assume that we are dealing with first order derivatives. If the derivatives up to order p are continuous and those of order $p + 1$ are not, then we can repeat the argument after differentiating the equation p times.

In the case under consideration the derivatives exist but are discontinuous. Since the collision term is continuous, it follows that the total discontinuity of the left-hand side of the equation is zero. If we denote by $[\![g]\!]$ the discontinuity of any function g, we have

$$[\![\partial_t f]\!] + \boldsymbol{\xi} \cdot [\![\partial_{\mathbf{x}} f]\!] = 0. \tag{9.7}$$

In addition, the differential in a tangential direction to the line, or surface of discontinuity, is continuous, because it can be computed using only the values on one side of the line or surface and these values are continuous. In other words, if $\tau(\mathbf{x}, t) = 0$ is the equation of a line, or surface of discontinuity, whenever $dt, d\mathbf{x}$ are such that

$$\partial_t \tau dt + \partial_{\mathbf{x}} \tau \cdot d\mathbf{x} = 0, \tag{9.8}$$

it must also be true that

$$[\![\partial_t f]\!] dt + [\![\partial_{\mathbf{x}} f]\!] \cdot d\mathbf{x} = 0. \tag{9.9}$$

Then it follows that the coefficients of the various differentials in the above relations must be proportional or, in other words, that a function $\lambda(\mathbf{x}, t)$ must exist, such that

$$[\![\partial_t f]\!] = \lambda \partial_t \tau; \qquad [\![\partial_{\mathbf{x}} f]\!] = \lambda \partial_{\mathbf{x}} \tau. \tag{9.10}$$

Hence, assuming $\lambda \neq 0$ (otherwise there is no discontinuity) and inserting (9.10) into (9.7), we obtain

$$\partial_t \tau + \boldsymbol{\xi} \cdot \partial_{\mathbf{x}} \tau = 0. \tag{9.11}$$

Hence $\tau = \tau(\mathbf{x} - \boldsymbol{\xi} t, \boldsymbol{\xi})$ and we obtain again that the discontinuities can occur only on sets built out of points which move according to (9.1). The discussion is similar to the case of the discontinuities of f. The considerations in the steady case also remain true. (The fact that the sets of possible discontinuities for the function and the derivatives are the same is a general statement holding for any equation or system in which the higher order derivatives appear only in linear terms.)

The speed of propagation is thus arbitrary, provided it is less than c, because the discontinuities can travel with any value taken by the molecular speed.

A word must be added about the damping of the discontinuities. We shall restrict ourselves to the case of discontinuities of the function f. Since the gain term tends to smooth out the irregularities of f (when they are $\boldsymbol{\xi}$-dependent, which is usually the case; see the above discussion on moments), we can assume that in most cases the discontinuities of f will evolve according to

$$\frac{d}{dt} [\![f]\!] = -\nu(f) [\![f]\!], \tag{9.12}$$

where d/dt indicates the derivative along the trajectory of a particle and $\nu(f)$ (the collision frequency) is the factor multiplying f in the collision term (which can be assumed to be continuous as we have done for the gain term). Then

$$[\![f]\!] = [\![f_0]\!] \exp\left(- \int_0^t \nu(\mathbf{x} + \boldsymbol{\xi} s) ds \right). \tag{9.13}$$

Thus we see that the discontinuities are damped on the scale of the time between collisions (inverse of the collision frequency ν). In the steady case, the previous formula must be slightly modified: s becomes a convenient parameter, $\mathbf{x}$ is replaced by $\mathbf{x}_0$, a point where $[\![f_0]\!]$ is evaluated and $t = |\mathbf{x}-\mathbf{x}_0|/|\boldsymbol{\xi}|$ is the value of s corresponding to another point $\mathbf{x}$ on the straight line in the direction of $\boldsymbol{\xi}$, where $[\![f]\!]$ is evaluated. We remark that the discontinuities are thus damped on a distance of the order of a mean free path. The discontinuities at $\boldsymbol{\xi}=0$ disappear suddenly; this applies, in particular, to the discontinuities at $\xi_{\mathbf{n}}$ on flat boundaries.

Problems

9.2.1 Verify that (9.10) follows if (9.9) holds whenever (9.8) does.

9.2.2 Check that (9.13) follows from (9.12).

9.3 Small oscillations

A tensorial equation of the type

$$\mathbf{A} = \hat{\mathbf{A}} \exp(-\alpha \mathbf{x}\cdot\mathbf{n}) \cos(K\mathbf{x}\cdot\mathbf{n} - \omega t + \delta), \tag{9.14}$$

represents a plane harmonic wave. In (9.14) $\hat{\mathbf{A}}$ denotes the amplitude of the wave, α the absorption coefficient, K the wave number, ω the circular frequency, δ the phase difference and $\mathbf{n}$ the unity vector in the direction of the propagation of the wave. If we introduce the complex amplitude $\tilde{\mathbf{A}} = \hat{\mathbf{A}} \exp i\delta$ and the complex wave number $\kappa = K + i\alpha$, we can write (9.14) as

$$\mathbf{A} = \Re\left\{\tilde{\mathbf{A}} \exp i(\kappa\mathbf{x}\cdot\mathbf{n} - \omega t)\right\} = \Re\left\{\tilde{\mathbf{A}} \exp\left(-i\kappa^\alpha x_\alpha\right)\right\}, \tag{9.15}$$

where $\kappa^0 = \omega/c$ and $\kappa^i = \kappa n^i$. As in Chapter 1, $\Re\{\}$ and $\Im\{\}$ represent the real and the imaginary parts of the quantity in the braces, respectively.

The phase speed and the absorption coefficient are given in terms of the complex wave number as

$$v_{ph} = \frac{\omega}{\Re\{\kappa\}}, \qquad \alpha = \Im\{\kappa\}. \tag{9.16}$$

In the next sections we shall analyze the propagation of small amplitudes by considering the Boltzmann equation and the balance equations of a continuum theory.

9.3.1 Boltzmann equation

Let us consider the waves produced by mechanical or thermal oscillations, assuming that the amplitude of the wave is so small that the distribution function is close to a Maxwell–Jüttner distribution function $f^{(0)}$ with constant parameters. Here we follow a paper by one of the authors [5], to show that the speed of propagation is less than the speed of light and the amplitude undergoes a damping.

We look for solutions of the Boltzmann equation without external forces

$$p^\alpha \frac{\partial f}{\partial x_\alpha} = Q(f,f), \tag{9.17}$$

– where $Q(f,f)$ is the bilinear quantity defined in (5.119) – of the form

$$f = f^{(0)}(1+\phi). \tag{9.18}$$

The quantity ϕ is treated as an infinitesimal, i.e., terms of degree higher than first in ϕ are neglected. Hence ϕ satisfies the linearized Boltzmann equation

$$p^\alpha \frac{\partial \phi}{\partial x_\alpha} = \mathcal{L}\phi, \tag{9.19}$$

where $\mathcal{L}$ is the so-called linearized collision operator

$$\mathcal{L}\phi = \frac{2}{f^{(0)}} Q(f^{(0)}\phi, f^{(0)}) = \int f_*^{(0)}(\phi'_* + \phi' - \phi_* - \phi) F\sigma d\Omega \frac{d^3p_*}{p_{*0}}. \tag{9.20}$$

We consider the Hilbert space $H = L^2(f^{(0)}d^3p/p_0)$ with the scalar product

$$(g,\phi) = \int \bar{g}\phi f^{(0)} \frac{d^3p}{p_0}, \tag{9.21}$$

where the bar over the function g denotes complex conjugation. If we take the scalar product of (9.20) with g we get

$$(g,\mathcal{L}\phi) = -\frac{1}{4}\int f^{(0)} f_*^{(0)} (\bar{g}'_* + \bar{g}' - \bar{g}_* - \bar{g})(\phi'_* + \phi' - \phi_* - \phi) F\sigma d\Omega \frac{d^3p_*}{p_{*0}}\frac{d^3p}{p_0}, \tag{9.22}$$

by using the symmetry properties of the collision operator.

From (9.22) one can show that

$$\begin{cases} (g,\mathcal{L}\phi) = (\mathcal{L}g,\phi), \\ (\phi,\mathcal{L}\phi) \quad \text{is real}, \\ (\phi,\mathcal{L}\phi) \le 0. \end{cases} \tag{9.23}$$

The equality sign in the last relationship holds if and only if ϕ is a summational invariant.

Now we look for solutions of (9.19) of the form

$$\phi(x^\alpha, p^\alpha) = \tilde{\phi}(p_\alpha, \kappa_\alpha)\exp(-i\kappa_\alpha x^\alpha). \tag{9.24}$$

As mentioned in the previous section, we distinguish between two kinds of waves, according to whether we fix a vector $\boldsymbol{\kappa}$ with real valued components and look for a (generally complex) $\omega = c\kappa_0$ for which a solution exists or we fix a real ω (positive, without loss of generality) and look for a vector $\boldsymbol{\kappa}$ (generally with complex components) for which a solution exists. In the first case we speak of free waves, because we can think of a periodic state in space, created in some way at $t = 0$ and evolving at subsequent times (with damping), whereas in the second case we speak of forced waves, because the oscillation is maintained with a fixed frequency by a suitable boundary condition. We shall concentrate on the second case, because it is the easiest to be investigated experimentally. In addition, we simplify the discussion by assuming that the wave has a fixed direction of propagation, that we take as the x-axis and let $\kappa^1 = \kappa$. Then we have:

$$-i\kappa_\alpha p^\alpha \tilde{\phi} = \mathcal{L}\tilde{\phi}, \quad \text{or} \quad -i\frac{\omega}{c}p^0\tilde{\phi} + i\kappa p^1\tilde{\phi} = \mathcal{L}\tilde{\phi}. \tag{9.25}$$

Now by taking the scalar product of (9.25) with $\tilde{\phi}$ it follows that

$$-i\frac{\omega}{c}(\tilde{\phi}, p^0\tilde{\phi}) + i\kappa(\tilde{\phi}, p^1\tilde{\phi}) = (\tilde{\phi}, \mathcal{L}\tilde{\phi}). \tag{9.26}$$

All the expressions $(\cdot, \cdot)$ are real (this is trivial for the expressions appearing in the left-hand side; for $(\tilde{\phi}, \mathcal{L}\tilde{\phi})$ see (9.23)). Hence, taking the real and imaginary parts of (9.26) and letting κ_R and κ_I denote the real and imaginary parts of κ, we obtain:

$$-\kappa_I(\tilde{\phi}, p^1\tilde{\phi}) = (\tilde{\phi}, \mathcal{L}\tilde{\phi}), \qquad \frac{\omega}{c}(\tilde{\phi}, p^0\tilde{\phi}) = \kappa_R(\tilde{\phi}, p^1\tilde{\phi}). \tag{9.27}$$

The second relation gives for the phase speed $v_{ph} = \omega/\kappa_R$:

$$v_{ph} = c\frac{(\tilde{\phi}, p^1\tilde{\phi})}{(\tilde{\phi}, p^0\tilde{\phi})}. \tag{9.28}$$

In order to analyze (9.28) we recall that the momentum four-vector p^α has a constant length so that

$$(p^0)^2 = m^2c^2 + |\mathbf{p}|^2. \tag{9.29}$$

By considering the energy as a positive quantity – i.e., $p^0 > 0$ – one can infer from (9.29) that

$$p^0 > mc, \qquad \text{and} \qquad |p^i| < p^0, \quad i = 1, 2, 3. \tag{9.30}$$

The relation (9.28) together with (9.30) immediately show that the phase speed can be arbitrary but less than c if we concentrate enough particles on the tail with, say, $p^1 > 0$ and sufficiently large. If, on the contrary, we produce a disturbance by letting the basic Maxwell–Jüttner distribution have slightly oscillating parameters, we can expect a speed of the order of $\sqrt{kT_0/m}$ i.e., of the ordinary adiabatic speed of sound for a monatomic gas $v_s = \sqrt{5kT_0/(3m)}$.

The first equation of the system (9.27) also gives an important piece of information; the imaginary part κ_I is positive in the direction of propagation. In fact, according to (9.28), the wave propagates toward increasing or decreasing values of x according to whether $(\tilde{\phi}, p^1\tilde{\phi})$ is positive or negative. Then the statement follows, because $(\tilde{\phi}, \mathcal{L}\tilde{\phi}) \leq 0$. This means that the wave decreases in amplitude in the direction of propagation (Problem 9.3.1.2).

There are two typical problems for which this theory is of interest, i.e., the propagation of sound and shear waves. One may think of a plate oscillating in the direction of its normal or in its own plane, with an assigned frequency ω; a periodic disturbance propagates through a gas filling the region at one side of the plate.

The computation of the relation between ω and k – the so-called dispersion relation – is not a completely easy matter. If the frequency is low, then one can apply the Navier–Stokes equations, but if the frequency becomes sufficiently high this approach fails even at ordinary densities, because ω^{-1} can become of the order of the time between collisions.

On the basis of an analogy with the non-relativistic case, we can conjecture that traditional moment methods also fail at high frequencies. The use of kinetic models gives interesting results (see [5], [6], [7] and [8], and here we shall follow the work by Cercignani and Majorana [7]. We start by recalling that the Anderson and Witting model equation (8.17) of the Boltzmann equation reads

$$p^\alpha \frac{\partial f}{\partial x^\alpha} = -\frac{U_L^\alpha p_\alpha}{c^2 \tau}(f - f^{(0)}). \tag{9.31}$$

The one-particle distribution function is written as

$$f(x^\alpha, p^\alpha) = f_0^{(0)}[1 + \phi(x^\alpha, p^\alpha)], \quad \text{with} \quad f_0^{(0)} = \exp(A_0 + B_0^\alpha p_\alpha), \tag{9.32}$$

where $\phi(x^\alpha, p^\alpha)$ is the perturbation of f with $f_0^{(0)}$ denoting the Maxwell–Jüttner distribution function of the unperturbed relativistic gas. Further the Maxwell–Jüttner distribution function can be approximated by

$$f^{(0)}(x^\alpha, p^\alpha) \approx f_0^{(0)}[1 + A'(x^\alpha) + B'_\alpha(x^\beta)p^\alpha]. \tag{9.33}$$

Above $A'(x^\alpha)$ and $B'_\alpha(x^\beta)$ denote the deviations from the constant values A_0 and B_0^α, respectively.

We choose a local Lorentz rest frame where $(U_L^\alpha) = (c, \mathbf{0})$, insert the representations (9.32) and (9.33) into the model equation (9.31) and obtain

$$p^\alpha \frac{\partial \phi}{\partial x^\alpha} = \frac{p^0}{c\tau}(A' + B'_\alpha p^\alpha - \phi). \tag{9.34}$$

According to (8.13) and (8.14) the following constraint must hold for the right-hand side of (9.31):

$$\int p^0 f_0^{(0)}(A' + B'_\alpha p^\alpha - \phi)\psi_{\mathcal{A}} \frac{d^3p}{p_0} = 0, \quad \text{with} \quad \mathcal{A} = 0, \ldots, 4, \tag{9.35}$$

where $\psi_{\mathcal{A}}$ denotes the summational invariants $\psi_\alpha = cp_\alpha$ and $\psi_4 = c$.

The solution of (9.34) we are looking for corresponds to a plane wave of the form

$$\phi = \tilde{\phi}\exp(-i\kappa_\alpha x^\alpha), \tag{9.36}$$

and the insertion of the above representation into (9.34) leads to

$$-i\kappa_\alpha p^\alpha \phi = \frac{p^0}{c\tau}(A' + B'_\alpha p^\alpha - \phi). \tag{9.37}$$

As in the previous case analyzed above, we assume that the wave has a fixed direction of propagation, the x-axis, so that we have: $\kappa^0 = \omega/c$, $\kappa^1 = \kappa$ and $\kappa^2 = \kappa^3 = 0$. Moreover by assuming that $p^0 \neq ic\tau\kappa_\alpha p^\alpha$ we can determine ϕ from (9.37), yielding

$$\phi = \frac{p^0(A' + B'_\alpha p^\alpha)}{p^0(1 - i\omega\tau) + ic\tau\kappa p^1}. \tag{9.38}$$

Once ϕ is known, we can insert it into the constraint (9.35) and get

$$\int f_0^{(0)} p^0 (A' + B'_\alpha p^\alpha)\frac{\xi}{1-\xi}\psi_A \frac{d^3p}{p_0} = 0, \tag{9.39}$$

where we have introduced the abbreviation

$$\xi = i\left(1 - \frac{\kappa c}{\omega}\frac{p^1}{p^0}\right)\omega\tau. \tag{9.40}$$

The constraint (9.39) represents a linear homogeneous system for A' and B'_α which can be written as

$$A'\mathcal{I} + B'_\alpha \mathcal{I}^\alpha = 0, \qquad A'\mathcal{I}^\alpha + B'_\beta \mathcal{I}^{\alpha\beta} = 0. \tag{9.41}$$

Above the $\mathcal{I}$'s denote the integrals

$$\mathcal{I} = \int f_0^{(0)} p^0 \frac{\xi}{1-\xi}\frac{d^3p}{p_0}, \qquad \mathcal{I}^\alpha = \int f_0^{(0)} p^0 p^\alpha \frac{\xi}{1-\xi}\frac{d^3p}{p_0}, \tag{9.42}$$

$$\mathcal{I}^{\alpha\beta} = \int f_0^{(0)} p^0 p^\alpha p^\beta \frac{\xi}{1-\xi}\frac{d^3p}{p_0}. \tag{9.43}$$

The linear homogeneous system (9.41) has a non-trivial solution if and only if the determinant of the coefficients of A' and B'_α vanishes which implies that

$$\begin{bmatrix} \mathcal{I} & \mathcal{I}^0 & \mathcal{I}^1 & 0 & 0 \\ \mathcal{I}^0 & \mathcal{I}^{00} & \mathcal{I}^{01} & 0 & 0 \\ \mathcal{I}^1 & \mathcal{I}^{01} & \mathcal{I}^{11} & 0 & 0 \\ 0 & 0 & 0 & \mathcal{I}^{22} & 0 \\ 0 & 0 & 0 & 0 & \mathcal{I}^{33} \end{bmatrix} = 0. \tag{9.44}$$

The above equation is equivalent to the conditions

$$\begin{bmatrix} \mathcal{I} & \mathcal{I}^0 & \mathcal{I}^1 \\ \mathcal{I}^0 & \mathcal{I}^{00} & \mathcal{I}^{01} \\ \mathcal{I}^1 & \mathcal{I}^{01} & \mathcal{I}^{11} \end{bmatrix} = 0, \qquad \text{or} \qquad \mathcal{I}^{22} = \mathcal{I}^{33} = 0. \tag{9.45}$$

From now on we shall restrict ourselves to the case of small frequencies where $|\xi| \ll 1$ so that

$$\frac{\xi}{1-\xi} = \sum_{n=0}^{\infty} \xi^{n+1} \approx i\left(1 - \frac{\kappa c}{\omega}\frac{p^1}{p^0}\right)\omega\tau - \left(1 - \frac{\kappa c}{\omega}\frac{p^1}{p^0}\right)^2 (\omega\tau)^2 + \cdots, \tag{9.46}$$

i.e., we shall truncate the series at the second term.

We proceed now to evaluate the $\mathcal{I}$'s integrals. First we note that $\mathcal{I}^2 = \mathcal{I}^3 = \mathcal{I}^{02} = \mathcal{I}^{03} = \mathcal{I}^{12} = \mathcal{I}^{13} = \mathcal{I}^{23} = 0$. Further by evaluating the remaining integrals in a local Lorentz rest frame and by considering the expansion for ξ given by (9.46) we get

$$\mathcal{I} = n\omega\tau\left\{i - \omega\tau\left[1 + \left(\frac{\kappa c}{\omega}\right)^2\left(\frac{m\lambda}{kp\tau\zeta^3 G^2} + \frac{1}{\zeta G}\right)\right]\right\}, \tag{9.47}$$

$$\mathcal{I}^0 = \frac{p\omega\tau}{c}\left\{i\frac{e}{kT} - \omega\tau\left[\frac{e}{kT} + \left(\frac{\kappa c}{\omega}\right)^2\right]\right\}, \quad \mathcal{I}^1 = -\frac{p\omega\tau}{c}\left(\frac{\kappa c}{\omega}\right)[i - 2\omega\tau], \tag{9.48}$$

$$\mathcal{I}^{00} = \frac{pkT\omega\tau}{c^2}\left\{(i-\omega\tau)\left(\zeta^2 + 3\frac{h_E}{kT}\right) - \omega\tau\frac{h_E}{kT}\left(\frac{\kappa c}{\omega}\right)^2\right\}, \quad \mathcal{I}^{01} = \frac{h_E}{c}\mathcal{I}^1, \tag{9.49}$$

$$\mathcal{I}^{11} = \frac{pkT\omega\tau}{c^2}\left\{(i-\omega\tau)\frac{h_E}{kT} - \frac{3\mu\omega}{p}\left(\frac{\kappa c}{\omega}\right)^2\right\}, \tag{9.50}$$

$$\mathcal{I}^{22} = \frac{pkT\omega\tau}{c^2}\left\{(i-\omega\tau)\frac{h_E}{kT} - \frac{\mu\omega}{p}\left(\frac{\kappa c}{\omega}\right)^2\right\}. \tag{9.51}$$

In the above equations the coefficients of shear viscosity μ and thermal conductivity λ are given by (8.86) and (8.83), respectively.

Let us analyze the first condition of (9.45), which furnishes the dispersion relation:

$$\mathcal{A}_1\left(\frac{\kappa c}{\omega}\right)^6 + \mathcal{A}_2\left(\frac{\kappa c}{\omega}\right)^4 + \mathcal{A}_3\left(\frac{\kappa c}{\omega}\right)^2 + \mathcal{A}_4 = 0, \tag{9.52}$$

where the coefficients $\mathcal{A}_1$ through $\mathcal{A}_4$ are given by

$$\mathcal{A}_1 = \frac{3\omega^3\tau m\lambda\mu}{\zeta^2 G k p^2}, \tag{9.53}$$

$$\mathcal{A}_2 = \frac{3\omega^2 m\lambda\mu}{\zeta^2 G^2 k p^2}(\zeta+3G)(\omega\tau - i) + \frac{\omega m\lambda}{\zeta kp}[3\omega\tau(i-\omega\tau)+1] + \frac{3\omega^2\tau\mu c_p}{\zeta G k p}(\omega\tau - i), \tag{9.54}$$

$$\mathcal{A}_3 = \left[\frac{3\omega c_v \mu}{kp} + \frac{\omega m\lambda}{\zeta G k p}(\zeta + 3G)\right][(\omega\tau)^2 - 2i\omega\tau - 1]$$

$$-\frac{c_p}{k}[3(\omega\tau)^3 - 6i(\omega\tau)^2 - 4\omega\tau + i], \tag{9.55}$$

$$\mathcal{A}_4 = \frac{\zeta G c_v}{k}[(\omega\tau)^3 - 3\omega\tau(1 + i\omega\tau) + i]. \tag{9.56}$$

The solution of the dispersion relation (9.52) for small frequencies read

$$\frac{\kappa v_s}{\omega} \approx 1 + \frac{i}{2}\left(\frac{c_v}{c_p}\right)^2\left(\frac{v_s}{c}\right)^2 \omega\tau \left[3\zeta G + \frac{m\lambda}{\mu c_v}(\zeta + 3G) - \frac{m\lambda\zeta G^2}{\mu c_p} - \frac{\zeta G c_p \tau p}{\mu c_v}\right], \tag{9.57}$$

where the adiabatic sound speed v_s was introduced:

$$v_s = \sqrt{\frac{c^2 c_p k T}{c_v h_E}}. \tag{9.58}$$

In the limiting case where $\zeta \gg 1$ – which refers to a non-relativistic gas – we have

$$v_s = \sqrt{\frac{5kT}{3m}}, \qquad \frac{\kappa v_s}{\omega} \approx 1 + \frac{3i}{5}\omega\tau, \tag{9.59}$$

i.e., the adiabatic sound speed reduces to that of a monatomic gas. Furthermore the attenuation coefficient differs from that found by using a Navier–Stokes and Fourier continuum theory (see next section) since as was previously stated the ratio between the transport coefficients here is $\lambda/\mu = 5k/(2m)$ instead of $\lambda/\mu = 15k/(4m)$. By using the latter value for this ratio we get that the factor of the attenuation coefficient becomes $7/10$ instead of $3/5$, which agrees with the Navier–Stokes and Fourier continuum theory.

If we consider $\zeta \ll 1$ we get the following results for an ultra-relativistic gas:

$$v_s = \frac{c}{\sqrt{3}}, \qquad \frac{\kappa v_s}{\omega} \approx 1 + \frac{i}{2}\omega\tau, \tag{9.60}$$

i.e., the adiabatic sound speed of an ultra-relativistic gas is of the order of the speed of light.

There exists another solution of the dispersion relation (9.52) for small frequencies which correspond to the so-called thermal or heat mode and where the complex wave number is given by

$$\kappa = \frac{1+i}{\sqrt{2}}\sqrt{\frac{n c_p \omega}{\lambda}} = \begin{cases} \frac{1+i}{\sqrt{3}}\sqrt{\frac{m\omega}{\tau k T}}, & \text{for} \quad \zeta \gg 1, \\ \frac{\sqrt{3}(1+i)}{\sqrt{2}c}\sqrt{\frac{\omega}{\tau}}, & \text{for} \quad \zeta \ll 1. \end{cases} \tag{9.61}$$

Furthermore the second condition of (9.45) furnishes the following expression for the complex wave number for small frequencies:

$$\kappa = \frac{1+i}{c\sqrt{2}}\sqrt{\frac{nh_E\omega}{\mu}} = \begin{cases} \frac{1+i}{\sqrt{2}}\sqrt{\frac{m\omega}{\tau kT}}, & \text{for} \quad \zeta \gg 1, \\ \frac{\sqrt{5}(1+i)}{\sqrt{2}c}\sqrt{\frac{\omega}{\tau}}, & \text{for} \quad \zeta \ll 1, \end{cases} \tag{9.62}$$

which represents the so-called shear mode.

Cercignani and Majorana [7] have also analyzed the dispersion relation at high frequencies and concluded that there exists no solution of the dispersion relation for frequencies sufficiently high.

Problems

9.3.1.1 Check that $(\phi, \mathcal{L}\phi)$ is real and negative.

9.3.1.2 Check that the linearized waves are damped in the direction of propagation.

9.3.1.3 Determine the integrals (9.47) through (9.51) by using the methods of integration described in previous chapters.

9.3.1.4 Obtain the solution (9.57) of the dispersion relation for small frequencies.

9.3.1.5 Obtain the expressions for v_s and $\kappa v_s/\omega$ in the limiting cases of a non-relativistic (9.59) and ultra-relativistic (9.60) gas.

9.3.1.6 Show that the complex wave number for the heat and shear modes are given by (9.61) and (9.62), respectively.

9.3.2 Continuum-like theories

The problem concerning the propagation of plane harmonic waves of small amplitudes through a relativistic gas at rest has been investigated by many authors. Among others we cite Weinberg [30] and Guichelaar et al. [17] who studied this problem by using a five-field theory which takes into account the fields of particle number density, temperature and four-velocity and employed the Navier–Stokes and Fourier constitutive equations. Anderson and Payne [1] have also considered a five-field theory but they used the Burnett constitutive equations which were obtained from a kinetic model of the Boltzmann equation. The same problem was analyzed by Müller [24], Israel and Stewart [18], de Groot et al. [16], Boillat [2], Seccia and Strumia [25] by using a fourteen-field theory where the fields of particle number density, temperature, four-velocity, dynamic pressure, pressure deviator and heat flux were taken into account, while Kranyš [20] employed a thirteen-field theory where the field of the dynamic pressure, i.e., the non-equilibrium part of the pressure, was not considered. For a non-relativistic gas Greenspan [15] has done a careful analysis on the dispersion and absorption of plane harmonic waves and compared the solutions that follow from a thirteen-field theory and from those that follow from a five-field theory with Navier–Stokes–Fourier and Burnett constitutive equations.

Here we follow [12] and extend Greenspan's analysis to a relativistic gas. To that end we rely on a five-field theory with Navier–Stokes–Fourier and Burnett constitutive equations and on a fourteen-field theory, and as a special case of the latter on a thirteen-field theory. In all cases the constitutive equations are derived from the Boltzmann equation in which a constant differential cross-section is taken into account, i.e., we consider a relativistic gas of hard-sphere particles. In the non-relativistic limit of the five- and thirteen-field theories the results of Greenspan are recovered, but the case of a fourteen-field theory is also analyzed. Further, the differences between expressions for the dispersion and absorption of plane waves in the non-relativistic and in the ultra-relativistic limit by using five-, thirteen- and fourteen-field theories are given.

Navier–Stokes–Fourier constitutive equations

We suppose that the plane harmonic wave is propagating through a viscous heat conducting relativistic gas at rest with constant particle number density n_0 and temperature T_0, i.e.,

$$n = n_0 + \tilde{n}\exp i(\kappa\mathbf{x}\cdot\mathbf{n} - \omega t), \qquad T = T_0 + \tilde{T}\exp i(\kappa\mathbf{x}\cdot\mathbf{n} - \omega t), \tag{9.63}$$

$$(U^\alpha) = \left(\frac{c}{\sqrt{1-[\tilde{v}\exp i(\kappa\mathbf{x}\cdot\mathbf{n} - \omega t)/c]^2}}, \frac{\tilde{\mathbf{v}}\exp i(\kappa\mathbf{x}\cdot\mathbf{n} - \omega t)}{\sqrt{1-[\tilde{v}\exp i(\kappa\mathbf{x}\cdot\mathbf{n} - \omega t)/c]^2}}\right). \tag{9.64}$$

The amplitudes $\tilde{n}$, $\tilde{T}$ and $\tilde{\mathbf{v}}$ are considered to be small so that then products can be neglected. The representation (9.64) for the four-velocity satisfies the condition of constant length $U^\alpha U_\alpha = c^2$.

As it was seen in previous chapters, a viscous heat conducting relativistic gas has the following constitutive equations for the dynamic pressure ϖ, pressure deviator $p^{\langle\alpha\beta\rangle}$ and heat flux q^α:

$$\varpi = -\eta\nabla_\alpha U^\alpha, \quad p^{\langle\alpha\beta\rangle} = 2\mu\nabla^{\langle\alpha}U^{\beta\rangle}, \quad q^\alpha = \lambda\left(\nabla^\alpha T - \frac{T}{nh_E}\nabla^\alpha p\right). \tag{9.65}$$

Here we shall use expressions (5.89) through (5.91) for the transport coefficients of bulk viscosity η, thermal conductivity λ and shear viscosity μ that correspond to a gas of hard-sphere particles.

Further we refer to Section 6.4 and write down the linearized balance equations for the particle number density, momentum density and energy density as

$$Dn + n\nabla^\alpha U_\alpha = 0, \tag{9.66}$$

$$\frac{nh_E}{c^2}DU^\alpha = \nabla^\alpha(p+\varpi) - \nabla_\beta p^{\langle\alpha\beta\rangle} - \frac{1}{c^2}Dq^\alpha, \tag{9.67}$$

$$nDe = -p\nabla^\alpha U_\alpha - \nabla_\alpha q^\alpha. \tag{9.68}$$

We substitute (9.63) through (9.65) into the linearized balance equations (9.66) through (9.68) and by using the relationships that are valid in the linear case,

$$D = U^\alpha \partial_\alpha = \frac{\partial}{\partial t}, \qquad \nabla^\alpha = \left(\eta^{\alpha\beta} - \frac{1}{c^2}U^\alpha U^\beta\right)\partial_\beta \equiv \left(0, -\frac{\partial}{\partial x^i}\right), \tag{9.69}$$

we get a system of equations for the amplitudes that reads

$$\begin{bmatrix} A_{11} & A_{12} & A_{13} \\ A_{21} & A_{22} & A_{23} \\ A_{31} & A_{32} & A_{33} \end{bmatrix} \begin{bmatrix} \tilde{n} \\ \tilde{v}_{\|} \\ \tilde{T} \end{bmatrix} = \begin{bmatrix} 0 \\ 0 \\ 0 \end{bmatrix}, \tag{9.70}$$

$$\left(\frac{n_0 h_0}{c^2}\omega + i\mu_0 \kappa^2\right)\mathbf{v}_\perp = 0. \tag{9.71}$$

In the above equations we have introduced the notation $\tilde{v}_{\|} = \tilde{\mathbf{v}}\cdot\mathbf{n}$ and $\tilde{\mathbf{v}}_\perp = (\mathbf{n}\times\tilde{\mathbf{v}})$ which refer to the longitudinal and transversal components of the amplitude of the velocity, respectively. Further the index zero indicates a reference state where $T = T_0$ and $n = n_0$ and the elements of the matrix in (9.70) read

$$\begin{cases} A_{11} = \omega, \quad A_{12} = -n_0\kappa, \quad A_{13} = 0, \quad A_{21} = -T_0 k\kappa\left(1 - i\frac{\lambda_0 T_0 \omega}{c^2 n_0 h_0}\right), \\ A_{22} = \frac{n_0 h_0 \omega}{c^2} + i\kappa^2\left(\eta_0 + \frac{4}{3}\mu_0\right), \quad A_{23} = -n_0 k\kappa\left(1 + i\frac{\lambda_0 e_0 \omega}{n_0 k c^2 h_0}\right), \\ A_{31} = -i\frac{\lambda_0 T_0^2 k\kappa^2}{n_0 h_0}, \quad A_{32} = -n_0 k T_0 \kappa, \quad A_{33} = n_0 c_v^0 \omega + i\frac{\lambda_0 e_0 \kappa^2}{h_0}. \end{cases} \tag{9.72}$$

A non-trivial solution of (9.71) is given by

$$\kappa = \frac{1+i}{c\sqrt{2}}\sqrt{\frac{n_0 h_0 \omega}{\mu_0}}, \tag{9.73}$$

which represents the dispersion relation of a shear wave (see (9.62)).

The system of equations (9.70) has a non-trivial solution if and only if the determinant of the coefficients for the amplitudes vanishes. This condition implies the dispersion relation for the longitudinal waves:

$$\left(\frac{\kappa}{\omega}\right)^4\left[\left(\eta_0 + \frac{4}{3}\mu_0\right)\lambda_0 \frac{e_0}{h_0}\omega^2 + i\lambda_0 n_0 T_0 k\omega\right] - \left(\frac{\kappa}{\omega}\right)^2\left[i\left(\eta_0 + \frac{4}{3}\mu_0\right)n_0 c_v^0 \omega \right.$$

$$\left. + i\frac{\lambda_0 n_0}{c^2 h_0}(e_0^2 + c_v^0 T_0^2 k)\omega - n_0^2 T_0 k c_p^0\right] - \frac{n_0^2 h_0 c_v^0}{c^2} = 0. \tag{9.74}$$

Here we have used the relationship between the heat capacities $c_p^0 = c_v^0 + k$ which is also valid for relativistic gases.

For small frequencies we can expand the ratio κ/ω in the following power series in ω:

$$\frac{\kappa}{\omega} = a_0 + a_1\omega + a_2\omega^2 + \cdots . \tag{9.75}$$

If we insert (9.75) into (9.74) and equate the coefficients of the same power of ω to zero, we can write the phase speed v_{ph} and the absorption coefficient α up to terms in ω^2 as

$$\begin{aligned}\frac{v_{ph}}{v_s} = 1 + \frac{1}{2}\left(\frac{\omega}{n_0 v_s^2}\right)^2 \Bigg\{ \frac{\lambda_0}{4c_p^0}\Bigg[\frac{c^2}{h_0}\left(\eta_0 + \frac{4}{3}\mu_0\right) + \frac{\lambda_0}{c_v^0}\Bigg(\frac{c_p^0 - c_v^0}{c_p^0} - \frac{2kT_0}{h_0} \\ + \frac{kT_0^2 c_p^0}{h_0^2}\Bigg)\Bigg]\left[\frac{3c_p^0 - 7c_v^0}{c_v^0} - 6\frac{kT_0 c_p^0}{c_v^0 h_0} + 3\frac{kT_0^2 c_p^{0\,2}}{c_v^0 h_0^2}\right] + \frac{c^2}{4h_0}\left(\eta_0 + \frac{4}{3}\mu_0\right) \\ \times \left[\frac{3c^2}{h_0}\left(\eta_0 + \frac{4}{3}\mu_0\right) + \frac{\lambda_0}{c_v^0}\left(\frac{7c_p^0 - 3c_v^0}{c_p^0} - \frac{10kT_0}{h_0} + \frac{3kT_0^2 c_p^0}{h_0^2}\right)\right]\Bigg\},\end{aligned} \tag{9.76}$$

$$\alpha = \frac{\omega^2}{2v_s^3 n_0}\left[\frac{c^2}{h_0}\left(\eta_0 + \frac{4}{3}\mu_0\right) + \frac{\lambda_0}{c_v^0}\left(\frac{c_p^0 - c_v^0}{c_p^0} - \frac{2kT_0}{h_0} + \frac{kT_0^2 c_p^0}{h_0^2}\right)\right]. \tag{9.77}$$

The phase speed (9.76) depends on the circular frequency ω and in this case we say that the wave exhibits a dispersion. The adiabatic sound speed

$$v_s = \sqrt{\frac{kT_0 c_p^0 c^2}{h_0 c_v^0}}, \tag{9.78}$$

in the limiting case of low temperatures where $\zeta \gg 1$, reduces to

$$v_s = v_s^{nr}\left[1 - \frac{7}{4\zeta} + \cdots\right], \qquad v_s^{nr} = \left(\frac{5}{3}\frac{k}{m}T_0\right)^{\frac{1}{2}}. \tag{9.79}$$

The leading term in (9.79) is the well-known adiabatic sound speed in a non-relativistic gas. For high temperatures where $\zeta \ll 1$ we have from (9.78)

$$v_s = v_s^{ur}\left[1 - \frac{1}{24}\zeta^2 + \cdots\right], \qquad v_s^{ur} = \frac{c}{\sqrt{3}}, \tag{9.80}$$

that is, the ultra-relativistic adiabatic sound speed is of the order of the speed of light.

For a relativistic gas with constant differential cross-section the absorption coefficient (9.77) in the non-relativistic limit ($\zeta \gg 1$) is

$$\alpha = \frac{7}{10}\frac{\omega^2 \tau_n}{v_s^{nr}}\left[1 + \frac{135}{112\zeta} + \cdots\right], \tag{9.81}$$

where the relaxation time in the non-relativistic regime τ_n was chosen as a function of the leading term of the coefficient of shear viscosity (see (5.94))

$$\tau_n = \frac{\mu_n}{p}, \qquad \text{with} \qquad \mu_n = \frac{5}{64\sigma}\left(\frac{mkT}{\pi}\right)^{\frac{1}{2}}. \tag{9.82}$$

The absorption coefficient (9.77) in the ultra-relativistic limiting case ($\zeta \ll 1$) for a relativistic gas with constant differential cross-section reduces to

$$\alpha = \frac{1}{2}\frac{\omega^2\tau_u}{v_s^{ur}}\left[1 + \frac{1}{20}\zeta^2 + \cdots\right]. \tag{9.83}$$

The relaxation time in the ultra-relativistic regime τ_u was chosen also as a function of the leading term of the coefficient of shear viscosity (see (5.100))

$$\tau_u = \frac{\mu_u}{p}, \qquad \text{with} \qquad \mu_u = \frac{3}{10\pi}\frac{kT}{c\sigma}. \tag{9.84}$$

We note that in both cases the absorption coefficient is proportional to the square of the circular frequency.

It is also instructive to analyze the dispersion relation in the two limiting cases, i.e., non-relativistic and ultra-relativistic. In the non-relativistic limiting case the transport coefficients reduce to

$$\mu_n = \tau_n p = \frac{5}{64\sigma}\left(\frac{mkT}{\pi}\right)^{\frac{1}{2}}, \qquad \lambda_n = \frac{15k}{4m}\mu_n, \qquad \eta_n \to 0, \tag{9.85}$$

and the dispersion relation reads

$$\left(\frac{\kappa v_s^{nr}}{\omega}\right)^4\left[\frac{6}{5}(\tau_n\omega)^2 + \frac{9}{10}i(\tau_n\omega)\right] - \left(\frac{\kappa v_s^{nr}}{\omega}\right)^2\left[\frac{23}{10}i(\tau_n\omega) - 1\right] - 1 = 0. \tag{9.86}$$

The first terms in the power series of $\kappa v_s^{nr}/\omega$ as a function of $(\tau_n\omega)$ are given by

$$\frac{\kappa v_s^{nr}}{\omega} = 1 + \frac{7}{10}i(\tau_n\omega) - \frac{141}{200}(\tau_n\omega)^2 - \frac{1559}{2000}i(\tau_n\omega)^3 + \cdots. \tag{9.87}$$

The above expression is the same as that obtained by Greenspan [15].

In the ultra-relativistic limiting case the transport coefficients are given by

$$\mu_u = \tau_u p = \frac{3}{10\pi}\frac{kT}{c\sigma}, \qquad \lambda_u = \frac{5c^2}{3T}\mu_u, \qquad \eta_u \to 0, \tag{9.88}$$

and the expression for the dispersion relation reduces to

$$\left(\frac{\kappa v_s^{ur}}{\omega}\right)^4\left[\frac{5}{4}(\tau_u\omega)^2 + \frac{5}{4}i(\tau_u\omega)\right] - \left(\frac{\kappa v_s^{ur}}{\omega}\right)^2\left[\frac{9}{4}i(\tau_u\omega) - 1\right] - 1 = 0. \tag{9.89}$$

In this case the power series of $\kappa v_s^{ur}/\omega$ in terms of $(\tau_u\omega)$ is

$$\frac{\kappa v_s^{ur}}{\omega} = 1 + \frac{1}{2}i(\tau_u\omega) - \frac{3}{8}(\tau_u\omega)^2 - \frac{5}{16}i(\tau_u\omega)^3 + \cdots. \tag{9.90}$$

Burnett constitutive equations

In Section 6.5.2 we obtained the linearized Burnett equations for the dynamic pressure ϖ, heat flux q^α and pressure deviator $p^{\langle\alpha\beta\rangle}$ which we reproduce below:

$$\varpi = -\eta\nabla_\alpha U^\alpha + \eta_1\nabla_\alpha\nabla^\alpha T - \eta_2\nabla_\alpha\nabla^\alpha p, \tag{9.91}$$

$$q^\alpha = \lambda\left(\nabla^\alpha T - \frac{T}{nh}\nabla^\alpha p\right) + \lambda_1\nabla^\alpha\nabla_\beta U^\beta - \lambda_2\nabla_\beta\nabla^{\langle\alpha}U^{\beta\rangle}, \tag{9.92}$$

$$p^{\langle\alpha\beta\rangle} = 2\mu\nabla^{\langle\alpha}U^{\beta\rangle} + \mu_1\nabla^{\langle\alpha}\nabla^{\beta\rangle}T - \mu_2\nabla^{\langle\alpha}\nabla^{\beta\rangle}p. \tag{9.93}$$

The transport coefficients η_1 through μ_2 for an interaction potential of hard-sphere particles are also given in the referred section.

If we insert the Burnett constitutive equations (9.91) through (9.93) together with the plane harmonic representations (9.63) and (9.64) into the balance equations (9.66) through (9.68), we get a system of equations for the amplitudes of the longitudinal waves of the form (9.70) where the elements of the matrix are now given by

$$\begin{cases} A_{11} = \omega, \qquad A_{12} = -n_0\kappa, \qquad A_{13} = 0, \\ A_{21} = -T_0k\kappa\left[1 - i\frac{\lambda_0T_0\omega}{c^2n_0h_0} + \left(\eta_2^0 + \frac{2}{3}\mu_2^0\right)\kappa^2\right], \\ A_{22} = \frac{n_0h_0\omega}{c^2} + i\kappa^2\left(\eta_0 + \frac{4}{3}\mu_0\right) + \frac{\kappa^2\omega}{c^2}\left(\lambda_1^0 - \frac{2}{3}\lambda_2^0\right), \\ A_{23} = -n_0k\kappa\left[1 + i\frac{\lambda_0e_0\omega}{n_0kc^2h_0} - \left(\eta_1^0 + \frac{2}{3}\mu_1^0\right)\frac{\kappa^2}{kn_0} + \left(\eta_2^0 + \frac{2}{3}\mu_2^0\right)\kappa^2\right], \\ A_{31} = -i\frac{\lambda_0T_0^2k\kappa^2}{n_0h_0}, \qquad A_{32} = -n_0kT_0\kappa\left[1 + \left(\lambda_1^0 - \frac{2}{3}\lambda_2^0\right)\frac{\kappa^2}{kTn_0}\right], \\ A_{33} = n_0c_v^0\omega + i\frac{\lambda_0e_0\kappa^2}{h_0}. \end{cases} \tag{9.94}$$

Here we shall analyze only the dispersion relation in the non-relativistic and ultra-relativistic limiting cases. In the non-relativistic limit case $\zeta \gg 1$ and the Burnett transport coefficients reduce to

$$\eta_1 \to 0, \quad \eta_2 \to 0, \quad \lambda_1 = \frac{15\mu_n^2}{4mn}, \quad \lambda_2 = \frac{3\mu_n^2}{mn}, \quad \mu_1 = \frac{3\mu_n^2}{Tmn}, \quad \mu_2 = \frac{2\mu_n^2}{pmn}. \tag{9.95}$$

The dispersion relation in this case is written as

$$\left(\frac{\kappa v_s^{nr}}{\omega}\right)^6\left[\frac{21}{125}(\tau_n\omega)^4 - \frac{18}{25}i(\tau_n\omega)^3\right] - \left(\frac{\kappa v_s^{nr}}{\omega}\right)^4\left[\frac{97}{50}(\tau_n\omega)^2 + \frac{9}{10}i(\tau_n\omega)\right]$$
$$+ \left(\frac{\kappa v_s^{nr}}{\omega}\right)^2\left[\frac{23}{10}i(\tau_n\omega) - 1\right] + 1 = 0. \tag{9.96}$$

The power series of $\kappa v_s^{nr}/\omega$ as a function of $(\tau_n\omega)$ is given by

$$\frac{\kappa v_s^{nr}}{\omega} = 1 + \frac{7}{10}i(\tau_n\omega) - \frac{43}{40}(\tau_n\omega)^2 - \frac{4203}{2000}i(\tau_n\omega)^3 + \cdots. \tag{9.97}$$

The above expression was also obtained by Greenspan [15] and we note that the dispersion term associated with $(\tau_n\omega)^2$ and the correction to the absorption coefficient, associated with $(\tau_n\omega)^3$, differ from the corresponding terms in the Navier–Stokes and Fourier theory (9.87).

The Burnett transport coefficients in the ultra-relativistic limit case ($\zeta \ll 1$) reduce to

$$\eta_1 \to 0, \quad \eta_2 \to 0, \quad \lambda_1 \to 0, \quad \lambda_2 = \frac{5\mu_u^2c^2}{6p}, \quad \mu_1 = \frac{5\mu_u^2c^2}{3pT}, \quad \mu_2 = \frac{7\mu_u^2c^2}{6p^2}, \tag{9.98}$$

so that the dispersion relation becomes

$$\left(\frac{\kappa v_s^{ur}}{\omega}\right)^6\left[\frac{5}{12}(\tau_u\omega)^4+\frac{15}{8}i(\tau_u\omega)^3\right]+\left(\frac{\kappa v_s^{ur}}{\omega}\right)^4\left[\frac{7}{3}(\tau_u\omega)^2+\frac{5}{4}i(\tau_u\omega)\right]$$
$$+\left(\frac{\kappa v_s^{ur}}{\omega}\right)^2\left[\frac{5}{12}(\tau_u\omega)^2-\frac{9}{4}i(\tau_u\omega)+1\right]-1=0. \tag{9.99}$$

In this case $\kappa v_s^{ur}/\omega$ expanded into a power series of $(\tau_u\omega)$ reads

$$\frac{\kappa v_s^{ur}}{\omega}=1+\frac{1}{2}i(\tau_u\omega)-\frac{9}{8}(\tau_u\omega)^2-\frac{95}{48}i(\tau_u\omega)^3+\cdots. \tag{9.100}$$

We note that the terms that correspond to the dispersion and the correction to the absorption coefficient also differ from those of the Navier–Stokes and Fourier theory (9.90).

Fourteen-field theory

Now we shall analyze the same problem of a plane harmonic wave which is propagating through a relativistic gas which is at rest, but we consider that the gas is described by the fourteen fields of particle number density, temperature, four-velocity, dynamic pressure, heat flux and pressure deviator. Here we are also interested in longitudinal waves propagating in the x-direction where the fields are represented by a plane harmonic wave of the form

$$n = n_0 + \tilde{n}\exp i(\kappa x-\omega t), \qquad T = T_0+\tilde{T}\exp i(\kappa x-\omega t), \tag{9.101}$$

$$(U^\alpha)=\left(\frac{c}{\sqrt{1-[\tilde{v}\exp i(\kappa x-\omega t)/c]^2}}, \frac{\tilde{v}\exp i(\kappa x-\omega t)}{\sqrt{1-[\tilde{v}\exp i(\kappa x-\omega t)/c]^2}},0,0\right), \tag{9.102}$$

$$\varpi=\tilde{\mathcal{P}}\exp i(\kappa x-\omega t),\ q^x=\tilde{\mathcal{Q}}\exp i(\kappa x-\omega t),\ p^{\langle xx\rangle}=\tilde{\mathcal{R}}\exp i(\kappa x-\omega t). \tag{9.103}$$

The amplitudes $\tilde{n}$, $\tilde{T}$, $\tilde{v}$, $\tilde{\mathcal{P}}$, $\tilde{\mathcal{Q}}$ and $\tilde{\mathcal{R}}$ are also considered to be small so that their products can be neglected.

The linearized fourteen-field equations were obtained in Section 6.4 and we reproduce them below:

$$Dn+n\nabla^\alpha U_\alpha=0, \tag{9.104}$$

$$\frac{nh}{c^2}DU^\alpha = \nabla^\alpha(p+\varpi) - \nabla_\beta p^{\langle\alpha\beta\rangle} - \frac{1}{c^2}Dq^\alpha, \tag{9.105}$$

$$nDe = -p\nabla_\beta U^\beta - \nabla_\beta q^\beta. \tag{9.106}$$

$$\frac{C_2}{2}D\varpi + \frac{1}{2}(m^2+C_1)Dn - \frac{\zeta}{2T}nC_1'DT - 5\frac{C_3}{c^2}\nabla^\alpha q_\alpha$$
$$+\frac{1}{6}(nm^2+5nC_1)\nabla^\alpha U_\alpha = -\frac{3}{c^2}B_1\varpi, \tag{9.107}$$

$$5C_3Dq^\alpha - \frac{c^4}{6}\left[(m^2-C_1)\nabla^\alpha n + \frac{\zeta}{T}nC_1'\nabla^\alpha T - C_2\nabla^\alpha\varpi\right] - c^2C_4\nabla_\beta p^{\langle\alpha\beta\rangle}$$
$$-\frac{c^2}{6}(nm^2+5nC_1)DU^\alpha = -B_2q^\alpha, \tag{9.108}$$

$$C_4Dp^{\langle\alpha\beta\rangle} + 2C_3\nabla^{\langle\alpha}q^{\beta\rangle} + \frac{c^2}{3}(nm^2-nC_1)\nabla^{\langle\alpha}U^{\beta\rangle} = B_3p^{\langle\alpha\beta\rangle}. \tag{9.109}$$

The scalar coefficients, C's and B's in the above equations, are given in Section 6.3.

We insert now the plane harmonic wave representations (9.101) through (9.103) into the field equations (9.104) through (9.109) and get, by neglecting all non-linear terms, a system of equations for the amplitudes that reads

$$\begin{bmatrix} A_{11} & A_{12} & A_{13} & A_{14} & A_{15} & A_{16} \\ A_{21} & A_{22} & A_{23} & A_{24} & A_{25} & A_{26} \\ A_{31} & A_{32} & A_{33} & A_{34} & A_{35} & A_{36} \\ A_{41} & A_{42} & A_{43} & A_{44} & A_{45} & A_{46} \\ A_{51} & A_{52} & A_{53} & A_{54} & A_{55} & A_{56} \\ A_{61} & A_{62} & A_{63} & A_{64} & A_{65} & A_{66} \end{bmatrix} \begin{bmatrix} \tilde{n} \\ \tilde{v} \\ \tilde{T} \\ \tilde{\mathcal{P}} \\ \tilde{\mathcal{Q}} \\ \tilde{\mathcal{R}} \end{bmatrix} = \begin{bmatrix} 0 \\ 0 \\ 0 \\ 0 \\ 0 \\ 0 \end{bmatrix}. \tag{9.110}$$

The elements of the above matrix are given by

$$\begin{cases} A_{11} = \omega,\ A_{12} = -n_0\kappa,\ A_{13} = A_{14} = A_{15} = A_{16} = 0,\ A_{21} = -kT_0\kappa, \\ A_{22} = \frac{n_0h_0\omega}{c^2},\ A_{23} = -n_0k\kappa,\ A_{24} = -\kappa,\ A_{25} = \frac{\omega}{c^2},\ A_{26} = -\kappa, \\ A_{31} = 0,\ A_{32} = -n_0kT_0\kappa,\ A_{33} = n_0c_v^0\omega,\ A_{34} = 0,\ A_{35} = -\kappa, \\ A_{36} = 0,\ A_{41} = \frac{(m^2+C_1)\omega}{2},\ A_{42} = -\frac{n_0(m^2+5C_1)\kappa}{6},\ A_{43} = -\frac{\zeta n_0C_1'\omega}{2T}, \\ A_{44} = \frac{C_2\omega}{2} + 3i\frac{B_1}{c^2},\ A_{45} = \frac{5C_3\kappa}{c^2},\ A_{46} = 0,\ A_{51} = \frac{c^4(C_1-m^2)\kappa}{6}, \\ A_{52} = -\frac{n_0c^2(m^2+5C_1)\omega}{6},\ A_{53} = -\frac{c^4\zeta n_0C_1'\kappa}{6T},\ A_{54} = \frac{c^4C_2\kappa}{6}, \\ A_{55} = 5C_3\omega + iB_2,\ A_{56} = c^2C_4\kappa,\ A_{61} = A_{63} = A_{64} = 0, \\ A_{62} = \frac{2n_0c^2(m^2-C_1)\kappa}{9},\ A_{65} = \frac{4C_3\kappa}{3},\ A_{66} = C_4\omega - iB_3. \end{cases} \tag{9.111}$$

In the following we shall analyze in more detail the non-relativistic and the ultra-relativistic limiting cases. We begin with the non-relativistic case and write the first terms of the ζ-expansions of the coefficients C's and B's for $\zeta \gg 1$ as

$$C_1 = m^2\left(1+\frac{6}{\zeta}+\frac{15}{\zeta^2}\right), \qquad C_2 = -\frac{m^2}{kT\zeta}\left(6+\frac{63}{\zeta}+\frac{693}{4}\frac{1}{\zeta^2}\right), \tag{9.112}$$

$$C_3 = -m\left(\frac{2}{5}+\frac{7}{5}\frac{1}{\zeta}+\frac{21}{20}\frac{1}{\zeta^2}\right), \qquad C_4 = m\left(1+\frac{7}{2}\frac{1}{\zeta}+\frac{35}{8}\frac{1}{\zeta^2}\right), \tag{9.113}$$

$$c_v = \frac{3}{2}k\left(1+\frac{5}{2}\frac{1}{\zeta}-\frac{15}{4}\frac{1}{\zeta^2}\right), \qquad h = mc^2\left(1+\frac{5}{2}\frac{1}{\zeta}+\frac{15}{8}\frac{1}{\zeta^2}\right), \tag{9.114}$$

$$B_1 = -\frac{2m}{3\tau_n}\left(1+\frac{95}{16}\frac{1}{\zeta}+\frac{4457}{512}\frac{1}{\zeta^2}\right), \; B_2 = -\frac{2m}{3\tau_n}\left(1+\frac{111}{16}\frac{1}{\zeta}+\frac{513}{512}\frac{1}{\zeta^2}\right), \tag{9.115}$$

$$B_3 = -\frac{m}{\tau_n}\left(1+\frac{15}{16}\frac{1}{\zeta}+\frac{923}{1536}\frac{1}{\zeta^2}\right). \tag{9.116}$$

If we use (9.112) through (9.116) we find that the dispersion relation, which follows from (9.110) in the non-relativistic limiting case by considering the fourteen-field equations, reduces to

$$\left(\frac{\kappa v_s^{nr}}{\omega}\right)^4\left[\frac{567}{100}i(\tau_n\omega)^3-\frac{477}{100}(\tau_n\omega)^2-\frac{9}{10}i(\tau_n\omega)\right]-\left(\frac{\kappa v_s^{nr}}{\omega}\right)^2\left[\frac{441}{50}i(\tau_n\omega)^3\right.$$

$$\left.-\frac{342}{25}(\tau_n\omega)^2-\frac{63}{10}i(\tau_n\omega)+1\right]+\frac{9}{4}i(\tau_n\omega)^3-\frac{21}{4}(\tau_n\omega)^2-4i(\tau_n\omega)+1=0. \tag{9.117}$$

From the above equation one can obtain that the power series expansion of $\kappa v_s^{nr}/\omega$ as a function of $(\tau_n\omega)$ is

$$\frac{\kappa v_s^{nr}}{\omega} = 1+\frac{7}{10}i(\tau_n\omega)-\frac{43}{40}(\tau_n\omega)^2-\frac{743}{400}i(\tau_n\omega)^3+\cdots. \tag{9.118}$$

We note that the dispersion term associated with $(\tau_n\omega)^2$ is the same as that of the Burnett equation (9.97) while the correction to the absorption coefficient, associated with $(\tau_n\omega)^3$, differs.

It is also interesting to find the limit of the phase speed when the circular frequency tends to infinity. While in the Navier–Stokes–Fourier and Burnett approximations the phase speed in this limit tends to infinity – which is the so-called paradox of heat conduction – one can get from (9.117) that for $\omega \to \infty$ two positive values for the phase speed are possible:

$$\frac{v_{ph}^{nr}}{v_s^{nr}} \to \left(\frac{7}{9}\mp\sqrt{\frac{118}{567}}\right)^{-\frac{1}{2}} \approx \begin{cases}1.76341,\\ 0.90022.\end{cases} \tag{9.119}$$

These two limits were obtained by Boillat [2] and Seccia and Strumia [25] and are known as the first and the second sound.

Let us analyze the ultra-relativistic limiting case where the first terms of the ζ-expansions of the coefficients C's and B's for $\zeta \ll 1$ are given by

$$C_1 = 24\left(\frac{kT}{c^2}\right)^2\left(1+\frac{1}{6}\zeta^2\right), \; C_2 = -144\frac{kT}{c^4\zeta^2}\left[1-\left(\frac{9}{4}+3\ln\frac{\zeta}{2}+3\gamma\right)\zeta^2\right], \tag{9.120}$$

$$C_3 = -2\frac{kT}{c^2}\left(1+\frac{1}{8}\zeta^2\right), \qquad C_4 = 6\frac{kT}{c^2}\left(1+\frac{1}{24}\zeta^2\right), \tag{9.121}$$

$$c_v = 3k\left(1-\frac{\zeta^2}{6}\right), \qquad h = 4kT\left(1+\frac{1}{8}\zeta^2\right), \tag{9.122}$$

$$B_1 = -\frac{48kT}{5\zeta^2\tau_u c^2}\left[1-\left(\frac{13}{6}+3\ln\frac{\zeta}{2}+3\gamma\right)\zeta^2\right], \tag{9.123}$$

$$B_2 = -\frac{12kT}{5c^2\tau_u}\left(1+\frac{\zeta^2}{8}\right), \quad B_3 = -\frac{4kT}{\tau_u c^2}\left(1+\frac{3}{40}\zeta^2\right). \tag{9.124}$$

The dispersion relation that follows from (9.110) in the ultra-relativistic limiting case reduces to

$$\left(\frac{\kappa v_s^{ur}}{\omega}\right)^4\left[\frac{225}{16}i(\tau_u\omega)^3-\frac{35}{4}(\tau_u\omega)^2-\frac{5}{4}i(\tau_u\omega)\right]-\left(\frac{\kappa v_s^{ur}}{\omega}\right)^2\left[\frac{175}{8}i(\tau_u\omega)^3\right.$$

$$\left.-\frac{145}{6}(\tau_u\omega)^2-\frac{25}{3}i(\tau_u\omega)+1\right]+\frac{125}{16}i(\tau_u\omega)^3-\frac{145}{12}(\tau_u\omega)^2-\frac{73}{12}i(\tau_u\omega)+1=0, \tag{9.125}$$

while the power series expansion of $\kappa v_s^{ur}/\omega$ as a function of $(\tau_u\omega)$ is given by

$$\frac{\kappa v_s^{ur}}{\omega} = 1+\frac{1}{2}i(\tau_u\omega)-\frac{9}{8}(\tau_u\omega)^2-\frac{113}{48}i(\tau_u\omega)^3+\cdots. \tag{9.126}$$

If we compare the above expression with the corresponding one for the Burnett equation (9.100) we note, as in the non-relativistic case, that the dispersion is the same while the correction to the absorption coefficient differs.

In the limit where the circular frequency tends to infinity we have also two phase speeds that correspond to the first and second sound

$$\frac{v_{ph}^{ur}}{v_s^{ur}} \to \begin{cases} 3/\sqrt{5} \approx 1.34164, \\ 1. \end{cases} \tag{9.127}$$

The two above limits were also found by Boillat [2] and Seccia and Strumia [25].

Thirteen-field theory

The dispersion relation for the thirteen-field theory, in which one does not take into account the field of dynamic pressure, can be obtained from the system of equations for the amplitudes (9.110) by taking off the line corresponding to the elements A_{4i} and the column corresponding to the elements A_{i4}. In the non-relativistic limiting case the dispersion relation for the thirteen fields is

$$\left(\frac{\kappa v_s^{nr}}{\omega}\right)^4\left[\frac{81}{50}(\tau_n\omega)^2+\frac{9}{10}i(\tau_n\omega)\right]-\left(\frac{\kappa v_s^{nr}}{\omega}\right)^2\left[\frac{117}{25}(\tau_n\omega)^2\right.$$

$$\left.+\frac{24}{5}i(\tau_n\omega)-1\right]+\frac{3}{2}(\tau_n\omega)^2+\frac{5}{2}i(\tau_n\omega)-1=0. \tag{9.128}$$

The power series expansion of $\kappa v_s^{nr}/\omega$ as a function of $(\tau_n\omega)$ is

$$\frac{\kappa v_s^{nr}}{\omega}=1+\frac{7}{10}i(\tau_n\omega)-\frac{43}{40}(\tau_n\omega)^2-\frac{527}{400}i(\tau_n\omega)^3+\cdots, \tag{9.129}$$

and we note that only the correction to the absorption coefficient is different when we compare (9.129) with (9.118) of the fourteen-field theory. The above expression is the same as that found by Greenspan [15].

The two values for the phase speed, i.e., the first and the second sound, when the circular frequency tends to infinity, agree with those found by Kranyš [20]:

$$\frac{v_{ph}^{nr}}{v_s^{nr}}\to\left(\frac{13}{9}\mp\sqrt{\frac{94}{81}}\right)^{-\frac{1}{2}}\approx\begin{cases}1.65029,\\0.62973.\end{cases} \tag{9.130}$$

The dispersion relation in the ultra-relativistic limiting case for the thirteen-field theory reads

$$\left(\frac{\kappa v_s^{ur}}{\omega}\right)^4\left[\frac{85}{24}(\tau_u\omega)^2+\frac{5}{4}i(\tau_u\omega)\right]-\left(\frac{\kappa v_s^{ur}}{\omega}\right)^2\left[\frac{15}{2}(\tau_u\omega)^2\right.$$

$$\left.+\frac{35}{6}i(\tau_u\omega)-1\right]+\frac{25}{8}(\tau_u\omega)^2+\frac{43}{12}i(\tau_u\omega)-1=0, \tag{9.131}$$

while the power series expansion of $\kappa v_s^{ur}/\omega$ as a function of $(\tau_u\omega)$ is

$$\frac{\kappa v_s^{ur}}{\omega}=1+\frac{1}{2}i(\tau_u\omega)-\frac{9}{8}(\tau_u\omega)^2-\frac{133}{148}i(\tau_u\omega)^3+\cdots. \tag{9.132}$$

The same conclusion can be drawn here as in the non-relativistic case, i.e., only the correction to the absorption coefficient in the above equation is different from that of the fourteen-field theory (9.126).

In the limit when the circular frequency tends to infinity the phase speeds of the first and second sound tend to

$$\frac{v_{ph}^{ur}}{v_s^{ur}} \rightarrow \left(\frac{18}{17} \mp \sqrt{\frac{69}{289}}\right)^{-\frac{1}{2}} \approx \begin{cases} 1.32430, \\ 0.80388. \end{cases} \tag{9.133}$$

Only the value of the first sound agrees with that found by Kranyš [20]. For the second sound Kranyš [20] has obtained that $v_{ph}^{ur}/v_s^{ur} \approx 0.696$.

Phase speeds and attenuation coefficients

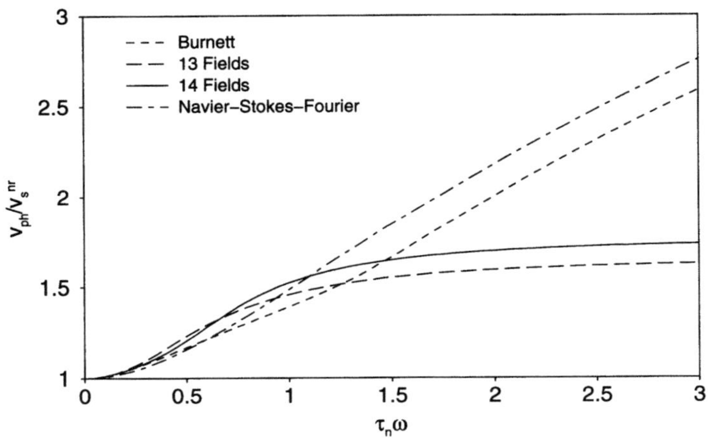

Figure 9.1: Phase speeds as functions of the circular frequency in the non-relativistic limiting case.

In this section we compare all phase speeds and attenuation coefficients obtained from different theories – that is Navier–Stokes–Fourier, Burnett, thirteen- and fourteen-field theories – by plotting these quantities as functions of the circular frequency.

In Figure 1 the ratio between the phase speeds and the speed of sound is plotted as functions of the product between the relaxation time and the circular frequency for a gas in the non-relativistic regime. For a gas in the ultra-relativistic regime the same quantities are plotted in Figure 2. As was previously stated the phase speeds calculated by using a five-field theory with Navier–Stokes–Fourier and Burnett constitutive equations increase with the increase in circular frequency, and the phase speeds tend to infinity for high values of the circular frequency (paradox of heat conduction). For the non-relativistic limiting case the phase speed that corresponds to the Navier–Stokes–Fourier solution is larger than the phase speed of the Burnett solution, while in the ultra-relativistic case there is an opposite

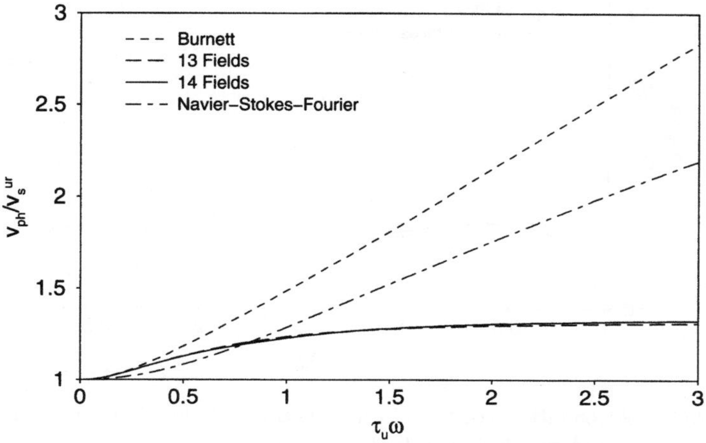

Figure 9.2: Phase speeds as functions of the circular frequency in the ultra-relativistic limiting case.

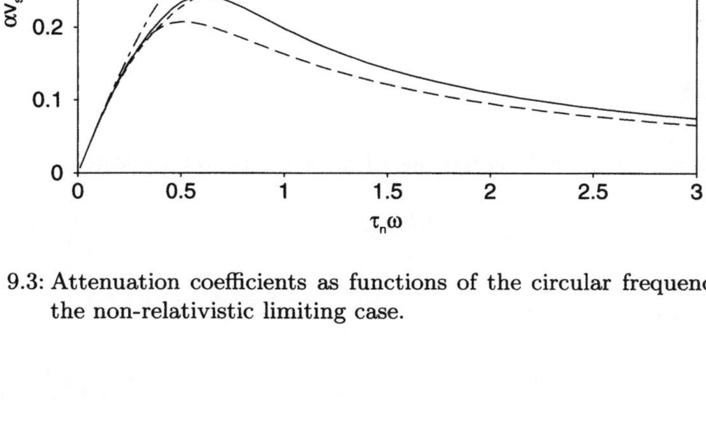

Figure 9.3: Attenuation coefficients as functions of the circular frequency in the non-relativistic limiting case.

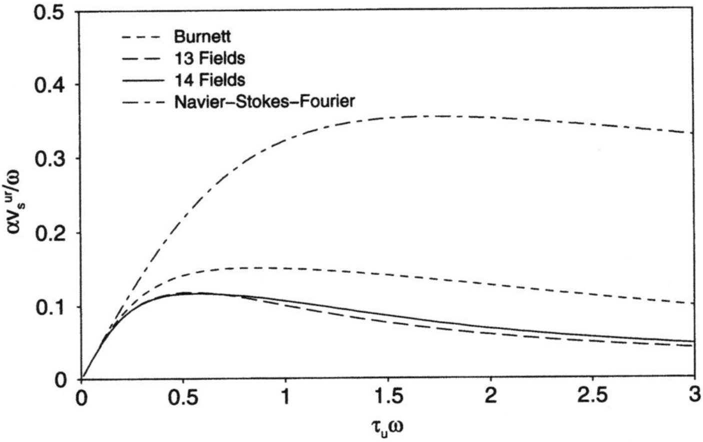

Figure 9.4: Attenuation coefficients as functions of the circular frequency in the ultra-relativistic limiting case.

behavior, i.e., the Burnett solution has a larger phase speed than that of the Navier–Stokes–Fourier solution. As was pointed out the phase speeds obtained from the thirteen- and fourteen-field theories tend to finite values for high values of the circular frequency (see (9.119), (9.127), (9.130) and (9.133)). We note also that the phase speeds which correspond to the thirteen- and fourteen-field theories in the ultra-relativistic limiting case practically coincide.

The behavior of the attenuation coefficients as functions of the circular frequency is shown in Figures 3 and 4 for the non-relativistic and ultra-relativistic limiting cases, respectively. From these figures we infer that for both cases the Navier–Stokes–Fourier solution furnishes the largest value for the attenuation coefficient while the thirteen-field solution gives the lowest value for the attenuation coefficient.

Problems

9.3.2.1 Consider the field equations of an Eulerian relativistic gas

$$Dn + n\nabla^\alpha U_\alpha = 0, \qquad \frac{nh_E}{c^2} DU^\alpha = \nabla^\alpha p, \qquad nDe = -p\nabla^\alpha U_\alpha,$$

where $h_E = e + p/n$ and

$$n = 4\pi(mc)^3 \frac{g_s}{h^3} J_{21}, \qquad e = mc^2 \frac{J_{22}}{J_{21}}, \qquad p = \frac{4\pi}{3} m^4 c^5 \frac{g_s}{h^3} J_{40}.$$

Show that the adiabatic sound speed of a degenerate relativistic gas reads

$$v_s = \sqrt{\zeta \frac{[J^\bullet_{23}(J^\bullet_{40})^2 - 2J^\bullet_{22}J^\bullet_{41}J^\bullet_{40} + J^\bullet_{21}(J^\bullet_{41})^2]}{3[J^\bullet_{21}J^\bullet_{23} - (J^\bullet_{22})^2]J^\bullet_{41}} \frac{kT}{m}},$$

where $J^\bullet_{nm}(\zeta, \mu_E)$ denotes the partial derivative of $J_{nm}(\zeta, \mu_E)$ with respect to $\mu_E/(kT)$.

9.3.2.2 Obtain the expressions for the phase speed (9.76) and attenuation coefficient (9.77).

9.3.2.3 Show that the dispersion relation in the non-relativistic and ultra-relativistic limiting cases are given respectively by: a) equations (9.86) and (9.90) for a five-field theory with Navier–Stokes and Fourier constitutive equations; b) equations (9.97) and (9.99) for a five-field theory with Burnett constitutive equations; c) equations (9.117) and (9.125) for a fourteen-field theory; d) equations (9.128) and (9.131) for a thirteen-field theory.

9.3.2.4 Obtain the limits of the phase speed in the non-relativistic and ultra-relativistic limiting cases when the frequency tends to infinity: a) equations (9.119) and (9.127) for a fourteen-field theory; b) equations (9.130) and (9.133) for a thirteen-field theory.

9.4 Shock waves

9.4.1 Continuum theory

We begin by writing the balance equations of the particle four-flow (2.73) and of the energy-momentum tensor (2.74) in the integral form as

$$\int (\partial_\alpha N^\alpha) d^4x = \int N^\alpha e_\alpha d^3x = 0, \tag{9.134}$$

$$\int (\partial_\beta T^{\alpha\beta}) d^4x = \int T^{\alpha\beta} e_\beta d^3x = 0. \tag{9.135}$$

The functions on the left-hand side of (9.134) and (9.135) are integrated over an arbitrary four-dimensional volume, while those on the right-hand side are integrated over a closed three-dimensional hypersurface with unit normal vector e_α. If we suppose that the three-dimensional hypersurface is a shell of thickness ϵ which envelopes a discontinuity surface at rest with three-dimensional unit normal vector n_i, we get from (9.134) and (9.135), in the limit when ϵ tends to zero, the jump equations

$$[\![N^i]\!] n_i = 0, \qquad [\![T^{\alpha i}]\!] n_i = 0. \tag{9.136}$$

$[\![\phi]\!] = \phi_+ - \phi_-$ represents the jump of the function ϕ across the surface of discontinuity, such that ϕ_+ is the value of the function ϕ on the side of the surface in the direction of the unit normal vector and ϕ_- the value of ϕ on the side of the surface opposite to the direction of the unit normal vector.

We recall that for a non-viscous and non-conducting fluid the energy-momentum tensor reduces to

$$T^{\alpha\beta} = -p\Delta^{\alpha\beta} + \frac{en}{c^2}U^\alpha U^\beta = -p\eta^{\alpha\beta} + \frac{nh_E}{c^2}U^\alpha U^\beta, \tag{9.137}$$

while the particle four-flow is written as

$$N^\alpha = nU^\alpha. \tag{9.138}$$

Without loss of generality we shall consider a coordinate system where the surface of discontinuity is perpendicular to the x^1- axis, i.e., $n_i = \delta_i^1$. Hence if we insert (9.137) and (9.138) into the jump equations (9.136) and take into account that $U_\pm^\alpha = (\gamma_\pm c, \gamma_\pm \mathbf{v}_\pm)$, where $\gamma = 1/\sqrt{1-v^2/c^2}$, we get

$$n_+\gamma_+v_+ = n_-\gamma_-v_-, \tag{9.139}$$

$$-p_+\eta^{\alpha 1} + \frac{n_+h_+}{c^2}\gamma_+v_+U_+^\alpha = -p_-\eta^{\alpha 1} + \frac{n_-h_-}{c^2}\gamma_-v_-U_-^\alpha. \tag{9.140}$$

In the above equations we have introduced the notation $v_\pm^1 \equiv v_\pm$.

The spatial and temporal components of (9.140) can be written respectively as

$$p_+ + \frac{n_+h_+}{c^2}(\gamma_+v_+)^2 = p_- + \frac{n_-h_-}{c^2}(\gamma_-v_-)^2, \tag{9.141}$$

$$n_+h_+\gamma_+^2v_+ = n_-h_-\gamma_-^2v_-. \tag{9.142}$$

Equations (9.139), (9.141) and (9.142) are the well-known Rankine–Hugoniot relativistic equations.

Now we shall calculate the velocities ahead of and behind the surface of discontinuity which characterizes the shock wave. For that end we multiply (9.140) by $U_{+\alpha}$ and get

$$(n_+e_+ + p_-)\gamma_+v_+ = \frac{n_-h_-}{c^2}\gamma_-v_-U_-^\alpha U_{+\alpha}, \tag{9.143}$$

while the multiplication of (9.140) by $U_{-\alpha}$ furnishes

$$\frac{n_+h_+}{c^2}\gamma_+v_+U_+^\alpha U_{-\alpha} = (n_-e_- + p_+)\gamma_-v_-. \tag{9.144}$$

We now eliminate $U_+^\alpha U_{-\alpha}$ from the two above equations, yielding

$$\frac{\gamma_+}{\gamma_-}\left(\frac{v_+}{v_-}\right)^2 = \frac{n_-(n_-e_- + p_+)}{n_+(n_+e_+ + p_-)}, \tag{9.145}$$

since according to (9.139) and (9.142) we have that

$$\gamma_+h_+ = \gamma_-h_-. \tag{9.146}$$

From (9.141), (9.145) and (9.146) we can write the velocities ahead of and behind the shock wave as

$$\frac{v_+}{c} = \sqrt{\frac{(n_-e_- + p_+)(p_- - p_+)}{(n_+e_+ + p_-)(n_-e_- - n_+e_+)}}, \tag{9.147}$$

$$\frac{v_-}{c} = \sqrt{\frac{(n_+e_+ + p_-)(p_- - p_+)}{(n_-e_- + p_+)(n_-e_- - n_+e_+)}}. \tag{9.148}$$

An alternative form to write (9.147) is

$$\frac{v_+\gamma_+}{c} = \sqrt{\frac{p_- - p_+}{h_+n_+\left(1 - \frac{n_+e_+ + p_-}{n_-e_- + p_+}\right)}}. \tag{9.149}$$

One can also build the ratio between the velocities ahead of and behind the shock wave, yielding

$$\frac{v_+}{v_-} = \frac{n_-e_- + p_+}{n_+e_+ + p_-}. \tag{9.150}$$

We could infer from equations (9.147) and (9.148) that: for a compression wave we have that $p_- > p_+$, hence (9.147) indicates that $n_-e_- > n_+e_+$, i.e., the energy density behind the shock wave is larger then the energy density ahead of it. Further (9.149) shows that we must have $(e_-n_- + p_+) > (e_+n_+ + p_-)$, hence we conclude from (9.150) that $v_+ > v_-$, i.e., the velocity ahead of the shock wave is larger than the velocity behind it.

The non-relativistic limit of the velocities (9.147) and (9.148) are

$$v_+ = \sqrt{\frac{n_-(p_- - p_+)}{mn_+(n_- - n_+)}}, \qquad v_- = \sqrt{\frac{n_+(p_- - p_+)}{mn_-(n_- - n_+)}}, \tag{9.151}$$

so that their ratio is given by

$$n_-v_- = n_+v_+. \tag{9.152}$$

The ultra-relativistic limit of the velocities (9.147) and (9.148) read

$$\frac{v_+}{c} = \sqrt{\frac{3n_-e_- + n_+e_+}{3(3n_+e_+ + n_-e_-)}}, \qquad \frac{v_-}{c} = \sqrt{\frac{3n_+e_+ + n_-e_-}{3(3n_-e_- + n_+e_+)}}. \tag{9.153}$$

It is also interesting to analyze the case when the shock wave is so strong that the energy density behind it n_-e_- becomes so large that the velocities in (9.153) tend to

$$v_+ \to c, \qquad v_- \to \frac{c}{3}, \tag{9.154}$$

that is, the velocity ahead of the shock wave tends to the speed of light whereas the velocity behind it is one third of the speed of light.

The structure of shock waves by using the balance equations of a continuum theory was analyzed among others by Koch [19], Cercignani and Majorana [9] and Majorana and Muscato [22]. In the next section we shall follow the work of Cercignani and Majorana [9] and analyze the structure of shock waves by using the Boltzmann equation in the limiting case where the upstream temperature T_- is close to zero. For that end we need the expressions for the velocities behind and ahead of the shock wave when $T_- \to 0$, which we proceed to derive.

We begin by writing (9.141) as

$$\frac{p_+}{n_+\gamma_+v_+} + \frac{h_+\gamma_+v_+}{c^2} = \frac{p_-}{n_-\gamma_-v_-} + \frac{h_-\gamma_-v_-}{c^2}. \tag{9.155}$$

In the limit $\zeta_- \to \infty$ we have that $G_- \equiv G(\zeta_-) \to 1$ and (9.146) reduces to

$$\gamma_- = \gamma_+ G_+, \tag{9.156}$$

while in this limit (9.155) becomes

$$\frac{\gamma_- v_-}{c} = \frac{c}{\zeta_+\gamma_+v_+} + \frac{G_+\gamma_+v_+}{c}. \tag{9.157}$$

From the above equation one can build the following expression

$$1 + \left(\frac{\gamma_- v_-}{c}\right)^2 = \gamma_-^2 \overset{(9.156)}{=} \gamma_+^2 G_+^2 = 1 + \frac{2G_+}{\zeta_+} + \left(\frac{G_+\gamma_+v_+}{c}\right)^2 + \left(\frac{c}{\zeta_+\gamma_+v_+}\right)^2, \tag{9.158}$$

which can be solved for v_+, yielding

$$\frac{v_+}{c} = \frac{1}{\zeta_+\sqrt{\left(G_+ - \frac{1}{\zeta_+}\right)^2 - 1}}. \tag{9.159}$$

The expression for v_- follows from (9.156):

$$\frac{v_-}{c} = \sqrt{1 + \frac{(v_+/c)^2 - 1}{G_+^2}}. \tag{9.160}$$

We note from (9.159) and (9.160) that the velocities ahead of and behind the shock wave in the limiting case where $T_- \to 0$ are only functions of the parameter ζ_+.

Problems

9.4.1.1 Obtain the expressions (9.147) and (9.148) for the velocities ahead of and behind the shock wave.

9.4.1.2 Show that in the non-relativistic and ultra-relativistic limiting cases (9.147) and (9.148) reduce to (9.151) and (9.153).

9.4.1.3 Obtain (9.155) from (9.141).

9.4.2 Boltzmann equation

So far the nonlinear nature of the Boltzmann equation has played only a little role in the solutions which have been discussed in this chapter. We are now going to discuss the simplest problem dominated by nonlinearity, the problem of the shock wave structure. We look for traveling waves of the form $f = f(x - at, \boldsymbol{\xi})$ where a is a constant with the condition that f tends to two different Maxwell–Jüttner distributions $f_\pm^{(0)}$ when $x \to \pm\infty$. Then

$$(\xi_1 - a)f' = Q(f, f), \tag{9.161}$$

where f' exceptionally denotes the derivative with respect to the first argument of f. The name of traveling wave is due to the fact that we deal with solutions which travel at a constant speed a without changing their shape.

It is clear that the constant a plays no significant role in the solution; sometimes it is used to impose some extra condition (e.g., that the bulk velocity goes to zero when $x \to \infty$). Here a difference arises between the classical and the relativistic case. In the classical case we must perform a Galilei transformation, in the relativistic case a Lorentz transformation. In either case we can replace f' by $\partial f/\partial x$, and we can find an observer, with respect to whom the Boltzmann equation can be written as:

$$\xi_1 \frac{\partial f}{\partial x} = Q(f, f); \quad f \to f_\pm^{(0)} \quad \text{when} \quad x \to \pm\infty \tag{9.162}$$

in both the classical and the relativistic case. In the latter case, however, the transformation is possible if and only if $|a| \le c$ (see Section 2.7).

In (9.162) the Maxwell–Jüttner distribution function is written as

$$f_\pm^{(0)} = \frac{n_\pm \zeta_\pm}{4\pi m^3 c^3 K_2(\zeta_\pm)} \exp\left(-\frac{p_\alpha U_\pm^\alpha}{kT_\pm}\right). \tag{9.163}$$

It is conventional to take $v_+ > 0$ (and hence also $v_- > 0$; see the Rankine–Hugoniot equation (9.139)). Thus the fluid flows from $-\infty$ (upstream) to $+\infty$ (downstream).

The parameters in the two Maxwell–Jüttner distributions cannot be chosen at will. In fact, conservation of particle four-flow and energy-momentum tensor imply that the flows of these quantities must be constant and, in particular, they must be the same at $\pm\infty$, i.e.,

$$N^1 = c\int p^1 f_-^{(0)} \frac{d^3p}{p_0} = c\int p^1 f_+^{(0)} \frac{d^3p}{p_0}, \tag{9.164}$$

$$T^{1\alpha} = c\int p^1 p^\alpha f_-^{(0)} \frac{d^3p}{p_0} = c\int p^1 p^\alpha f_+^{(0)} \frac{d^3p}{p_0}. \tag{9.165}$$

These conditions imply the Rankine–Hugoniot equations (9.139), (9.141) and (9.142) which relate the upstream and downstream values of particle number density, velocity and temperature in an ideal compressible relativistic fluid.

The problem that we have just defined is the problem of the shock wave structure. In kinetic theory, a shock wave cannot be classified among the discontinuities, as is the case for an Euler fluid. The same situation occurs for the Navier–Stokes equations. In both cases we find that the solution is smooth but changes very rapidly through a thin layer. The solution for the non-relativistic Navier–Stokes equations is not hard to find [13], but it is in a sense contradictory, unless the shock is very weak (the strength of a shock can be defined in several ways, e.g., by the ratio r between the downstream and upstream pressures). In fact, unless r is close to unity, the shock layer turns out to be just a few mean free paths thick, and we expect the Navier–Stokes equations to be invalid on this scale.

Before studying the problem of the shock wave structure we remark that, since we can arbitrarily shift the origin along the x-axis without changing the Boltzmann equation, the solution is defined up to a constant. The latter can be fixed by putting the origin at a special point, which could be the point where the particle number density takes on the arithmetic mean of the downstream and upstream values, or where its derivative takes on the maximum value.

We start our study by sketching the theory of weak shock waves. In this case the parameter $s = r - 1$ is small and shall be denoted by ϵ, which thus measures the difference between upstream and downstream values. In this case we can expect the transition between these values to be not steep and the space derivative to be of the order of ϵ. This is not *a priori* evident, but can be justified by starting with some undefined expansion parameter and then identifying it later with ϵ on the basis of the results.

It is possible to apply an expansion similar to the Hilbert expansion discussed in Chapter 5, with one (simplifying) exception: the zeroth order term, a Maxwell–Jüttner distribution, can be assumed to be uniform, since the upstream and downstream parameters are assumed to differ by terms of order ϵ. Thus we are led to writing the Boltzmann equation with a parameter ϵ in front of the space derivative of equation (9.161) and looking for a solution of our problem in the form of a power series in ϵ. It turns out that a solution exists if and only if the bulk velocity in the unperturbed Maxwell–Jüttner distribution equals the speed of sound. For the non-relativistic case see [4]; the details of the relativistic case do not appear to have been worked out.

The first analytic approach to the problem of shock structure in the non-relativistic case is due to Mott-Smith [23], who introduced it in 1954 (a similar approach had been suggested by Tamm [28] in the Soviet Union). The method consists in approximating the distribution function by a linear combination of the upstream and downstream Maxwell–Jüttner distributions with x-dependent coefficients $a_\pm(x)$. Then the conservation equations yield

$$a_+ n_+ \gamma_+ v_+ + a_- n_- \gamma_- v_- = n_\pm \gamma_\pm v_\pm, \tag{9.166}$$

$$a_+ \left[p_+ + \frac{n_+ h_+}{c^2}(\gamma_+ v_+)^2\right] + a_- \left[p_- + \frac{n_- h_-}{c^2}(\gamma_- v_-)^2\right]$$

$$= p_\pm + \frac{n_\pm h_\pm}{c^2}(\gamma_\pm v_\pm)^2, \tag{9.167}$$

$$a_+ n_+ h_+ \gamma_+^2 v_+ + a_- n_- h_- \gamma_-^2 v_- = n_\pm h_\pm \gamma_\pm^2 v_\pm. \tag{9.168}$$

It is clear that these equations are satisfied if, and only if, $a_+(x) + a_-(x) = 1$ (provided the Rankine–Hugoniot conditions are satisfied) (Problem 9.4.2.2). Hence we can write $a_+(x) = a(x)$ and $a_-(x) = 1 - a(x)$. An additional equation is needed to determine $a = a(x)$.

A possible choice for the extra relation is to take the moment equation obtained by multiplying the Boltzmann equation by a monomial in the momentum components and integrating. This calculation does not seem to have been performed in the relativistic case. The relativistic case was considered by Cercignani and Majorana [9] just in the limiting case of an upstream temperature T_- close to zero. We remark that, under this assumption, the upstream Maxwell–Jüttner distribution degenerates into a delta function when $T_- \to 0$ (infinitely strong shock). Here we shall follow their paper, where the fluid is assumed to be a cold hard sphere gas, i.e., a gas with constant differential cross-section σ.

For a non-relativistic gas this extreme case was studied by Grad [14] who suggested that the limit of the shock profile for $r \to \infty$ exists (at least for collision operators with a finite collision frequency) and is given by a multiple of the delta function centered at the upstream bulk velocity plus a comparatively smooth function for which it is not hard to derive an equation. The latter seems more complicated than the Boltzmann equation itself, but the presumed smoothness of its solution should allow a simple approximate solution to be obtained. The simplest choice for the smooth remainder is a Maxwellian distribution [14], the parameters of which are determined by the conservation equations. Thus the case studied by Cercignani and Majorana [9] can be considered as the relativistic version of Grad's approach. Essentially one uses the relations derived so far in the sketch of the relativistic version of Mott-Smith's approach (in the limiting case $T_- \to 0$); the missing equation is provided by equating the delta function terms in the left- and right-hand sides of the Boltzmann equation. We proceed to determine $a(x)$ by following the work of Cercignani and Majorana [9].

We start by assuming that the one-particle distribution function is of the form

$$f = [1 - a(x)]f_-^{(0)} + a(x)f_+^{(0)}. \tag{9.169}$$

Next we write the Maxwell–Jüttner distribution function $f_-^{(0)}$ as

$$f_-^{(0)} = \frac{n_- \zeta_-}{4\pi m^3 c^3 K_2(\zeta_-) e^{\zeta_-}} \exp\left[-\frac{\zeta_-}{mc^2}(p_\alpha U_-^\alpha - mc^2)\right], \tag{9.170}$$

and by making use of the asymptotic expansion for the modified Bessel function

$$\lim_{\zeta_- \to +\infty} K_2(\zeta_-) = \sqrt{\frac{\pi}{2\zeta_-}} \frac{1}{e^{\zeta_-}}, \tag{9.171}$$

we get that the limit of the Maxwell–Jüttner distribution function when the upstream temperature goes to zero is

$$\lim_{T_-\to 0+} f_-^{(0)} = \begin{cases} +\infty, & \text{if} \quad p_\alpha U_-^\alpha - mc^2 = 0, \\ 0, & \text{if} \quad p_\alpha U_-^\alpha - mc^2 > 0. \end{cases} \tag{9.172}$$

In the following we shall make use of the result:

$$p_\alpha U_-^\alpha - mc^2 = 0 \quad \text{is equivalent to} \quad p^\alpha = mU_-^\alpha. \tag{9.173}$$

In order to prove the above conjecture we introduce the four-vector $X^\alpha = p^\alpha - mU_-^\alpha$ and take the scalar product $X^\alpha X_\alpha = -2m(p_\alpha U_-^\alpha - mc^2)$. If $(9.173)_1$ holds, $X^\alpha X_\alpha = 0$ and X^α is a null four-vector which implies $(9.173)_2$. Furthermore if $(9.173)_2$ holds, then $p_\alpha U_-^\alpha = mc^2$ which implies $(9.173)_1$.

Hence the Maxwell–Jüttner distribution function $f_-^{(0)}$ reduces to a delta function

$$\lim_{T_-\to 0+} f_-^{(0)} \to \frac{n_-}{m^3c^3}\delta(p^\alpha - mU_-^\alpha), \tag{9.174}$$

which is defined such that, for an arbitrary function $\mathcal{F}(p^\beta)$,

$$\int \mathcal{F}(p^\beta)\delta(p^\alpha - mU_-^\alpha)\frac{d^3p}{p_0} = \mathcal{F}(mU_-^\alpha). \tag{9.175}$$

From the above considerations we can write the one-particle distribution function (9.169) as

$$f = \frac{n_-}{m^3c^3}\delta(p^\alpha - mU_-^\alpha)[1 - a(x)] + a(x)f_+^{(0)}. \tag{9.176}$$

We insert the one-particle distribution function (9.176) into the Boltzmann equation and obtain by equating the delta function terms in the left- and right-hand sides of the Boltzmann equation

$$p^1\frac{da}{dx} = 4\pi m a(1-a)\sigma \int f_{*+}^{(0)}\sqrt{(U_-^\alpha p_{*\alpha})^2 - m^2c^4}\,\frac{d^3p_*}{p_{*0}}, \tag{9.177}$$

or

$$\frac{da}{dx} = \alpha a(1-a), \tag{9.178}$$

where

$$\alpha = \frac{4\pi\sigma}{\gamma_- v_-}\int f_{*+}^{(0)}\sqrt{(U_-^\alpha p_{*\alpha})^2 - m^2c^4}\,\frac{d^3p_*}{p_{*0}}. \tag{9.179}$$

The above equation easily integrates to

$$a(x) = \frac{e^{\alpha x}}{e^{\alpha x} + 1}, \tag{9.180}$$

and an arbitrary constant has been set equal to zero, since this amounts to fixing the position of the center of the shock layer (which, as we know, is arbitrary). Hence one can obtain every quantity which is defined in terms of the distribution function, for example the particle number density and the bulk velocity reads

$$n\gamma = \frac{n_+\gamma_+ e^{\alpha x} + n_-\gamma_-}{e^{\alpha x}+1}, \tag{9.181}$$

$$v = \frac{v_- v_+ (e^{\alpha x}+1)}{e^{\alpha x} v_- + v_+}. \tag{9.182}$$

The other physical quantities can be easily computed (Problem 9.4.2.3).

A significant quantity related to the structure of shock waves is the shock thickness L defined by

$$L = \frac{|v_+ - v_-|}{|dv/dx|_{\max}}. \tag{9.183}$$

It is easy to verify that $L = 4/\alpha$ so that, if one introduces the mean free path of the upstream gas $\ell_- = 1/(n_-\sigma)$, it is possible to build the dimensionless quantity

$$\frac{\ell_-}{L} = \frac{\alpha}{4n_-\sigma}. \tag{9.184}$$

The integral (9.179) is evaluated in a new frame where the upstream four-velocity has components given by $(u_-^\alpha) = (c, \mathbf{0})$ whereas the downstream four-velocity reads $(u_+^\alpha) = (u^0, u^1, 0, 0)$. By making use of the conditions $U_-^\alpha U_{+\alpha} = u_-^\alpha u_{+\alpha}$ and $u_+^\alpha u_{+\alpha} = c^2$ one can get

$$\frac{u^0}{c} = \gamma_+\gamma_-\left(1 - \frac{v_+ v_-}{c^2}\right), \qquad \frac{u^1}{c} = \gamma_+\gamma_-\left|\frac{v_+}{c} - \frac{v_-}{c}\right|. \tag{9.185}$$

In this frame the integral (9.179) reduces to

$$\alpha = \frac{n_-\sigma\zeta_+}{\gamma_+ v_+ m^3 c^3 K_2(\zeta_+)} \int \exp\left(-\frac{p_{*0}u^0 - p_*^1 u^1}{kT_+}\right)\sqrt{c^2 p_{*0}^2 - m^2c^4}\,\frac{d^3p_*}{p_{*0}}. \tag{9.186}$$

Now by introducing spherical coordinates and by changing the variables of integration – such that $p_*^1 = mcy\cos\theta$ where $|\mathbf{p}_*| = mcy$ and $p_{*0} = mc\sqrt{y^2+1}$ – one can obtain by integration over the angular variables

$$\alpha = \frac{4\pi c^2 n_-\sigma}{\gamma_+ v_+ K_2(\zeta_+)u^1}\int_0^\infty \exp\left(-\zeta_+\frac{u^0}{c}\sqrt{y^2+1}\right)\sinh\left(\zeta_+\frac{u^1}{c}y\right)\frac{y^2dy}{\sqrt{y^2+1}}. \tag{9.187}$$

The above integral is tabulated in [9].

For the case $\zeta_+ \ll 1$ one can get an explicit expression for the ratio ℓ_-/L. For that end we introduce a new variable of integration $Y = \zeta_+ y$ and write (9.187) as

$$\alpha = \frac{4\pi c^2 n_-\sigma}{\gamma_+ v_+ K_2(\zeta_+)u^1\zeta_+^2}\int_0^\infty \exp\left(-\frac{u^0}{c}\sqrt{Y^2+\zeta_+^2}\right)\sinh\left(\frac{u^1}{c}Y\right)\frac{Y^2dY}{\sqrt{Y^2+\zeta_+}}$$

ζ_+	v_+/c	v_-/c	ℓ_-/L
0.05	0.33324	0.99993	533.533
0.10	0.33297	0.99973	267.339
0.50	0.32582	0.99336	56.696
1.00	0.31067	0.97606	32.397
1.50	0.29444	0.95262	25.239
1.80	0.28512	0.93709	23.130

Table 9.1: v_+/c, v_-/c and ℓ_-/L as functions of ζ_+

$$\approx \frac{2\pi c^2 n_- \sigma}{\gamma_+ v_+ u^1} \int_0^\infty \exp\left(-\frac{u^0}{c}Y\right) \sinh\left(\frac{u^1}{c}Y\right) Y dY = \frac{4\pi n_- \sigma u^0}{\gamma_+ v_+}, \tag{9.188}$$

since $\lim_{\zeta_+ \to 0} \zeta_+^2 K_2(\zeta_+) = 2$ (see Problem 3.2.1.2).

The final expression for the ratio ℓ_-/L that follows from (9.184) and (9.188) reads

$$\frac{\ell_-}{L} = \pi\gamma_- \left(\frac{c}{v_+} - \frac{v_-}{c}\right), \tag{9.189}$$

which is a good approximation for $0 < \zeta_+ \leq 1.80$.

In the same paper [9] a comparison is presented between the results of the kinetic and hydrodynamic models. The results turn out to be surprisingly close. A tentative explanation [9] is that relativistic shocks are to be considered as weak, in a sense, because the difference in the speeds upstream and downstream is relatively small. Another explanation may be found in the crude approximation of Grad's method for strong shock waves.

In Table 9.1 we give the values of v_+/c, v_-/c and ℓ_-/L for some values of ζ_+ in the range $0.05 \leq \zeta_+ \leq 1.80$ calculated by using equations (9.159), (9.160) and (9.189), respectively. We note that in the limiting case where $\zeta_+ \to 0$ we have: $v_+/c \to 1/\sqrt{3}$, $v_-/c \to 1$ and $\ell_-/L \to \infty$, i.e., the shock thickness tends to zero. The results given here for the ratio ℓ_-/L are two times larger than those given in [9], since here the differential cross-section is one-half the one defined in [9].

Problems

9.4.2.1 Compute the first and second order approximation to a weak shock wave (see [4] for the non-relativistic case).

9.4.2.2 Show that the relations (9.166) through (9.168) are satisfied if, and only if, $a_+(x) + a_-(x) = 1$ (provided the Rankine–Hugoniot conditions (9.139), (9.141) and (9.142) are satisfied).

9.4.2.3 Compute the pressure and the energy per particle in an infinitely strong shock.

9.4.2.4 Obtain the expression (9.177) by insertion of the one-particle distribution function (9.176) into the Boltzmann equation and equating the delta function terms in the left- and right-hand sides of the Boltzmann equation.

9.4.2.5 Show that the shock thickness is given by $L = 4/\alpha$.

9.4.2.6 Obtain the expressions for u^0/c and u^1/c given in (9.185) from the conditions $U_-^\alpha U_{+\alpha} = u_-^\alpha u_{+\alpha}$ and $u_+^\alpha u_{+\alpha} = c^2$.

References

[1] J. L. Anderson, A. C. Payne Jr., The relativistic Burnett equations and sound propagation, *Physica* **85A**, 261–286 (1976).

[2] G. Boillat, Wave velocities in relativistic extended thermodynamics, in *Kinetic theory and extended thermodynamics* pp. 55–60, eds. I. Müller and T. Ruggeri, (Pitagora, Bologna, 1987).

[3] G. Boillat, T. Ruggeri, Maximum wave velocity in the moments system of a relativistic gas, *Continuum Mech. Thermodyn.* **11**, 107–111 (1999).

[4] C. Cercignani, Bifurcation problems in fluid mechanics, *Meccanica* **5**, 7–16 (1970)

[5] C. Cercignani, Speed of propagation of infinitesimal disturbances in a relativistic gas, *Phys. Rev. Lett.* **50**, 1122–1124 (1983).

[6] C. Cercignani and A. Majorana, Propagation of infinitesimal disturbances in a gas according to a relativistic model, *Meccanica* **19**, 175–181 (1984).

[7] C. Cercignani and A. Majorana, Analysis of thermal, sound and shear waves according to a relativistic kinetic model, *Phys. Fluids* **28**, 1673–1683 (1985).

[8] C. Cercignani, Solutions of a linearized kinetic model for an ultrarelativistic gas, *Jour. Stat. Phys.* **42**, 601–620 (1986).

[9] C. Cercignani and A. Majorana, Structure of shock waves in relativistic simple gases, *Phys. Fluids* **31**, 1064–1068 (1988).

[10] C. Cercignani, Propagation phenomena in classical and relativistic rarefied gases, *Transp. Theory Stat. Phys.* **29**, 607–614 (2000).

[11] C. Cercignani, *Rarefied gas dynamics*, (Cambridge University Press, Cambridge, 2000).

[12] C. Cercignani and G. M. Kremer, Dispersion and absorption of plane harmonic waves in a relativistic gas, *Continuum Mech. Thermodyn.* **13**, 171–182 (2001).

[13] D. Gilbarg and D. Paolucci, The structure of shock waves in the continuum theory of fluids, *J. Rational Mech. Anal.* **2**, 617–642 (1953).

[14] H. Grad, Singular and non-uniform limits of solutions of the Boltzmann equation in *Transport Theory* pp. 269–308, eds. R. Bellman, G. Birkhoff and I. Abu-Shumays (American Mathematical Society, Providence, 1969).

[15] M. Greenspan, Translational dispersion in gases, in *Dispersion and absorption of sound by molecular processes* pp. 73–100, ed. D. Sette (Academic Press, New York, 1963).

[16] S. R. de Groot, W. A. van Leeuwen, Ch. G. van Weert, *Relativistic kinetic theory* (North-Holland, Amsterdam, 1980).

[17] J. Guichelaar, W. A. van Leeuwen and S. R. de Groot, On the relativistic kinetic gas theory VII. The propagation and absorption of sound, *Physica* **59**, 97–108 (1972).

[18] W. Israel, J. M. Stewart, Transient relativistic thermodynamics and kinetic theory, *Ann. Phys.* **118**, 341–372 (1979).

[19] P. A. Koch, Relativistic shock structure, *Phys. Rev.* **140 A**, 1161–1165 (1965).

[20] M. Kranyš, Phase and signal velocities of waves in dissipative media. Special relativistic theory, *Arch. Rational Mech. Anal.* **48**, 274–301 (1972).

[21] A. Majorana, Analytical solutions of hte Rankine–Hugoniot relations for a relativistic simple gas, *Nuovo Cimento* **98 B**, 111–118 (1987).

[22] A. Majorana, O. Muscato, Shock structure in an ultrarelativistic gas, *Meccanica* **25**, 77–82 (1990).

[23] H. Mott-Smith, The solution of the Boltzmann equation for a shock wave, *Phys. Rev.* **82**, 885–892 (1951).

[24] I. Müller, Zur Ausbreitungsgeschwindigkeit von Störungen in kontinuierlichen Media, *Dissertation RWTH* (Aachen, 1966).

[25] L. Seccia, A. Strumia, Wave propagation in relativistic extended thermodynamics, *Continuum Mech. Thermodyn.* **2**, 151–161 (1990).

[26] Y. Sone, On a new kind of boundary layer over a convex solid boundary in rarefied gas, *Kyoto University Research Report* **24** (1972).

[27] J. L. Synge, *The relativistic gas* (North-Holland, Amsterdam, 1957).

[28] I. E. Tamm, On the thickness of shock waves of large intensity, *Tr. Fiz. Inst. Akad. Nauk SSSR* **29**, 239–249 (1965) (in Russian).

[29] A. H. Taub, Relativistic Rankine–Hugoniot equations, *Phys. Rev.* **74**, 328–334 (1948).

[30] S. Weinberg, Entropy generation and the survival of protogalaxies in the expanding universe, *Ap. J.* **168**, 175–194 (1971).

Chapter 10

Tensor Calculus in General Coordinates

10.1 Introduction

In this chapter we shall introduce the mathematical tools that are needed for dealing with tensors in non-Cartesian coordinates. These tools are useful in ordinary three-dimensional space and in special relativity, but become essential in general relativity, as we shall see in the next chapter. We begin this chapter by introducing the definitions of the transformation rules of the components of tensors and tensor densities. Further we introduce the concept of affine connection which is important in defining the differentiation of tensors in general coordinates, especially the covariant derivative and the absolute derivative of a four-vector. The definition and the properties of the spatial metric tensor are also given in this chapter. As an application we analyze the equation of motion of a mass point in special relativity by using general coordinates.

10.2 Tensor components in general coordinates

Let us consider the four-dimensional space of special relativity with coordinates x^μ ($\mu = 0, 1, 2, 3$) (but what we say applies to the ordinary three-dimensional case as well, with obvious changes) and a coordinate transformation (not necessarily linear but with non-vanishing Jacobian)

$$x'^\mu = x'^\mu(x^0, x^1, x^2, x^3), \tag{10.1}$$

with inverse transformation given by

$$x^\mu = x^\mu(x'^0, x'^1, x'^2, x'^3). \tag{10.2}$$

The differentials are transformed according to

$$dx'^\mu = \frac{\partial x'^\mu}{\partial x^\nu} dx^\nu, \qquad dx^\mu = \frac{\partial x^\mu}{\partial x'^\nu} dx'^\nu, \tag{10.3}$$

and we have the identities

$$\frac{\partial x'^{\mu}}{\partial x^{\nu}}\frac{\partial x^{\nu}}{\partial x'^{\kappa}} = \delta^{\mu}_{\kappa}, \qquad \frac{\partial x^{\mu}}{\partial x'^{\nu}}\frac{\partial x'^{\nu}}{\partial x^{\kappa}} = \delta^{\mu}_{\kappa}. \tag{10.4}$$

With respect to the coordinate transformations (10.1) and (10.2) the contravariant and covariant laws of transformation of four-vectors are defined by

$$A'^{\mu} = \frac{\partial x'^{\mu}}{\partial x^{\nu}} A^{\nu}, \qquad A'_{\mu} = \frac{\partial x^{\nu}}{\partial x'^{\mu}} A_{\nu}, \tag{10.5}$$

respectively. The gradient of a scalar field ϕ is a covariant four-vector since

$$\frac{\partial \phi}{\partial x'^{\mu}} = \frac{\partial x^{\nu}}{\partial x'^{\mu}}\frac{\partial \phi}{\partial x^{\nu}}. \tag{10.6}$$

The differentials of the coordinates are an example of contravariant components (because of (10.3)) of a vector. One must be careful here because some authors use (see, e.g., Wald [9]) dx^{μ} to denote *four-vectors* rather than the *four components of a vector*. In order to underline the difference, we use a boldface notation $\mathbf{dx}^{\mu}$ to denote the vector of a basis; then a vector $\mathbf{A}$ with covariant components A_{μ} equals $A_{\mu}\mathbf{dx}^{\mu}$, and $\mathbf{dx}^{\mu}$ form the basis to represent vectors in terms of *covariant* components. One must be careful in distinguishing $A_{\mu}\mathbf{dx}^{\mu}$ and $A_{\mu}dx^{\mu}$; the latter is a scalar, the former a vector; this is easy here, because we have used a boldface symbol for vectors, but it may be harder when no special typographical trick underlines the difference.

There is a similar basis $\partial/\partial\mathbf{x}^{\nu}$ for contravariant vectors; in this case the lack of boldface notation does not lead to confusion, since the symbol $\partial/\partial x^{\nu}$ has a different meaning only if it acts on some function. In terms of this basis one can represent a vector $\mathbf{A}$ with contravariant components A^{μ} as $A^{\mu}\partial/\partial\mathbf{x}^{\mu}$. We shall not use this notation in this book, but we have inserted these remarks to facilitate comparison.

We remark that the rule is self consistent: if we transform from one system of coordinates to another, and then to a third one, we obtain the same components as if we had transformed the components from the first system to the third directly, thanks to the chain rule.

We can generalize the transformation rule to tensors of any order. Let us restrict ourselves to tensors of the second order, since the generalization is obvious. The contravariant components of a very simple tensor are given by $A^{\mu}B^{\nu}$ where A^{μ} and B^{ν} are components of a four-vector. This suggests the rule

$$T'^{\mu\nu} = \frac{\partial x'^{\mu}}{\partial x^{\tau}}\frac{\partial x'^{\nu}}{\partial x^{\rho}} T^{\tau\rho}, \qquad T'_{\mu\nu} = \frac{\partial x^{\tau}}{\partial x'^{\mu}}\frac{\partial x^{\rho}}{\partial x'^{\nu}} T_{\tau\rho}. \tag{10.7}$$

It is easy to conclude that this is the only possible rule, when we realize that the contravariant and covariant components of any tensor can be written as a linear combination of a finite number of products of four-vector components.

The square of the magnitude of the vector is given by $A'^{\mu} A'_{\mu}$ which equals $A^{\mu} A_{\mu}$, thanks to (10.4).

Whereas the upper and lower indices are not required when we deal with vectors and tensors in the ordinary three-dimensional space, and can be regarded as a nuisance in special relativity (related to the presence of the minus sign in the line element) when we use the usual coordinates (Cartesian coordinates and time), they become a necessity when dealing with general coordinates.

As an example, let us consider the familiar spherical coordinates in three dimensions:

$$x'^1 = \rho = \sqrt{(x^1)^2 + (x^2)^2 + (x^3)^2}, \tag{10.8}$$

$$x'^2 = \vartheta = \arccos\left\{\frac{x^3}{\sqrt{(x^1)^2 + (x^2)^2 + (x^3)^2}}\right\}, \tag{10.9}$$

$$x'^3 = \varphi = \arccos\left\{\frac{x^1}{\sqrt{(x^1)^2 + (x^2)^2}}\right\}, \tag{10.10}$$

where ϑ takes the values between 0 and π and φ between 0 and π for $x^2 > 0$ and between π and 2π for $x^2 < 0$. Then the Jacobian matrix $J = ((\partial x'^i/\partial x^k))$ is given by

$$J = \begin{pmatrix} x^1/\rho & x^2/\rho & x^3/\rho \\ x^1 x^3/(\rho^2 \mathcal{X}) & x^2 x^3/(\rho^2 \mathcal{X}) & -\mathcal{X}/\rho^2 \\ -x^2/\mathcal{X}^2 & x^1/\mathcal{X}^2 & 0 \end{pmatrix}, \tag{10.11}$$

where we have introduced the abbreviation $\mathcal{X} \equiv \sqrt{(x^1)^2 + (x^2)^2}$.

The inverse Jacobian matrix $J' = ((\partial x^k/\partial x'^i))$ is given by

$$J' = \begin{pmatrix} \sin\vartheta\cos\varphi & \rho\cos\vartheta\cos\varphi & -\rho\sin\vartheta\sin\varphi \\ \sin\vartheta\sin\varphi & \rho\cos\vartheta\sin\varphi & \rho\sin\vartheta\cos\varphi \\ \cos\vartheta & -\rho\sin\vartheta & 0 \end{pmatrix}. \tag{10.12}$$

Here each matrix is expressed in terms of the variables with respect to which the differentiation has been performed. Clearly, however, in order to verify (10.4) one must express both matrices in terms of the same variables. If we express J in terms of the spherical coordinates, we get

$$J = \begin{pmatrix} \sin\vartheta\cos\varphi & \sin\vartheta\sin\varphi & \cos\vartheta \\ \cos\vartheta\cos\varphi/\rho & \cos\vartheta\sin\varphi/\rho & -\sin\vartheta/\rho \\ -\sin\varphi/(\rho\sin\vartheta) & \cos\varphi/(\rho\sin\vartheta) & 0 \end{pmatrix}. \tag{10.13}$$

It is now easy to verify that the product JJ' is the identity matrix.

Let us now compute the contravariant and covariant components of a vector $\mathbf{v}$ with respect to the spherical coordinates, according to the rule (10.5). We obtain:

$$v'^1 = v^1 \sin\vartheta\cos\varphi + v^2 \sin\vartheta\sin\varphi + v^3\cos\vartheta, \tag{10.14}$$

$$v'^2 = \frac{v^1}{\rho}\cos\vartheta\cos\varphi + \frac{v^2}{\rho}\cos\vartheta\sin\varphi - \frac{v^3}{\rho}\sin\vartheta, \tag{10.15}$$

$$v'^3 = -\frac{v^1}{\rho}\frac{\sin\varphi}{\sin\vartheta} + \frac{v^2}{\rho}\frac{\cos\varphi}{\sin\vartheta}, \tag{10.16}$$

$$v'_1 = v_1\sin\vartheta\cos\varphi + v_2\sin\vartheta\sin\varphi + v_3\cos\vartheta, \tag{10.17}$$

$$v'_2 = v_1\rho\cos\vartheta\cos\varphi + v_2\rho\cos\vartheta\sin\varphi - v_3\rho\sin\vartheta, \tag{10.18}$$

$$v'_3 = -v_1\rho\sin\vartheta\sin\varphi + v_2\rho\sin\vartheta\cos\varphi. \tag{10.19}$$

In the above equations $v^1 \equiv v_1$, $v^2 \equiv v_2$ and $v^3 \equiv v_3$ are the Cartesian components of the vector $\mathbf{v}$. We remark two things. First, the components do not have the same physical dimensions: if $\mathbf{v}$ is a velocity, then the first contravariant component has the dimensions of a velocity, but the other two have the dimensions of a velocity divided by a length; the first covariant component has the dimensions of a velocity, but the other two have the dimensions of a velocity multiplied by a length. The two circumstances are not unrelated because if we compute the magnitude of $\mathbf{v}$, we must obtain something having the dimensions of a velocity. Second, the components do not coincide with the usual components obtained by projecting along the tangent to the coordinate curves (otherwise, our first remark would not be true).

We consider now the relation between the covariant and contravariant components in general. If the unprimed variables are those of a usual Minkowski space (orthogonal Cartesian coordinates and time), we have

$$A^\nu = \eta^{\nu\sigma}A_\sigma, \quad A_\nu = \eta_{\nu\sigma}A^\sigma. \tag{10.20}$$

Hence if we use general coordinates, it follows that

$$A'^\mu = \frac{\partial x'^\mu}{\partial x^\nu}\eta^{\nu\sigma}A_\sigma = \frac{\partial x'^\mu}{\partial x^\nu}\eta^{\nu\sigma}\frac{\partial x'^\tau}{\partial x^\sigma}A'_\tau, \tag{10.21}$$

where both formulas in (10.5) have been used.

Thus we have found

$$A'^\mu = g'^{\mu\tau}A'_\tau, \tag{10.22}$$

where

$$g'^{\mu\tau} = \eta^{\nu\sigma}\frac{\partial x'^\mu}{\partial x^\nu}\frac{\partial x'^\tau}{\partial x^\sigma}. \tag{10.23}$$

It is clear from the remarks made above that a similar formula holds for a tensor

$$T'^{\mu\nu} = g'^{\mu\tau}g'^{\nu\sigma}T'_{\tau\sigma}. \tag{10.24}$$

It is also clear that the new quantities $g^{\mu\tau}$ are the contravariant components of a second order tensor (the *metric* tensor). It is associated with the tensor $\eta^{\mu\nu}$ and reduces to it when we use the simple space-time coordinates used in Chapter 1.

The corresponding covariant components are easily obtained. They must serve similar purpose, i.e., to start from the covariant components to obtain the covariant tensor. This is one of the departure points to obtain it. Another one is from (10.24) applied to $g'^{\mu\nu}$ itself which reads

$$g'^{\mu\nu} = g'^{\mu\tau} g'^{\nu\sigma} g'_{\tau\sigma}. \tag{10.25}$$

This shows that if the determinant of $g'^{\mu\nu}$ does not vanish (and it does not because of (10.23), unless the Jacobian of the transformation vanishes, something that we excluded), we have

$$g'^{\nu\sigma} g'_{\tau\sigma} = \delta^{\nu}_{\tau}. \tag{10.26}$$

Thus the contravariant components of the metric tensor, $g'_{\tau\sigma}$, are the elements of the inverse matrix. From this, or directly, it follows that

$$g'_{\tau\sigma} = \eta_{\nu\mu} \frac{\partial x^{\mu}}{\partial x'^{\tau}} \frac{\partial x^{\nu}}{\partial x'^{\sigma}}. \tag{10.27}$$

It is easy to verify that these are indeed the elements of the inverse of the matrix with elements (10.23).

The line element can be expressed in general coordinates x'^{μ} in terms of the metric tensor $g_{\mu\nu}$ such that

$$ds'^2 = g'_{\mu\nu} dx'^{\mu} dx'^{\nu}. \tag{10.28}$$

This follows either directly or by transforming the expression $dx'_{\mu} dx'^{\mu}$.

The general rule to transform mixed tensors is

$$T'^{\mu\sigma}_{\ \ \nu} = \frac{\partial x'^{\mu}}{\partial x^{\tau}} \frac{\partial x^{\rho}}{\partial x'^{\nu}} \frac{\partial x'^{\sigma}}{\partial x^{\epsilon}} T^{\tau\epsilon}_{\ \ \rho}. \tag{10.29}$$

So far we have distinguished pseudo-Cartesian frames from the others, by affecting the quantities referring to the latter by a prime. Henceforth, unless explicitly declared, we shall treat all the systems of coordinates in the same way and hence primed and unprimed quantities refer to two different coordinate systems of general nature.

The transformation law of the metric tensor follows from the general rule

$$g_{\mu\nu} = g'_{\sigma\tau} \frac{\partial x'^{\sigma}}{\partial x^{\mu}} \frac{\partial x'^{\tau}}{\partial x^{\nu}}. \tag{10.30}$$

The determinant of (10.30) leads to

$$g = \left| \frac{\partial x'}{\partial x} \right|^2 g', \qquad \text{or} \qquad \left| \frac{\partial x}{\partial x'} \right| g = \left| \frac{\partial x'}{\partial x} \right| g', \tag{10.31}$$

where $|\partial x'/\partial x|$ is the Jacobian of the transformation (10.1) and we have introduced $-g \equiv \det((g_{\mu\nu}))$. Further we have used the relationship

$$\left| \frac{\partial x}{\partial x'} \right| = \left| \frac{\partial x'}{\partial x} \right|^{-1}, \tag{10.32}$$

that follows from the determinants of the identities (10.4).

Let us calculate the transformation law of the Levi–Civita tensor defined in (1.62). According to (1.66) we have

$$g\epsilon_{\mu\nu\kappa\lambda} = \epsilon^{\alpha\beta\gamma\delta} g_{\mu\alpha} g_{\nu\beta} g_{\kappa\gamma} g_{\lambda\delta}$$

$$\overset{(10.30)}{=} \epsilon^{\alpha\beta\gamma\delta} g'_{\sigma\tau} \frac{\partial x'^{\sigma}}{\partial x^{\mu}} \frac{\partial x'^{\tau}}{\partial x^{\alpha}} g'_{\epsilon\theta} \frac{\partial x'^{\epsilon}}{\partial x^{\nu}} \frac{\partial x'^{\theta}}{\partial x^{\beta}} g'_{\rho\pi} \frac{\partial x'^{\rho}}{\partial x^{\kappa}} \frac{\partial x'^{\pi}}{\partial x^{\gamma}} g'_{\zeta\eta} \frac{\partial x'^{\zeta}}{\partial x^{\lambda}} \frac{\partial x'^{\eta}}{\partial x^{\delta}}$$

$$\overset{(1.66)}{=} \left|\frac{\partial x'}{\partial x}\right| \epsilon^{\tau\theta\pi\eta} g'_{\sigma\tau} g'_{\epsilon\theta} g'_{\rho\pi} g'_{\zeta\eta} \frac{\partial x'^{\sigma}}{\partial x^{\mu}} \frac{\partial x'^{\epsilon}}{\partial x^{\nu}} \frac{\partial x'^{\rho}}{\partial x^{\kappa}} \frac{\partial x'^{\zeta}}{\partial x^{\lambda}}$$

$$\overset{(1.66)}{=} \left|\frac{\partial x'}{\partial x}\right| g' \epsilon_{\sigma\epsilon\rho\zeta} \frac{\partial x'^{\sigma}}{\partial x^{\mu}} \frac{\partial x'^{\epsilon}}{\partial x^{\nu}} \frac{\partial x'^{\rho}}{\partial x^{\kappa}} \frac{\partial x'^{\zeta}}{\partial x^{\lambda}}. \tag{10.33}$$

The above equation together with $(10.31)_2$ lead to

$$\sqrt{g'}\epsilon_{\mu\nu\kappa\lambda} = \sqrt{g}\epsilon_{\sigma\epsilon\rho\zeta} \frac{\partial x^{\sigma}}{\partial x'^{\mu}} \frac{\partial x^{\epsilon}}{\partial x'^{\nu}} \frac{\partial x^{\rho}}{\partial x'^{\kappa}} \frac{\partial x^{\zeta}}{\partial x'^{\lambda}}, \tag{10.34}$$

that is $\sqrt{g}\epsilon_{\mu\nu\kappa\lambda}$ transforms like a covariant tensor. Due to relationship (1.65) we have that $\epsilon^{\mu\nu\kappa\lambda}/\sqrt{g}$ transforms like a contravariant tensor.

A transformation law of the type

$$T'^{\mu_1\ldots\mu_p}_{\ \ \nu_1\ldots\nu_q} = \left|\frac{\partial x}{\partial x'}\right|^N \frac{\partial x'^{\mu_1}}{\partial x^{\sigma_1}} \cdots \frac{\partial x'^{\mu_p}}{\partial x^{\sigma_p}} \frac{\partial x^{\tau_1}}{\partial x'^{\nu_1}} \cdots \frac{\partial x^{\tau_q}}{\partial x'^{\nu_q}} T^{\sigma_1\ldots\sigma_p}_{\ \ \tau_1\ldots\tau_q}, \tag{10.35}$$

defines a tensor density of weight N. By the use of $(10.31)_1$ the above equation can be written as

$$g'^{-N/2} T'^{\mu_1\ldots\mu_p}_{\ \ \nu_1\ldots\nu_q} = \frac{\partial x'^{\mu_1}}{\partial x^{\sigma_1}} \cdots \frac{\partial x'^{\mu_p}}{\partial x^{\sigma_p}} \frac{\partial x^{\tau_1}}{\partial x'^{\nu_1}} \cdots \frac{\partial x^{\tau_q}}{\partial x'^{\nu_q}} g^{-N/2} T^{\sigma_1\ldots\sigma_p}_{\ \ \tau_1\ldots\tau_q}, \tag{10.36}$$

and one concludes that $g^{-N/2}T^{\sigma_1\ldots\sigma_p}_{\tau_1\ldots\tau_q}$ transforms like an ordinary tensor. Hence we have that $\epsilon_{\mu\nu\sigma\tau}$ is a tensor density of weight -1, g is a scalar density of weight 2, while $\epsilon^{\mu\nu\sigma\tau}$ is a tensor density of weight 1.

The volume element d^4x transforms according to

$$d^4x' = \left\|\frac{\partial x'}{\partial x}\right\| d^4x \qquad \text{or} \qquad \sqrt{g'}d^4x' = \sqrt{g}d^4x, \tag{10.37}$$

where $||\partial x'/\partial x||$ is the modulus of the Jacobian. Hence $\sqrt{g}d^4x$ is an invariant volume element and d^4x is a scalar density of weight -1.

Problems

10.2.1 Show that the transformation of the components of a four-vector from one system to another and then to a third one is the same as if we had transformed the components of a four-vector from the first system to the third.

10.2.2 Show that the contravariant components of any tensor can be written as a combination of a finite number of products of four-vector components.

10.2.3 Show that $A'_\mu A'^\mu = A_\mu A^\mu$.

10.2.4 Show that the product of the Jacobians JJ' given by (10.12) and (10.13) is the identity matrix.

10.2.5 Show that the matrix $((g'_{\mu\nu}))$ has its inverse given by $((g'^{\mu\nu}))$.

10.2.6 Prove that $dx'_\mu dx'^\mu = g'_{\mu\nu} dx'^\mu dx'^\nu = \eta_{\mu\nu} dx^\mu dx^\nu = ds^2$.

10.2.7 Show that the general rule to transform the mixed tensor $T'^{\mu\sigma}_{\ \ \nu}$ is given by (10.29).

10.2.8 Show that the transformation law of the metric tensor is given by (10.30).

10.3 Affine connection

Tensors do not exhaust the tools of tensor analysis in general coordinates. Another important concept is that of affine connection, which plays an important role in the differentiation of vectors and tensors in general coordinates. To explain its role, we remark that in the ordinary three-dimensional space as well as in Minkowski space it is easy to define parallel vectors. If we consider two points P and Q, a vector at P and a vector at Q will be parallel if they have the same components with respect to a Cartesian (or pseudo-Cartesian) frame. This definition does not work in general coordinates. In fact, let us take a constant vector field with components A_ν with respect to a system of Cartesian (or pseudo-Cartesian) coordinates x^ν, and let A'_ν be the components of the same vector with respect to a general system of coordinates x'^ν; the derivatives of A_ν with respect to the coordinates x^ν will vanish, but this will not be the case for the derivatives of A'_ν with respect to the coordinates x'^ν. In fact, if we differentiate (10.5) with respect to x'^σ, taking into account the fact that A^ν is constant, we obtain

$$\frac{\partial A'_\mu}{\partial x'^\sigma} = \frac{\partial^2 x^\nu}{\partial x'^\sigma \partial x'^\mu} A_\nu = \frac{\partial^2 x^\nu}{\partial x'^\sigma \partial x'^\mu} \frac{\partial x'^\tau}{\partial x^\nu} A'_\tau = \Gamma'^\tau_{\sigma\mu} A'_\tau. \tag{10.38}$$

The object which we have introduced,

$$\Gamma'^\mu_{\sigma\nu} = \frac{\partial^2 x^\tau}{\partial x'^\sigma \partial x'^\nu} \frac{\partial x'^\mu}{\partial x^\tau} = \Gamma'^\mu_{\nu\sigma}, \tag{10.39}$$

is called the affine connection relative to the coordinate transformation (10.1). The geometrical meaning of this object is a compensation for the change of the components of a derivative in order to form again a tensor, as we shall see soon, and hence to provide objective statements in any coordinate system.

In the following we shall proceed to analyze the transformation law of this affine connection and its relationship with the metric tensor.

Let $\bar{x}^\mu$ be another coordinate system; then the following relations hold:

$$\frac{\partial x^\eta}{\partial \bar{x}^\nu} = \frac{\partial x^\eta}{\partial x'^\tau}\frac{\partial x'^\tau}{\partial \bar{x}^\nu}, \quad \text{and} \quad \frac{\partial \bar{x}^\lambda}{\partial x^\rho} = \frac{\partial \bar{x}^\lambda}{\partial x'^\epsilon}\frac{\partial x'^\epsilon}{\partial x^\rho}. \tag{10.40}$$

If we differentiate $(10.40)_1$ with respect to $\bar{x}^\sigma$ we get

$$\frac{\partial^2 x^\eta}{\partial \bar{x}^\sigma \partial \bar{x}^\nu} = \frac{\partial^2 x^\eta}{\partial x'^\rho \partial x'^\tau}\frac{\partial x'^\rho}{\partial \bar{x}^\sigma}\frac{\partial x'^\tau}{\partial \bar{x}^\nu} + \frac{\partial x^\eta}{\partial x'^\tau}\frac{\partial^2 x'^\tau}{\partial \bar{x}^\sigma \partial \bar{x}^\nu}. \tag{10.41}$$

The multiplication of the above equation by $\partial \bar{x}^\mu / \partial x^\eta$ leads to

$$\begin{aligned}
\bar{\Gamma}^\mu_{\sigma\nu} \equiv \frac{\partial^2 x^\eta}{\partial \bar{x}^\sigma \partial \bar{x}^\nu}\frac{\partial \bar{x}^\mu}{\partial x^\eta} &= \frac{\partial^2 x^\eta}{\partial x'^\rho \partial x'^\tau}\frac{\partial x'^\rho}{\partial \bar{x}^\sigma}\frac{\partial x'^\tau}{\partial \bar{x}^\nu}\frac{\partial \bar{x}^\mu}{\partial x^\eta} + \frac{\partial x^\eta}{\partial x'^\tau}\frac{\partial^2 x'^\tau}{\partial \bar{x}^\sigma \partial \bar{x}^\nu}\frac{\partial \bar{x}^\mu}{\partial x^\eta} \\
&\overset{(10.40)_2}{=} \frac{\partial^2 x^\eta}{\partial x'^\rho \partial x'^\tau}\frac{\partial x'^\rho}{\partial \bar{x}^\sigma}\frac{\partial x'^\tau}{\partial \bar{x}^\nu}\frac{\partial \bar{x}^\mu}{\partial x'^\lambda}\frac{\partial x'^\lambda}{\partial x^\eta} + \frac{\partial^2 x'^\tau}{\partial \bar{x}^\sigma \partial \bar{x}^\nu}\frac{\partial \bar{x}^\mu}{\partial x'^\tau} \\
&= \Gamma'^\lambda_{\rho\tau}\frac{\partial x'^\rho}{\partial \bar{x}^\sigma}\frac{\partial x'^\tau}{\partial \bar{x}^\nu}\frac{\partial \bar{x}^\mu}{\partial x'^\lambda} + \frac{\partial^2 x'^\tau}{\partial \bar{x}^\sigma \partial \bar{x}^\nu}\frac{\partial \bar{x}^\mu}{\partial x'^\tau}.
\end{aligned} \tag{10.42}$$

This is one of the forms of the transformation law for an affine connection from one coordinate system to another. Due to the second term on the right-hand side of (10.42) the affine connection is not a tensor. The adjective "affine" is frequently used to mean "linear but not homogeneous"; we see that the components of a tensor with respect to a coordinate system are linear homogeneous functions of the components with respect to another system. This is not the case for $\Gamma^\lambda_{\rho\tau}$ because of the extra term which is independent of the components of the object under consideration; hence the transformation is linear but not homogeneous and the adjective "affine" is justified. The name "connection" is justified by the fact that $\Gamma^\lambda_{\rho\tau}$ connects the values of the components of a vector or a tensor at different (infinitesimally close) points. In fact, the reason why we need this object is that in order to form the derivative of a vector, we need to compare the values of the vector at two (infinitesimally close) points; in Cartesian coordinates this is not a problem because the change of components indicates the change undergone by the vector. Not so in general coordinates, as is clear when using spherical coordinates; a constant vector has different values of the components at different points. The affine connection tells us how to correct the change due to this circumstance and to obtain the change in the vector as opposed to the change in the components.

If we now differentiate the identity

$$\frac{\partial x'^\tau}{\partial \bar{x}^\sigma}\frac{\partial \bar{x}^\mu}{\partial x'^\tau} = \delta^\mu_\sigma \tag{10.43}$$

with respect to $\bar{x}^\nu$, it follows that

$$\frac{\partial^2 x'^\tau}{\partial \bar{x}^\nu \partial \bar{x}^\sigma}\frac{\partial \bar{x}^\mu}{\partial x'^\tau} = -\frac{\partial x'^\tau}{\partial \bar{x}^\sigma}\frac{\partial x'^\eta}{\partial \bar{x}^\nu}\frac{\partial^2 \bar{x}^\mu}{\partial x'^\eta \partial x'^\tau}. \tag{10.44}$$

This last equation is used to write the transformation law (10.42) in the alternative form

$$\bar{\Gamma}^{\mu}_{\sigma\nu} = \Gamma'^{\lambda}_{\rho\tau} \frac{\partial x'^{\rho}}{\partial \bar{x}^{\sigma}} \frac{\partial x'^{\tau}}{\partial \bar{x}^{\nu}} \frac{\partial \bar{x}^{\mu}}{\partial x'^{\lambda}} - \frac{\partial^2 \bar{x}^{\mu}}{\partial x'^{\eta} \partial x'^{\tau}} \frac{\partial x'^{\tau}}{\partial \bar{x}^{\sigma}} \frac{\partial x'^{\eta}}{\partial \bar{x}^{\nu}}. \tag{10.45}$$

According to (10.30) the metric tensor in two coordinate system x'^{μ} and $\bar{x}^{\mu}$ are related by

$$\bar{g}_{\mu\nu} = g'_{\sigma\tau} \frac{\partial x'^{\sigma}}{\partial \bar{x}^{\mu}} \frac{\partial x'^{\tau}}{\partial \bar{x}^{\nu}}. \tag{10.46}$$

If we differentiate (10.46) with respect to $\bar{x}^{\lambda}$ we get

$$\frac{\partial \bar{g}_{\mu\nu}}{\partial \bar{x}^{\lambda}} = \frac{\partial g'_{\sigma\tau}}{\partial x'^{\epsilon}} \frac{\partial x'^{\epsilon}}{\partial \bar{x}^{\lambda}} \frac{\partial x'^{\sigma}}{\partial \bar{x}^{\mu}} \frac{\partial x'^{\tau}}{\partial \bar{x}^{\nu}} + g'_{\sigma\tau} \left(\frac{\partial^2 x'^{\sigma}}{\partial \bar{x}^{\lambda} \partial \bar{x}^{\mu}} \frac{\partial x'^{\tau}}{\partial \bar{x}^{\nu}} + \frac{\partial x'^{\sigma}}{\partial \bar{x}^{\mu}} \frac{\partial^2 x'^{\tau}}{\partial \bar{x}^{\lambda} \partial \bar{x}^{\nu}} \right). \tag{10.47}$$

Now by using a cyclic permutation of the indices μ, ν, λ in (10.47) we can build the relation

$$\frac{1}{2} \bar{g}^{\kappa\nu} \left(\frac{\partial \bar{g}_{\mu\nu}}{\partial \bar{x}^{\lambda}} + \frac{\partial \bar{g}_{\lambda\nu}}{\partial \bar{x}^{\mu}} - \frac{\partial \bar{g}_{\mu\lambda}}{\partial \bar{x}^{\nu}} \right) = \frac{\partial^2 x'^{\eta}}{\partial \bar{x}^{\lambda} \partial \bar{x}^{\mu}} \frac{\partial \bar{x}^{\kappa}}{\partial x'^{\eta}}$$

$$+ \frac{1}{2} g'^{\eta\tau} \left(\frac{\partial g'_{\sigma\tau}}{\partial x'^{\epsilon}} + \frac{\partial g'_{\epsilon\tau}}{\partial x'^{\sigma}} - \frac{\partial g'_{\sigma\epsilon}}{\partial x'^{\tau}} \right) \frac{\partial x'^{\epsilon}}{\partial \bar{x}^{\lambda}} \frac{\partial x'^{\sigma}}{\partial \bar{x}^{\mu}} \frac{\partial \bar{x}^{\kappa}}{\partial x'^{\eta}}. \tag{10.48}$$

In the above equation we have used the relationship for the contravariant components of the metric tensor

$$\bar{g}^{\kappa\nu} = \frac{\partial \bar{x}^{\kappa}}{\partial x'^{\eta}} \frac{\partial \bar{x}^{\nu}}{\partial x'^{\rho}} g'^{\eta\rho}. \tag{10.49}$$

We thus see, by comparing (10.42) with (10.48) that there is an affine connection which can be built from the derivatives of the metric tensor

$$\Gamma^{\kappa}_{\mu\lambda} = \frac{1}{2} g^{\kappa\nu} \left(\frac{\partial g_{\mu\nu}}{\partial x^{\lambda}} + \frac{\partial g_{\lambda\nu}}{\partial x^{\mu}} - \frac{\partial g_{\mu\lambda}}{\partial x^{\nu}} \right). \tag{10.50}$$

The affine connection $\Gamma^{\kappa}_{\mu\lambda}$ is also known as the Christoffel symbol of the second kind. Further one can show from (10.50) that

$$\frac{\partial g_{\mu\sigma}}{\partial x^{\lambda}} = g_{\kappa\sigma} \Gamma^{\kappa}_{\mu\lambda} + g_{\kappa\mu} \Gamma^{\kappa}_{\sigma\lambda}. \tag{10.51}$$

It is now easy to see that the two affine connections (10.39) and (10.50) must coincide; in fact if we use a pseudo-Cartesian system of coordinates x'^{λ} such that $g'_{\kappa\sigma} = \eta_{\kappa\sigma}$ is constant, (10.48) reduces to (10.39) except for the fact that the coordinates are x'^{λ} and $\bar{x}^{\lambda}$ rather than x^{λ} and x'^{λ}.

There is also a relation between the affine connection and the derivative of the determinant of the metric tensor $g \equiv -\det((g_{\mu\nu}))$. In order to deduce such a relation we need to know the expression for the derivative of the metric tensor which we proceed to derive. If we identify the tensor $T_{\mu\nu}$ with the metric tensor $g_{\mu\nu}$ in $(1.66)_2$ and differentiate this expression with respect to x^ϵ we get

$$-24\frac{\partial g}{\partial x^\epsilon} = \epsilon^{\alpha\beta\gamma\delta}\epsilon^{\mu\nu\kappa\lambda}\left[\frac{\partial g_{\mu\eta}}{\partial x^\epsilon}g^{\eta\tau}g_{\tau\alpha}g_{\nu\beta}g_{\kappa\gamma}g_{\lambda\delta} + g_{\mu\alpha}\frac{\partial g_{\nu\eta}}{\partial x^\epsilon}g^{\eta\tau}g_{\tau\beta}g_{\kappa\gamma}g_{\lambda\delta}\right.$$

$$\left.+g_{\mu\alpha}g_{\nu\beta}\frac{\partial g_{\kappa\eta}}{\partial x^\epsilon}g^{\eta\tau}g_{\tau\gamma}g_{\lambda\delta} + g_{\mu\alpha}g_{\nu\beta}g_{\kappa\gamma}\frac{\partial g_{\lambda\eta}}{\partial x^\epsilon}g^{\eta\tau}g_{\tau\delta}\right]$$

$$\overset{(1.66)_1}{=} g^{\eta\tau}\epsilon^{\mu\nu\kappa\lambda}g\left[\epsilon_{\tau\nu\kappa\lambda}\frac{\partial g_{\mu\eta}}{\partial x^\epsilon} + \epsilon_{\mu\tau\kappa\lambda}\frac{\partial g_{\nu\eta}}{\partial x^\epsilon} + \epsilon_{\mu\nu\tau\lambda}\frac{\partial g_{\kappa\eta}}{\partial x^\epsilon} + \epsilon_{\mu\nu\kappa\tau}\frac{\partial g_{\lambda\eta}}{\partial x^\epsilon}\right]$$

$$\overset{(1.65)_1}{=} -24g\frac{\partial g_{\mu\nu}}{\partial x^\epsilon}g^{\mu\nu}. \tag{10.52}$$

Hence we have from (10.52) that the derivative of the determinant of the metric tensor is given by

$$\frac{\partial \ln g}{\partial x^\epsilon} = \frac{\partial g_{\mu\nu}}{\partial x^\epsilon}g^{\mu\nu}. \tag{10.53}$$

Now by contracting the Christoffel symbol (10.50) we get

$$\Gamma^\nu_{\mu\nu} = \frac{1}{2}g^{\nu\sigma}\frac{\partial g_{\nu\sigma}}{\partial x^\mu} \overset{(10.53)}{=} \frac{\partial \ln\sqrt{g}}{\partial x^\mu}. \tag{10.54}$$

This formula will be useful in the derivation of the covariant divergence of a four-vector.

Problems

10.3.1 Show that by performing a cyclic permutation of the indices μ, ν and λ in (10.47), the equation (10.48) follows.

10.3.2 Show that (10.51) holds.

10.3.3 Show that the two affine connections (10.39) and (10.50) coincide by using a pseudo-Cartesian system of coordinates x'^λ such that $g'_{\kappa\sigma} = \eta_{\kappa\sigma}$ is constant.

10.3.4 Show that (10.53) holds by performing the calculations indicated in (10.52).

10.4 Covariant differentiation

As we saw in the previous section, if we differentiate a four-vector we do not get a tensor. Indeed the differentiation of $(10.5)_1$ with respect to x'^σ leads to

$$\frac{\partial A'^\mu}{\partial x'^\sigma} = \frac{\partial x'^\mu}{\partial x^\eta}\frac{\partial x^\tau}{\partial x'^\sigma}\frac{\partial A^\eta}{\partial x^\tau} + \frac{\partial^2 x'^\mu}{\partial x^\eta \partial x^\tau}\frac{\partial x^\tau}{\partial x'^\sigma}A^\eta. \tag{10.55}$$

Due to the second term on the right-hand side of (10.55), as we know, the derivative of a four-vector does not transform like a tensor.

On the other hand, for the coordinates x'^μ and x^μ we obtain from the transformation law of the affine connection (10.45):

$$\frac{\partial^2 x'^\mu}{\partial x^\eta \partial x^\tau}\frac{\partial x^\tau}{\partial x'^\sigma} = \Gamma^\lambda_{\rho\eta}\frac{\partial x^\rho}{\partial x'^\sigma}\frac{\partial x'^\mu}{\partial x^\lambda} - \Gamma'^\mu_{\sigma\nu}\frac{\partial x'^\nu}{\partial x^\eta}. \tag{10.56}$$

We introduce (10.56) into the second term on the right-hand side of (10.55) and get after some manipulations

$$\frac{\partial A'^\mu}{\partial x'^\sigma} + \Gamma'^\mu_{\sigma\nu}A'^\nu = \frac{\partial x'^\mu}{\partial x^\eta}\frac{\partial x^\tau}{\partial x'^\sigma}\left[\frac{\partial A^\eta}{\partial x^\tau} + \Gamma^\eta_{\tau\nu}A^\nu\right]. \tag{10.57}$$

We conclude from the above equation that

$$A^\mu{}_{;\nu} \equiv \frac{\partial A^\mu}{\partial x^\nu} + \Gamma^\mu_{\nu\sigma}A^\sigma \tag{10.58}$$

transforms like a tensor. This quantity is called covariant derivative.

This derivative is hidden in the usual Lagrange equation for the motion of a particle when we use general coordinates (we use here x_μ rather than the traditional q_μ and Greek letters for the indices in order to stress the analogy):

$$\frac{d}{dt}\frac{\partial T}{\partial \dot{x}^\sigma} - \frac{\partial T}{\partial x^\sigma} = Q_\sigma \tag{10.59}$$

where Q_σ is the generalized component of the force acting on the particle and T is the kinetic energy, given by

$$T = \frac{1}{2}g_{\mu\nu}\frac{dx^\mu}{dt}\frac{dx^\nu}{dt}. \tag{10.60}$$

From the above equation it follows

$$\frac{\partial T}{\partial \dot{x}^\sigma} = g_{\mu\sigma}\frac{dx^\mu}{dt}, \tag{10.61}$$

$$\frac{\partial T}{\partial x^\sigma} = \frac{1}{2}\frac{\partial g_{\mu\nu}}{\partial x^\sigma}\frac{dx^\mu}{dt}\frac{dx^\nu}{dt}. \tag{10.62}$$

Hence performing the derivation with respect to time in (10.61) and rearranging, we obtain from (10.59) together with (10.62)

$$\frac{\partial}{\partial x^\nu}\left(\frac{dx^\tau}{dt}\right)\frac{dx^\nu}{dt} + \frac{1}{2}g^{\tau\sigma}\left(\frac{\partial g_{\mu\sigma}}{\partial x^\nu} + \frac{\partial g_{\sigma\nu}}{\partial x^\mu} - \frac{\partial g_{\mu\nu}}{\partial x^\sigma}\right)\frac{dx^\mu}{dt}\frac{dx^\nu}{dt} = Q^\tau \tag{10.63}$$

the left-hand side is clearly related to the covariant derivative and can be written

$$\left(\frac{dx^\tau}{dt}\right)_{;\nu}\frac{dx^\nu}{dt} = Q^\tau \tag{10.64}$$

where (10.58) with dx^μ/dt in place of A^μ has been used.

The covariant derivative of a covariant four-vector A_μ can be obtained in a similar manner, yielding

$$A_{\mu;\nu} \equiv \frac{\partial A_\mu}{\partial x^\nu} - \Gamma^\sigma_{\mu\nu} A_\sigma. \tag{10.65}$$

The covariant derivative of a covariant and of a mixed tensor are easily found by remarking that $A_\mu B_\nu$ is a tensor and any second order tensor can be written as a product of a finite number of products of this kind. Then the rule to differentiate the aforementioned tensors is found by starting from the rule to differentiate a product (which we require to be satisfied by the new derivative) and are given respectively by

$$T_{\mu\nu;\sigma} = \frac{\partial T_{\mu\nu}}{\partial x^\sigma} - \Gamma^\lambda_{\mu\sigma} T_{\lambda\nu} - \Gamma^\lambda_{\nu\sigma} T_{\mu\lambda}, \qquad T^\mu_{\nu;\sigma} = \frac{\partial T^\mu_\nu}{\partial x^\sigma} + \Gamma^\mu_{\sigma\lambda} T^\lambda_\nu - \Gamma^\lambda_{\nu\sigma} T^\mu_\lambda. \tag{10.66}$$

Moreover, by using (10.51) it follows that the covariant derivative of the metric tensor is zero, i.e.,

$$g_{\mu\nu;\sigma} = \frac{\partial g_{\mu\nu}}{\partial x^\sigma} - \Gamma^\lambda_{\sigma\nu} g_{\mu\lambda} - \Gamma^\lambda_{\sigma\mu} g_{\nu\lambda} = 0. \tag{10.67}$$

The relationship $g^{\mu\nu}_{\ ;\sigma} = 0$ implies that the covariant derivative of the Kronecker symbol is zero ($\delta^\mu_{\nu;\sigma} = 0$) since it can be expressed as a product of metric tensors.

The covariant derivative of a scalar ϕ is the gradient

$$\phi_{;\mu} = \frac{\partial \phi}{\partial x^\mu}, \tag{10.68}$$

while the covariant divergence of a four-vector follows from (10.58) yielding

$$A^\mu{}_{;\mu} = \frac{\partial A^\mu}{\partial x^\mu} + \Gamma^\mu_{\mu\sigma} A^\sigma \overset{(10.54)}{=} \frac{\partial A^\mu}{\partial x^\mu} + \frac{\partial \ln \sqrt{g}}{\partial x^\mu} A^\mu = \frac{1}{\sqrt{g}} \frac{\partial \sqrt{g} A^\mu}{\partial x^\mu}. \tag{10.69}$$

The covariant curl can be obtained from (10.65)

$$A_{\mu;\nu} - A_{\nu;\mu} = \frac{\partial A_\mu}{\partial x^\nu} - \frac{\partial A_\nu}{\partial x^\mu}, \tag{10.70}$$

which is the usual definition of the curl.

For a tensor $T^{\mu\nu}$ the contravariant components of the divergence are given by

$$T^{\mu\nu}{}_{;\nu} = \frac{\partial T^{\mu\nu}}{\partial x^\nu} + \Gamma^\nu_{\nu\lambda} T^{\mu\lambda} + \Gamma^\mu_{\nu\lambda} T^{\lambda\nu} \overset{(10.54)}{=} \frac{1}{\sqrt{g}} \frac{\partial \sqrt{g} T^{\mu\nu}}{\partial x^\nu} + \Gamma^\mu_{\nu\lambda} T^{\lambda\nu}. \tag{10.71}$$

Further for antisymmetric tensors the following relationship holds:

$$T^{\mu\nu}{}_{;\nu} = \frac{1}{\sqrt{g}} \frac{\partial \sqrt{g} T^{\mu\nu}}{\partial x^\nu}, \qquad \text{for} \qquad T^{\mu\nu} = -T^{\nu\mu}, \tag{10.72}$$

since $\Gamma^\mu_{\nu\lambda}$ is symmetric in the indices ν and λ.

From (10.71) one can obtain the covariant components of the divergence of a tensor $T_\sigma{}^\nu$. Indeed if we multiply (10.71) by $g_{\mu\sigma}$ it follows by performing some rearrangements that

$$T_\sigma{}^\nu{}_{;\nu} - T^{\mu\nu} g_{\mu\sigma;\nu} = \frac{1}{\sqrt{g}} \frac{\partial \sqrt{g} T_\sigma{}^\nu}{\partial x^\nu} - T^{\mu\nu} \frac{\partial g_{\mu\sigma}}{\partial x^\nu} + g_{\mu\sigma} \Gamma^\mu_{\nu\lambda} T^{\lambda\nu}. \tag{10.73}$$

By using the identity (10.67), equation (10.73) reduces to

$$T_\sigma{}^\nu{}_{;\nu} = \frac{1}{\sqrt{g}} \frac{\partial \sqrt{g} T_\sigma{}^\nu}{\partial x^\nu} - g_{\mu\tau} \Gamma^\tau_{\nu\sigma} T^{\mu\nu}, \tag{10.74}$$

which is the final expression for the covariant components of the divergence of a tensor $T_\sigma{}^\nu$. For symmetric tensors (10.74) can be written as

$$T_\sigma{}^\nu{}_{;\nu} = \frac{1}{\sqrt{g}} \frac{\partial \sqrt{g} T_\sigma{}^\nu}{\partial x^\nu} - \frac{1}{2} T^{\mu\nu} \frac{\partial g_{\mu\nu}}{\partial x^\sigma}. \tag{10.75}$$

Since the d'Alembertian is the divergence of the gradient of a scalar function ϕ, it follows from (10.68) and (10.69) that its covariant expression is

$$\Box \phi = \frac{1}{\sqrt{g}} \frac{\partial}{\partial x^\mu} \left(\sqrt{g}\, g^{\mu\nu} \frac{\partial \phi}{\partial x^\nu} \right). \tag{10.76}$$

If we add the cyclic permutations of the covariant derivative of an antisymmetric covariant tensor we get

$$T_{\mu\nu;\sigma} + T_{\nu\sigma;\mu} + T_{\sigma\mu;\nu} = \frac{\partial T_{\mu\nu}}{\partial x^\sigma} + \frac{\partial T_{\nu\sigma}}{\partial x^\mu} + \frac{\partial T_{\sigma\mu}}{\partial x^\nu}, \quad \text{for} \quad T_{\mu\nu} = -T_{\nu\mu}, \tag{10.77}$$

by the use of the definition of the covariant derivative $(10.66)_1$.

We consider now a contravariant four-vector $A^\mu(\lambda)$ which is defined over a curve $x^\mu = x^\mu(\lambda)$ where λ is an invariant parameter of the curve. We write the transformation rule $(10.5)_1$ as

$$A'^\mu(\lambda) = \frac{\partial x'^\mu}{\partial x^\nu} A^\nu(\lambda) \tag{10.78}$$

and differentiate the above expression with respect to the parameter λ, yielding

$$\frac{dA'^\mu(\lambda)}{d\lambda} = \frac{\partial x'^\mu}{\partial x^\nu} \frac{dA^\nu(\lambda)}{d\lambda} + \frac{\partial^2 x'^\mu}{\partial x^\nu \partial x^\sigma} \frac{\partial x^\sigma}{\partial \lambda} A^\nu(\lambda). \tag{10.79}$$

Equation (10.79) shows that the differentiation of a covariant four-vector with respect to the parameter λ is not a tensor. However if we use the relationship that follows from (10.56),

$$\frac{\partial^2 x'^\mu}{\partial x^\sigma \partial x^\nu} = \Gamma^\tau_{\sigma\nu} \frac{\partial x'^\mu}{\partial x^\tau} - \Gamma'^\mu_{\tau\eta} \frac{\partial x'^\tau}{\partial x^\sigma} \frac{\partial x'^\eta}{\partial x^\nu}, \tag{10.80}$$

to eliminate from (10.79) the term $\partial^2 x'^\mu/(\partial x^\nu \partial x^\sigma)$, we get

$$\frac{dA'^\mu(\lambda)}{d\lambda} + \Gamma'^\mu_{\nu\sigma}\frac{dx'^\nu}{d\lambda}A'^\sigma(\lambda) = \frac{\partial x'^\mu}{\partial x^\nu}\left[\frac{dA^\nu(\lambda)}{d\lambda} + \Gamma^\nu_{\sigma\tau}\frac{dx^\sigma}{d\lambda}A^\tau(\lambda)\right]. \tag{10.81}$$

Hence the quantity

$$\frac{\delta A^\mu(\lambda)}{\delta\lambda} \equiv \frac{dA^\mu(\lambda)}{d\lambda} + \Gamma^\mu_{\nu\sigma}\frac{dx^\nu}{d\lambda}A^\sigma(\lambda) \tag{10.82}$$

transforms like a contravariant four-vector. This quantity is called absolute derivative of the vector with components A^μ along the curve; when it vanishes, the vector is said to be dragged along in a parallel fashion, since this is equivalent to having constant components in Cartesian coordinates.

An alternative expression for the absolute derivative (10.82) is

$$\frac{\delta A^\mu(\lambda)}{\delta\lambda} = \left[\frac{\partial A^\mu}{\partial x^\nu} + \Gamma^\mu_{\nu\sigma}A^\sigma\right]\frac{dx^\nu}{d\lambda} = A^\mu{}_{;\nu}\frac{dx^\nu}{d\lambda}, \tag{10.83}$$

an expression which has already appeared in the particular case of the Lagrange equations (10.64).

For the covariant components $A_\mu(\lambda)$ it holds that

$$\frac{\delta A_\mu(\lambda)}{\delta\lambda} = A_{\mu;\nu}\frac{dx^\nu}{d\lambda}. \tag{10.84}$$

If $\delta A^\mu(\lambda)/\delta\lambda$ vanishes we have that

$$\frac{dA^\mu(\lambda)}{d\lambda} + \Gamma^\mu_{\sigma\nu}\frac{dx^\sigma}{d\lambda}A^\nu(\lambda) = 0 \tag{10.85}$$

and we say that the four-vector $A^\mu(\lambda)$ suffers a parallel displacement along the curve $x^\mu = x^\mu(\lambda)$. This is in agreement with the example we gave before: the Lagrange equation when the generalized force vanishes tells us that the velocity vector suffers a parallel displacement along the trajectory, i.e., it is a constant vector (uniform motion). The general statement is clearly justified by the fact that if (10.85) holds, and we use pseudo-Cartesian coordinates ($\Gamma^\mu_{\sigma\nu} = 0$), then the components A^μ are constant and hence the corresponding vector is parallel (in the usual sense) to itself in the various points along the curve.

Problems

10.4.1 Obtain (10.57) from (10.55) and (10.56).

10.4.2 Show that the Lagrange equation (10.59) can be written as (10.63) by performing the time derivative of (10.61) and using (10.62).

10.4.3 Show that the covariant derivatives $T_{\mu\nu;\sigma}$ and $T^\mu_{;\sigma}$ are given by (10.66).

10.4.4 Show that the relationship $\Gamma^{\sigma}_{\mu\nu}T^{\mu\nu} = 0$ holds for any antisymmetric tensor $T^{\mu\nu} = -T^{\nu\mu}$.

10.4.5 Show that a) for any symmetric tensor $T^{\mu\nu} = T^{\nu\mu}$ (10.75) holds and b) for any antisymmetric tensor $T^{\mu\nu} = -T^{\nu\mu}$ (10.77) holds.

10.4.6 Obtain the transformation law (10.81) of the absolute derivative of $A^{\mu}(\lambda)$.

10.5 Spatial metric tensor

Let e^{μ} be a unit four-vector directed along the time coordinate with contravariant and covariant components given by

$$(e^{\mu}) = \left(\frac{1}{\sqrt{g_{00}}}, 0, 0, 0\right), \qquad e_{\mu} = g_{\mu\nu}e^{\nu} = \frac{g_{\mu 0}}{\sqrt{g_{00}}}, \tag{10.86}$$

and such that $e^{\mu}e_{\mu} = 1$.

We use the definition of e^{μ} and introduce the projection of an arbitrary four-vector A^{μ} onto the time coordinate

$$A^{\mu}_{\parallel} = (A_{\nu}e^{\nu})e^{\mu}, \qquad \text{with} \qquad A^{0}_{\parallel} = \frac{A_0}{g_{00}}, \qquad A^{i}_{\parallel} = 0. \tag{10.87}$$

Further we define the projection $A^{\mu}_{\perp}$ onto an orthogonal plane to the time coordinate through the decomposition

$$A^{\mu} = A^{\mu}_{\parallel} + A^{\mu}_{\perp}. \tag{10.88}$$

$A^{\mu}_{\perp}$ is orthogonal to the unit four-vector, i.e., $A^{\mu}_{\perp}e_{\mu} = 0$ and its components are

$$A^{0}_{\perp} = A^{0} - \frac{A_0}{g_{00}} = -\frac{g_{0i}}{g_{00}}A^{i}, \qquad A^{i}_{\perp} = A^{i}; \tag{10.89}$$

that is the spatial coordinates of $A^{\mu}_{\perp}$ are the spatial coordinates of A^{μ}.

In order to determine the spatial metric tensor we calculate

$$\begin{aligned} g_{\mu\nu}A^{\mu}_{\perp}A^{\nu}_{\perp} &\overset{(10.88)}{=} g_{\mu\nu}(A^{\mu} - A^{\mu}_{\parallel})(A^{\nu} - A^{\nu}_{\parallel}) \\ &\overset{(10.87)_1}{=} g_{\mu\nu}\left[A^{\mu}A^{\nu} - (A^{\sigma}e_{\sigma})(A^{\tau}e_{\tau})e^{\mu}e^{\nu}\right] \\ &\overset{(10.86)_2}{=} \left(g_{\mu\nu} - \frac{g_{\mu 0}g_{\nu 0}}{g_{00}}\right)A^{\mu}A^{\nu} = \left(g_{ij} - \frac{g_{i0}g_{j0}}{g_{00}}\right)A^{i}A^{j}. \end{aligned} \tag{10.90}$$

The right-hand side of (10.90) is equal to minus the square of the modulus of the spatial components of A^{μ}. Hence we have

$$|\mathbf{A}|^2 = \gamma^{*}_{ij}A^{i}A^{j}, \qquad \gamma^{*}_{ij} = -g_{ij} + \frac{g_{i0}g_{j0}}{g_{00}}, \tag{10.91}$$

where γ^*_{ij} is the spatial metric tensor. We note that in a Minkowski space γ^*_{ij} reduces to $-\eta_{ij}$.

Let us relate the determinant of the spatial metric tensor γ^* with the determinant of the metric tensor g. First we write the determinant of the spatial metric tensor as

$$\gamma^* = \begin{vmatrix} \gamma^*_{11} & \gamma^*_{12} & \gamma^*_{13} \\ \gamma^*_{21} & \gamma^*_{22} & \gamma^*_{23} \\ \gamma^*_{31} & \gamma^*_{32} & \gamma^*_{33} \end{vmatrix} = -\frac{1}{g_{00}} \begin{vmatrix} -g_{00} & 0 & 0 & 0 \\ -g_{10} & \gamma^*_{11} & \gamma^*_{12} & \gamma^*_{13} \\ -g_{20} & \gamma^*_{21} & \gamma^*_{22} & \gamma^*_{23} \\ -g_{30} & \gamma^*_{31} & \gamma^*_{32} & \gamma^*_{33} \end{vmatrix}. \tag{10.92}$$

By using the property that the determinant does not change when we add to a row a linear combination of the other rows, we can write (10.92) as

$$\gamma^* = -\frac{1}{g_{00}} \begin{vmatrix} -g_{00} & -\frac{g_{10}g_{00}}{g_{00}} & -\frac{g_{20}g_{00}}{g_{00}} & -\frac{g_{30}g_{00}}{g_{00}} \\ -g_{10} & \gamma^*_{11} - \frac{g_{10}g_{10}}{g_{00}} & \gamma^*_{12} - \frac{g_{10}g_{20}}{g_{00}} & \gamma^*_{13} - \frac{g_{10}g_{30}}{g_{00}} \\ -g_{20} & \gamma^*_{21} - \frac{g_{20}g_{10}}{g_{00}} & \gamma^*_{22} - \frac{g_{20}g_{20}}{g_{00}} & \gamma^*_{23} - \frac{g_{20}g_{30}}{g_{00}} \\ -g_{30} & \gamma^*_{31} - \frac{g_{30}g_{10}}{g_{00}} & \gamma^*_{32} - \frac{g_{30}g_{20}}{g_{00}} & \gamma^*_{33} - \frac{g_{30}g_{30}}{g_{00}} \end{vmatrix}$$

$$\overset{(10.91)_2}{=} \frac{(-1)^5}{g_{00}} \begin{vmatrix} g_{00} & g_{01} & g_{02} & g_{03} \\ g_{10} & g_{11} & g_{12} & g_{13} \\ g_{20} & g_{21} & g_{22} & g_{23} \\ g_{30} & g_{31} & g_{32} & g_{33} \end{vmatrix}. \tag{10.93}$$

Hence it follows from (10.93) with $-g = \det(g_{\mu\nu})$ that

$$\gamma^* g_{00} = g. \tag{10.94}$$

In the three-dimensional curvilinear coordinate system the determinant of the spatial metric tensor γ^* plays the same role as the determinant of the metric tensor g in a four-dimensional curvilinear coordinate system. For example, the element of spatial volume is $\sqrt{\gamma^*}dx^1dx^2dx^3$, the permutation symbol is $\sqrt{\gamma^*}\epsilon^{ijk}$ while the divergence of a spatial vector $\mathbf{A}$ is given by

$$\text{div}\mathbf{A} = \frac{1}{\sqrt{\gamma^*}} \frac{\partial\sqrt{\gamma^*}A^i}{\partial x^i}. \tag{10.95}$$

Problems

10.5.1 Show that the components of $A^\mu_\perp$ are given by (10.89).

10.5.2 Obtain the expression given in (10.90).

10.6 Special relativity in general coordinates

It is sometimes useful to use general coordinates in special relativity in order to understand certain phenomena in a clearer way. Let us, as an example, consider the equations of motion of a mass point

$$m\frac{dU^\alpha}{d\tau} = K^\alpha, \tag{10.96}$$

where K^α is the Minkowski force. In general coordinates we can replace $dU^\alpha/d\tau$ by the absolute derivative $\delta U^\alpha/\delta\tau$. Then we obtain

$$m\frac{\delta U^\alpha}{\delta\tau} = m\left(\frac{dU^\alpha}{d\tau} + \Gamma^\alpha_{\mu\lambda}U^\mu U^\lambda\right) = K^\alpha, \tag{10.97}$$

or

$$m\frac{d(g_{\beta\alpha}U^\alpha)}{d\tau} + m\left(g_{\beta\alpha}\Gamma^\alpha_{\mu\lambda} - \frac{\partial g_{\mu\beta}}{\partial x^\lambda}\right)U^\mu U^\lambda = K_\beta. \tag{10.98}$$

However from (10.50) we have that

$$g_{\beta\alpha}\Gamma^\alpha_{\mu\lambda} - \frac{\partial g_{\mu\beta}}{\partial x^\lambda} = \frac{1}{2}\left(-\frac{\partial g_{\mu\beta}}{\partial x^\lambda} + \frac{\partial g_{\beta\lambda}}{\partial x^\mu} - \frac{\partial g_{\mu\lambda}}{\partial x^\beta}\right) \tag{10.99}$$

and hence only one partial derivative survives in (10.98), which becomes:

$$m\frac{d(g_{\beta\alpha}U^\alpha)}{d\tau} - \frac{m}{2}\frac{\partial g_{\mu\lambda}}{\partial x^\beta}U^\mu U^\lambda = K_\beta, \tag{10.100}$$

or also

$$m\left(g_{\beta\alpha}\frac{dU^\alpha}{d\tau}\right) + m\left(\frac{\partial g_{\beta\lambda}}{\partial x^\mu} - \frac{1}{2}\frac{\partial g_{\mu\lambda}}{\partial x^\beta}\right)U^\mu U^\lambda = K_\beta. \tag{10.101}$$

This rearrangement is useful because (10.100) and (10.101) do not require the use of the contravariant components of the metric tensor and hence are more handy when performing explicit calculations.

When using general coordinates in special relativity, one must be aware that the coordinates can have a well-defined mathematical meaning, but also that they can become singular at some points of space-time and hence give meaningless results beyond these points.

Let us start by an easy example. We keep the usual time coordinate but use cylindrical coordinates for ordinary space. Then we have

$$ds^2 = (cd\tau)^2 = (cdt)^2 - [(d\rho)^2 + \rho^2(d\theta)^2 + (dz)^2]. \tag{10.102}$$

These coordinates do not produce particular problems, except the usual singularity at $\rho = 0$. Let us now transform to other space-time coordinates, by letting $\theta' = \theta - \omega t$, with ω constant. We get

$$ds^2 = (cdt)^2\left(1 - \frac{\rho^2\omega^2}{c^2}\right) - 2\frac{\rho^2\omega}{c}d\theta'(cdt) - [(d\rho)^2 + \rho^2(d\theta')^2 + (dz)^2]. \tag{10.103}$$

Then we let $\theta' = \arctan(y/x)$, $\rho = \sqrt{x^2+y^2}$, $z = z$ and obtain

$$ds^2 = (cdt)^2\left[1 - \frac{\omega^2(x^2+y^2)}{c^2}\right] + 2\frac{\omega y}{c}dx(cdt) - 2\frac{\omega x}{c}dy(cdt)$$

$$-[(dx)^2 + (dy)^2 + (dz)^2)], \tag{10.104}$$

i.e., the covariant components of the metric tensor read

$$g_{00} = \left[1 - \frac{\omega^2(x^2+y^2)}{c^2}\right], \qquad g_{01} = 2\frac{\omega y}{c}, \tag{10.105}$$

$$g_{02} = -2\frac{\omega x}{c}, \qquad g_{03} = 0, \qquad g_{ij} = -\delta_{ij}, \tag{10.106}$$

while its contravariant components are given by

$$g^{00} = \frac{1}{1+3\omega^2(x^2+y^2)/c^2}, \quad g^{01} = \frac{2\omega y/c}{1+3\omega^2(x^2+y^2)/c^2}, \tag{10.107}$$

$$g^{02} = -\frac{2\omega x/c}{1+3\omega^2(x^2+y^2)/c^2}, \quad g^{11} = -\frac{1+\omega^2(3x^2-y^2)/c^2}{1+3\omega^2(x^2+y^2)/c^2}, \tag{10.108}$$

$$g^{12} = -\frac{4\omega^2 xy/c^2}{1+3\omega^2(x^2+y^2)/c^2}, \quad g^{22} = -\frac{1+\omega^2(3y^2-x^2)/c^2}{1+3\omega^2(x^2+y^2)/c^2}, \tag{10.109}$$

$$g^{03} = g^{13} = g^{23} = 0, \qquad g^{33} = -1. \tag{10.110}$$

The advantage of using these coordinates is that they differ from the usual Minkowski coordinates by a transformation which looks like a solid body rotation about the z-axis. It can be shown, however, that solid body rotations are meaningless in special relativity. This is revealed by the expression (10.104), in which the component g_{00} vanishes when $\rho\omega = c$. This is not surprising; if we could have a solid body rotation, we could pass the speed of light by going sufficiently far from the rotation axis. The singularity appears exactly at the point where the speed of light equals the hypothetical speed of a point of the rotating body. Thus we cannot attach a direct physical meaning to the results we obtain by using these coordinates (i.e., we cannot interpret them as lengths or times) unless we go to the classical limit.

We see that using non-Minkowskian coordinates produces an apparent force

$$\mathcal{K}_\beta = -m\left(\frac{\partial g_{\beta\lambda}}{\partial x^\mu} - \frac{1}{2}\frac{\partial g_{\mu\lambda}}{\partial x^\beta}\right)U^\mu U^\lambda. \tag{10.111}$$

Let us compute the spatial components of this force in the case of the above coordinates:

$$\mathcal{K}_i = \frac{m}{2}\frac{\partial g_{00}}{\partial x^i}U^0U^0 + m\frac{\partial g_{j0}}{\partial x^i}U^jU^0 - m\frac{\partial g_{i0}}{\partial x^j}U^jU^0. \tag{10.112}$$

Hence $\mathcal{K}_3 = 0$ and

$$\mathcal{K}_1 = -m\frac{\omega^2 x}{c^2}U^0U^0 - 4m\frac{\omega}{c}U^2U^0, \tag{10.113}$$

$$\mathcal{K}_2 = -m\frac{\omega^2 y}{c^2}U^0U^0 + 4m\frac{\omega}{c}U^1U^0. \tag{10.114}$$

Since for low speeds $(U^\mu) \approx (c, v_x, v_y, v_z)$, when we pass to the contravariant components - more appropriate to deal with velocity and forces - we have (apart from terms negligible for low rotation speeds where $w/c \ll 1$ and $g^{\mu\nu} \to \eta^{\mu\nu}$)

$$\mathcal{K}^1 = m\omega^2 x + 4m\omega v_y, \tag{10.115}$$

$$\mathcal{K}^2 = m\omega^2 y - 4m\omega v_x, \tag{10.116}$$

i.e., we recover the centrifugal and Coriolis forces.

Problems

10.6.1 Obtain (10.101) from (10.97).

10.6.2 Show that ds^2 in cylindrical coordinates is given by (10.102).

10.6.3 By using the transformations of coordinates $\theta' = \theta - \omega t$ in (10.102) and afterwards $\theta' = \arctan(y/x)$, $\rho = \sqrt{x^2 + y^2}$, $z = z$, obtain (10.104).

10.6.4 Obtain the contravariant components of the metric tensor given in (10.107) through (10.110).

10.6.5 Show that a) the covariant space components of the apparent force are given by (10.113) and (10.114) and b) the contravariant components of this force reduce to (10.115) and (10.116) for low speeds.

References

[1] P. A. M. Dirac, *General theory of relativity* (Princeton University Press, Princeton, 1996).

[2] T. Fliessbach, *Allgemeine Relativitätstheorie* (B. I. Wissenschaftsverlag, Mannheim, 1990).

[3] V. Fock, *The theory of space time and gravitation* (Pergamon Press, London, 1959).

[4] L. D. Landau and E. M. Lifshitz, *The classical theory of fields*, 4th ed. (Pergamon Press, Oxford, 1980).

[5] A. Lichnerowicz, *Théories relativistes de la gravitation et de l'électromagnétisme* (Masson, Paris, 1955).

[6] C. Møller, *The theory of relativity,* (Oxford University Press, London, 1952).

[7] W. Pauli, *Theory of relativity* (Dover, New York, 1981).

[8] E. Schrödinger, *Space-time structure* (Cambridge Univiversity Press, Cambridge, 1950).

[9] R. M. Wald, *General relativity* (The University of Chicago Press, Chicago, 1984).

[10] S. Weinberg, *Gravitation and cosmology. Principles and applications of the theory of relativity* (Wiley, New York, 1972).

Chapter 11

Riemann Spaces and General Relativity

11.1 Introduction

In this chapter we shall introduce the mathematical tools that are needed for the development of the theory of general relativity. They generalize to non-Euclidean spaces the tools that we developed in the last chapter for the tensor calculus in general coordinates. We begin with the characterization of a Riemannian space and the introduction of the Riemann–Christoffel curvature tensor, the Ricci tensor, the curvature scalar and the Bianchi identities. After stating the physical principles of the general relativity we analyze mechanics, electrodynamics and fluid dynamics in the presence of gravitational fields. We derive Einstein's field equations and find the solution corresponding to a weak gravitational field and the Schwarzschild solution. We present also the solutions of Einstein's field equations that correspond to the evolution of the cosmic scale factor in a universe described by the Robertson–Walker metric in the radiation and in the matter dominated periods.

11.2 Tensors in Riemannian spaces

A Riemannian space is a manifold where the line element in terms of general coordinates x^μ is given by

$$ds^2 = g_{\mu\nu} dx^\mu dx^\nu, \tag{11.1}$$

where the coefficients $g_{\mu\nu}$, assigned functions of the coordinates, are the components of the metric tensor and the concept of parallel displacement is given through the Christoffel symbols

$$\Gamma^\kappa_{\mu\lambda} = \frac{1}{2} g^{\kappa\nu} \left(\frac{\partial g_{\mu\nu}}{\partial x^\lambda} + \frac{\partial g_{\lambda\nu}}{\partial x^\mu} - \frac{\partial g_{\mu\lambda}}{\partial x^\nu} \right), \tag{11.2}$$

exactly in the same way as in a Euclidean (or Minkowski) space when we use general coordinates. Thus, in a sense, these spaces are particular cases of a Riemann space. What characterizes these special cases is the fact that we can find a coordinate system such that the metric tensor is constant.

We remark that the name "Riemann space" (or "Riemann manifold") is frequently restricted to the case of a positive definite metric; otherwise the space is called pseudo-Riemann space. When the signature of the metric is the same as a Minkowski space, the term "Lorentzian space" is also used. Here we shall not make these fine distinctions, since no confusion is possible.

We can introduce the absolute differentiation in a Riemann space exactly as we did when dealing with tensor calculus in general coordinates. Nothing new occurs if we restrict ourselves to first order derivatives. In the same way, nothing unusual is found if we do not displace vectors beyond a first order neighborhood, where the square of the displacement can be neglected. In other words, we cannot discover any difference between a Riemann and a Euclidean space if we consider the values of $g_{\mu\nu}$ and $\Gamma^{\kappa}_{\mu\lambda}$ at a point. Here and henceforth, when it is not necessary to emphasize the difference, we shall include the case of the Minkowski space when we say "Euclidean space".

Yet, it is easy to give examples of Riemannian spaces where the Euclidean geometry does not hold in the special case of two dimensions. The surface of an ordinary sphere is a two-dimensional Riemann space with the line element

$$ds^2 = a^2(d\theta)^2 + a^2\sin^2\theta(d\phi)^2, \tag{11.3}$$

where a is the sphere radius. It is well known that Euclidean geometry does not hold on the surface of the sphere. The equivalent of the straight lines are here the largest circles, and it is possible to construct a triangle with two and even three right angles.

We must distinguish local and global violations of Euclidean geometry: all the theorems of this geometry hold on the surface of an infinite cylinder, provided we do not consider too big objects: the only axiom which is violated is the fact that the straight line has an infinite length, but many consequences of this axiom remain true in a sufficiently small neighborhood. Parallel transport along a closed path, e.g., always brings the vector to its initial value. A subtler case is the surface of the cone, even if we stay away from the vertex, which is a singular point: parallel transport of a vector may not bring the vector to its initial value. This is almost always the case for parallel transport on the surface of the sphere.

In order to investigate the origin of these different behaviors, we must clearly investigate the derivatives of $\Gamma^{\kappa}_{\mu\lambda}$.

11.3 Curvature tensor

We start from the law of transformation of the affine connection established in Chapter 10 (see (10.56)):

$$\frac{\partial^2 x'^{\mu}}{\partial x^{\sigma}\partial x^{\nu}} = \Gamma^{\tau}_{\sigma\nu}\frac{\partial x'^{\mu}}{\partial x^{\tau}} - \Gamma'^{\mu}_{\tau\eta}\frac{\partial x'^{\tau}}{\partial x^{\sigma}}\frac{\partial x'^{\eta}}{\partial x^{\nu}}. \tag{11.4}$$

In order to learn something about the second derivatives of the affine connection, we differentiate (11.4) with respect to x^λ, yielding

$$\frac{\partial^3 x'^\mu}{\partial x^\lambda \partial x^\sigma \partial x^\nu} = \frac{\partial \Gamma^\tau_{\sigma\nu}}{\partial x^\lambda}\frac{\partial x'^\mu}{\partial x^\tau} + \Gamma^\tau_{\sigma\nu}\frac{\partial^2 x'^\mu}{\partial x^\lambda \partial x^\tau} - \frac{\partial \Gamma'^\mu_{\tau\eta}}{\partial x'^\rho}\frac{\partial x'^\rho}{\partial x^\lambda}\frac{\partial x'^\tau}{\partial x^\sigma}\frac{\partial x'^\eta}{\partial x^\nu}$$

$$-\Gamma'^\mu_{\tau\eta}\frac{\partial^2 x'^\tau}{\partial x^\lambda \partial x^\sigma}\frac{\partial x'^\eta}{\partial x^\nu} - \Gamma'^\mu_{\tau\eta}\frac{\partial x'^\tau}{\partial x^\sigma}\frac{\partial^2 x'^\eta}{\partial x^\lambda \partial x^\nu}. \tag{11.5}$$

If we subtract (11.5) from the equation obtained by interchanging the indices λ and ν in (11.5), we get

$$0 = \left(\frac{\partial \Gamma^\tau_{\sigma\nu}}{\partial x^\lambda} - \frac{\partial \Gamma^\tau_{\sigma\lambda}}{\partial x^\nu}\right)\frac{\partial x'^\mu}{\partial x^\tau} + \Gamma^\tau_{\sigma\nu}\frac{\partial^2 x'^\mu}{\partial x^\lambda \partial x^\tau} - \Gamma^\tau_{\sigma\lambda}\frac{\partial^2 x'^\mu}{\partial x^\nu \partial x^\tau}$$

$$-\frac{\partial \Gamma'^\mu_{\tau\eta}}{\partial x'^\rho}\left(\frac{\partial x'^\rho}{\partial x^\lambda}\frac{\partial x'^\eta}{\partial x^\nu} - \frac{\partial x'^\varrho}{\partial x^\nu}\frac{\partial x'^\eta}{\partial x^\lambda}\right)\frac{\partial x'^\tau}{\partial x^\sigma}$$

$$-\Gamma'^\mu_{\tau\eta}\left(\frac{\partial^2 x'^\tau}{\partial x^\lambda \partial x^\sigma}\frac{\partial x'^\eta}{\partial x^\nu} - \frac{\partial^2 x'^\tau}{\partial x^\nu \partial x^\sigma}\frac{\partial x'^\eta}{\partial x^\lambda}\right). \tag{11.6}$$

Now by eliminating the second derivatives by the use of (11.4), it follows after some manipulations that

$$\frac{\partial \Gamma^\tau_{\sigma\nu}}{\partial x^\lambda} - \frac{\partial \Gamma^\tau_{\sigma\lambda}}{\partial x^\nu} + \Gamma^\epsilon_{\sigma\nu}\Gamma^\tau_{\lambda\epsilon} - \Gamma^\epsilon_{\sigma\lambda}\Gamma^\tau_{\nu\epsilon} = \frac{\partial x'^\rho}{\partial x^\lambda}\frac{\partial x'^\epsilon}{\partial x^\sigma}\frac{\partial x'^\eta}{\partial x^\nu}\frac{\partial x^\tau}{\partial x'^\mu}\left[\frac{\partial \Gamma'^\mu_{\epsilon\eta}}{\partial x'^\rho}\right.$$

$$\left.-\frac{\partial \Gamma'^\mu_{\epsilon\rho}}{\partial x'^\eta} + \Gamma'^\theta_{\epsilon\eta}\Gamma'^\mu_{\rho\theta} - \Gamma'^\theta_{\epsilon\rho}\Gamma'^\mu_{\eta\theta}\right]. \tag{11.7}$$

We conclude from (11.7) that

$$R^\tau{}_{\sigma\nu\lambda} = \frac{\partial \Gamma^\tau_{\sigma\nu}}{\partial x^\lambda} - \frac{\partial \Gamma^\tau_{\sigma\lambda}}{\partial x^\nu} + \Gamma^\epsilon_{\sigma\nu}\Gamma^\tau_{\lambda\epsilon} - \Gamma^\epsilon_{\sigma\lambda}\Gamma^\tau_{\nu\epsilon} \tag{11.8}$$

is a mixed tensor. This quantity is called the Riemann–Christoffel curvature tensor[1]. The contracted forms

$$R_{\mu\nu} = R^\tau{}_{\mu\tau\nu}, \qquad \text{and} \qquad R = g^{\mu\nu}R_{\mu\nu} \tag{11.9}$$

are called Ricci tensor and curvature scalar, respectively.

[1]The convention adopted here for the Riemann–Christoffel curvature tensor in connection of the signature of the metric tensor is the same (among others) as the books by Adler et al [1] and Fock [6], but it differs from the books by Landau and Lifshitz [8], Misner et al [10] and Weinberg [16]. For more details on this subject one is referred to the inner cover of the book by Misner et al [10].

In order to see better the physical meaning of the Riemann tensor, we consider the change undergone by a vector when it is parallel transported along a curve $\mathcal{C}$:

$$\Delta A^{\mu} = \int_{\mathcal{C}} \delta A^{\mu}. \tag{11.10}$$

The change δA^{μ} is given by (10.85) and hence

$$\Delta A^{\mu} = -\int_{\mathcal{C}} \Gamma^{\mu}_{\alpha\beta} A^{\alpha} dx^{\beta}. \tag{11.11}$$

We remark that, in analogy with the proof of Stokes' theorem, we can stretch a surface $\mathcal{S}$ over the closed curve and divide it into small parallelograms, which eventually shrink around a point to become infinitesimal. It is sufficient to examine the case of an infinitesimal neighborhood of a point $\overline{P}$.

We also need to define A^{μ} for points which do not lie on the curve. We can assume that it is parallel transported there as well. The parallel transport is an operation which is defined explicitly only in a first order neighborhood; this is sufficient to conclude that the ordinary derivatives of A^{μ} at $\overline{P}$ are given by

$$\frac{\partial \overline{A}^{\alpha}}{\partial \overline{x}^{\sigma}} = -\overline{\Gamma}^{\alpha}_{\sigma\beta} \overline{A}^{\beta}. \tag{11.12}$$

Then if $\overline{P}$ has coordinates $\overline{x}^{\beta}$, any other point in the neighborhood will have coordinates $x^{\beta} = \overline{x}^{\beta} + y^{\beta}$, where y^{β} can be treated as infinitesimal, i.e., we can neglect terms of higher order. We then have

$$\begin{aligned} \Gamma^{\mu}_{\alpha\beta} A^{\alpha} &= \overline{\Gamma}^{\mu}_{\alpha\beta} \overline{A}^{\alpha} + \frac{\partial \overline{\Gamma}^{\mu}_{\alpha\beta}}{\partial \overline{x}^{\sigma}} \overline{A}^{\alpha} y^{\sigma} + \overline{\Gamma}^{\mu}_{\alpha\beta} \frac{\partial \overline{A}^{\alpha}}{\partial \overline{x}^{\sigma}} y^{\sigma} + \mathcal{O}(y^{\mu} y^{\nu}) \\ &\overset{(11.12)}{=} \overline{\Gamma}^{\mu}_{\alpha\beta} \overline{A}^{\alpha} + \frac{\partial \overline{\Gamma}^{\mu}_{\alpha\beta}}{\partial \overline{x}^{\sigma}} \overline{A}^{\alpha} y^{\sigma} - \overline{\Gamma}^{\mu}_{\alpha\beta} \overline{\Gamma}^{\alpha}_{\sigma\nu} \overline{A}^{\nu} y^{\sigma} + \mathcal{O}(y^{\mu} y^{\nu}), \end{aligned} \tag{11.13}$$

where $\mathcal{O}(y^{\mu} y^{\nu})$ denotes the terms of second and higher orders in y^{μ}.

The term not containing y^{σ} integrates to zero because

$$\int_{\mathcal{C}} dx^{\beta} = 0. \tag{11.14}$$

Thus we are left with just the contributions which are linear in y^{σ}; since they are integrated along an infinitesimal curve, we find that the total change is of second order. Hence (11.11), thanks to (11.13) and (11.14), reads

$$\Delta A^{\mu} = -\left[\frac{\partial \overline{\Gamma}^{\mu}_{\nu\beta}}{\partial \overline{x}^{\sigma}} - \overline{\Gamma}^{\mu}_{\alpha\beta} \overline{\Gamma}^{\alpha}_{\sigma\nu} \right] \overline{A}^{\nu} \int_{\mathcal{C}} y^{\sigma} dy^{\beta} + \mathcal{O}(y^{\mu} y^{\nu}). \tag{11.15}$$

We have also that

$$\int_{\mathcal{C}} y^{\sigma} dy^{\beta} = \frac{1}{2} \int_{\mathcal{C}} (y^{\sigma} dy^{\beta} - y^{\beta} dy^{\sigma}) = \mathcal{A}^{\beta\sigma}. \tag{11.16}$$

Here $\mathcal{A}^{\beta\sigma}$ is the area of the projection of the surface $\mathcal{S}$ stretched over $\mathcal{C}$ onto the "plane" described by the coordinates y^β and y^σ; we can talk of a plane because we are in an infinitesimal neighborhood of a point. The expression shown above indicates that $\mathcal{A}^{\beta\sigma}$ is an antisymmetric tensor of second order (the coordinate planes are oriented and an exchange of coordinates changes the orientation of the curve and hence the sign of the area). Hence the meaningful part of the factor multiplying it is obtained by making it antisymmetric with respect to the indices β and σ. Thus the final form of (11.15) is

$$\Delta A^\mu = -\left[\frac{\partial \overline{\Gamma}^\mu_{\nu\beta}}{\partial \overline{x}^\sigma} - \frac{\partial \overline{\Gamma}^\mu_{\nu\sigma}}{\partial \overline{x}^\beta} + \overline{\Gamma}^\mu_{\alpha\sigma}\overline{\Gamma}^\alpha_{\beta\nu} - \overline{\Gamma}^\mu_{\alpha\beta}\overline{\Gamma}^\alpha_{\sigma\nu}\right]\overline{A}^\nu \mathcal{A}^{\beta\sigma} + \mathcal{O}(y^\mu y^\nu), \qquad (11.17)$$

or according to (11.8) we have

$$\Delta A^\mu = -\overline{R}^\mu{}_{\nu\beta\sigma}\overline{A}^\nu \mathcal{A}^{\beta\sigma} + \mathcal{O}(y^\mu y^\nu) \qquad (11.18)$$

where $\overline{R}^\mu{}_{\nu\beta\sigma}$ is the Riemann tensor evaluated at $\overline{P}$.

This formula shows that even if the Riemann tensor vanishes in a finite neighborhood of $\mathcal{C}$, the parallel transport along this closed curve may produce a different vector, if the Riemann tensor does not vanish everywhere on the surface $\mathcal{S}$. We remark that there is not an analogue formula for a finite area because there is not a simple way of dragging along the vector by parallelism inside the curve at a finite distance. We remark however that if in the region between two closed curves $\mathcal{C}_1$ and $\mathcal{C}_2$ on some surface $\mathcal{S}$ of our manifold the Riemann tensor vanishes, then because of the formula just proved, we have

$$\Delta_{\mathcal{C}_1} A^\mu = \Delta_{\mathcal{C}_2} A^\mu, \qquad (11.19)$$

where $\Delta_{\mathcal{C}_i} A^\mu$ $(i = 1, 2)$ denotes the change of the vector when displaced along the curves $\mathcal{C}_i$ $(i = 1, 2)$. This formula explains why the transport of a vector around a curve on a conical surface produces a different vector if the curve encircles the vertex. The latter is a singular point, but we might smooth it to something similar to the surface of a hemisphere, without changing the rest of the surface and hence the transport along a curve far away from the vertex. The curvature in the part which replaces the vertex influences, so to speak, the transport far away. This is the geometric analogue of the Bohm–Aharonov effect in quantum mechanics where a magnetic field produces effects at points where it vanishes. There is no magic in this; indeed, the fact that the Riemann tensor (the magnetic field, respectively) vanishes, implies that we can locally take a system of coordinates in which the Christoffel symbols (the vector potential, respectively) vanish, but this choice cannot be made global along curves surrounding regions of non-vanishing curvature (of non-vanishing magnetic field, respectively).

We can express the Riemann–Christoffel curvature tensor in terms of derivatives of the metric tensor. Let us proceed: first we write from (11.2)

$$g_{\kappa\sigma}\Gamma^\kappa_{\mu\lambda} = \frac{1}{2}\left(\frac{\partial g_{\mu\sigma}}{\partial x^\lambda} + \frac{\partial g_{\lambda\sigma}}{\partial x^\mu} - \frac{\partial g_{\mu\lambda}}{\partial x^\sigma}\right). \qquad (11.20)$$

If we add (11.20) to the equation that follows from (11.20) by interchanging μ and σ, we get

$$\frac{\partial g_{\mu\sigma}}{\partial x^\lambda} = g_{\kappa\sigma}\Gamma^\kappa_{\mu\lambda} + g_{\kappa\mu}\Gamma^\kappa_{\sigma\lambda}. \tag{11.21}$$

Further by differentiating the identity $\delta^\nu_\mu = g^{\nu\sigma} g_{\mu\sigma}$ with respect to x^λ we find that

$$g_{\mu\sigma}\frac{\partial g^{\nu\sigma}}{\partial x^\lambda} = -g^{\nu\sigma}\frac{\partial g_{\mu\sigma}}{\partial x^\lambda} \overset{(11.21)}{=} -g^{\nu\sigma}(g_{\kappa\sigma}\Gamma^\kappa_{\mu\lambda} + g_{\kappa\mu}\Gamma^\kappa_{\sigma\lambda}). \tag{11.22}$$

Next we obtain from (11.8) by the use of (11.2) and (11.20) that the covariant components of the Riemann–Christoffel curvature tensor can be written, by performing some manipulations, as

$$R_{\mu\sigma\nu\lambda} = g_{\tau\mu}R^\tau{}_{\sigma\nu\lambda} = \frac{1}{2}\left[\frac{\partial^2 g_{\mu\nu}}{\partial x^\lambda \partial x^\sigma} - \frac{\partial^2 g_{\sigma\nu}}{\partial x^\lambda \partial x^\mu} - \frac{\partial^2 g_{\mu\lambda}}{\partial x^\nu \partial x^\sigma} + \frac{\partial^2 g_{\sigma\lambda}}{\partial x^\nu \partial x^\mu}\right]$$

$$+g_{\tau\mu}\frac{\partial g^{\tau\eta}}{\partial x^\lambda} g_{\kappa\eta}\Gamma^\kappa_{\sigma\nu} - g_{\tau\mu}\frac{\partial g^{\tau\eta}}{\partial x^\nu} g_{\kappa\eta}\Gamma^\kappa_{\lambda\sigma} + g_{\tau\mu}\left(\Gamma^\epsilon_{\sigma\nu}\Gamma^\tau_{\lambda\epsilon} - \Gamma^\epsilon_{\sigma\lambda}\Gamma^\tau_{\nu\epsilon}\right). \tag{11.23}$$

Finally by using (11.22) the above equation reduces to

$$R_{\mu\sigma\nu\lambda} = \frac{1}{2}\left[\frac{\partial^2 g_{\mu\nu}}{\partial x^\lambda \partial x^\sigma} - \frac{\partial^2 g_{\sigma\nu}}{\partial x^\lambda \partial x^\mu} - \frac{\partial^2 g_{\mu\lambda}}{\partial x^\nu \partial x^\sigma} + \frac{\partial^2 g_{\sigma\lambda}}{\partial x^\nu \partial x^\mu}\right]$$

$$+g_{\kappa\eta}\left(\Gamma^\kappa_{\mu\nu}\Gamma^\eta_{\sigma\lambda} - \Gamma^\kappa_{\mu\lambda}\Gamma^\eta_{\sigma\nu}\right), \tag{11.24}$$

which is the final expression for the covariant components of the Riemann–Christoffel curvature tensor in terms of the derivatives of the metric tensor.

The following properties of $R_{\mu\sigma\nu\lambda}$ can be obtained from (11.24):

$$\begin{cases} \text{1. Symmetry: } R_{\mu\sigma\nu\lambda} = R_{\nu\lambda\mu\sigma}, \\ \text{2. Cyclicity: } R_{\mu\sigma\nu\lambda} + R_{\mu\nu\lambda\sigma} + R_{\mu\lambda\sigma\nu} = 0, \\ \text{3. Antisymmetry: } R_{\mu\sigma\nu\lambda} = -R_{\sigma\mu\nu\lambda} = -R_{\mu\sigma\lambda\nu} = R_{\sigma\mu\lambda\nu}. \end{cases} \tag{11.25}$$

Let now A_μ be a covariant four-vector; then the covariant derivative of $A_{\mu;\nu}$ is given by

$$A_{\mu;\nu;\sigma} \overset{(10.66)_1}{=} \frac{\partial A_{\mu;\nu}}{\partial x^\sigma} - \Gamma^\tau_{\nu\sigma}A_{\mu;\tau} - \Gamma^\tau_{\mu\sigma}A_{\tau;\nu} \overset{(10.65)}{=} \frac{\partial^2 A_\mu}{\partial x^\nu \partial x^\sigma} - \frac{\partial \Gamma^\tau_{\mu\nu}}{\partial x^\sigma}A_\tau$$

$$-\Gamma^\tau_{\mu\nu}\frac{\partial A_\tau}{\partial x^\sigma} - \Gamma^\tau_{\nu\sigma}\left(\frac{\partial A_\mu}{\partial x^\tau} - \Gamma^\epsilon_{\mu\tau}A_\epsilon\right) - \Gamma^\tau_{\mu\sigma}\left(\frac{\partial A_\tau}{\partial x^\nu} - \Gamma^\epsilon_{\tau\nu}A_\epsilon\right). \tag{11.26}$$

If we subtract (11.26) from the equation obtained from it by interchanging the indices ν and σ and use (11.8), we get

$$A_{\mu;\nu;\sigma} - A_{\mu;\sigma;\nu} = -R^\tau{}_{\mu\nu\sigma}A_\tau. \tag{11.27}$$

Further if $T_{\mu\nu}$ is a covariant tensor it holds that

$$T_{\mu\nu;\sigma;\tau} - T_{\mu\nu;\tau;\sigma} = -R^{\kappa}{}_{\mu\sigma\tau}T_{\kappa\nu} - R^{\kappa}{}_{\nu\sigma\tau}T_{\mu\kappa}. \tag{11.28}$$

We shall now derive the so-called Bianchi identities which are related with the covariant derivative of the Riemann–Christoffel curvature tensor. The covariant derivative of (11.27) with respect to κ yields

$$A_{\mu;\nu;\sigma;\kappa} - A_{\mu;\sigma;\nu;\kappa} = -R^{\tau}{}_{\mu\nu\sigma;\kappa}A_{\tau} - R^{\tau}{}_{\mu\nu\sigma}A_{\tau;\kappa}. \tag{11.29}$$

Now we get by addition of (11.29) with the two equations obtained by permuting the indices ν, σ, κ:

$$\begin{aligned}(A_{\mu;\nu;\sigma;\kappa} - A_{\mu;\nu;\kappa;\sigma}) + (A_{\mu;\sigma;\kappa;\nu} - A_{\mu;\sigma;\nu;\kappa}) + (A_{\mu;\kappa;\nu;\sigma} - A_{\mu;\kappa;\sigma;\nu})\\ + (R^{\tau}{}_{\mu\nu\sigma}A_{\tau;\kappa} + R^{\tau}{}_{\mu\sigma\kappa}A_{\tau;\nu} + R^{\tau}{}_{\mu\kappa\nu}A_{\tau;\sigma})\\ = -(R^{\tau}{}_{\mu\nu\sigma;\kappa} + R^{\tau}{}_{\mu\sigma\kappa;\nu} + R^{\tau}{}_{\mu\kappa\nu;\sigma})A_{\tau},\end{aligned} \tag{11.30}$$

where some rearrangements were made. On the other hand, one can obtain from (11.28) that

$$A_{\mu;\nu;\sigma;\kappa} - A_{\mu;\nu;\kappa;\sigma} = -R^{\tau}{}_{\mu\sigma\kappa}A_{\tau;\nu} - R^{\tau}{}_{\nu\sigma\kappa}A_{\mu;\tau}. \tag{11.31}$$

If we add (11.31) with the two equations that follow by permuting the indices ν, σ, κ and use the cyclic property $(11.25)_2$, we get that the left-hand side of (11.30) vanishes and it follows that

$$\begin{cases} R^{\tau}{}_{\mu\nu\sigma;\kappa} + R^{\tau}{}_{\mu\sigma\kappa;\nu} + R^{\tau}{}_{\mu\kappa\nu;\sigma} = 0, \quad \text{or} \\ R_{\tau\mu\nu\sigma;\kappa} + R_{\tau\mu\sigma\kappa;\nu} + R_{\tau\mu\kappa\nu;\sigma} = 0, \end{cases} \tag{11.32}$$

since A_{τ} is arbitrary. Equations (11.32) are the Bianchi identities. If we multiply $(11.32)_2$ with $g^{\tau\nu}$ and note that $g^{\tau\nu}_{;\kappa} = 0$, we get

$$R_{\mu\sigma;\kappa} + R^{\tau}{}_{\mu\sigma\kappa;\tau} - R_{\mu\kappa;\sigma} = 0. \tag{11.33}$$

Finally, multiplying (11.33) by $g^{\mu\sigma}$ leads to

$$R_{;\kappa} - 2R^{\mu}{}_{\kappa;\mu} = 0 \qquad \text{or} \qquad \left(R^{\mu\kappa} - \frac{1}{2}Rg^{\mu\kappa}\right)_{;\mu} = 0. \tag{11.34}$$

The above equation will be used to derive Einstein's field equations.

Problems

11.3.1 Through the elimination of the second derivatives of x'^{μ} from (11.6) by using (11.4), obtain (11.7).

11.3.2 Check that (11.16) holds and make the conventions on the sign of $\mathcal{A}^{\beta\sigma}$ explicit.

11.3.3 Obtain (11.24) from (11.8).

11.3.4 Show that the components of the Riemann–Christoffel curvature tensor has the properties stated in (11.25), i.e., symmetry, cyclicity and antisymmetry.

11.3.5 Show that the Riemann–Christoffel curvature tensor has only 20 linearly independent components. (Hint: Make use of the properties of symmetry, cyclicity and antisymmetry given in (11.25).)

11.3.6 By using the following relationship

$$T_{\mu\nu;\sigma;\tau} = \frac{\partial T_{\mu\nu;\sigma}}{\partial x^\tau} - \Gamma^\lambda_{\mu\tau} T_{\lambda\nu;\sigma} - \Gamma^\lambda_{\nu\tau} T_{\mu\lambda;\sigma} - \Gamma^\lambda_{\sigma\tau} T_{\mu\nu;\lambda} \, ,$$

check that (11.28) holds.

11.3.7 Obtain the Bianchi identities (11.32) by following the procedure described above.

11.3.8 Obtain (11.34) by contracting twice the Bianchi identities given in (11.32).

11.4 Physical principles of general relativity

The general theory of relativity was formulated by Einstein (see the collection of papers in [3]) in its final form in 1916. Based on the experiments of Eötvös on the equivalence of gravitational and inertial masses, Einstein proposed the principle of equivalence which was an extension of the postulate of relativity to non-inertial systems. This principle states that the physical laws should take the same form in a local system of reference in the presence of gravity as they do in an inertial system of reference in the absence of gravity. This is suggested by the fact that apparent forces, such as centrifugal or Coriolis force, can be described in special relativity, as we have seen in Chapter 10, by writing the equations in general coordinates. Since gravity is locally indistinguishable from an apparent force (and can be annihilated by balancing it with an apparent force as inside an orbiting satellite) because it is proportional to the inertial mass of each body, one can simply use general coordinates in a Riemannian space to obtain gravity as a consequence of geometry.

A variant version of the principle of equivalence is the principle of covariance which was also identified in the literature as the mathematical formulation of the principle of equivalence (see Pauli [12] pp. 149–150). The principle of covariance states that the physical laws are expressed by equations that have the same form in all systems of reference connected by arbitrary coordinate transformations. In other words we may say that the equations expressing physical laws are covariant with respect to arbitrary coordinate transformations. In the next section we shall use this principle to determine the equations of mechanics and electromagnetism in the presence of gravitational fields.

It is clear that one can always write any equation in tensor form with the help of some extra vector field, e.g., the four-velocity of a medium; this trick is excluded when we want to deal with the gravitational field itself. In this case, when we say that the equations must be covariant, the gravitational field must be described in a way that just involves the geometry of the manifold and not these extra fields; the latter may be introduced to describe the sources of the field in a phenomenological way, as we shall see later.

11.5 Mechanics in gravitational fields

11.5.1 Four-velocity

In order to determine the components of the four-velocity $U^\mu = dx^\mu/d\tau$, let us find the relationship between the elements of the proper time $d\tau$ and of the coordinate time $dt = dx^0/c$. For this purpose we write the interval between two events as

$$\begin{aligned} ds^2 = (cd\tau)^2 &= g_{00}(dx^0)^2 + 2g_{0i}dx^0dx^i + g_{ij}dx^idx^j \\ &\overset{(10.91)_2}{=} g_{00}(dx^0)^2 + 2g_{0i}dx^0dx^i - \gamma^*_{ij}dx^idx^j + \frac{g_{0i}g_{0j}}{g_{00}}dx^idx^j, \end{aligned} \tag{11.35}$$

by using the spatial metric tensor γ^*_{ij}. If we divide (11.35) by $(cdt)^2$ and introduce the velocity and the speed of the particle defined through

$$v^i = \frac{dx^i}{dt}, \qquad v = \sqrt{\gamma^*_{ij}\frac{dx^i}{dt}\frac{dx^j}{dt}}, \tag{11.36}$$

respectively, we get after some rearrangements

$$\Gamma \equiv \frac{dt}{d\tau} = \frac{1}{\sqrt{g_{00}\left(1 + \frac{g_{0i}}{g_{00}}\frac{v^i}{c}\right)^2 - \frac{v^2}{c^2}}}. \tag{11.37}$$

Note that in a Minkowski space, Γ reduces to γ which is defined by (1.35).

Now we use (11.37) and write the time and space contravariant components of the four-velocity as

$$U^0 = \frac{dx^0}{d\tau} = \Gamma c, \quad U^i = \frac{dx^i}{d\tau} = \frac{dx^i}{dt}\frac{dt}{d\tau} = \Gamma v^i, \quad \text{or} \quad (U^\mu) = (\Gamma c, \Gamma v^i). \tag{11.38}$$

The covariant components follows from

$$U_\mu = g_{\mu\nu}U^\nu = \Gamma c g_{\mu 0} + \Gamma v^i g_{\mu i}, \tag{11.39}$$

or by writing the time and spatial components explicitly:

$$(U_\mu) = (\Gamma c g_{00} + \Gamma v^i g_{0i}, \Gamma c g_{i0} + \Gamma v^j g_{ij}). \tag{11.40}$$

From (11.38) and (11.40) one can show that $U^\mu U_\mu = c^2$.

Problems

11.5.1.1 Show that the ratio $\Gamma = dt/d\tau$ is given by (11.37) and that Γ reduces to (1.35), by considering a Minkowski space.

11.5.1.2 By using the equations (11.38) and (11.40) show that the relationship $U^{\mu}U_{\mu} = c^2$ holds.

11.5.2 Equations of motion

The equation of motion of a free particle in special relativity is given by (see (1.100))

$$\frac{dU^{\alpha}}{d\tau} = 0 \qquad \text{or} \qquad \frac{dU^{\alpha}}{ds} = 0, \tag{11.41}$$

where U^{α} is the four-velocity and $d\tau = ds/c$ is the proper time. Equation $(11.41)_2$ represents the equation of a geodesic and according to the principle of covariance its form in an arbitrary coordinate system is given by

$$\frac{\delta U^{\mu}}{\delta s} = 0, \tag{11.42}$$

where $\delta U^{\mu}/\delta s$ is the absolute derivative (10.82) and U^{μ} suffers a parallel displacement along a curve with parameter s. With $U^{\mu} = dx^{\mu}/d\tau$ we get from (11.42) together with (10.82)

$$\frac{d^2x^{\mu}}{d\tau^2} + \Gamma^{\mu}_{\sigma\nu}\frac{dx^{\sigma}}{d\tau}\frac{dx^{\nu}}{d\tau} = 0. \tag{11.43}$$

In special relativity, this equation describes the motion of a free particle with respect to a non-inertial observer, as we have seen in Chapter 10, or an inertial observer using non-Cartesian space coordinates.

Thus we simply extend (11.43) to the case of a gravitational field; the only difference will be that we cannot get rid of the field globally (as for apparent forces in special relativity) but only locally. In other words, our space will be a Riemann space.

In the following we shall get from (11.43) the Newton equation of motion of a particle in the presence of a gravitational field. Let us proceed: we consider that the particle is moving with a small velocity in a stationary weak gravitational field so that we can write

$$v^i \ll c, \qquad \text{or} \qquad \frac{dx^i}{d\tau} \ll \frac{dx^0}{d\tau}. \tag{11.44}$$

In this approximation we have that

$$\Gamma^{\mu}_{\sigma\nu}\frac{dx^{\sigma}}{d\tau}\frac{dx^{\nu}}{d\tau} \approx \Gamma^{\mu}_{00}\left(\frac{dx^0}{d\tau}\right)^2 = c^2\Gamma^{\mu}_{00}\left(\frac{dt}{d\tau}\right)^2. \tag{11.45}$$

For stationary gravitational fields the time derivatives of $g^{\mu\nu}$ vanish and the affine connection (11.2) reduces to

$$\Gamma^{\mu}_{00} = -\frac{1}{2} g^{\mu\nu} \frac{\partial g_{00}}{\partial x^{\nu}}. \tag{11.46}$$

Further for weak gravitational fields we may approximate

$$g^{\mu\nu} \approx \eta^{\mu\nu} + h^{\mu\nu}, \qquad \text{with} \qquad |h^{\mu\nu}| \ll 1. \tag{11.47}$$

That is $g^{\mu\nu}$ is the metric of a Minkowski space $\eta^{\mu\nu}$ plus a small perturbation $h^{\mu\nu}$. If we insert (11.47) into the expression for Γ^{μ}_{00} and neglect all non-linear terms in $h^{\mu\nu}$ and its derivatives, it follows that

$$\Gamma^{\mu}_{00} = -\frac{1}{2}(\eta^{\mu\nu} + h^{\mu\nu}) \frac{\partial h_{00}}{\partial x^{\nu}} \approx -\frac{1}{2} \eta^{\mu\nu} \frac{\partial h_{00}}{\partial x^{\nu}}, \tag{11.48}$$

which reduces to

$$\Gamma^{0}_{00} = 0, \qquad \Gamma^{i}_{00} = \frac{1}{2} \vec{\nabla}^{i} h_{00}, \tag{11.49}$$

since the gravitational field is considered to be stationary. Hence we have from (11.43), (11.45) and (11.49)

$$\frac{d^2 t}{d\tau^2} = 0, \qquad \text{i.e.,} \qquad \frac{dt}{d\tau} = \text{constant}, \tag{11.50}$$

$$\frac{d^2 \mathbf{x}}{d\tau^2} = -\frac{c^2}{2} \vec{\nabla} h_{00} \left(\frac{dt}{d\tau} \right)^2, \qquad \text{or by (11.50)} \qquad \frac{d^2 \mathbf{x}}{dt^2} = -\frac{c^2}{2} \vec{\nabla} h_{00}. \tag{11.51}$$

On the other hand, the Newton equation of motion of a particle in the presence of a gravitational field is given by

$$\frac{d^2 \mathbf{x}}{dt^2} = -\vec{\nabla} \Phi, \qquad \text{with} \qquad \Phi(r) = -\frac{GM}{r}, \tag{11.52}$$

where $\Phi(r)$ is the gravitational potential at a distance r from a spherical distribution of mass M and $G = 6.664 \times 10^{-11}$ Nm^2/kg^2 is the usual gravitational constant. If we compare $(11.51)_2$ with $(11.52)_1$ we conclude that

$$h_{00} = \frac{2\Phi}{c^2} + \text{constant}, \qquad \text{i.e.,} \qquad h_{00} = \frac{2\Phi}{c^2}, \tag{11.53}$$

the constant being zero if we assume that for large distances from the spherical distribution of mass $g^{\mu\nu}$ must be reduced to the Minkowski metric $\eta^{\mu\nu}$. Hence we have that

$$g_{00} = 1 + \frac{2\Phi}{c^2}. \tag{11.54}$$

In Section 11.9 we shall discuss how to calculate the other components of the metric tensor $g^{\mu\nu}$ in a first approximation. Let us now estimate the correction $2\Phi/c^2$ at the surface of some bodies:

a) Earth: $M_\oplus \approx 5.97 \times 10^{24}$ kg; $R_\oplus \approx 6.38 \times 10^6$ m; $2|\Phi|/c^2 \approx 1.4 \times 10^{-9}$;
b) Sun: $M_\odot \approx 1.99 \times 10^{30}$ kg; $R_\odot \approx 6.96 \times 10^8$ m; $2|\Phi|/c^2 \approx 4.3 \times 10^{-6}$;
c) White dwarf: $M \approx 1.02 M_\odot$; $R \approx 5.4 \times 10^6$ m; $2|\Phi|/c^2 \approx 5.6 \times 10^{-4}$;
d) Neutron star: $M \approx M_\odot$; $R \approx 2 \times 10^4$ m; $2|\Phi|/c^2 \approx 1.5 \times 10^{-1}$;
e) Black hole: $M \approx 3M_\odot$; $R \approx 3 \times 10^3$ m; $2|\Phi|/c^2 \approx 2.97$.

From the above estimates we can conclude that for neutron stars and black holes it is not possible to use (11.54) since for those cases the approximation $(11.47)_2$ does not hold.

The equation of motion of a particle of rest mass m subjected to a force is given in special relativity by (1.100). According to the principle of covariance the equation of motion in general relativity will be

$$m\frac{\delta U^\mu}{\delta \tau} = K^\mu, \tag{11.55}$$

where K^μ is the Minkowski force. Another way to write (11.55) is

$$m\frac{d^2x^\mu}{d\tau^2} = K^\mu - \Gamma^\mu_{\sigma\nu}\frac{dx^\sigma}{d\tau}\frac{dx^\nu}{d\tau}, \tag{11.56}$$

where the last term on the right-hand side of (11.56), if we choose coordinates which reduce to those of a Lorentzian frame in the absence of a gravitational field, represents a gravitational force, in a way completely analogous to the apparent force that we discussed in Section 10.6 of previous chapter.

Problems

11.5.2.1 Check that for small velocities (11.45) holds.

11.5.2.2 Show that for weak gravitational fields where (11.47) holds the components of Γ^μ_{00} reduce to (11.49).

11.6 Electrodynamics in gravitational fields

In order to write the Maxwell equations in a covariant form we first note that the electromagnetic field tensor can be written in terms of the covariant derivative of the four-potential A_μ, since by (1.150) and (10.70) we have that

$$F_{\mu\nu} = \frac{\partial A_\mu}{\partial x^\nu} - \frac{\partial A_\nu}{\partial x^\mu} = A_{\mu;\nu} - A_{\nu;\mu}. \tag{11.57}$$

Further, due to the fact that $F_{\mu\nu}$ is an antisymmetric tensor, the relationship (10.77) must hold, that is

$$F_{\mu\nu;\sigma} + F_{\nu\sigma;\mu} + F_{\sigma\mu;\nu} = \frac{\partial F_{\mu\nu}}{\partial x^\sigma} + \frac{\partial F_{\nu\sigma}}{\partial x^\mu} + \frac{\partial F_{\sigma\mu}}{\partial x^\nu}, \tag{11.58}$$

which represents the pair of Maxwell equations $(1.146)_2$.

For the other pair of Maxwell equations we apply the covariant principle for (1.145) and write

$$F^{\mu\nu}{}_{;\nu} = -\mu_0 c J^\mu, \qquad \text{or by (10.72)} \qquad \frac{1}{\sqrt{g}}\frac{\partial \sqrt{g} F^{\mu\nu}}{\partial x^\nu} = -\mu_0 c J^\mu, \tag{11.59}$$

where J^μ is the current four-vector.

The energy-momentum tensor of the electromagnetic field (1.162) is represented as

$$T^{\mu\nu}_{\text{em}} = \epsilon_0 \left(F^\mu{}_\sigma F^{\sigma\nu} + \frac{1}{4} g^{\mu\nu} F^{\sigma\tau} F_{\sigma\tau} \right) \tag{11.60}$$

while its balance equation (1.161), obtained by the covariant principle, is

$$T^{\mu\nu}_{\text{em};\nu} = -\frac{1}{c} F^{\mu\nu} J_\nu, \qquad \text{or by (10.71)} \qquad \frac{1}{\sqrt{g}}\frac{\partial \sqrt{g} T^{\mu\nu}_{\text{em}}}{\partial x^\nu} = -\frac{1}{c} F^{\mu\nu} J_\nu - \Gamma^\mu_{\nu\lambda} T^{\lambda\nu}_{\text{em}}. \tag{11.61}$$

Further the equation of motion (1.154) of a particle with electric charge q in terms of the absolute derivative (10.82) is

$$\frac{\delta p^\mu}{\delta t} = \frac{dp^\mu}{d\tau} + \Gamma^\mu_{\nu\sigma} U^\nu p^\sigma = \frac{q}{c} F^{\mu\nu} U_\nu. \tag{11.62}$$

Let us now analyze the continuity equation for the charge density (1.142) which can be written according to the principle of covariance as

$$J^\mu{}_{;\mu} = 0 \qquad \text{or by (10.69)} \qquad \frac{1}{\sqrt{g}}\frac{\partial \sqrt{g} J^\mu}{\partial x^\mu} = 0. \tag{11.63}$$

The electric charge in a volume element $\sqrt{\gamma^*} dx^1 dx^2 dx^3$ of a three-dimensional space is given by $dq = \varrho_q \sqrt{\gamma^*} dx^1 dx^2 dx^3$. For the determination of the electric current four-vector we construct, as in Section 1.5.1, the four-vector

$$dq\, dx^\mu = \varrho_q \sqrt{\gamma^*} dx^1 dx^2 dx^3 d\tau \frac{dx^\mu}{d\tau} \overset{(11.37)}{=} \varrho_q \frac{\sqrt{\gamma^*}}{\sqrt{g}} \frac{1}{c\Gamma} \frac{dx^\mu}{d\tau} (\sqrt{g} d^4 x). \tag{11.64}$$

Since $(\sqrt{g} d^4 x)$ is a scalar invariant,

$$J^\mu = \frac{\varrho_q}{\Gamma} \sqrt{\frac{\gamma^*}{g}} \frac{dx^\mu}{d\tau} = \frac{\varrho_q}{\Gamma} \sqrt{\frac{\gamma^*}{g}} U^\mu \tag{11.65}$$

is the electric current four-vector. The components of the electric current four-vector can be obtained from (11.65) and $(11.38)_3$, yielding

$$(J^\mu) = \left(c\varrho_q \sqrt{\frac{\gamma^*}{g}}, \varrho_q \sqrt{\frac{\gamma^*}{g}} v^i \right) \overset{(10.94)}{=} \left(\frac{c\varrho_q}{\sqrt{g_{00}}}, \frac{\varrho_q v^i}{\sqrt{g_{00}}} \right). \tag{11.66}$$

Now we insert (11.66) into (11.63) and get the three-dimensional form of the continuity equation for the charge density,

$$\frac{\partial \varrho_q\sqrt{\gamma^*}}{\partial t}+\frac{\partial \varrho_q\sqrt{\gamma^*}v^i}{\partial x^i}=0, \quad \text{or by (10.95)} \quad \frac{1}{\sqrt{\gamma^*}}\frac{\partial \varrho_q\sqrt{\gamma^*}}{\partial t}+\operatorname{div}\mathbf{I}=0, \quad (11.67)$$

where $\mathbf{I}=\varrho_q\sqrt{\gamma^*}\mathbf{v}$ is the electric current density.

Problem

11.6.1 Show that the electric current four-vector (11.66) and the continuity equation for the charge density (11.67) reduce respectively to (1.141) and (1.138) by considering the metric of a Minkowski space.

11.7 Perfect fluids

We recall that in a locally inertial Lorentz rest frame the components of the energy-momentum tensor and of the particle four-flow of a perfect fluid are given by

$$\begin{cases} T_{\mathrm{R}}^{00}=ne, \quad T_{\mathrm{R}}^{0i}=T_{\mathrm{R}}^{i0}=0, \quad T_{\mathrm{R}}^{ij}=-p\eta^{ij}, \\ N_R^0=nc, \quad N_R^i=0, \end{cases} \tag{11.68}$$

where n, p and e are measured by an observer in this frame.

In an arbitrary frame the energy-momentum tensor reads

$$T^{\mu\nu}=\frac{\partial x^\mu}{\partial x_R^\alpha}\frac{\partial x^\nu}{\partial x_R^\beta}T^{\alpha\beta}, \tag{11.69}$$

where x_R^α are the coordinates of the locally inertial Lorentz rest frame. The insertion of $(11.68)_1$ into (11.69) leads to

$$T^{\mu\nu}=ne\frac{\partial x^\mu}{\partial x_R^0}\frac{\partial x^\nu}{\partial x_R^0}-p\eta^{ij}\frac{\partial x^\mu}{\partial x_R^i}\frac{\partial x^\nu}{\partial x_R^j}. \tag{11.70}$$

Since the relationship between the metric tensors is given by (10.23) we have that

$$g^{\mu\nu}=\frac{\partial x^\mu}{\partial x_R^\alpha}\frac{\partial x^\nu}{\partial x_R^\beta}\eta^{\alpha\beta}=\frac{\partial x^\mu}{\partial x_R^0}\frac{\partial x^\nu}{\partial x_R^0}+\eta^{ij}\frac{\partial x^\mu}{\partial x_R^i}\frac{\partial x^\nu}{\partial x_R^j}. \tag{11.71}$$

If we multiply (11.71) by p and add the resulting equation with (11.70) we get

$$T^{\mu\nu}=(ne+p)\frac{\partial x^\mu}{\partial x_{\mathrm{R}}^0}\frac{\partial x^\nu}{\partial x_{\mathrm{R}}^0}-pg^{\mu\nu}. \tag{11.72}$$

On the other hand, the four-velocity U^μ can be written as

$$U^\mu=\frac{dx^\mu}{d\tau}=\frac{\partial x^\mu}{\partial x_R^0}\frac{dx_R^0}{d\tau}+\frac{\partial x^\mu}{\partial x_R^i}\frac{dx_R^i}{d\tau}. \tag{11.73}$$

Due to the fact that in a proper coordinate system it holds that

$$\frac{dx^0_R}{d\tau} = c, \qquad \text{and} \qquad \frac{dx^i_R}{d\tau} = 0, \tag{11.74}$$

the expression for the four-velocity (11.73) reduces to

$$U^\mu = c\frac{\partial x^\mu}{\partial x^0_R}. \tag{11.75}$$

Hence we have from (11.72) and (11.75) that the energy-momentum tensor for a perfect fluid in an arbitrary frame is given by

$$T^{\mu\nu} = (ne + p)\frac{U^\mu U^\nu}{c^2} - pg^{\mu\nu}. \tag{11.76}$$

By using the same methodology one can calculate the particle four-flow of a perfect fluid in an arbitrary frame yielding

$$N^\mu = \frac{\partial x^\mu}{\partial x^\alpha_{\rm R}} N^\alpha \overset{(11.68)_2}{=} \frac{\partial x^\mu}{\partial x^0_{\rm R}} cn \overset{(11.75)}{=} nU^\mu. \tag{11.77}$$

Let us analyze the balance equation for the particle four-flow. According to the principle of covariance we can write $\partial_\alpha N^\alpha = 0$ as

$$(nU^\mu)_{;\mu} = 0, \qquad \text{or by (10.69)} \qquad \frac{1}{\sqrt{g}}\frac{\partial\sqrt{g}nU^\mu}{\partial x^\mu} = 0. \tag{11.78}$$

The balance equation for the particle four-flow $(11.78)_1$ becomes

$$\frac{dn}{d\tau} + nU^\mu{}_{;\mu} = 0, \tag{11.79}$$

if one uses the relationship

$$n_{;\mu}U^\mu = \frac{\partial n}{\partial x^\mu}\frac{dx^\mu}{d\tau} = \frac{dn}{d\tau}. \tag{11.80}$$

By applying the principle of covariance for the energy-momentum tensor balance equation $\partial_\beta T^{\alpha\beta} = 0$ it follows that

$$T^{\mu\nu}{}_{;\nu} = 0, \qquad \text{or by (11.76)} \qquad \left[(ne + p)\frac{U^\mu U^\nu}{c^2} - pg^{\mu\nu}\right]_{;\nu} = 0. \tag{11.81}$$

Further $(11.81)_2$ is reduced to

$$\frac{d(ne + p)}{d\tau}\frac{U^\mu}{c^2} + \frac{(ne + p)}{c^2}\frac{\delta U^\mu}{\delta\tau} + (ne + p)\frac{U^\mu}{c^2}U^\nu_{;\nu} - \frac{\partial p}{\partial x_\mu} = 0, \tag{11.82}$$

by the use of the relationships $g^{\mu\nu}{}_{;\nu} = 0$, (10.83) and (11.80). The balance equation for the energy per particle of a perfect fluid can be obtained through the multiplication of (11.82) by U_μ, yielding

$$n\frac{de}{d\tau} + pU^\mu{}_{;\mu} = 0. \tag{11.83}$$

In the above equation we have used the balance equation for the particle four-flow (11.79) and the relationship $U_\mu \delta U^\mu/\delta\tau = 0$ which is a consequence of the constraint $U^\mu U_\mu = c^2$. If the derivative $d(ne)/d\tau$ is eliminated from (11.82) by the use of (11.79) and (11.83), the balance equation for the momentum density of a perfect fluid follows:

$$\frac{nh_E}{c^2}\frac{\delta U^\mu}{\delta\tau} = \frac{\partial p}{\partial x_\mu} - \frac{U^\mu}{c^2}\frac{dp}{d\tau}, \tag{11.84}$$

where $h_E = e + p/n$ is the enthalpy per particle.

Let us find a solution of the balance equation for the momentum density corresponding to a perfect fluid in hydrostatic equilibrium and in the presence of a weak gravitational static field. In this case $v^i = 0$ and we have from (11.37) and (11.38)

$$U^0 = \frac{c}{\sqrt{g_{00}}}, \qquad U^i = 0. \tag{11.85}$$

The absolute derivative of U^μ becomes

$$\frac{\delta U^\mu}{\delta\tau} = U^0\left[\frac{\partial U^\mu}{\partial x^0} + \Gamma^\mu_{00}U^0\right], \tag{11.86}$$

by the use of (10.83). The affine connection Γ^μ_{00} that follows from (11.2) is

$$\Gamma^\mu_{00} = -\frac{1}{2}g^{\mu i}\frac{\partial g_{00}}{\partial x^i}, \tag{11.87}$$

since all time derivatives must vanish. Insertion of (11.86) together with (11.87) into (11.84) and again by neglecting all time derivatives and taking into account weak gravitational static fields where (11.47) holds, we have that

$$-\frac{1}{2}\frac{nh_E}{c^2}(U^0)^2\vec{\nabla}g_{00} = \vec{\nabla}p, \quad \text{or by } (11.85)_1 \quad -\vec{\nabla}\left(\ln\sqrt{g_{00}}\right) = \frac{1}{nh_E}\vec{\nabla}p. \tag{11.88}$$

By considering only the rest energy in the limit of a non-relativistic gas and for weak gravitational static fields, we have

$$h_E = mc^2, \qquad g_{00} = 1 + \frac{2\Phi}{c^2}. \tag{11.89}$$

In this case we can use the approximation $\ln\sqrt{g_{00}} \approx \Phi/c^2$ and write $(11.88)_2$ as

$$\vec{\nabla}p = -\varrho\vec{\nabla}\Phi, \tag{11.90}$$

where $\varrho = mn$ denotes the mass density of the perfect fluid. Equation (11.90) is the usual hydrostatic equation of a fluid in the non-relativistic theory.

On the other hand, in the ultra-relativistic limit (see Section 3.1) $ne = 3p$ so that $nh_E = 4p$ and we obtain by integrating $(11.88)_2$:

$$\left(\frac{p}{\mathcal{C}}\right)^{\frac{1}{4}} = \left(\frac{1}{g_{00}}\right)^{\frac{1}{2}}, \quad \text{or by } (11.89)_2 \quad p \approx \mathcal{C}\left(1 - \frac{4\Phi}{c^2}\right), \tag{11.91}$$

where $\mathcal{C}$ is a constant. Since the pressure must vanish outside the fluid, we refer to Weinberg [16] to infer from $(11.91)_2$ that it is impossible to have a hydrostatic equilibrium in a finite ultra-relativistic perfect fluid.

Problems

11.7.1 Show that the balance equation for the particle four-flow (11.78) can be written, thanks to (11.38), as

$$\frac{\partial\sqrt{g}\,\Gamma n}{\partial t} + \frac{\partial\sqrt{g}\,\Gamma n v^i}{\partial x^i} = 0.$$

Further show that it reduces to (4.1) in a Minkowski space.

11.7.2 Show that the spatial components of the balance equation for the momentum density (11.84) can be written as

$$\frac{nh_E}{c^2}\Gamma^2\frac{\delta v^i}{\delta t} = \frac{\partial p}{\partial x_i} - \frac{v^i}{c}\frac{\partial p}{\partial x_0}.$$

Further show that it reduces to (4.8) in a Minkowski space. (Hint: Write $U^i = U^0 v^i/c$ and eliminate $\delta U^0/\delta\tau$ by using the temporal component of the balance equation for the momentum density.)

11.7.3 Check that (11.88) follows from (11.84) by considering a perfect fluid in hydrostatic equilibrium in the presence of a weak gravitational static field.

11.8 Einstein's field equations

We shall proceed to infer Einstein field equations that are fundamental laws of the gravitational field in the general theory of relativity. The starting point is the Poisson equation of the Newtonian theory of gravitation

$$\nabla^2\Phi = 4\pi G\varrho, \tag{11.92}$$

which we should obtain for weak static gravitational fields generated by non-relativistic matter. To achieve this aim we recall that the energy-momentum tensor of a perfect fluid in special relativity is given by

$$T^{\alpha\beta} = (ne + p)\frac{U^\alpha U^\beta}{c^2} - p\eta^{\alpha\beta}. \tag{11.93}$$

In the non-relativistic limit its components reduce to

$$T^{00} = ne \approx nmc^2 = \varrho c^2, \tag{11.94}$$

$$T^{0i} = T^{i0} \approx nh_E\frac{v^i}{c} \approx \varrho c^2\frac{v^i}{c}, \qquad \text{that is} \qquad \frac{T^{0i}}{T^{00}} \approx \frac{v^i}{c} \ll 1, \tag{11.95}$$

$$T^{ij} \approx -p\eta^{ij} + mnc^2\frac{v^iv^j}{c^2}, \qquad \text{that is} \qquad \frac{T^{ij}}{T^{00}} \approx \frac{kT}{mc^2}\eta^{ij} + \frac{v^iv^j}{c^2} \ll 1, \tag{11.96}$$

since $(U^\mu) = (c, v^i)$, $p = nkT$ and $\zeta = mc^2/kT \gg 1$. Hence in the non-relativistic limit we can approximate the energy-momentum tensor of a perfect fluid as

$$(T^{\alpha\beta}) = \begin{pmatrix} \varrho c^2 & 0 & 0 & 0 \\ 0 & 0 & 0 & 0 \\ 0 & 0 & 0 & 0 \\ 0 & 0 & 0 & 0 \end{pmatrix}. \tag{11.97}$$

We use (11.54) and (11.97) and write the Poisson equation (11.92) for weak static gravitational fields as

$$\nabla^2 g_{00} = \frac{8\pi G}{c^4}T_{00}. \tag{11.98}$$

To generalize (11.98) we search for an equation of the form

$$G_{\mu\nu} = \frac{8\pi G}{c^4}T_{\mu\nu}, \tag{11.99}$$

such that for weak static gravitational fields, (11.99) reduces to (11.98). The quantity $G_{\mu\nu}$ must fulfill the following conditions:

1. $G_{\mu\nu}$ must be a symmetric Riemannian tensor such that $G^{\mu\nu}{}_{;\nu} = 0$, since $T_{\mu\nu}$ is a symmetric tensor that satisfies $T^{\mu\nu}{}_{;\nu} = 0$;
2. $G_{\mu\nu}$ must be linear in the second derivatives of the metric tensor $g_{\mu\nu}$ and no higher order derivatives of this tensor must occur;
3. For weak static gravitational fields we must have that

$$G_{00} \approx \nabla^2 g_{00}. \tag{11.100}$$

The only symmetric tensor that satisfies the second condition above can be written as a linear combination of the Ricci tensor $R_{\mu\nu}$ and of the curvature scalar R, i.e.,[2]

$$G_{\mu\nu} = aR_{\mu\nu} + bRg_{\mu\nu} \tag{11.101}$$

where a and b are two constants which will be determined in the following.

[2]We do not consider the cosmological term $\Lambda g_{\mu\nu}$ where Λ is the cosmological constant, which does not vanish even in a Minkowski space.

We apply the first condition to (11.101) and get by the use of (11.34) that $b = -a/2$. Hence (11.101) becomes

$$G_{\mu\nu} = a\left(R_{\mu\nu} - \frac{1}{2}Rg_{\mu\nu}\right) \overset{(11.99)}{=} \frac{8\pi G}{c^4}T_{\mu\nu}. \tag{11.102}$$

In order to obtain the other constant, the contraction of (11.102) leads to

$$-aR = \frac{8\pi G}{c^4}T^{\mu}_{\mu}. \tag{11.103}$$

The non-relativistic limit of the above equation is

$$-aR \approx \frac{8\pi G}{c^4}T^0_0 = \frac{8\pi G}{c^4}g^{0\nu}T_{\nu 0} \approx \frac{8\pi G}{c^4}g^{00}T_{00}, \tag{11.104}$$

since the non-relativistic conditions $|T^i_i| \ll |T^0_0|$, $|T_{0i}| \ll |T_{00}|$ hold. On the other hand, the temporal component of (11.102) is

$$a\left(R_{00} - \frac{1}{2}Rg_{00}\right) = \frac{8\pi G}{c^4}T_{00}, \quad \text{or by (11.104)} \quad aR_{00} = \frac{4\pi G}{c^4}T_{00}, \tag{11.105}$$

due to the fact that for weak gravitational static fields $g^{00}g_{00} \approx 1$.

The temporal component of the Ricci tensor can be obtained from (11.8) and (11.9), yielding

$$R_{00} \approx \frac{\partial \Gamma^{\tau}_{0\tau}}{\partial x^0} - \frac{\partial \Gamma^{\tau}_{00}}{\partial x^{\tau}}, \tag{11.106}$$

by neglecting the quadratic terms in the affine connections. Hence for weak static gravitational fields (11.106) reduces to

$$R_{00} = -\frac{\partial \Gamma^i_{00}}{\partial x^i} \overset{(11.2)}{=} \frac{1}{2}g^{ij}\frac{\partial^2 g_{00}}{\partial x^i \partial x^j} \approx -\frac{1}{2}\eta^{ij}\frac{\partial^2 g_{00}}{\partial x^i \partial x^j} = -\frac{1}{2}\nabla^2 g_{00}. \tag{11.107}$$

Now we have from $(11.105)_2$ and (11.107) that

$$a\nabla^2 g_{00} = -\frac{8\pi G}{c^4}T_{00}. \tag{11.108}$$

If we compare (11.98) with (11.108) we conclude that $a = -1$. Further this condition and (11.102) lead to the Einstein field equations

$$R_{\mu\nu} - \frac{1}{2}Rg_{\mu\nu} = -\frac{8\pi G}{c^4}T_{\mu\nu}. \tag{11.109}$$

An equivalent form of the above equation is

$$R_{\mu\nu} = -\frac{8\pi G}{c^4}\left(T_{\mu\nu} - \frac{1}{2}T^{\sigma}_{\sigma}g_{\mu\nu}\right), \tag{11.110}$$

since the trace of (11.109) provides that the curvature scalar is given by

$$R = \frac{8\pi G}{c^4}T^{\sigma}_{\sigma}. \tag{11.111}$$

Problems

11.8.1 Prove that the only symmetric tensor that is linear in the second derivatives of the metric tensor, and does not contain higher order derivatives of the latter, can be written as a linear combination of the Ricci tensor and of the curvature scalar, provided we disregard the cosmological term $\Lambda g_{\mu\nu}$. (Hint: See the appendix of the book of Weyl [17].)

11.8.2 Check the approximation (11.104).

11.8.3 Show that for weak gravitational static fields $g_{00}g^{00} \approx 1$. (Hint: $g^{\mu\nu}g_{\nu\sigma} = \delta^{\mu}_{\sigma}$.)

11.9 Solution for weak fields

In this section we shall find a solution of Einstein's field equations (11.109) or (11.110) for weak gravitational fields. Since (11.110) are non-linear partial differential equations for the metric tensor $g_{\mu\nu}$, for weak gravitational fields they reduce to a system of linear partial differential equations which we shall proceed to analyze.

For weak gravitational fields we may approximate the metric tensor $g_{\mu\nu}$ by (11.47) and the Ricci tensor $R_{\mu\nu}$, calculated from (11.24), becomes

$$R_{\mu\kappa} = g^{\lambda\nu}R_{\lambda\mu\nu\kappa} \approx \frac{1}{2}\eta^{\lambda\nu}\left[\frac{\partial^2 h_{\lambda\nu}}{\partial x^\kappa \partial x^\mu} - \frac{\partial^2 h_{\lambda\kappa}}{\partial x^\nu \partial x^\mu} - \frac{\partial^2 h_{\mu\nu}}{\partial x^\kappa \partial x^\lambda} + \frac{\partial^2 h_{\mu\kappa}}{\partial x^\nu \partial x^\lambda}\right]$$

$$= \frac{1}{2}\Box h_{\mu\kappa} - \frac{1}{2}\frac{\partial}{\partial x^\mu}\left[\frac{\partial}{\partial x^\nu}\left(h^\nu_\kappa - \frac{1}{2}h^\sigma_\sigma\delta^\nu_\kappa\right)\right] - \frac{1}{2}\frac{\partial}{\partial x^\kappa}\left[\frac{\partial}{\partial x^\nu}\left(h^\nu_\mu - \frac{1}{2}h^\sigma_\sigma\delta^\nu_\mu\right)\right], \tag{11.112}$$

where $\Box$ is the d'Alembertian.

We consider now an infinitesimal coordinate transformation

$$x^\mu = x'^\mu + \epsilon\varphi^\mu(x') \qquad \text{such that} \qquad |\epsilon| \ll 1. \tag{11.113}$$

According to (10.30) the metric tensor in the new coordinates will be given by

$$g'_{\mu\nu} = g_{\sigma\tau}\frac{\partial x^\sigma}{\partial x'^\mu}\frac{\partial x^\tau}{\partial x'^\nu} \approx g_{\mu\nu} + \epsilon\left(\frac{\partial \varphi_\mu}{\partial x'^\nu} + \frac{\partial \varphi_\nu}{\partial x'^\mu}\right). \tag{11.114}$$

By considering that $g'_{\mu\nu} = \eta_{\mu\nu} + h'_{\mu\nu}$ we have that

$$h'_{\mu\nu} = h_{\mu\nu} + \epsilon\left(\frac{\partial \varphi_\mu}{\partial x'^\nu} + \frac{\partial \varphi_\nu}{\partial x'^\mu}\right). \tag{11.115}$$

From the above considerations it is easy to show that the transformation (11.115) is a sort of gauge transformation, since it leaves the linearized Ricci tensor invariant.

Hence, as in Section 1.5.3, we can choose a suitable function φ_μ such that the gauge condition

$$\frac{\partial}{\partial x^\nu}\left(h^\nu_\kappa - \frac{1}{2}h^\sigma_\sigma \delta^\nu_\kappa\right) = 0 \tag{11.116}$$

is satisfied.

If we take into account the constraint (11.116) and the expression for the Ricci tensor (11.112), we get that the linearized Einstein field equations reduce to

$$\Box\, h_{\mu\nu} = -\frac{16\pi G}{c^4}\mathcal{T}_{\mu\nu}, \qquad \text{with} \qquad \mathcal{T}_{\mu\nu} = T_{\mu\nu} - \frac{1}{2}T^\sigma_\sigma \eta_{\mu\nu}. \tag{11.117}$$

Equation (11.117) represents an inhomogeneous wave equation whose solution is given in terms of a retarded potential (see (1.192)):

$$h_{\mu\nu}(\mathbf{x},t) = -\frac{4G}{c^4}\int \frac{\mathcal{T}_{\mu\nu}(\mathbf{x}', t - |\mathbf{x}-\mathbf{x}'|/c)}{|\mathbf{x}-\mathbf{x}'|} dx'^1 dx'^2 dx'^3. \tag{11.118}$$

In the above equation we have neglected the solution of the homogeneous wave equation because we have assumed that there is no incoming gravitational wave.

For a static distribution of mass where (11.94) through (11.97) hold we get from $(11.117)_2$ that

$$\mathcal{T}_{00} = \frac{1}{2}\varrho c^2, \qquad \mathcal{T}_{ij} = -\frac{1}{2}\varrho c^2 \eta_{ij}, \tag{11.119}$$

since $T^\sigma_\sigma = \varrho c^2$. Hence the components of $h_{\mu\nu}$ can be written as

$$h_{00}(\mathbf{x},t) = \frac{2\Phi}{c^2}, \qquad h_{ij}(\mathbf{x},t) = -\frac{2\Phi}{c^2}\eta_{ij}, \tag{11.120}$$

where Φ is the Newtonian potential

$$\Phi(\mathbf{x}) = -G\int \frac{\varrho(\mathbf{x}')}{|\mathbf{x}-\mathbf{x}'|} dx'^1 dx'^2 dx'^3. \tag{11.121}$$

The components of the metric tensor $g_{\mu\nu} = \eta_{\mu\nu} + h_{\mu\nu}$ follow from (11.120) yielding

$$g_{00} = 1 + \frac{2\Phi}{c^2}, \qquad g_{ij} = \left(1 - \frac{2\Phi}{c^2}\right)\eta_{ij}, \tag{11.122}$$

whereas the line element $ds^2 = g_{\mu\nu}dx^\mu dx^\nu$ becomes

$$ds^2 = \left(1 + \frac{2\Phi}{c^2}\right)(dx^0)^2 - \left(1 - \frac{2\Phi}{c^2}\right)[(dx^1)^2 + (dx^2)^2 + (dx^3)^2]. \tag{11.123}$$

Problems

11.9.1 Show that for a weak gravitational field the Ricci tensor reduces to (11.112).

11.9.2 Show that the metric tensor under the infinitesimal coordinate transformation (11.113) is given by (11.114).

11.9.3 Prove that one can choose a coordinate system with the transformation (11.115) in a such a way that (11.116) holds.

11.10 Exact solutions of Einstein's field equations

Exact solutions of Einstein's field equations are very rare. The most widely known is due to Schwarzschild and describes the gravitational field in the neighborhood of a point mass. Since we expect that this solution has a sort of symmetry which resembles the spherical symmetry in ordinary space, we adopt coordinates which have the same name as spherical coordinates, r, θ, ϕ and write the line element in the form

$$ds^2 = A(r)(cdt)^2 - B(r)(dr)^2 - C(r)r^2[(d\theta)^2 + \sin^2\theta(d\phi)^2]. \tag{11.124}$$

Here $A(r)$, $B(r)$, and $C(r)$ are three functions of r to be presently determined. When we take $A = B = C = 1$, we obtain the line element of special relativity in spherical coordinates.

The element given by (11.124) is the most general one, which seems to comply with our idea of static spherical symmetry. The coordinate r is still somehow arbitrary, because if we replace it by an arbitrary, differentiable function of r, we obtain a different line element which has, however, the same form as (11.124). We can exploit this circumstance to get rid of one of the three functions $A(r)$, $B(r)$, and $C(r)$. The two simplest choices are $B = C$ and $C = 1$. The second one leads to a simpler calculation and so we rewrite (11.124) in the form

$$ds^2 = A(r)(cdt)^2 - B(r)(dr)^2 - r^2[(d\theta)^2 + \sin^2\theta(d\phi)^2]. \tag{11.125}$$

In order to simplify the calculations, it is expedient to let

$$A = e^F; \quad B = e^H, \tag{11.126}$$

where F and H are two new functions of r. Our coordinates are $x^0 = ct$, $x^1 = r$, $x^2 = \theta$, and $x^3 = \phi$ and the metric tensor is diagonal with the components

$$g_{00} = e^F; \quad g_{11} = -e^H; \quad g_{22} = -r^2; \quad g_{33} = -r^2\sin^2\theta. \tag{11.127}$$

The determinant is the product of the leading diagonal, $-e^{F+H}r^4\sin^2\theta$ and the contravariant components are

$$g^{00} = e^{-F}; \quad g^{11} = -e^{-H}; \quad g^{22} = -r^{-2}; \quad g^{33} = -r^{-2}\sin^{-2}\theta. \tag{11.128}$$

Since the tensors we are dealing with are diagonal, the expression for the Christoffel symbols can be rewritten without a sum on a repeated index. The total number of (different) components of the affine connection is 40, but one easily checks that 31 of them vanish and the only non-zero components are

$$\Gamma^0_{10} = \frac{1}{2}F', \qquad \Gamma^1_{00} = \frac{1}{2}e^{F-H}F', \tag{11.129}$$

$$\Gamma^1_{11} = \frac{1}{2}H', \qquad \Gamma^2_{12} = \Gamma^3_{13} = \frac{1}{r}, \tag{11.130}$$

$$\Gamma^1_{22} = -re^{-H}, \qquad \Gamma^3_{23} = \cot\theta, \tag{11.131}$$

$$\Gamma^1_{33} = -r\sin^2\theta e^{-H}, \qquad \Gamma^2_{33} = -\sin\theta\cos\theta, \tag{11.132}$$

where the prime denotes differentiation with respect to r. This produces an enormous simplification in the calculation of the components of the contracted Riemann–Christoffel curvature tensor, which turns out to be diagonal (this is immediate except for one non-diagonal component R_{12}, which can be shown to vanish only after some calculations). The diagonal components are

$$R_{00} = e^{F-H}\left(-\frac{1}{2}F'' + \frac{1}{4}F'H' - \frac{1}{4}F'^2 - \frac{F'}{r}\right), \tag{11.133}$$

$$R_{11} = \frac{1}{2}F'' - \frac{1}{4}F'H' + \frac{1}{4}F'^2 - \frac{H'}{r}, \tag{11.134}$$

$$R_{22} = e^{-H}\left[1 + \frac{r}{2}(F' - H')\right] - 1, \tag{11.135}$$

$$R_{33} = \sin^2\theta\left\{e^{-H}\left[1 + \frac{r}{2}(F' - H')\right] - 1\right\}. \tag{11.136}$$

When we equate these to zero (as appropriate outside $r = 0$, where we expect a singularity), the last component gives an equation, which is automatically satisfied as a consequence of the previous one. After suppressing the exponential in the first equation, we see that the first two equations coincide, except for their last terms. Hence we obtain, by subtraction, $F' = -H'$ and since we impose the condition that at space infinity the geometry becomes Minkowskian ($F = H = 0$) we conclude that $F = -H$. The third equation $R_{22} = 0$, then reduces to

$$e^F(1 + rF') = 1. \tag{11.137}$$

Now it is convenient to reintroduce A and B according to (11.126). Then $B = 1/A$ and

$$A + rA' = 1, \tag{11.138}$$

which integrates immediately to give

$$A = 1 - \frac{2GM}{rc^2}, \tag{11.139}$$

where the integration constant has been chosen to agree with (11.54) and (11.52).

Hence the line element of the Schwarzschild solution can be written as

$$ds^2 = \left(1 - \frac{2GM}{rc^2}\right)(cdt)^2 - \frac{1}{\left(1 - \frac{2GM}{rc^2}\right)}(dr)^2 - r^2[(d\theta)^2 + \sin^2\theta(d\phi)^2], \quad (11.140)$$

which is singular for $r = 2GM/c^2$.

We remark that there are two singularities; one at $r = 0$ and another one at $r = 2GM/c^2$. The first is expected; the second is not. The radius $r_S = 2GM/c^2$ where the second singularity occurs is called the Schwarzschild radius. Something peculiar occurs there; there is however no physical singularity, because the Riemann tensor is finite and smooth at $r = r_S$. It is a *coordinate singularity*; i.e., we can make a change of coordinates which causes the singularity in the metric to disappear. The required change of coordinates was first investigated by Kruskal [7] in 1960. The new coordinates T and X are defined by

$$X^2 - T^2 = \left(\frac{rc^2}{2GM} - 1\right)e^{rc^2/(2GM)}, \quad (11.141)$$

$$\ln\left(\frac{X+T}{X-T}\right) = 2\text{artanh}\frac{T}{X} = \frac{c^3 t}{2GM}, \quad (11.142)$$

which shows that, since $r > 0$, X and T undergo the restriction

$$T^2 - X^2 < 1. \quad (11.143)$$

Then the metric becomes

$$ds^2 = \frac{32G^3M^3e^{-rc^2/(2GM)}}{rc^6}[(dT)^2 - (dX)^2] - r^2[(d\theta)^2 + \sin^2\theta(d\phi)^2], \quad (11.144)$$

where, of course, r must be expressed in terms of T and X, i.e., $r = r(T, X)$.

The Schwarzschild radius is no longer a locus of singularities. There is a feature, however, that shows the peculiar behavior of space-time inside the Schwarzschild sphere: a radially infalling observer, when crossing this radius (the surface $X = T$ in the Kruskal coordinates) enters a region, whence he cannot escape. Actually he will fall into the singularity $r = 0$ where $X = \sqrt{T^2 - 1}$ in a finite proper time and, if, when falling in this region, he tries to send a light signal, the signal will not be able to pass $r = r_S$ and will eventually fall into $r = 0$. For this reason the region $r < r_S$ is called a *black hole*.

We remark that the specularly symmetric region with $X < 0$ has a similar, but time-reversed behavior; for obvious reasons, its part where $r > r_S$ is called a *white hole*. This region must be considered as unphysical because no time signal can be sent between the two regions $X > 0$ and $X < 0$.

Another interesting case is obtained when we look for a spherically symmetric solution of Einstein's field equations when the energy-momentum tensor corresponds to that of a perfect fluid,

$$T_{\sigma\nu} = (\varrho c^2 + p)\frac{U_\sigma U_\nu}{c^2} - p g_{\sigma\nu}, \quad (11.145)$$

by considering $ne \approx nmc^2 = \varrho c^2$. In order to be compatible with the static symmetry we have that the four-velocity reduces to $(U_\mu) = (c\sqrt{g_{00}}, \mathbf{0})$.

Einstein's field equations (11.109) then become

$$\frac{B'}{rB^2} + \frac{1}{r^2}\left(1 - \frac{1}{B}\right) = \frac{8\pi G}{c^2}\varrho, \tag{11.146}$$

$$\frac{A'}{rAB} - \frac{1}{r^2}\left(1 - \frac{1}{B}\right) = \frac{8\pi G}{c^4}p, \tag{11.147}$$

$$\frac{1}{2\sqrt{AB}}\frac{d}{dr}\left[\frac{A'}{\sqrt{AB}}\right] + \frac{A'}{2rAB} - \frac{B'}{2rB^2} = \frac{8\pi G}{c^4}p, \tag{11.148}$$

by using (11.126), (11.133) through (11.136) and (11.145).

The first equation (11.146) only involves B and can be rewritten in the form

$$\frac{1}{r^2}\frac{d}{dr}\left[r\left(1 - \frac{1}{B}\right)\right] = \frac{8\pi G}{c^2}\varrho, \tag{11.149}$$

which integrates immediately to

$$B(r) = \left[1 - \frac{2m(r)G}{rc^2}\right]^{-1}, \tag{11.150}$$

where, if a denotes a constant,

$$m(r) = 4\pi\int_0^r \varrho(r')r'^2 dr' + a. \tag{11.151}$$

Since we no longer have a singularity at $r = 0$, a must be zero. If we want r to be a space coordinate, we must have $r \geq 2m(r)$.

The solution we are calculating represents a space-time filled with a fluid. If we want the fluid to be only inside a sphere of radius R, then $\varrho = 0$ for $r > R$ and the expression for B only holds for $r \leq R$; for $r > R$ a Schwarzschild solution for a point mass holds, with a total mass

$$M = m(R) = 4\pi\int_0^R \varrho(r)r^2 dr. \tag{11.152}$$

This expression is formally identical to the expression for the total mass in non-relativistic mechanics, but this analogy is partly misleading, since the total *proper mass* should be

$$M_p = \int \varrho\sqrt{\gamma^*}dr\, d\theta\, d\phi = 4\pi\int_0^R \frac{\varrho(r)r^2 dr}{\sqrt{1 - 2m(r)/r}}. \tag{11.153}$$

The difference between M and M_p can be interpreted as the gravitational binding energy.

We must now find A; in order to stress the analogy with the Newtonian case (see Chapter 3), we let

$$A = e^{2\Phi^\star} \tag{11.154}$$

and rewrite (11.147) as

$$\frac{d(c^2\Phi^\star)}{dr} = \frac{m(r)Gc^2 + 4\pi G r^3 p}{r[rc^2 - 2m(r)G]}. \tag{11.155}$$

This is the analogue of the Poisson equation for the Newtonian potential $c^2\Phi^\star$, to which it reduces if we let $p = 0$.

The third equation (11.148), after some algebra and elimination of $\Phi^\star$ reduces to a fluid equilibrium equation, known as the *Tolman–Oppenheimer–Volkoff equation:*

$$\frac{dp}{dr} = -(p + \varrho c^2)\left\{\frac{m(r)Gc^2 + 4\pi G r^3 p}{rc^2[rc^2 - 2m(r)G]}\right\}. \tag{11.156}$$

Thus we have found a solution for an arbitrary equation of state, $p = p(\varrho)$. A particular case is offered by the case of an incompressible fluid, for which $\varrho = \varrho_0$. Then

$$m(r) = \frac{4}{3}\pi r^3 \varrho_0, \qquad (r \le R). \tag{11.157}$$

Equation (11.156) can then be integrated, with the boundary condition $p(R) = 0$, to yield

$$p = \varrho_0 c^2 \left\{\frac{\sqrt{1 - 2MGr^2/(R^3c^2)} - \sqrt{1 - 2MG/(Rc^2)}}{3\sqrt{1 - 2MG/(Rc^2)} - \sqrt{1 - 2MGr^2/(R^3c^2)}}\right\}. \tag{11.158}$$

Thus the pressure at $r = 0$ (*central pressure*) required to keep a star in uniform density at equilibrium is

$$p = \varrho_0 c^2 \left\{\frac{1 - \sqrt{1 - 2MG/(Rc^2)}}{3\sqrt{1 - 2MG/(Rc^2)} - 1}\right\}. \tag{11.159}$$

The Newtonian value of the right-hand side, $\varrho_0 MG/(2R)$, is obtained by letting $R \to \infty$. If we do not take the limit, however, we see that there is a value $R = 9MG/(4c^2)$ for which p_c tends to infinity. If $R < 9MG/(4c^2)$ the star cannot exist. This result can be extended to any equation of state.

Problems

11.10.1 Show that the line element

$$ds^2 = A(r)(cdt)^2 - B(r)(dr)^2 - C(r)r^2(d\theta)^2 + D(r)r^2\sin^2\theta(d\phi)^2,$$

reduces to (11.125) by considering first a spherical symmetry so that $D(r) = C(r)$ and then by changing the variable r so that $C(r) = 1$. (Hint: For the first step consider $dr = dt = 0$, choose two values of θ and impose the isotropy in θ and ϕ.)

11.10.2 Obtain the expressions (11.129) through (11.132) for the components of the affine conection.

11.10.3 Show that the diagonal components of the Ricci tensor are given by (11.133) through (11.136) and that the component R_{12} vanishes.

11.10.4 Show that the line element in the Kruskal coordinates (11.141) and (11.142) is given by (11.144).

11.10.5 Show that Einstein's field equations (11.109) reduce to (11.146) through (11.148) by using the equations (11.126), (11.133) through (11.136) and (11.145).

11.10.6 Show that the equation of Tolman–Oppenheimer–Volkoff given in (11.156) follows from (11.148).

11.10.7 Integrate the Tolman–Oppenheimer–Volkoff equation (11.156) with the boundary condition $p(R) = 0$ and obtain (11.158).

11.11 Robertson–Walker metric

The so-called cosmological principle is based on the assumption that the universe is spatially homogeneous and isotropic. The metric that describes such a kind of universe, known as the Robertson–Walker metric, has the form

$$ds^2 = (cdt)^2 - \kappa(t)^2 \left\{ \frac{(dr)^2}{1 - \varepsilon r^2} + r^2[(d\theta)^2 + \sin^2\theta(d\phi)^2] \right\}. \tag{11.160}$$

In the above equation ε may assume the values $0, +1$ or -1, $\kappa(t)$ is an unknown function of time which has a dimension of length and r is a dimensionless quantity.

The components and the determinant $g = -\det((g_{\mu\nu}))$ of the metric tensor $g_{\mu\nu}$ for the Robertson–Walker metric (11.160) with respect to the coordinates $x^0 = ct$, $x^1 = r$, $x^2 = \theta$ and $x^3 = \phi$ are

$$g_{00} = g^{00} = 1, \quad g_{11} = \frac{-\kappa^2}{1 - \varepsilon r^2} = \frac{1}{g^{11}}, \quad g_{22} = -\kappa^2 r^2 = \frac{1}{g^{22}}, \tag{11.161}$$

$$g_{33} = -\kappa^2 r^2 \sin^2\theta = \frac{1}{g^{33}}, \qquad g = \frac{\kappa^6 r^4 \sin^2\theta}{1 - \varepsilon r^2}. \tag{11.162}$$

The corresponding non-zero Christoffel symbols obtained from (11.2), (11.161) and (11.162) read

$$\Gamma^0_{11} = \frac{\dot{\kappa}\kappa}{1 - \varepsilon r^2}, \quad \Gamma^0_{22} = \dot{\kappa}\kappa r^2, \quad \Gamma^0_{33} = \dot{\kappa}\kappa r^2 \sin^2\theta, \quad \Gamma^1_{01} = \Gamma^2_{02} = \Gamma^3_{03} = \frac{\dot{\kappa}}{\kappa}, \tag{11.163}$$

$$\Gamma^1_{11} = \frac{\varepsilon r}{1-\varepsilon r^2}, \quad \Gamma^1_{22} = -r(1-\varepsilon r^2), \quad \Gamma^1_{33} = -r(1-\varepsilon r^2)\sin^2\theta, \tag{11.164}$$

$$\Gamma^2_{12} = \Gamma^3_{13} = \frac{1}{r}, \quad \Gamma^2_{33} = -\sin\theta\cos\theta, \quad \Gamma^3_{23} = \cot\theta, \tag{11.165}$$

where the dot denotes the derivative with respect to the coordinate $x^0 = ct$.

Once the Christoffel symbols are known the non-vanishing components of the Ricci tensor $R_{\mu\nu} = R^{\tau}_{\mu\tau\nu}$ and the curvature scalar $R = g^{\mu\nu}R_{\mu\nu}$ can be obtained from (11.8) and (11.163) through (11.165). Hence it follows that

$$R_{00} = 3\frac{\ddot{\kappa}}{\kappa}, \quad R_{11} = -\frac{\ddot{\kappa}\kappa + 2\dot{\kappa}^2 + 2\varepsilon}{1-\varepsilon r^2}, \quad R_{22} = -(\ddot{\kappa}\kappa + 2\dot{\kappa}^2 + 2\varepsilon)r^2, \tag{11.166}$$

$$R_{33} = R_{22}\sin^2\theta, \qquad R = 6\left(\frac{\ddot{\kappa}}{\kappa} + \frac{\dot{\kappa}^2}{\kappa^2} + \frac{\varepsilon}{\kappa^2}\right). \tag{11.167}$$

From the knowledge of the curvature scalar and of the components of the Ricci tensor it is straightforward to obtain the non-vanishing components of the Einstein tensor $G_{\mu\nu} = R_{\mu\nu} - Rg_{\mu\nu}/2$:

$$G_{00} = -3\left(\frac{\dot{\kappa}^2}{\kappa^2} + \frac{\varepsilon}{\kappa^2}\right), \qquad G_{11} = \frac{2\ddot{\kappa}\kappa + \dot{\kappa}^2 + \varepsilon}{1-\varepsilon r^2}, \tag{11.168}$$

$$G_{22} = G_{11}r^2(1-\varepsilon r^2), \qquad G_{33} = G_{22}\sin^2\theta. \tag{11.169}$$

Problems

11.11.1 Show that the non-zero Christoffel symbols for the Robertson–Walker metric are given by (11.163) through (11.165).

11.11.2 Obtain the expression for the curvature scalar $(11.167)_2$.

11.11.1 Geometrical meaning

In the following we refer to Misner et al. [10] and Weinberg [16] and give the geometrical meaning of the Robertson–Walker metric. First we shall consider a surface which is a three-dimensional sphere in a four-dimensional Euclidean space with Cartesian coordinates w, x, y, z. This surface is described by the equation

$$w^2 + x^2 + y^2 + z^2 = \kappa^2, \tag{11.170}$$

and a line element $d\sigma$ in this space is given by

$$d\sigma^2 = dw^2 + dx^2 + dy^2 + dz^2. \tag{11.171}$$

If we consider the spherical coordinates

$$x = \kappa\sin\chi\sin\theta\cos\phi, \qquad y = \kappa\sin\chi\sin\theta\sin\phi, \tag{11.172}$$

$$z = \kappa \sin\chi \cos\theta, \qquad w = \kappa \cos\chi, \tag{11.173}$$

we have that (11.171) can be written as

$$d\sigma^2 = \kappa^2\{(d\chi)^2 + \sin^2\chi[(d\theta)^2 + \sin^2\theta(d\phi)^2]\}. \tag{11.174}$$

By introducing a new variable $r = \sin\chi$ the line element (11.174) reduces to

$$d\sigma^2 = \kappa^2\left\{\frac{(dr)^2}{1-r^2} + r^2[(d\theta)^2 + \sin^2\theta(d\phi)^2]\right\}, \tag{11.175}$$

which corresponds to the spatial line element of the Robertson–Walker metric (11.160) for $\varepsilon = +1$.

Let us consider now a surface which is a three-dimensional hyperboloid in a four-dimensional Minkowski space of signature $(+1, -1, -1, -1)$. The equation of this surface with pseudo-Cartesian coordinates w, x, y, z is

$$w^2 - x^2 - y^2 - z^2 = \kappa^2, \tag{11.176}$$

while the line element $d\sigma$ in this space reads

$$d\sigma^2 = dw^2 - dx^2 - dy^2 - dz^2. \tag{11.177}$$

We introduce the hyperbolic coordinates

$$x = \kappa \sinh\chi \sin\theta \cos\phi, \qquad y = \kappa \sinh\chi \sin\theta \sin\phi, \tag{11.178}$$

$$z = \kappa \sinh\chi \cos\theta, \qquad w = \kappa \cosh\chi, \tag{11.179}$$

so that (11.177) reduces to

$$d\sigma^2 = \kappa^2\{(d\chi)^2 + \sinh^2\chi[(d\theta)^2 + \sin^2\theta(d\phi)^2]\}. \tag{11.180}$$

If a new variable $r = \sinh\chi$ is introduced, the line element (11.180) is written as

$$d\sigma^2 = \kappa^2\left\{\frac{(dr)^2}{1+r^2} + r^2[(d\theta)^2 + \sin^2\theta(d\phi)^2]\right\}. \tag{11.181}$$

We infer from the above equation that it corresponds to the spatial line element of the Robertson–Walker metric (11.160) for $\varepsilon = -1$.

Finally we consider a line element given by

$$d\sigma^2 = dx^2 + dy^2 + dz^2, \tag{11.182}$$

which represents a flat three-dimensional Euclidean space. In spherical coordinates

$$x = \kappa r \sin\theta \cos\phi, \quad y = \kappa r \sin\theta \sin\phi, \quad z = \kappa r \cos\theta, \tag{11.183}$$

and the line element (11.182) becomes

$$d\sigma^2 = \kappa^2\{(dr)^2 + r^2[(d\theta)^2 + \sin^2\theta(d\phi)^2]\}. \tag{11.184}$$

If we compare (11.184) with the spatial element of the Robertson–Walker metric (11.160) we infer that it corresponds to the case where $\varepsilon = 0$.

We can also calculate the curvature scalar for the three cases described above. For that end we use the result that an N-dimensional isotropic space with $N \geq 3$ has a constant scalar curvature R which can be represented by

$$R = -N(N-1)K^{(N)}, \tag{11.185}$$

where $K^{(N)}$ is a constant curvature of the N-dimensional space (for more details one is referred to the book by Weinberg [16], pages 382–383). Here we have a three-dimensional space characterized by the line elements (11.175), (11.181) and (11.184) whose non-vanishing components of the spatial metric tensor read

$$\gamma^*_{11} = \frac{\kappa^2}{1-\varepsilon r^2}, \qquad \gamma^*_{22} = \kappa^2 r^2, \qquad \gamma^*_{33} = \kappa^2 r^2 \sin^2\theta. \tag{11.186}$$

In this case the scalar curvature becomes

$$R = -\frac{6\varepsilon}{\kappa^2}, \qquad \text{so that} \qquad K^{(3)} = \frac{\varepsilon}{\kappa^2}. \tag{11.187}$$

For $\varepsilon = +1$ the universe has a positive spatial curvature and $\kappa(t)$ may be identified as the radius of the universe. For $\varepsilon = -1$ we have a universe with a negative spatial curvature while for $\varepsilon = 0$ the spatial curvature is zero. For $\varepsilon = -1$ and for $\varepsilon = 0$, $\kappa(t)$ is interpreted as a cosmic scale factor (Weinberg [16] page 413) and this interpretation can also be extended to the case $\varepsilon = +1$.

Problems

11.11.1.1 Obtain (11.174) from (11.171) by considering the spherical coordinates (11.172) and (11.173).

11.11.1.2 Show that the scalar curvature for the three-dimensional space characterized by the metric tensor (11.186) is given by $(11.187)_1$.

11.11.2 Determination of the energy density

The universe at large scale can be modeled as a mixture of two perfect fluids, namely matter and radiation. Matter refers to galaxies, cosmic rays and intergalactic gas, while radiation is associated with electromagnetic, neutrino and gravitational radiations.

According to (11.76) the energy-momentum tensor of a perfect fluid is given by

$$T^{\mu\nu} = (\epsilon + p)\frac{U^\mu U^\nu}{c^2} - pg^{\mu\nu}. \tag{11.188}$$

The pressure p and the energy density $\epsilon = ne$ are written as a sum of two contributions due to matter and radiation:

$$p = p_m + p_r, \qquad \epsilon = \epsilon_m + \epsilon_r. \tag{11.189}$$

The indices m and r refer to the matter and radiation, respectively.

In the early universe we had a radiation dominated period where the pressure due to the matter could be neglected with respect to the radiation pressure. Nowadays we have a matter dominated period in which the matter can be modeled as a dust with a negligible pressure of the matter. Hence the only contribution to the pressure in all periods is the radiation pressure, which in terms of the energy density is given by

$$p = p_r = \frac{1}{3}\epsilon_r. \tag{11.190}$$

By taking into account the above reasonings, the energy-momentum tensor (11.188) – thanks to (11.189) and (11.190) – can be written as

$$T^{\mu\nu} = \left(\epsilon_m + \frac{4}{3}\epsilon_r\right)\frac{U^\mu U^\nu}{c^2} - \frac{1}{3}\epsilon_r g^{\mu\nu}. \tag{11.191}$$

Note that the energy-momentum tensor (11.191) is a function of the energy densities ϵ_m and ϵ_r.

Since we are considering that the universe is homogeneous and isotropic we have that the energy densities may depend only on time, i.e., $\epsilon_m = \epsilon_m(t)$ and $\epsilon_r = \epsilon_r(t)$. Further we choose the so-called comoving coordinate system in which the space coordinates r, θ and ϕ are constant along the world-lines. In the comoving coordinate system the four-velocity has components $(U^\mu) = (c, \mathbf{0})$ so that the space of coordinates r, θ and ϕ moves with the particles.

From the balance equation of the energy-momentum tensor $T^{\mu\nu}{}_{;\nu} = 0$, together with the representation (11.191), we get

$$-\frac{1}{3}\frac{\partial \epsilon_r}{\partial x^\nu}g^{\mu\nu} + \frac{1}{\sqrt{g}}\frac{\partial}{\partial x^\nu}\left[\sqrt{g}\left(\epsilon_m + \frac{4}{3}\epsilon_r\right)\frac{U^\mu U^\nu}{c^2}\right]$$
$$+\Gamma^\mu_{\nu\lambda}\left(\epsilon_m + \frac{4}{3}\epsilon_r\right)\frac{U^\lambda U^\nu}{c^2} = 0, \tag{11.192}$$

by taking into account that $g^{\mu\nu}{}_{;\nu} = 0$. The spatial components of (11.192) are identically zero, while the temporal component reduces to

$$3\kappa^2\dot{\kappa}\left(\frac{4}{3}\epsilon_r + \epsilon_m\right) + \kappa^3(\dot{\epsilon}_r + \dot{\epsilon}_m) = 0, \tag{11.193}$$

thanks to $(11.161)_1$, $(11.162)_2$ and $\Gamma^\mu_{00} = 0$.

If we assume that the exchange between the energy densities of matter and radiation is only important in the first few seconds at the beginning of the evolution of the universe, we can decouple (11.193) into two equations, one for the energy density of the radiation and another for the energy density of the matter, which read

$$4\kappa^2\dot{\kappa}\epsilon_r + \kappa^3\dot{\epsilon}_r = 0, \qquad 3\kappa^2\dot{\kappa}\epsilon_m + \kappa^3\dot{\epsilon}_m = 0. \tag{11.194}$$

The solutions of the two differential equations above are

$$\epsilon_r = \frac{C_1}{\kappa^4(t)}, \qquad \epsilon_m = \frac{C_2}{\kappa^3(t)}, \tag{11.195}$$

where C_1 and C_2 are two constants of integration. Hence the energy density of radiation scales as $\kappa(t)^{-4}$, while the energy density of the matter scales as $\kappa(t)^{-3}$. To determine the energy densities of radiation and matter completely one has to know $\kappa(t)$, which can be found from the Einstein field equations, and this will be the subject of the next section.

Problems

11.11.2.1 Obtain (11.192) from the balance equation for the energy-momentum tensor $T^{\mu\nu}{}_{;\nu} = 0$.

11.11.2.2 Show that the temporal component of the balance equation for the energy-momentum tensor of a perfect fluid in a comoving frame is given by

$$3\dot{\kappa}(\epsilon + p) + \kappa\dot{\epsilon} = 0.$$

11.11.2.3 Obtain from the balance equation of the particle four-flow $N^{\mu}{}_{;\mu} = 0$ that $n \propto 1/\kappa^3$.

11.11.2.4 By using the results of the previous problems show that: a) for radiation where $\epsilon = 3p$ holds we have $p \propto 1/\kappa^4$ and $T \propto 1/\kappa$; b) for matter where $\epsilon \approx \varrho c^2 + 3p/2$ holds we have $p \propto 1/\kappa^5$ and $T \propto 1/\kappa^2$.

11.11.3 Determination of $\kappa(t)$

In order to write down the system of equations that follow from the Einstein field equations

$$G_{\mu\nu} = -\frac{8\pi G}{c^4} T_{\mu\nu}, \tag{11.196}$$

we have to know the non-vanishing covariant components of the energy-momentum tensor (11.188) in a comoving system, which are given by

$$T_{00} = \epsilon, \quad T_{11} = \frac{p\kappa^2}{1 - \varepsilon r^2}, \quad T_{22} = p\kappa^2 r^2, \quad T_{33} = p\kappa^2 r^2 \sin^2\theta. \tag{11.197}$$

Now we use (11.168), (11.169), (11.196) and (11.197) to write the components G_{00} and G_{11} of the Einstein field equations as

$$\dot{\kappa}^2 + \varepsilon = \frac{8\pi G}{3c^4}\kappa^2\epsilon, \tag{11.198}$$

$$2\ddot{\kappa}\kappa + \dot{\kappa}^2 + \varepsilon = -\frac{8\pi G}{c^4}p\kappa^2, \quad \text{or by (11.198)} \quad \ddot{\kappa} = -\frac{4\pi G}{3c^4}(\epsilon + 3p)\kappa. \tag{11.199}$$

The components G_{22} and G_{33} lead to the same equation as that for the component G_{11}.

From $(11.199)_2$ we infer that $\ddot{\kappa} < 0$ since $p \geq 0$ and $\epsilon > 0$. When $\dot{\kappa} > 0$ the universe is expanding while for $\dot{\kappa} < 0$ the universe is contracting. For the integration of (11.198) we shall analyze two separate cases that correspond to a radiation dominated period and to a matter dominated period.

Radiation dominated period

By using the result that the radiation energy density is given by $(11.195)_1$ we get from (11.198)

$$\dot{\kappa}^2 - \frac{\mathcal{C}_1}{\kappa^2} + \varepsilon = 0, \tag{11.200}$$

where $\mathcal{C}_1 = 8\pi G C_1/(3c^4)$. The solutions of (11.200) by considering the initial condition $\kappa(0) = 0$ are:

i) Three-dimensional sphere ($\varepsilon = +1$)

$$\kappa(t) = \sqrt{\mathcal{C}_1}\sqrt{1 - \left(1 - \frac{ct}{\sqrt{\mathcal{C}_1}}\right)^2}; \tag{11.201}$$

ii) Flat three-dimensional space ($\varepsilon = 0$)

$$\kappa(t) = (4c^2\mathcal{C}_1)^{\frac{1}{4}}\sqrt{t}; \tag{11.202}$$

iii) Three-dimensional hyperboloid ($\varepsilon = -1$)

$$\kappa(t) = \sqrt{\mathcal{C}_1}\sqrt{\left(1 + \frac{ct}{\sqrt{\mathcal{C}_1}}\right)^2 - 1}. \tag{11.203}$$

Matter dominated period

In this case the matter energy density is given by $(11.195)_2$ and (11.198) reduces to

$$\dot{\kappa}^2 - \frac{\mathcal{C}_2}{\kappa} + \varepsilon = 0, \tag{11.204}$$

where $\mathcal{C}_2 = 8\pi G C_2/(3c^4)$. By considering the initial condition $\kappa(0) = 0$ we get that the solutions of (11.204) are given by:

i) Three-dimensional sphere ($\varepsilon = +1$)

$$\kappa(t) = \frac{\mathcal{C}_2}{2}(1 - \cos \mathcal{T}), \qquad \text{with} \qquad t = \frac{\mathcal{C}_2}{2c}(\mathcal{T} - \sin \mathcal{T}), \tag{11.205}$$

which is the parametric equation of a cycloid;

ii) Flat three-dimensional space ($\varepsilon = 0$)

$$\kappa(t) = \left(\frac{9c^2\mathcal{C}_2}{4}\right)^{\frac{1}{3}} t^{\frac{2}{3}}; \tag{11.206}$$

iii) Three-dimensional hyperboloid ($\varepsilon = -1$)

$$\kappa(t) = \frac{\mathcal{C}_2}{2}(\cosh \mathcal{T} - 1), \qquad \text{with} \qquad t = \frac{\mathcal{C}_2}{2c}(\sinh \mathcal{T} - \mathcal{T}). \tag{11.207}$$

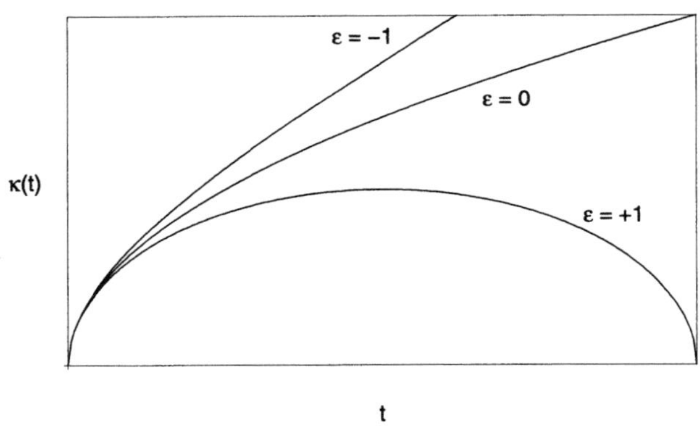

Figure 11.1: Cosmic scale factor κ as a function of time t.

Hence we have determined $\kappa(t)$ for the radiation and matter dominated periods. A graphic representation of the solutions (11.201) through (11.203) or (11.205) through (11.207) is given in Figure 11.1. Note that for $\varepsilon = +1$ the space is unbounded but finite, whereas for $\varepsilon = -1$ or $\varepsilon = 0$ the space is infinite.

Problems

11.11.3.1 Show that $(11.199)_2$ follows by differentiating (11.198) with respect to x^0 and by taking into account the equation of Problem 11.11.2.2.

11.11.3.2 For the radiation and matter dominated periods obtain the solutions for $\kappa(t)$ given by (11.201) through (11.203) and (11.205) through (11.207), respectively.

References

[1] R. Adler, M. Bazin and M. Schiffer, *Introduction to general relativity* 2nd. ed. (McGraw-Hill, New York, 1975).

[2] P. A. M. Dirac, *General theory of relativity* (Princeton University Press, Princeton, 1996).

[3] H. A. Lorentz, A. Einstein, H. Minkowski and H. Weyl, *The principle of relativity* (Dover, New York, 1952).

[4] A. Einstein, *The meaning of relativity* 5th. ed. (Princeton University Press, Princeton, 1974).

[5] T. Fliessbach, *Allgemeine Relativitätstheorie* (B. I. Wissenschaftsverlag, Mannheim, 1990).

[6] V. Fock, *The theory of space time and gravitation* (Pergamon Press, London, 1959).

[7] M. D. Kruskal, Maximal extension of Schwarzschild metric, *Phys. Rev.* **119**, 1743–1745 (1960).

[8] L. D. Landau and E. M. Lifshitz, *The classical theory of fields*, 4th ed. (Pergamon Press, Oxford, 1980).

[9] A. Lichnerowicz, *Théories relativistes de la gravitation et de l'électromagnétisme* (Masson, Paris, 1955).

[10] C. W. Misner, K. S. Thorne and J. A. Wheeler, *Gravitation* (W. H. Freeman, San Francisco, 1973).

[11] C. Møller, *The theory of relativity,* (Oxford University Press, London, 1952).

[12] W. Pauli, *Theory of relativity* (Dover, New York, 1981).

[13] E. Schrödinger, *Space-time Ssructure* (Cambridge University Press, Cambridge, 1950).

[14] H. Stephani, *General relativity* 2nd. ed. (Cambridge University Press, Cambridge, 1990).

[15] R. M. Wald, *General relativity* (The University of Chicago Press, Chicago, 1984).

[16] S. Weinberg, *Gravitation and cosmology. Principles and applications of the theory of relativity* (Wiley, New York, 1972).

[17] H. Weyl, *Space-time-matter* (Dover, New York, 1952).

Chapter 12

Boltzmann Equation in Gravitational Fields

12.1 Introduction

In this chapter a relativistic gas in the presence of a gravitational field is studied. First, in order to define properly the scalar invariant corresponding to the one-particle distribution function, the transformation laws of the volume elements in phase space are considered. The Boltzmann equation and the general equation of transfer in the presence of a gravitational field are derived and the constraints imposed on the fields by the Boltzmann equation in the analysis of the states of equilibrium are discussed. As an application, the dynamic pressure and the entropy production rate in a homogeneous and isotropic universe are calculated.

12.2 Transformation of volume elements

We recall that in a Minkowski space the transformations of the volume elements d^3x and d^3p were given by (see Section 2.1)

$$d^3x = \frac{1}{\gamma}d^3x', \quad \frac{d^3p}{p_0} = \frac{d^3p'}{p_0'}, \quad p_0' = \frac{p_0}{\gamma}, \quad \text{so that} \quad d^3xd^3p = d^3x'd^3p'. \tag{12.1}$$

Further we have introduced the one-particle distribution function, defined in terms of the space-time and momentum coordinates $f(x^\alpha, p^\alpha)$, which was taken equal to $f(\mathbf{x}, \mathbf{p}, t)$ since $p^0 = \sqrt{m^2c^2 + |\mathbf{p}|^2}$. The distribution function was defined as a scalar invariant such that $f(\mathbf{x}, \mathbf{p}, t)d^3xd^3p$ gives at time t the number of particles in the volume element d^3x about $\mathbf{x}$ with momenta in a range d^3p about $\mathbf{p}$.

In this section we shall determine the invariant element of volume in the phase space spanned by the space and momentum coordinates $(\mathbf{x}, \mathbf{p})$ in a Riemannian space in order to properly define the scalar invariant one-particle distribution function.

In a Riemannian space the constant length of the momentum four-vector $g_{\mu\nu}p^\mu p^\nu = m^2c^2$ implies that

$$p^0 = \frac{p_0 - g_{0i}p^i}{g_{00}}, \qquad \text{where} \qquad p_0 = \sqrt{g_{00}m^2c^2 + (g_{0i}g_{0j} - g_{00}g_{ij})\, p^i p^j}. \quad (12.2)$$

Let us first analyze the transformation of the volume element d^3p which is more involved. We begin by writing the transformation between the volume elements in terms of the Jacobian

$$d^3p' = |J|d^3p = \left|\frac{\partial(p'^1, p'^2, p'^3)}{\partial(p^1, p^2, p^3)}\right| d^3p. \quad (12.3)$$

For the coordinate transformation $x'^\mu = x'^\mu(x^0, x^1, x^2, x^3)$ the contravariant and the covariant coordinates of the momentum four-vector transform according to

$$p'^\mu = \frac{\partial x'^\mu}{\partial x^\nu} p^\nu, \qquad p_\mu = \frac{\partial x'^\nu}{\partial x^\mu} p'_\nu, \quad (12.4)$$

so that we can build from $(12.4)_1$ the relationship

$$\frac{\partial p'^i}{\partial p^j} = \frac{\partial x'^i}{\partial x^j} + \frac{\partial x'^i}{\partial x^0}\frac{\partial p^0}{\partial p^j} = \frac{\partial x'^i}{\partial x^j} - \frac{\partial x'^i}{\partial x^0}\frac{p_j}{p_0}. \quad (12.5)$$

Hence the Jacobian of the transformation can be written as

$$J = \begin{vmatrix} \frac{\partial x'^1}{\partial x^1} - \frac{\partial x'^1}{\partial x^0}\frac{p_1}{p_0} & \frac{\partial x'^1}{\partial x^2} - \frac{\partial x'^1}{\partial x^0}\frac{p_2}{p_0} & \frac{\partial x'^1}{\partial x^3} - \frac{\partial x'^1}{\partial x^0}\frac{p_3}{p_0} \\ \frac{\partial x'^2}{\partial x^1} - \frac{\partial x'^2}{\partial x^0}\frac{p_1}{p_0} & \frac{\partial x'^2}{\partial x^2} - \frac{\partial x'^2}{\partial x^0}\frac{p_2}{p_0} & \frac{\partial x'^2}{\partial x^3} - \frac{\partial x'^2}{\partial x^0}\frac{p_3}{p_0} \\ \frac{\partial x'^3}{\partial x^1} - \frac{\partial x'^3}{\partial x^0}\frac{p_1}{p_0} & \frac{\partial x'^3}{\partial x^2} - \frac{\partial x'^3}{\partial x^0}\frac{p_2}{p_0} & \frac{\partial x'^3}{\partial x^3} - \frac{\partial x'^3}{\partial x^0}\frac{p_3}{p_0} \end{vmatrix}$$

$$= \begin{vmatrix} 1 & \frac{p_1}{p_0} & \frac{p_2}{p_0} & \frac{p_3}{p_0} \\ \frac{\partial x'^1}{\partial x^0} & \frac{\partial x'^1}{\partial x^1} & \frac{\partial x'^1}{\partial x^2} & \frac{\partial x'^1}{\partial x^3} \\ \frac{\partial x'^2}{\partial x^0} & \frac{\partial x'^2}{\partial x^1} & \frac{\partial x'^2}{\partial x^2} & \frac{\partial x'^2}{\partial x^3} \\ \frac{\partial x'^3}{\partial x^0} & \frac{\partial x'^3}{\partial x^1} & \frac{\partial x'^3}{\partial x^2} & \frac{\partial x'^3}{\partial x^3} \end{vmatrix}, \quad (12.6)$$

where the last equality follows by the use of the property of lowering the order of the determinant. Moreover from the transformation $(12.4)_2$ one can obtain that

$$p_0 = \frac{\partial x'^0}{\partial x^0} p'_0 + \frac{\partial x'^i}{\partial x^0} p'_i, \qquad p_i = \frac{\partial x'^0}{\partial x^i} p'_0 + \frac{\partial x'^j}{\partial x^i} p'_j, \quad (12.7)$$

and the Jacobian (12.6) reduces to

$$J = \begin{vmatrix} \frac{\partial x'^0}{\partial x^0}\frac{p'_0}{p_0} + \frac{\partial x'^i}{\partial x^0}\frac{p'_i}{p_0} & \frac{\partial x'^0}{\partial x^1}\frac{p'_0}{p_0} + \frac{\partial x'^j}{\partial x^1}\frac{p'_j}{p_0} & \frac{\partial x'^0}{\partial x^2}\frac{p'_0}{p_0} + \frac{\partial x'^j}{\partial x^2}\frac{p'_j}{p_0} & \frac{\partial x'^0}{\partial x^3}\frac{p'_0}{p_0} + \frac{\partial x'^j}{\partial x^3}\frac{p'_j}{p_0} \\ \frac{\partial x'^1}{\partial x^0} & \frac{\partial x'^1}{\partial x^1} & \frac{\partial x'^1}{\partial x^2} & \frac{\partial x'^1}{\partial x^3} \\ \frac{\partial x'^2}{\partial x^0} & \frac{\partial x'^2}{\partial x^1} & \frac{\partial x'^2}{\partial x^2} & \frac{\partial x'^2}{\partial x^3} \\ \frac{\partial x'^3}{\partial x^0} & \frac{\partial x'^3}{\partial x^1} & \frac{\partial x'^3}{\partial x^2} & \frac{\partial x'^3}{\partial x^3} \end{vmatrix}$$

$$= \frac{p_0'}{p_0}\left|\frac{\partial x'}{\partial x}\right| \overset{(10.31)}{=} \frac{p_0'}{p_0}\sqrt{\frac{g}{g'}}. \tag{12.8}$$

Above we have used the property that the determinant does not change when we add to a row a linear combination of the other rows. From (12.3) and (12.8) follows the transformation law for the volume element d^3p:

$$\sqrt{g'}\frac{d^3p'}{p_0'} = \sqrt{g}\frac{d^3p}{p_0}. \tag{12.9}$$

For the transformation of the volume element d^3x we recall (Section 10.2) that $\sqrt{g}d^4x = \sqrt{g'}d^4x'$ is an invariant volume element and we say that d^4x is a scalar density of weight -1. As in Section 2.1 we choose the primed frame as a local rest system of inertia where $\mathbf{p}' = \mathbf{0}$. In this case the volume element $d^3x' \equiv dx'^1dx'^2dx'^3$ is an element of proper volume, $dx'^0 \equiv cd\tau$ is an element of proper time and $g' \equiv -\det((\eta_{\alpha\beta})) = 1$. Hence we write the transformation between the volume elements as

$$c\sqrt{g}dtdx^1dx^2dx^3 = c\sqrt{g'}d\tau dx'^1dx'^2dx'^3, \tag{12.10}$$

which implies that

$$d^3x = \sqrt{\frac{g'}{g}\frac{1}{\Gamma}}d^3x', \quad \text{with} \quad \Gamma \equiv \frac{dt}{d\tau} \overset{(11.37)}{=} \frac{1}{\sqrt{g_{00}\left(1+\frac{g_{0i}}{g_{00}}\frac{v^i}{c}\right)^2 - \frac{v^2}{c^2}}}. \tag{12.11}$$

If the primed frame is chosen as a local rest system of inertia where $\mathbf{p}' = \mathbf{0}$, we have from $(12.4)_1$ that

$$p^0 = \frac{\partial x^0}{\partial x'^0}p'^0 = \Gamma p'^0, \tag{12.12}$$

since along the trajectory of the volume element $\partial x^0/\partial x'^0 \equiv \Gamma$. By using the above relationship we can get from $(12.11)_1$ and (12.9) that the transformation law for the invariant element of volume in the phase space is given by

$$\frac{p^0}{p_0}gd^3xd^3p = \frac{p'^0}{p_0'}g'd^3x'd^3p' = d^3x'd^3p', \tag{12.13}$$

since $g' = 1$ and $p_0' = p'^0$.

We can define in a Riemannian space the one-particle distribution function $f(\mathbf{x}, \mathbf{p}, t)$ as the scalar invariant such that

$$dN = f(\mathbf{x}, \mathbf{p}, t)(p^\mu n_\mu)d\Sigma\frac{d^3p}{p_0}, \tag{12.14}$$

gives the number of particle world lines that cross the hypersurface element $d\Sigma$ with unit normal n_μ and with momentum four-vector contained in the cell d^3p/p_0

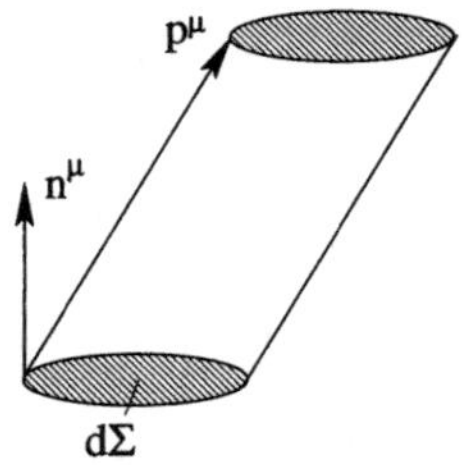

Figure 12.1: Tube of particle world lines

of the mass-shell. If the hypersurface element is chosen as the three-dimensional space on the surface x^0 = constant, we have

$$dN = f(\mathbf{x}, \mathbf{p}, t)\frac{p^0}{p_0} g d^3x d^3p, \tag{12.15}$$

which reduces to $dN = f(\mathbf{x}, \mathbf{p}, t)d^3xd^3p$ in a Minkowskian space.

Another invariant volume element to be used later in this chapter is

$$\frac{g'}{p'_0} d^3p' d^4x' = \frac{g}{p_0} d^3p d^4x, \tag{12.16}$$

which can be obtained from the product of (12.9) by the invariant $\sqrt{g}d^4x$.

Problems

12.2.1 Obtain the relationships (12.2) from the conditions $g_{\mu\nu}p^\mu p^\nu = m^2c^2$ and $p_0 = g_{0\mu}p^\mu$.

12.2.2 Show that $\partial p'^i/\partial p^j$ is given by (12.5).

12.2.3 Obtain (12.8) from (12.6).

12.3 Boltzmann equation

In Section 2.1 we derived the Boltzmann equation in a Minkowski space for a single non-degenerate relativistic gas. In the absence of external forces the Boltzmann equation reads (see (2.36))

$$p^\alpha \frac{\partial f}{\partial x^\alpha} = \int (f'_* f' - f_* f)\, F\, \sigma\, d\Omega \frac{d^3p_*}{p_{*0}}, \tag{12.17}$$

where σ is the invariant differential cross-section, $d\Omega$ a solid angle element that characterizes the binary collision between the particles of the gas and F is the invariant flux

$$F = \sqrt{(p_*^\mu p_\mu)^2 - m^4c^4} = \sqrt{(g_{\mu\nu}p_*^\mu p^\nu)^2 - m^4c^4}. \tag{12.18}$$

We shall derive now the appropriate form of the Boltzmann equation in the presence of a gravitational field and we begin with the analysis of its left-hand side. First we write the one-particle distribution function $f(\mathbf{x}, \mathbf{p}, t)$ as $f(x^\mu(\tau^*), p^i(\tau^*))$ where $\tau^* = \tau/m$ is an affine parameter along the world line of a particle of rest mass m with τ denoting the proper time. Hence the variation of the one-particle distribution function with respect to the affine parameter τ^* is given by

$$\frac{df(x^\mu(\tau^*), p^i(\tau^*))}{d\tau^*} = \frac{\partial f}{\partial x^\mu}\frac{dx^\mu}{d\tau^*} + \frac{\partial f}{\partial p^i}\frac{dp^i}{d\tau^*}. \tag{12.19}$$

According to (11.43) the equation of motion of a particle in the presence of a gravitational field is given by

$$\frac{dp^i}{d\tau^*} = -\Gamma^i_{\mu\nu}p^\mu p^\nu, \qquad \text{where} \qquad p^\mu = \frac{dx^\mu}{d\tau^*}. \tag{12.20}$$

Hence (12.19) is written as

$$\frac{df(x^\mu(\tau^*), p^i(\tau^*))}{d\tau^*} = p^\mu\frac{\partial f}{\partial x^\mu} - \Gamma^i_{\mu\nu}p^\mu p^\nu\frac{\partial f}{\partial p^i}, \tag{12.21}$$

so that the left-hand side of the Boltzmann equation is replaced by

$$p^\mu\frac{\partial f}{\partial x^\mu} \longrightarrow p^\mu\frac{\partial f}{\partial x^\mu} - \Gamma^i_{\mu\nu}p^\mu p^\nu\frac{\partial f}{\partial p^i}. \tag{12.22}$$

In the expression on the right-hand side of the Boltzmann equation we have to replace the invariant element d^3p_*/p_{*0} by $\sqrt{g}d^3p_*/p_{*0}$. Therefore the Boltzmann equation in the presence of a gravitational field reads

$$p^\mu\frac{\partial f}{\partial x^\mu} - \Gamma^i_{\mu\nu}p^\mu p^\nu\frac{\partial f}{\partial p^i} = \int (f_*'f' - f_*f)\, F\,\sigma\, d\Omega\sqrt{g}\frac{d^3p_*}{p_{*0}} = Q(f,f). \tag{12.23}$$

If the mass-shell condition $g_{\mu\nu}p^\mu p^\nu = m^2c^2$ is not taken into account, the Boltzmann equation for $f(x^\mu, p^\mu)$ reads

$$p^\mu\frac{\partial f}{\partial x^\mu} - \Gamma^\sigma_{\mu\nu}p^\mu p^\nu\frac{\partial f}{\partial p^\sigma} = \int (f_*'f' - f_*f)\, F\,\sigma\, d\Omega\sqrt{g}\frac{d^3p_*}{p_{*0}} = Q(f,f). \tag{12.24}$$

The above equation can be obtained from (12.23) as follows. First we write

$$\frac{\partial f}{\partial x^\mu} = \frac{\partial f}{\partial x^\mu} + \frac{\partial f}{\partial p^0}\frac{\partial p^0}{\partial x^\mu}, \qquad \frac{\partial f}{\partial p^i} = \frac{\partial f}{\partial p^i} + \frac{\partial f}{\partial p^0}\frac{\partial p^0}{\partial p^i}, \tag{12.25}$$

where in the left-hand sides of (12.25) the one-particle distribution function is considered a function of (x^μ, p^i), while in their right-hand sides f is a function of (x^μ, p^μ). From the mass-shell condition it follows that

$$\frac{\partial p^0}{\partial x^\mu} = -\frac{1}{p_0} p^\nu p_\kappa \Gamma^\kappa_{\mu\nu}, \qquad \frac{\partial p^0}{\partial p^i} = -\frac{p_i}{p_0}, \tag{12.26}$$

so that the insertion of (12.25) together with (12.26) into (12.23) leads to (12.24).

We consider a seven-dimensional phase space spanned by the coordinates (x^μ, p^i) with invariant element of volume given by

$$dF = \frac{g}{p_0} d^3p d^4x, \tag{12.27}$$

and introduce a seven-dimensional momentum p^A and a seven-dimensional gradient $\partial/(\partial x^A)$ defined by

$$(p^A) = \left(\frac{dx^\mu}{d\tau^*}, \frac{dp^i}{d\tau^*}\right) \equiv \left(p^\mu, -\Gamma^i_{\mu\nu} p^\mu p^\nu\right), \qquad \left(\frac{\partial}{\partial x^A}\right) = \left(\frac{\partial}{\partial x^\mu}, \frac{\partial}{\partial p^i}\right). \tag{12.28}$$

Liouville's theorem asserts that the density of the points in the phase space is constant along the trajectories in the phase space. This means that the density of the points in the phase space moves like an incompressible fluid. If we identify g/p_0 as the density of the points in the phase space spanned by (x^μ, p^i), we have according Liouville's theorem that the divergence of gp^A/p_0 must vanish, i.e.,

$$\frac{\partial}{\partial x^A}\left(\frac{g}{p_0} p^A\right) = 0, \quad \text{or} \quad \frac{\partial}{\partial x^\mu}\left(\frac{g}{p_0} p^\mu\right) + \frac{\partial}{\partial p^i}\left(-\frac{g}{p_0} \Gamma^i_{\mu\nu} p^\mu p^\nu\right) = 0. \tag{12.29}$$

We shall use the last condition in order to derive the transfer equation from the Boltzmann equation (12.23).

Problems

12.3.1 Obtain the relationships (12.26) from the mass-shell condition $g_{\mu\nu}p^\mu p^\nu = m^2c^2$.

12.3.2 Show by direct calculation that $(12.29)_2$ is valid.

12.4 Transfer equation

To obtain the transfer equation from the Boltzmann equation we multiply (12.23) by an arbitrary scalar function $\psi(x^\mu, p^i)$ and integrate over all values of $g/p_0 d^3p d^4x$, yielding

$$\int \psi \left[p^\mu \frac{\partial f}{\partial x^\mu} - \Gamma^i_{\mu\nu} p^\mu p^\nu \frac{\partial f}{\partial p^i}\right] \frac{g}{p_0} d^3p d^4x = \int \frac{\partial}{\partial x^\mu}\left[\psi p^\mu f \frac{g}{p_0}\right] d^3p d^4x$$

$$-\int p^\mu f\frac{\partial\psi}{\partial x^\mu}\frac{g}{p_0}d^3p d^4x-\int\frac{\partial}{\partial p^i}\left[\psi\Gamma^i_{\mu\nu}p^\mu p^\nu f\frac{g}{p_0}\right]d^3p d^4x$$

$$+\int\frac{\partial\psi}{\partial p^i}\Gamma^i_{\mu\nu}p^\mu p^\nu f\frac{g}{p_0}d^3p d^4x=\int\psi\left(f'_*f'-f_*f\right)F\,\sigma\,d\Omega\sqrt{g}\frac{d^3p_*}{p_{*0}}\frac{g}{p_0}d^3p d^4x,\tag{12.30}$$

where the condition $(12.29)_2$ has been used. The first term in the second equality above can be written, thanks to (10.54), as

$$\int\frac{\partial}{\partial x^\mu}\left[\psi p^\mu f\frac{g}{p_0}\right]d^3p d^4x=\int\left\{\frac{\partial}{\partial x^\mu}\left[\int\psi p^\mu f\sqrt{g}\frac{d^3p}{p_0}\right]\right.$$

$$\left.+\Gamma^\kappa_{\mu\kappa}\int\psi p^\mu f\sqrt{g}\frac{d^3p}{p_0}\right\}\sqrt{g}d^4x\overset{(10.69)}{=}\int\left[\int\psi p^\mu f\sqrt{g}\frac{d^3p}{p_0}\right]_{;\mu}\sqrt{g}d^4x.\tag{12.31}$$

The third term in the second equality in (12.30) vanishes since we can transform the volume integral in momentum space into an integral over an infinitely far surface where the one-particle distribution function tends to zero. Hence (12.30) reduces to

$$\int\left\{\left[\int\psi p^\mu f\sqrt{g}\frac{d^3p}{p_0}\right]_{;\mu}-\int p^\mu f\frac{\partial\psi}{\partial x^\mu}\sqrt{g}\frac{d^3p}{p_0}+\int\frac{\partial\psi}{\partial p^i}\Gamma^i_{\mu\nu}p^\mu p^\nu f\sqrt{g}\frac{d^3p}{p_0}\right.$$

$$\left.-\int\psi\left(f'_*f'-f_*f\right)F\,\sigma\,d\Omega\sqrt{g}\frac{d^3p_*}{p_{*0}}\sqrt{g}\frac{d^3p}{p_0}\right\}\sqrt{g}d^4x=0.\tag{12.32}$$

The final form of the transfer equation follows from (12.32) by considering that the integrand of this equation must vanish due to the fact that the integration over $\sqrt{g}d^4x$ is arbitrary. Hence we get

$$\left[\int\psi p^\mu f\sqrt{g}\frac{d^3p}{p_0}\right]_{;\mu}-\int p^\mu f\frac{\partial\psi}{\partial x^\mu}\sqrt{g}\frac{d^3p}{p_0}+\int\frac{\partial\psi}{\partial p^i}\Gamma^i_{\mu\nu}p^\mu p^\nu f\sqrt{g}\frac{d^3p}{p_0}$$

$$=\frac{1}{4}\int\left(\psi+\psi_*-\psi'-\psi'_*\right)\left(f'_*f'-f_*f\right)F\,\sigma\,d\Omega\sqrt{g}\frac{d^3p_*}{p_{*0}}\sqrt{g}\frac{d^3p}{p_0},\tag{12.33}$$

by using the symmetry properties of the collision term. Equation (12.33) represents the general equation of transfer for an arbitrary scalar function of a relativistic gas in the presence of a gravitational field.

For an arbitrary tensorial function $\psi^\nu(x^\mu,p^i)$ the general equation of transfer reads

$$\left[\int\psi^\nu p^\mu f\sqrt{g}\frac{d^3p}{p_0}\right]_{;\mu}-\int p^\mu f\psi^\nu{}_{;\mu}\sqrt{g}\frac{d^3p}{p_0}+\int\frac{\partial\psi^\nu}{\partial p^i}\Gamma^i_{\mu\sigma}p^\mu p^\sigma f\sqrt{g}\frac{d^3p}{p_0}$$

$$= \frac{1}{4}\int (\psi^\nu + \psi^\nu_* - \psi'^\nu - \psi'^\nu_*)\,(f'_*f' - f_*f)\; F\,\sigma\, d\Omega \sqrt{g}\frac{d^3p_*}{p_{*0}}\sqrt{g}\frac{d^3p}{p_0}. \qquad (12.34)$$

Now we are able to derive the balance equations for the particle four-flow N^μ, energy-momentum tensor $T^{\mu\nu}$ and entropy four-flow S^μ, defined by

$$N^\mu = c\int p^\mu f\sqrt{g}\frac{d^3p}{p_0}, \qquad T^{\mu\nu} = c\int p^\mu p^\nu f\sqrt{g}\frac{d^3p}{p_0}, \qquad (12.35)$$

$$S^\mu = -kc\int p^\mu f \ln\left(\frac{fh^3}{eg_s}\right)\sqrt{g}\frac{d^3p}{p_0}. \qquad (12.36)$$

Indeed by choosing in (12.33) and (12.34) the following values for ψ and ψ^ν,

$$\psi = c, \qquad \psi^\nu = cp^\nu, \qquad \psi = -kc\ln\left(\frac{fh^3}{eg_s}\right), \qquad (12.37)$$

we get the corresponding equations

$$N^\mu{}_{;\mu} = 0, \qquad T^{\mu\nu}{}_{;\nu} = 0, \qquad S^\mu{}_{;\mu} \geq 0\,, \qquad (12.38)$$

which represent the balance equations for the particle four-flow, energy-momentum tensor and entropy four-flow, respectively.

Problems

12.4.1 Show that the general equation of transfer for an arbitrary tensorial function ψ^ν is given by (12.34).

12.4.2 Obtain the balance equations (12.38) for the particle four-flow, energy-momentum tensor and entropy four-flow from the equations of transfer (12.33) and (12.34).

12.5 Equilibrium states

This section is based mainly on the work of Chernikov [3] and its aim is to investigate the equilibrium states of a relativistic gas in the presence of a gravitational field. We recall that the equilibrium distribution function (2.123) which refers to the Maxwell–Jüttner distribution function

$$f^{(0)} = \frac{g_s}{h^3}e^{\frac{\mu_E}{kT} - \frac{U^\alpha p_\alpha}{kT}}, \qquad (12.39)$$

makes the right-hand side of the Boltzmann equation (12.23) vanish identically, i.e.,

$$\int \left(f'^{(0)}_* f'^{(0)} - f^{(0)}_* f^{(0)}\right) F\,\sigma\, d\Omega\sqrt{g}\frac{d^3p_*}{p_{*0}} = 0. \qquad (12.40)$$

As in Section 2.10 we search for the restrictions that the left-hand side of the Boltzmann equation (12.23) imposes in the fields that appear in the equilibrium distribution function. For that end we insert (12.39) into (12.23) and get, after some calculations by using (12.2) and by introducing the covariant derivative,

$$p^\mu \left[\frac{\mu_E}{kT}\right]_{;\mu} - \frac{1}{2}p^\mu p^\nu \left\{\left[\frac{U_\mu}{kT}\right]_{;\nu} + \left[\frac{U_\nu}{kT}\right]_{;\mu}\right\} = 0. \tag{12.41}$$

The above equation can be written in terms of a polynomial equation for the spatial components of the four-momenta p^i. Since the components p^i are linearly independent and arbitrary, all coefficients of the polynomial equation must vanish and it follows that

$$\left[\frac{\mu_E}{kT}\right]_{;\mu} = 0, \qquad \text{or} \qquad [\ln \mu_E]_{;\mu} = [\ln T]_{;\mu}, \tag{12.42}$$

$$\left[\frac{U_\mu}{kT}\right]_{;\nu} + \left[\frac{U_\nu}{kT}\right]_{;\mu} = \begin{cases} 0, & \text{if } m \neq 0, \\ \varphi(x^\delta) g_{\mu\nu}, & \text{if } m = 0. \end{cases} \tag{12.43}$$

$\varphi(x^\delta)$ is an arbitrary function that depends on the space-time coordinates and we note that for a relativistic gas with particles of zero rest mass we have the condition $g_{\mu\nu}p^\mu p^\nu = 0$.

Let us analyze the case of a relativistic gas with particles of non-vanishing rest mass. If we multiply (12.43) by $U^\mu U^\nu$ it follows that

$$U^\mu T_{;\mu} = 0, \tag{12.44}$$

while the multiplication of the same equation by U^μ leads to

$$U^\mu U_{\nu;\mu} = \frac{c^2}{T} T_{;\nu}, \tag{12.45}$$

thanks to (12.44) and by noting that $U^\mu U_{\mu;\nu} = 0$. We infer from (12.45) that a relativistic gas with a gradient of temperature may be in a state of equilibrium only if the gradient of temperature is compensated by an acceleration $U^\mu U_{\nu;\mu}$ of the gas. From (12.44) we conclude that in equilibrium the temperature of a relativistic gas may not depend on time since in a comoving frame we have that $\partial T/\partial t = 0$. The same conclusion is valid, thanks to (12.42), for the chemical potential in equilibrium, i.e., $\partial \mu_E/\partial t = 0$. Further let us write (12.43) for $m \neq 0$ as

$$\left[\frac{U_\mu}{kT}\right]_{;\nu} + \left[\frac{U_\nu}{kT}\right]_{;\mu} = g_{\sigma\mu}\left[\frac{U^\sigma}{kT}\right]_{;\nu} + g_{\sigma\nu}\left[\frac{U^\sigma}{kT}\right]_{;\mu}$$

$$= g_{\sigma\nu}\frac{\partial}{\partial x^\mu}\left(\frac{U^\sigma}{kT}\right) + g_{\sigma\mu}\frac{\partial}{\partial x^\nu}\left(\frac{U^\sigma}{kT}\right) + \frac{U^\sigma}{kT}\frac{\partial g_{\mu\nu}}{\partial x^\sigma} = 0, \tag{12.46}$$

where we have used the definition of the covariant derivative (10.58) and the expression (10.51) for the Christoffel symbols. Equation (12.46) is a Killing equation and $U^\mu/(kT)$ is identified as a Killing vector. If we choose a reference frame where the Killing vector has components $(U^\mu/(kT)) = (U^0/(kT), \mathbf{0})$ with $U^0/(kT)$ being constant, we get from (12.46) that

$$\frac{\partial g_{\mu\nu}}{\partial x^0} = 0. \tag{12.47}$$

Hence (12.39) represents an equilibrium distribution function for a relativistic gas with particles of non-vanishing rest mass only if the metric tensor does not depend on time, and this excludes e.g., the Robertson–Walker metric (see Section 11.11).

We turn now to a relativistic gas with particles of vanishing rest mass. In this case instead of (12.44) through (12.46) we have

$$-\frac{2}{kT^2}U^\mu T_{;\mu} = \varphi, \qquad U^\mu U^\nu{}_{;\mu} = \frac{c^2}{T}\left(g^{\mu\nu} - \frac{1}{c^2}U^\mu U^\nu\right)T_{;\mu}, \tag{12.48}$$

$$g_{\sigma\nu}\frac{\partial}{\partial x^\mu}\left(\frac{U^\sigma}{kT}\right) + g_{\sigma\mu}\frac{\partial}{\partial x^\nu}\left(\frac{U^\sigma}{kT}\right) + \frac{U^\sigma}{kT}\frac{\partial g_{\mu\nu}}{\partial x^\sigma} = -\frac{2}{kT^2}U^\sigma T_{;\sigma} g_{\mu\nu}. \tag{12.49}$$

Equation $(12.48)_1$ identifies φ with the component of the temperature gradient parallel to the four-velocity. The interpretation of $(12.48)_2$ is the same as that for a relativistic gas with particles of non-vanishing rest mass. From the last equation above (12.49) it is not possible to infer that (12.47) holds, i.e., the metric tensor in this case may depend on the time coordinate.

In the following we shall determine the distribution function in equilibrium for some particular cases.

a) Relativistic gas in absence of a gravitational field

In this case the metric tensor is that of a Minkowski space $g_{\mu\nu} = \eta_{\mu\nu}$ and according to Section 2.10, the Killing equation has a solution of the form (see (2.166)):

$$\frac{U^\mu}{kT} = \Omega^{\mu\nu}x_\nu + \Omega^\mu, \tag{12.50}$$

where $\Omega^{\mu\nu} = -\Omega^{\nu\mu}$ is a constant antisymmetric tensor and Ω^μ a constant four-vector.

If we make the choice

$$\Omega^0 > 0, \quad \Omega^i = 0, \quad \Omega^{0i} = 0, \quad \Omega^{12} = \frac{\omega}{c}\Omega^0, \quad \Omega^{13} = \Omega^{23} = 0, \tag{12.51}$$

where ω has the dimension of an angular frequency, we have that the components of the Killing vectors reduce to

$$\frac{U^0}{kT} = \Omega^0, \qquad \frac{U^1}{kT} = -\frac{\omega}{c}\Omega^0 x^2, \qquad \frac{U^2}{kT} = \frac{\omega}{c}\Omega^0 x^1, \qquad \frac{U^3}{kT} = 0. \tag{12.52}$$

From (12.52) one can obtain the components of the velocity of the gas, namely

$$\frac{U^1}{U^0}=\frac{dx^1}{cdt}=-\frac{\omega x^2}{c},\qquad \frac{U^2}{U^0}=\frac{dx^2}{cdt}=\frac{\omega x^1}{c},\qquad \frac{U^3}{U^0}=\frac{dx^3}{cdt}=0. \tag{12.53}$$

The above equations represent a uniform rotation of the relativistic gas with constant angular velocity ω about the x^3-axis.

Due to the relationship

$$\eta_{\mu\nu}\frac{U^\mu}{kT}\frac{U^\nu}{kT}=\left(\frac{c}{kT}\right)^2=(\Omega^0)^2\left\{1-\left(\frac{\omega}{c}\right)^2[(x^1)^2+(x^2)^2]\right\}\geq 0, \tag{12.54}$$

it follows that the solution is valid in the interior of the cylinder with radius r given by

$$r^2=(x^1)^2+(x^2)^2\leq\left(\frac{c}{\omega}\right)^2. \tag{12.55}$$

Further from (12.54) one can obtain that the temperature field becomes

$$T(r)=\frac{T(0)}{\sqrt{1-\frac{\omega^2r^2}{c^2}}},\qquad\text{where}\qquad T(0)=\frac{c}{k\Omega^0} \tag{12.56}$$

is the temperature at the axis of the cylinder. We may infer from (12.56) that the temperature field increases when the distance from the axis is increased.

The Maxwell–Jüttner distribution function (12.39) in this case reads

$$f^{(0)}=\frac{g_s}{h^3}\exp\left[\frac{\mu_E}{kT}-\frac{c\sqrt{m^2c^2+|\mathbf{p}|^2}+\omega(x^2p^1-x^1p^2)}{kT(0)}\right]. \tag{12.57}$$

Note that according to $(12.42)_1$ $\mu_E/(kT)$ is a constant.

b) Relativistic gas in a Schwarzschild metric

We study now the case of a spherically symmetric static gravitational field whose line element is given by (see Section 11.10)

$$ds^2=A(r)(cdt)^2-B(r)(dr)^2-r^2[(d\theta)^2+\sin^2\theta(d\phi)^2]. \tag{12.58}$$

If we consider that the components of the Killing vectors are given by

$$\frac{U^0}{kT}=\Omega^0=\text{constant}>0,\qquad\frac{U^1}{kT}=\frac{U^2}{kT}=\frac{U^3}{kT}=0, \tag{12.59}$$

we get that equation (12.46) is identically satisfied.

Moreover the constraint $g_{\mu\nu}U^\mu U^\nu=c^2$ leads to

$$T=\frac{c}{k\Omega^0\sqrt{A(r)}}. \tag{12.60}$$

We recall (see (11.139)) that the solution of the Einstein field equations leads to

$$A(r) = 1 - \frac{2GM}{rc^2}, \tag{12.61}$$

so that the temperature field (12.60) can be written as

$$T(r) = T(R)\frac{\sqrt{1-\frac{2GM}{Rc^2}}}{\sqrt{1-\frac{2GM}{rc^2}}}. \tag{12.62}$$

In the above equation $T(R)$ is the temperature at the surface of the spherical distribution of mass. We can infer from (12.62) that the temperature field decreases with increasing r.

The Maxwell–Jüttner distribution function in this case is given by

$$f^{(0)} = \frac{g_s}{h^3}\exp\left[\frac{\mu_E}{kT} - p_0\Omega^0\right]. \tag{12.63}$$

c) Relativistic gas in a Robertson–Walker metric

We consider a relativistic gas with particles of vanishing rest mass in a space which is described by the line element of the Robertson–Walker metric (11.160)

$$ds^2 = (cdt)^2 - \kappa(t)^2\left\{\frac{(dr)^2}{1-\varepsilon r^2} + r^2[(d\theta)^2 + \sin^2\theta(d\phi)^2]\right\}. \tag{12.64}$$

In a comoving frame where $(U^\mu) = (c, \mathbf{0})$ holds, the component $\mu = \nu = 0$ of (12.49) is identically satisfied, while the components $\mu = \nu = 1, 2,$ or 3 lead to

$$\frac{\partial \ln \kappa}{\partial x^0} = -\frac{\partial \ln T}{\partial x^0}. \tag{12.65}$$

We conclude from (12.65) that in equilibrium a relativistic gas with particles of vanishing rest mass has a temperature field that may depend on the time. Further the temperature is given by the inverse of the cosmic scale factor $\kappa(t)$ of the Robertson–Walker metric, i.e.,

$$T(t) = \frac{C}{\kappa(t)}, \tag{12.66}$$

where C is a constant of integration.

In this case the Maxwell–Jüttner distribution function reads

$$f^{(0)} = \frac{g_s}{h^3}\exp\left[\frac{\mu_E}{kT} - \frac{cp_0\kappa(t)}{kC}\right]. \tag{12.67}$$

Problems

12.5.1 Show that the left-hand side of the Boltzmann equation (12.23) for the equilibrium distribution function $f^{(0)}$ leads to (12.41).

12.5.2 Write (12.41) in terms of the spatial components of the four-momenta p^i and conclude from the polynomial equation in p^i that (12.42) and (12.43) hold.

12.5.3 Obtain the conditions (12.44), (12.45) and (12.48) from (12.43).

12.6 Boltzmann equation in a spherically symmetric gravitational field

This section is based on the work of Rein and Rendall [7] and its aim is to write the left-hand side of the Boltzmann equation in Cartesian coordinates for a relativistic gas in presence of a non-static spherically symmetric gravitational field. The results of this section will be used in the next chapter in the analysis of the Vlasov–Einstein system.

The line element for a non-static spherically symmetric gravitational field follows from (12.58) and we reproduce it here by introducing $A(r,t) = e^{F(r,t)}$ and $B(r,t) = e^{H(r,t)}$:

$$ds^2 = e^{F(r,t)}(cdt)^2 - e^{H(r,t)}(dr)^2 - r^2[(d\theta)^2 + \sin^2\theta(d\phi)^2]. \tag{12.68}$$

In Cartesian coordinates

$$x^1 = r\sin\theta\cos\phi, \qquad x^2 = r\sin\theta\sin\phi, \qquad x^3 = r\cos\phi, \tag{12.69}$$

the line element (12.68) can be written as

$$ds^2 = e^{F(r,t)}(cdt)^2 - \left[\delta_{ij} + (e^{H(r,t)} - 1)\delta_{ik}\delta_{jl}\frac{x^k x^l}{r^2}\right] dx^i dx^j$$

$$= e^{F(r,t)}(cdt)^2 + g_{ij}dx^i dx^j. \tag{12.70}$$

For transformation of the Boltzmann equation we shall need expressions of the Christoffel symbols $\Gamma^i_{\mu\nu}$, which are given by

$$\Gamma^i_{00} = \frac{F'}{2}e^{F-H}\frac{x^i}{r}, \qquad \Gamma^i_{0j} = \frac{\dot{H}}{2}\delta_{jk}\frac{x^i x^k}{r^2}, \tag{12.71}$$

$$\Gamma^i_{jk} = (1 - e^{-H})\left[\frac{x^i}{r^2}\delta_{jk} - \delta_{jm}\delta_{kn}\frac{x^i x^m x^n}{r^4}\right] + \frac{H'}{2}\delta_{jm}\delta_{kn}\frac{x^i x^m x^n}{r^3}, \tag{12.72}$$

where the prime and the dot denote the derivative with respect to r and x^0, respectively.

We follow Rein and Rendall and define a frame on $\mathcal{R}^3$ characterized by the vectors

$$e_a^i = \delta_a^i + (e^{-H/2} - 1)\delta_{aj}\frac{x^i x^j}{r^2}, \qquad \text{such that} \qquad g_{ij}e_a^i e_b^j = \delta_{ab}. \tag{12.73}$$

The index a refers to the frame while the index i to the coordinates. Further a dimensionless vector v^i is introduced such that

$$\frac{p^i}{mc} = v^a e_a^i, \qquad \text{with} \qquad v^i = \frac{p^i}{mc} + (e^{H/2} - 1)\delta_{jk}\frac{x^i x^k}{r^2}\frac{p^j}{mc}. \tag{12.74}$$

In terms of the dimensionless vector v^i the components of the momentum four-vector p^μ read

$$\frac{p^0}{mc} = e^{-F/2}\sqrt{1+|\mathbf{v}|^2}, \qquad \frac{p^i}{mc} = v^i + (e^{-H/2} - 1)\delta_{jk}\frac{x^i x^k}{r^2}v^j, \tag{12.75}$$

where $|\mathbf{v}|^2 = \delta_{ij}v^i v^j$ and $r^2 = \delta_{ij}x^i x^j$.

We make a change of variables and instead of $(\mathbf{x}, \mathbf{p}, t)$ we use $(\mathbf{x}, \mathbf{v}, t)$ as new variables, so that the derivatives of the distribution function that appear in the Boltzmann equation (12.23) become

$$p^0 \left.\frac{\partial f}{\partial x^0}\right|_{\mathbf{p}} = p^0 \left[\left.\frac{\partial f}{\partial x^0}\right|_{\mathbf{v}} + \frac{\partial f}{\partial v^i}\frac{\dot{H}}{2}\delta_{jk}\frac{x^j v^k x^i}{r^2}\right], \tag{12.76}$$

$$p^i \left.\frac{\partial f}{\partial x^i}\right|_{\mathbf{p}} = p^i \left\{\left.\frac{\partial f}{\partial x^i}\right|_{\mathbf{v}} + \frac{\partial f}{\partial v^j}\left[\frac{H'}{2}\delta_{ik}\delta_{lm}\frac{x^j x^k x^l v^m}{r^3} + (1 - e^{-H/2})\delta_i^j \delta_{kl}\frac{x^k v^l}{r^2}\right.\right.$$
$$\left.\left. + (e^{H/2} - 1)\delta_{ik}\frac{x^j v^k}{r^2} + (e^{-H/2} - e^{H/2})\delta_{ik}\frac{x^j x^k}{r^2}\delta_{mn}\frac{x^m v^n}{r^2}\right]\right\}, \tag{12.77}$$

$$\Gamma^i_{\mu\nu}p^\mu p^\nu \frac{\partial f}{\partial p^i} = \Gamma^i_{\mu\nu}\frac{p^\mu p^\nu}{mc}\frac{\partial f}{\partial v^j}\left[\delta_i^j + (e^{H/2} - 1)\delta_{ik}\frac{x^j x^k}{r^2}\right]. \tag{12.78}$$

If we insert (12.76) through (12.78) into the Boltzmann equation (12.23) we get, after some rearrangements and by dividing the resulting equation by p^0,

$$\frac{\partial f}{\partial x^0} + \frac{e^{F/2}}{\sqrt{1+|\mathbf{v}|^2}}\left[v^i + (e^{-H/2} - 1)\delta_{jk}\frac{x^i x^j v^k}{r^2}\right]\frac{\partial f}{\partial x^i}$$
$$+ \left[\frac{e^{F/2}}{\sqrt{1+|\mathbf{v}|^2}}\left(\frac{e^{-H/2} - 1}{r^2}\right)\delta_{jk}v^j(x^i v^k - x^k v^i)\right.$$
$$\left. - \frac{F'}{2}e^{(F-H)/2}\frac{x^i}{r}\sqrt{1+|\mathbf{v}|^2} - \frac{\dot{H}}{2}\delta_{jk}\frac{x^i x^j v^k}{r^2}\right]\frac{\partial f}{\partial v^i} = \frac{Q(f,f)}{p^0}. \tag{12.79}$$

The above expression can be put in a simplified form if we consider that the distribution function is spherically symmetric so that the following condition holds:

$$\delta_{jk}x^j(x^kv^i - x^iv^k)\frac{\partial f}{\partial x^i} = \delta_{jk}v^j(v^kx^i - v^ix^k)\frac{\partial f}{\partial v^i}. \tag{12.80}$$

Indeed if we take into account (12.80) the Boltzmann equation (12.79) reduces to

$$\frac{\partial f}{\partial x^0} + \frac{e^{(F-H)/2}}{\sqrt{1+|\mathbf{v}|^2}}v^i\frac{\partial f}{\partial x^i} - \left[\frac{F'}{2}e^{(F-H)/2}\sqrt{1+|\mathbf{v}|^2}\right.$$
$$\left. + \frac{\dot{H}}{2}\frac{(\mathbf{x}\cdot\mathbf{v})}{r}\right]\frac{x^i}{r}\frac{\partial f}{\partial v^i} = \frac{Q(f,f)}{p^0}, \tag{12.81}$$

where $(\mathbf{x}\cdot\mathbf{v}) = \delta_{ij}x^iv^j$.

It is also instructive to write the energy density $\epsilon = ne$ and the sum of the hydrostatic pressure with the dynamic pressure $\Pi = p + \varpi$ in terms of the integrals involving the dimensionless vector v^i. To this end we note first that the transformation between the elements d^3p and d^3v is given by

$$d^3p = |J|d^3v, \qquad \text{where} \qquad |J| = (mc)^3e^{-H/2}. \tag{12.82}$$

Further the determinant of the metric tensor $g = -\det((g_{\mu\nu}))$ with components (see (12.70))

$$g_{00} = e^F, \qquad g_{ij} = -\left[\delta_{ij} + (e^H - 1)\delta_{ik}\delta_{jl}\frac{x^kx^l}{r^2}\right], \tag{12.83}$$

reads $g = e^{F+H}$.

The energy density $\epsilon = ne$ follows from the projection

$$\epsilon = \frac{\partial x'^\mu}{\partial x^0}\frac{\partial x'^\nu}{\partial x^0}T_{\mu\nu}, \tag{12.84}$$

where the primed coordinates refer to (x^0, x^1, x^2, x^3) and the unprimed coordinates to (x^0, r, θ, ϕ). By considering the definition of the energy-momentum tensor $(12.35)_2$ and by taking into account (12.75) and (12.82) it follows that

$$\epsilon = m^4c^5\int\sqrt{1+|\mathbf{v}|^2}fd^3v. \tag{12.85}$$

The sum of the hydrostatic pressure with the dynamic pressure $\Pi = p + \varpi$ is calculated from the projection

$$\Pi = \frac{\partial x'^\mu}{\partial r}\frac{\partial x'^\nu}{\partial r}T_{\mu\nu}. \tag{12.86}$$

Applying the same procedure we have used to calculate the energy density yields

$$\Pi = m^4 c^5 \int \left(\frac{\mathbf{x}\cdot\mathbf{v}}{r}\right)^2 f \frac{d^3v}{\sqrt{1+|\mathbf{v}|^2}}. \tag{12.87}$$

We shall use the above results in the next chapter where the Vlasov–Einstein system is studied.

Problems

12.6.1 Show that $g_{ij}e_a^i e_b^j = \delta_{ab}$.

12.6.2 Obtain the relationships (12.76) through (12.78).

12.6.3 Show that the Boltzmann equation (12.23) by taking into account (12.76) through (12.78) reduces to (12.79).

12.6.4 Show that the Jacobian of the transformation between the elements d^3p and d^3v is given by $(12.82)_2$.

12.6.5 Obtain the expressions for the energy density (12.85) and for the sum of the hydrostatic pressure with the dynamic pressure (12.87).

12.7 Dynamic pressure in a homogeneous and isotropic universe

In this section we shall determine the dynamic pressure and the entropy production rate in a spatially homogeneous and isotropic universe. These topics were first discussed, to the best of our knowledge, by Weinberg [11] within the framework of a phenomenological theory. Bernstein [1] has also determined the expressions for the dynamic pressure and for the entropy production rate from the kinetic theory of gases. Here we shall follow the work of Bernstein but in order to get explicit results for the bulk viscosity we shall use the Anderson and Witting model (8.12) for the Boltzmann equation (12.23) which reads:

$$p^\mu \frac{\partial f}{\partial x^\mu} - \Gamma^i_{\mu\nu} p^\mu p^\nu \frac{\partial f}{\partial p^i} = -\frac{U_L^\mu p_\mu}{c^2 \tau}(f - f^{(0)}). \tag{12.88}$$

We consider the Robertson–Walker metric where the line element represents a flat three-dimensional space

$$ds^2 = (cdt)^2 - \kappa(t)^2[(dx)^2 + (dy)^2 + (dz)^2]. \tag{12.89}$$

In (12.89) $\kappa(t)$ is taken as a dimensionless quantity. The non-vanishing components of the metric tensor and of the Christoffel symbols are given by

$$g_{00} = 1, \quad g_{ij} = -\kappa^2\delta_{ij}, \quad g = \kappa^6, \quad \Gamma^0_{ij} = \dot\kappa\kappa\delta_{ij}, \quad \Gamma^i_{0j} = \frac{\dot\kappa}{\kappa}\delta^i_j, \tag{12.90}$$

where the dot denotes the derivative with respect to x^0. By neglecting the space gradients – since we are dealing with a spatially homogeneous and isotropic universe – and by considering a comoving frame where $(U_L^\mu) = (c, \mathbf{0})$ in a flat three-dimensional space, it follows that the Boltzmann equation (12.88) reduces to

$$\frac{\partial f}{\partial x^0} - 2\frac{\dot{\kappa}}{\kappa}p^i\frac{\partial f}{\partial p^i} = -\frac{1}{c\tau}(f - f^{(0)}). \tag{12.91}$$

We use the Chapman and Enskog method and search for a solution of the Boltzmann equation (12.91) of the form

$$f = f^{(0)}(1+\phi), \tag{12.92}$$

where $f^{(0)}$ is the Maxwell–Jüttner distribution function and ϕ is its deviation. We recall (see Section 12.4) that $f^{(0)}$ does not represent an equilibrium distribution function for relativistic gases with particles of non-vanishing rest mass when the metric tensor depends on the time coordinate. Furthermore in a comoving frame the Maxwell–Jüttner distribution function for a non-degenerate gas (see (3.27)) reads

$$f^{(0)} = \frac{n}{4\pi m^2 ckTK_2(\zeta)}e^{-\frac{cp_0}{kT}}, \qquad \zeta = \frac{mc^2}{kT}, \tag{12.93}$$

where according to $(12.2)_2$ and (12.90) we have

$$p_0 = \sqrt{m^2c^2 + \kappa^2|\mathbf{p}|^2}, \tag{12.94}$$

with $|\mathbf{p}|^2 = \delta_{ij}p^ip^j$.

We write the Boltzmann equation (12.91) as

$$\left[1 + c\tau\left(\frac{\partial}{\partial x^0} - 2\frac{\dot{\kappa}}{\kappa}p^i\frac{\partial}{\partial p^i}\right)\right]f = f^{(0)}, \tag{12.95}$$

which can be solved for f, yielding

$$f = \sum_{l=0}^{\infty}\left[-c\tau\left(\frac{\partial}{\partial x^0} - 2\frac{\dot{\kappa}}{\kappa}p^i\frac{\partial}{\partial p^i}\right)\right]^l f^{(0)}. \tag{12.96}$$

In the above equation we have used the series expansion for $(1-x)^{-1} = \sum_{l=0}^{\infty} x^l$ in the case when x is an operator. The series is convergent if and only if the norm of x is less than unity, which is certainly not the case for a differential operator. Thus, in general, the resulting expansion will not be convergent but only asymptotic for $\tau \to 0$. By considering the terms up to $c\tau$ we get from (12.96)

$$f = f^{(0)}(1+\phi) \approx f^{(0)} - c\tau\left(\frac{\partial f^{(0)}}{\partial x^0} - 2\frac{\dot{\kappa}}{\kappa}p^i\frac{\partial f^{(0)}}{\partial p^i}\right)$$

$$\overset{(12.93)}{=} f^{(0)}\left\{1 - c\tau\left[\frac{\dot{n}}{n} + \left(1 - \zeta\frac{K_3}{K_2}\right)\frac{\dot{T}}{T} + \frac{c}{kT}p_0\frac{\dot{T}}{T} + \frac{c}{kT}\frac{\dot{\kappa}}{\kappa}\frac{|\mathbf{p}|^2}{p_0}\right]\right\}. \tag{12.97}$$

Once we know the non-equilibrium distribution function (12.97) we can calculate the projection of the energy-momentum tensor in a comoving frame which corresponds to the sum of the hydrostatic pressure with the dynamic pressure, i.e.,

$$p+\varpi=-\frac{1}{3}\left(g_{\mu\nu}-\frac{U_{L\mu}U_{L\nu}}{c^2}\right)c\int p^\mu p^\nu f\sqrt{g}\frac{d^3p}{p_0}\overset{(12.90)}{=}\frac{c\kappa^5}{3}\int|\mathbf{p}|^2f\frac{d^3p}{p_0}. \tag{12.98}$$

To this end we introduce a new variable $y=p_0/(mc)$ such that we can write, from (12.94),

$$|\mathbf{p}|=\frac{mc}{\kappa}(y^2-1)^{\frac{1}{2}},\qquad \text{and}\qquad d|\mathbf{p}|=\frac{mc}{\kappa}\frac{ydy}{(y^2-1)^{\frac{1}{2}}}. \tag{12.99}$$

We insert (12.97) into (12.98) and take into account (12.99), yielding

$$p+\varpi=\frac{p\zeta^2}{3K_2(\zeta)}\int_1^\infty e^{-\zeta y}(y^2-1)^{\frac{3}{2}}\left\{1-c\tau\left[\frac{\dot n}{n}+\left(1-\zeta\frac{K_3}{K_2}\right)\frac{\dot T}{T}\right.\right.$$
$$\left.\left.+\zeta y\frac{\dot T}{T}+\zeta\frac{y^2-1}{y}\frac{\dot\kappa}{\kappa}\right]\right\}dy. \tag{12.100}$$

The subsequent integration of the above equation leads to the following constitutive equation for the dynamic pressure:

$$\varpi=-cp\tau\left\{\frac{\dot n}{n}+\frac{\dot T}{T}+\frac{\zeta^3}{3}\left[\frac{3K_3}{\zeta^2K_2}-\frac{1}{\zeta}+\frac{K_1}{K_2}-\frac{\mathrm{Ki}_1}{K_2}\right]\frac{\dot\kappa}{\kappa}\right\}. \tag{12.101}$$

thanks to (3.19) and (8.34).

The elimination of $\dot n$ and $\dot T$ from (12.101) proceeds in the same manner as in the Chapman and Enskog method, i.e., we use the balance equations of the particle four-flow and energy-momentum tensor of an Eulerian gas whose constitutive equations read

$$N^\mu=nU_L^\mu,\qquad T^{\mu\nu}=(ne+p)\frac{U_L^\mu U_L^\nu}{c^2}-pg^{\mu\nu}. \tag{12.102}$$

Insertion of the above representations into the balance equations $N^\mu{}_{;\mu}=0$ and $T^{\mu\nu}{}_{;\nu}=0$ lead to

$$\frac{1}{\sqrt{g}}\frac{\partial}{\partial x^\mu}\left(\sqrt{g}nU_L^\mu\right)=0, \tag{12.103}$$

$$-\frac{\partial p}{\partial x^\nu}g^{\mu\nu}+\frac{1}{\sqrt{g}}\frac{\partial}{\partial x^\nu}\left[\sqrt{g}(ne+p)\frac{U_L^\mu U_L^\nu}{c^2}\right]+(ne+p)\Gamma^\mu_{\lambda\nu}\frac{U_L^\lambda U_L^\nu}{c^2}=0. \tag{12.104}$$

In a comoving frame (12.103) becomes

$$\frac{\dot n}{n}=-3\frac{\dot\kappa}{\kappa}, \tag{12.105}$$

while the spatial components of (12.104) are identically satisfied due to the constraint that all quantities n, p, e and κ are only functions of the time coordinate. The temporal component of (12.104) can be written as

$$\frac{\dot{T}}{T} = -\frac{3k}{c_v}\frac{\dot{\kappa}}{\kappa}, \tag{12.106}$$

where c_v is the heat capacity per particle[1] (see $(3.34)_1$)

$$c_v = \frac{k}{K_2^2}\left(\zeta^2 K_2^2 + 5K_3K_2\zeta - K_3^2\zeta^2 - K_2^2\right). \tag{12.107}$$

The equations (12.105) and (12.106) are used to eliminate $\dot{n}$ and $\dot{T}$ from (12.101), yielding

$$\varpi = -cp\tau\zeta\left[\frac{3(K_3^2\zeta - 5K_3K_2 - K_2^2\zeta)}{\zeta^2K_2^2 + 5K_3K_2\zeta - K_3^2\zeta^2 - K_2^2}\right.$$
$$\left.+\frac{\zeta^2}{3}\left(\frac{3K_3}{\zeta^2K_2} - \frac{1}{\zeta} + \frac{K_1}{K_2} - \frac{\mathrm{Ki}_1}{K_2}\right)\right]\frac{\dot{\kappa}}{\kappa} \overset{(8.85)}{=} -3\eta c\frac{\dot{\kappa}}{\kappa}. \tag{12.108}$$

Hence we have identified the coefficient of proportionality between ϖ and $3c\dot{\kappa}/\kappa$ as the bulk viscosity η. If we compare (12.108) with the constitutive equation for the dynamic pressure, given in terms of the divergence of the four-velocity, i.e., $\varpi = -\eta U^\mu{}_{;\mu}$, we infer that here $3c\dot{\kappa}/\kappa$ plays the same role as $U^\mu{}_{;\mu}$. Furthermore due to the fact that the bulk viscosity is a positive quantity the dynamic pressure decreases when the universe is expanding ($\dot{\kappa} > 0$) while it increases when the universe is contracting ($\dot{\kappa} < 0$).

For the determination of the entropy production rate ς we use (8.16) and (12.36) to write ς as

$$\varsigma = \frac{k}{c\tau}\int U_L^\mu p_\mu (f - f^{(0)}) \ln\frac{f}{f^{(0)}}\sqrt{g}\frac{d^3p}{p_0}. \tag{12.109}$$

If we consider that the distribution function is given by (12.97), i.e., $f = f^{(0)}(1+\phi)$, we can use the approximation $\ln(1+\phi) \approx \phi$ valid for $|\phi| \ll 1$ to write (12.109) in a comoving frame as

$$\varsigma = \frac{k}{\tau}\int f^{(0)}\phi^2\sqrt{g}d^3p. \tag{12.110}$$

We insert now (12.93) and (12.97) – which correspond to the representations of $f^{(0)}$ and ϕ in a comoving frame respectively – into (12.110) and get by integrating the resulting equation

$$\varsigma = \frac{9\eta c^2}{T}\left(\frac{\dot{\kappa}}{\kappa}\right)^2. \tag{12.111}$$

Hence the entropy production rate is connected with the bulk viscosity. Weinberg [11] has derived this formula by using a phenomenological theory and has

[1] Here we have not introduced the abbreviation $G = K_3/K_2$.

also shown that the bulk viscosity alone could not explain the high entropy of the present microwave background radiation. For more details one is referred to Weinberg [10, 11].

Problems

12.7.1 Obtain the non-equilibrium distribution function (12.97).

12.7.2 Show that the dynamic pressure is given by (12.101) by considering the non-equilibrium distribution function (12.97).

12.7.3 Obtain the two evolution equations for n and T given by (12.105) and (12.106), respectively.

12.7.4 Obtain the expression (12.111) for the entropy production rate.

12.7.5 By considering the non-equilibrium distribution function (12.97), show that the pressure deviator and the heat flux vanish, i.e., $p^{\langle\mu\nu\rangle} = 0$ and $q^\mu = 0$.

References

[1] J. Bernstein, *Kinetic theory in the expanding universe* (Cambridge University Press, Cambridge, 1988).

[2] N. A. Chernikov, The relativistic gas in the gravitational field, *Acta Phys. Polon.* **23**, 629–645 (1963).

[3] N. A. Chernikov, Equilibrium distribution of the relativistic gas, *Acta Phys. Polon.* **26**, 1069–1092 (1964).

[4] J. Ehlers, General relativity and kinetic theory, in *Proceedings of the international school of physics, Enrico Fermi, course 47* , pp. 1–70, ed. R. K. Sachs (Academic Press, New York, 1971).

[5] W. Israel, The relativistic Boltzmann equation, in *General relativity*, pp. 201–241, ed. L. O'Raifeartaigh (Clarendon Press, Oxford, 1972).

[6] C. Møller, *The theory of relativity*, (Oxford University Press, London, 1952).

[7] G. Rein and A. D. Rendall, Global existence of solutions of the spherically symmetric Vlasov–Einstein system with small initial data, *Commun. Math. Phys.* **150**, 561–583 (1992).

[8] J. M. Stewart, *Non-equilibrium relativistic kinetic theory*, Lecture Notes in Physics, vol. 10 (Springer, Heidelberg, 1971).

[9] J. L. Synge, *The relativistic gas* (North-Holland, Amsterdam, 1957).

[10] S. Weinberg, *Gravitation and cosmology. Principles and applications of the theory of relativity* (Wiley, New York, 1972).

[11] S. Weinberg, Entropy generation and the survival of protogalaxies in the expanding universe, *Ap. J.* **168**, 175–194 (1971).

Chapter 13

The Vlasov Equation and Related Systems

13.1 Introduction

There are many situations in which the forces acting among the particles are long-range and thus the effect of collisions (short-range interactions) can be neglected with respect to the averaged action of several particles at distance. One well-known case arises with charged particles in a plasma or in a semiconductor; in the second case the short-range interactions are with the impurities (including thermal vibrations) of the lattice and usually cannot be neglected, but there are several situations in plasma physics where the mean electromagnetic field plays a much more important role than the interparticle collisions. In the non-relativistic case this is dealt with via the force term in the left-hand side of the collisionless Boltzmann equation; the electric field is assumed to possess a potential which is computed via the Poisson equation, where the charge density is determined by the integral of the distribution function with respect to momentum. This produces a nonlinear system which is called the Vlasov–Poisson system. If the electric field is expressed as an integral with respect to space coordinates via the Green's function and substituted into the collisionless Boltzmann equation, we obtain a non-linear equation, usually called the Vlasov equation. We shall call the Vlasov equation, generally speaking, the collisionless Boltzmann equation in which the force term depends on fields which in turn are determined by field equations containing the distribution function in the source term. When both the mean field and the collisions are taken into account (as in semiconductors) one talks of the Boltzmann–Vlasov equation.

Thus many systems related to the Vlasov equation may arise, depending on the choice of the field appearing in the force term; in the case of electric charges one may wish to include the effect of the magnetic field produced by their motion, if they travel at high speeds, and then the Poisson equation is naturally replaced by the Maxwell equations. In this way we lose Galilei invariance and, if we want to gain Lorentz invariance in its place, we are forced to introduce a relativistic formulation. We obtain in this way the (relativistic) Vlasov–Maxwell system.

The Vlasov equation for gravitating masses has long been studied by astrophysicists. Frequently then it is known as the Jeans equation, but, mathematically, it is identical with the Vlasov equation, except for the fact that only attraction (and not repulsion) may occur. We shall always use the name Vlasov equation, not paying any attention to possible historical objections, for the sake of uniformity in terminology.

When we consider general relativity, the gravitational field occurs in the Vlasov equation via the Christoffel symbols, and the coupling of the force with the distribution function is then via the Einstein field field equations and we talk of the Vlasov–Einstein equation.

Whereas interest in the Vlasov–Maxwell system is clearly well founded by its role in many applications, we should devote a few words to justify a study of the Vlasov–Einstein system. There is, of course, the purely mathematical interest of seeing how the properties of the related, simpler systems which have been mentioned above extend to this system, which is much more complicated. In addition, we remark that one of the central themes in general relativity is investigation of the gravitational collapse of matter fields and the formation of space-time singularities. On one hand, we have the interest of mathematicians in seeing how their methods work on the system describing coupling between the Vlasov equation, which describes the conservation of the number of particles in phase space, and the elliptic or hyperbolic equations in space-time provided by Einstein's field equations; on the other hand, we would like to see the properties of the singularities which may develop in a gravitational collapse when the model for matter, which is crucial, is chosen to be that provided by kinetic theory. Other models describe matter as a perfect fluid or dust (i.e., a fluid without pressure) and develop singularities even in the Newtonian case. Hence their use prevents a study of "real singularities" of space-time (black-holes) because the matter model itself breaks down and is not capable of describing the formation of the interesting singularities.

Interest in the matter was revived in 1992 by two independent papers: on one hand Pfaffelmoser [20] produced a global existence result for classical solutions of the initial value problem for the classical Vlasov–Poisson system and, on the other hand, Rein and Rendall [23] proved the first theorem of global existence of solutions for the Vlasov–Einstein system (in the case of spherical symmetry and small data).

The first of the aforementioned results indicated that any singularity in the solutions of the Vlasov–Einstein system must be due to relativistic effects, and generated the hope that the solutions of this system may really extend up to possible space-time singularities. The second result shows that the Vlasov–Einstein system does not produce naked singularities (at least in the case of spherical symmetry and small data). This must be contrasted with the work of Christodoulos [6] and Yodzis et al [34], who showed that dust or perfect fluids in general do develop singularities, even for small, spherically symmetric data. In addition, these singularities are naked, i.e., violate the so-called cosmic censorship hypothesis. We recall that a black hole is expected to contain a singularity where space-time curvature goes to infinity. The assumption that this is the only kind of acceptable singularity is called the cosmic censorship hypothesis; any other singularity is called naked because it is not covered from the sight of observers by the black hole.

All the above results, taken together, seem to indicate that the Vlasov model of a collisionless gas may be a more suitable model for a description of matter in general relativity, with a restoration of the standard picture where cosmic censorship holds. It is to be admitted that some authors [29] claim to have observed the formation of naked singularities on a computer, through a numerical solution of the Vlasov–Einstein system. Since, however, the same numerical method exhibits singularities for the classical Vlasov–Poisson system, this work must be regarded with some suspicion, in view of [20] and can only reinforce the motivations for a more thorough study of the Vlasov–Einstein system.

We finally remark that one might envisage a Vlasov–Maxwell–Einstein system embodying both gravity and electromagnetism. No work seems to have been devoted to this extremely complex system.

13.2 The Vlasov–Maxwell system

In the frame of special relativity, we can use Minkowskian coordinates and the collisionless Boltzmann equation (Vlasov equation henceforth) may be written as

$$p^\alpha \frac{\partial f}{\partial x^\alpha} + \frac{\partial (f K^\alpha)}{\partial p^\alpha} = 0, \tag{13.1}$$

where K^α is the Minkowski force, given by

$$K^\alpha = \frac{q}{c} F^{\alpha\beta} U_\beta, \tag{13.2}$$

q being the charge of each particle and $F^{\alpha\beta}$ the electromagnetic field tensor. If there are several species (ions, electrons, neutrals), then we must consider a mixture with several distribution functions and several Vlasov equations: then the factor q in front of the electromagnetic field tensor will be different in each Vlasov equation.

The Vlasov equation must be coupled with the Maxwell equations

$$\frac{\partial F^{\alpha\beta}}{\partial x^\beta} = \partial_\beta F^{\alpha\beta} = -\mu_0 c J^\alpha, \tag{13.3}$$

$$\epsilon_{\alpha\beta\gamma\delta} \frac{\partial F^{\gamma\delta}}{\partial x_\beta} = \epsilon_{\alpha\beta\gamma\delta} \partial^\beta F^{\gamma\delta} = 0, \quad \text{or} \quad \partial_\gamma F_{\alpha\beta} + \partial_\alpha F_{\beta\gamma} + \partial_\beta F_{\gamma\alpha} = 0. \tag{13.4}$$

Here J^α is not assigned, but is expressed in terms of the distribution function f as

$$J^\alpha = qc \int p^\alpha f \frac{d^3 p}{p_0}. \tag{13.5}$$

This clearly couples the system of Maxwell equations to the Vlasov equation and produces the Vlasov–Maxwell system. Again, if there are several species a modification is called for and a sum over the species with the appropriate charges and distribution functions should be introduced in the last equation.

13.3 The Vlasov–Einstein system

The Vlasov–Maxwell system is invariant with respect to Lorentz transformations, but not with respect to more general coordinate transformations. In order to allow more general coordinates or to describe a self-gravitating collisionless gas we need to use the Christoffel symbols. To reduce the length of the equations and also the specifications to be given we shall assume that the electromagnetic field (as well as any force other than gravitation) vanishes, and we write the Vlasov equation in the form

$$p^\mu \frac{\partial f}{\partial x^\mu} - \Gamma^\sigma_{\mu\nu} p^\mu p^\nu \frac{\partial f}{\partial p^\sigma} = 0. \tag{13.6}$$

We remark that when space-time is curved, the momentum space is a Minkowski vector space, tangent to space-time at each point, but the phase space is not the direct product of the two spaces; it is a sub-manifold of the so called *tangent bundle* TM of the space-time manifold M, an eight-dimensional manifold with coordinates (x^μ, p^ν). The true phase space is a seven-dimensional manifold, of course, because of the constraint $p^\mu p_\mu = m^2c^2$. We remark that the latter is more complicated in general relativity, because the indices are raised and lowered with the field $g_{\mu\nu}$, and it is better to write the constraint in the form

$$g_{\mu\nu} p^\mu p^\nu = m^2 c^2. \tag{13.7}$$

Thus the derivative appearing in the Vlasov equation is taken in a direction tangent to this manifold, along which the eight-vector

$$(2g_{\mu\nu} p^\nu, p^\nu p^\sigma \partial_\mu g_{\nu\sigma}), \tag{13.8}$$

and p^0 depends not only on p_i but on the space-time coordinates as well.

Examples of collision-free systems described by the Vlasov–Einstein system are galaxies (systems of stars), the system of all galaxies, and the photon gas known as *cosmic fireball.* On the other hand, if we want to deal with stellar matter, the more complicated description with collisions seems to be required.

The Vlasov equation must be coupled with the Einstein field equations

$$R_{\mu\nu} - \frac{1}{2} R g_{\mu\nu} = -\frac{8\pi G}{c^4} T_{\mu\nu}. \tag{13.9}$$

Here $T_{\mu\nu}$ is not assigned, but is expressed in terms of the distribution function f as

$$T_{\mu\nu} = c \int p_\mu p_\nu f \sqrt{g} \frac{d^3 p}{p_0}. \tag{13.10}$$

This clearly couples the system of Einstein field equations to the Vlasov equation and produces the Vlasov–Einstein system. Again, if there are several species a modification is called for and a sum over the species with the distribution functions should be introduced in the last equation. If an electromagnetic field is present, the corresponding energy-momentum tensor must be added; if the particles described by the distribution function are the sources of that electromagnetic field, we must couple the above system to the Maxwell equations and obtain the more complicated Vlasov–Maxwell–Einstein system, alluded to at the end of the introduction.

13.4 Steady Vlasov–Einstein system in case of spherical symmetry

We recall from the previous chapters[1] that the line element in space-time takes on a particularly simple form in the case of spherical symmetry:

$$ds^2 = A(r)(dt)^2 - B(r)(dr)^2 - r^2[(d\theta)^2 + \sin^2\theta(d\phi)^2]. \tag{13.11}$$

In order to simplify the calculations, we found it expedient to let

$$A = e^F; \quad B = e^H \tag{13.12}$$

where F and H are two new functions of r. Our coordinates are $x^0 = ct$, $x^1 = r$, $x^2 = \theta$, and $x^3 = \phi$ and the metric tensor is diagonal with the components

$$g_{00} = e^F; \quad g_{11} = -e^H; \quad g_{22} = -r^2; \quad g_{33} = -r^2\sin^2\theta. \tag{13.13}$$

The determinant $g = -\det((g_{\mu\nu}))$ is the product of the leading diagonal, so that $g = e^{F+H}r^4\sin^2\theta$ and the contravariant components are

$$g^{00} = e^{-F}; \quad g^{11} = -e^{-H}; \quad g^{22} = -r^{-2}; \quad g^{33} = -r^{-2}\sin^{-2}\theta. \tag{13.14}$$

We found that only nine Christoffel symbols do not vanish:

$$\Gamma^0_{10} = \frac{1}{2}F', \qquad \Gamma^1_{00} = \frac{1}{2}e^{F-H}F', \tag{13.15}$$

$$\Gamma^1_{11} = \frac{1}{2}H', \qquad \Gamma^2_{12} = \Gamma^3_{13} = \frac{1}{r}, \tag{13.16}$$

$$\Gamma^1_{22} = -re^{-H}, \qquad \Gamma^3_{23} = \cot\theta, \tag{13.17}$$

$$\Gamma^1_{33} = -r\sin^2\theta e^{-H}, \qquad \Gamma^2_{33} = -\sin\theta\cos\theta, \tag{13.18}$$

and the contracted Riemann–Christoffel curvature tensor is diagonal:

$$R_{00} = e^{F-H}\left(-\frac{1}{2}F'' + \frac{1}{4}F'H' - \frac{1}{4}F'^2 - \frac{F'}{r}\right), \tag{13.19}$$

[1] Here and in the subsequent sections the gravitational constant, the speed of light and the rest mass are taken to be unit, i.e., $G = m = c = 1$.

$$R_{11} = \frac{1}{2}F'' - \frac{1}{4}F'H' + \frac{1}{4}F'^2 - \frac{H'}{r}, \tag{13.20}$$

$$R_{22} = e^{-H}\left[1 + \frac{r}{2}(F' - H')\right] - 1, \tag{13.21}$$

$$R_{33} = \sin^2\theta\left\{e^{-H}\left[1 + \frac{r}{2}(F' - H')\right] - 1\right\}. \tag{13.22}$$

At variance with what we did in the previous chapter where we equated these components to zero except for $r = 0$, we must first compute the components of the Einstein tensor $G_{\mu\nu} = R_{\mu\nu} - \frac{1}{2}Rg_{\mu\nu}$ and equate the results to a (negative) constant times the corresponding components of the energy-momentum tensor. The scalar curvature $R = R_{\mu\nu}g^{\mu\nu}$ is given by (Problem 13.4.1)

$$R = e^{-H}\left(-F'' + \frac{1}{2}F'H' - \frac{1}{2}F'^2 - 2\frac{F' - H'}{r} - 2\frac{1 - e^H}{r^2}\right). \tag{13.23}$$

Then

$$G_{00} = e^{F-H}\left(-\frac{H'}{r} + \frac{1 - e^H}{r^2}\right), \tag{13.24}$$

$$G_{11} = -\frac{F'}{r} - \frac{1 - e^H}{r^2}, \tag{13.25}$$

$$G_{22} = -r^2e^{-H}\left(\frac{1}{2}F'' - \frac{1}{4}F'H' + \frac{1}{4}F'^2 + \frac{1}{2}\frac{F' - H'}{r}\right), \tag{13.26}$$

$$G_{33} = -r^2\sin^2\theta e^{-H}\left(\frac{1}{2}F'' - \frac{1}{4}F'H' + \frac{1}{4}F'^2 + \frac{1}{2}\frac{F' - H'}{r}\right). \tag{13.27}$$

In order to write the Vlasov–Einstein system, it is convenient to use the vector v^i defined in Section 12.6, so that we get from (12.81), and from the Einstein field equations (11.109) together with (12.85) and (12.87)

$$v^i\frac{\partial f}{\partial x^i} - \frac{F'}{2}(1 + |\mathbf{v}|^2)\frac{x^i}{r}\frac{\partial f}{\partial v^i} = 0, \tag{13.28}$$

$$e^{-H}(rH' - 1) + 1 - 8\pi r^2\int\sqrt{1 + |\mathbf{v}|^2}fd^3v = 0, \tag{13.29}$$

$$e^{-H}(rF' + 1) - 1 - 8\pi r^2\int\left(\frac{\mathbf{x}\cdot\mathbf{v}}{r}\right)^2 f\frac{d^3v}{\sqrt{1 + |\mathbf{v}|^2}} = 0. \tag{13.30}$$

We have just written the (00) and (11) components of the Einstein field equations. The (22) and (33) components can be shown to follow from the rest of the system (Problem 13.4.2). This is not surprising because the Vlasov equation implies the conservation equations for the number of particles and the momentum-energy and so does the system of Einstein field equations, because of the contracted Bianchi identities. Thus there are four relations which reduce the number of equations; because of the spherical symmetry, they make two field equations unnecessary.

It is possible to construct solutions of these equations, as done by several authors (see below). Here we shall follow the papers by Rein and Rendall [23] and [24] who gave a fairly general discussion of a wide class of solutions. They start from the remark that the trajectories of a particle in a given gravitational field of the given symmetry satisfy

$$\dot{\mathbf{x}} = \frac{\mathbf{v}}{\sqrt{1+|\mathbf{v}|^2}}; \quad \dot{\mathbf{v}} = -\frac{F'}{2}\sqrt{1+|\mathbf{v}|^2}\frac{\mathbf{x}}{r} \tag{13.31}$$

where the dot denotes differentiation with respect to a convenient parameter (which turns out to be t). The system (13.31) has two first integrals; in fact the two quantities

$$E = e^{F/2}\sqrt{1+|\mathbf{v}|^2}; \quad L = |\mathbf{x}\times\mathbf{v}|^2 \tag{13.32}$$

are constants when equations (13.31) are satisfied. These first integrals can be interpreted as conservation of energy and angular momentum.

Since f is constant along the same trajectories, an arbitrary function of E and L,

$$f = \Phi(E, L), \tag{13.33}$$

is a solution of the Vlasov equation, no matter what (regular) function of r the parameter F is.

If we look for an asymptotically flat metric, we must impose the conditions

$$\lim_{r\to\infty} F = \lim_{r\to\infty} H = 0; \quad H(0) = 0 \tag{13.34}$$

where the last condition requires that the origin is a regular point (at a regular point, the metric is locally Euclidean).

Before proceeding further, we show that Φ must vanish for large values of E if the resulting steady state is to have a finite mass, i.e., if the quantity

$$M = \int \sqrt{1+|\mathbf{v}|^2} f d^3v d^3x, \tag{13.35}$$

is finite. This quantity is usually called the ADM mass, after Arnowitt, Deser and Minzer, who provided a canonical formalism for general relativity which gives unique definitions of physical quantities (otherwise having the ambiguity associated with an arbitrary change of space-time coordinates).

We first remark that (13.29) implies

$$e^{-H} = 1 - \frac{2m(r)}{r} \tag{13.36}$$

where

$$m(r) = \int_{|\mathbf{x}|\le r} \sqrt{1+|\mathbf{v}|^2} f d^3v d^3x, \tag{13.37}$$

at least as long as $2m(r) < r$ (otherwise H should become complex). Substituting (13.36) in (13.30) we obtain that F is increasing with r at least for $r > M$ $(> m(r))$. To be precise,

$$F(r) = F(4M) + 2\int_{4M}^{r} e^{H}\left(\frac{m(s)}{s^2} + 4\pi s\int\left(\frac{\mathbf{x}\cdot\mathbf{v}}{s}\right)^2 f\frac{d^3v}{\sqrt{1+|\mathbf{v}|^2}}\right)ds. \quad (13.38)$$

Since we are assuming that the ADM mass M is finite, we conclude that $F_\infty = \lim_{r\to\infty} F$ exists. In addition $e^H \leq 2$ for $r > 4M$, and the pressure is always smaller than the energy density, i.e.,

$$\int\left(\frac{\mathbf{x}\cdot\mathbf{v}}{r}\right)^2 f\frac{d^3v}{\sqrt{1+|\mathbf{v}|^2}} < \int\sqrt{1+|\mathbf{v}|^2} f d^3v; \quad (13.39)$$

we then have

$$F(r) \leq F(4M) + \frac{3}{2}, \quad (13.40)$$

which proves that F_∞ is finite. With a change of variables and an integration over the angles, we obtain from (13.35)

$$M = 8\pi^2\int_0^\infty e^{-F(r)}\int_{e^{F(r)/2}}^{\infty}\int_0^{r^2(E^2e^{-F(r)}-1)}\frac{\Phi(E,L)E^2 dL\, dE\, dr}{\sqrt{E^2 - e^{F(r)}(1+L/r^2)}}. \quad (13.41)$$

Hence

$$M \geq 8\pi^2 e^{-F_\infty}\int_{e^{F_\infty}/2}^{\infty}\Phi(E,L)\left(\int_{\sqrt{L/(E^2e^{-F_\infty}-1)}}^{\infty}\frac{dr}{\sqrt{E^2 - e^{F(r)}}}\right)E^2 dL dE. \quad (13.42)$$

The integral with respect to r diverges for any $E^2 > e^{-F_\infty}$ and $L > 0$; this implies that Φ has to vanish for such arguments if we want M to be finite.

This still leaves room for many choices for Φ, but, following Rein and Rendall [24], we restrict ourselves to solutions of the form

$$f(\mathbf{x},\mathbf{U}) = \phi(E)L^\ell \quad (13.43)$$

with $\ell > -1/2$ and ϕ measurable. Since we are interested only in states with finite mass, we assume that $\phi(E)$ vanishes for all energies larger than some given $E_0 > 0$. In addition we assume that in each compact set on the positive real axis there is a $C \geq 0$ such that $0 \leq \phi(E) \leq C(E_0 - E)_+^k$, where $k > -1$ is a given exponent and x_+ equals x when x is positive and 0 when x is negative.

Substituting the *ansatz* for f in the field equations, we obtain

$$e^{-H}(rH' - 1) + 1 - 16\pi^2 c_{\ell,-1/2} r^{2\ell+2} H_\ell(e^{F/2}) = 0, \quad (13.44)$$

$$e^{-H}(rF' + 1) - 1 - 16\pi^2 c_{\ell,1/2} r^{2\ell+2} e^{-(\ell+2)F} h_{\ell+3/2}(e^{F/2}) = 0, \quad (13.45)$$

where

$$h_m = \int_u^\infty \phi(E)(E^2 - u^2)^m dE, \qquad (u > 0) \tag{13.46}$$

$$H_\ell(e^{F/2}) = e^{-(\ell+2)F} h_{\ell+3/2}(e^{F/2}) + e^{-(\ell+1)F} h_{\ell+1/2}(e^{F/2}), \tag{13.47}$$

and

$$c_{a,b} = \int_0^1 s^a(1-s)^b ds = \frac{\Gamma(a+1)\Gamma(b+1)}{\Gamma(a+b+2)}, \tag{13.48}$$

where Γ denotes the gamma function. The function h_m can be shown to be continuous for positive values of E if $m > -1$ and $k + m + 1 > 0$, and differentiable for positive values of E if $m > 0$ and $k + m > 0$, with $h'_m = -2mh_{m-1}$.

The system formed by (13.44) and (13.45) has a global, unique solution in r (no singularity for finite values of r), as we shall presently discuss. Let us assume that $H(0) = 0$ and $F(0)$ has some assigned real value F_0. The local existence of a differentiable solution in some interval $(0, \delta)$ follows from a contraction mapping argument (for details see [21]). In order to prove global existence, one must control the inequality $2m(r) < r/2$ [21]. Although nothing is said in the proof about the boundary conditions at infinity, since we know that a steady state has a finite mass, then F has a finite limit at infinity as we discussed above. Subtracting this limit from F and redefining E_0 accordingly gives a steady state with the same f, but satisfying the boundary condition at infinity; the boundary condition for H follows from (13.36) because M is assumed to be finite.

Rein and Rendall [24] prove the following

Theorem. Let k, ℓ be two real numbers such that

$$k > -1, \quad \ell > -\frac{1}{2}, \quad k + \ell + \frac{1}{2} > 0,, \quad k < \ell + \frac{3}{2},$$

let ϕ be measurable and such that $\phi \in L^\infty_{\text{loc}}(]0, E_0[)$, and let

$$\phi(E) = c(E_0 - E)_+^k + \mathrm{O}((E_0 - E)_+^{k+\delta}) \quad \text{as} \quad E \to E_0 \tag{13.49}$$

for some positive E_0, c, and δ. If (f, H, F) is a steady solution of the Vlasov–Einstein system, in the sense that $f = \phi(E)L^\ell$ with E and L defined in (13.32), and if H, $F \in C^1([0, \infty[)$ satisfy (13.44), (13.45), then the steady state has compact support and a finite value of M as defined in (13.35).

It is clear that all the properties discussed above apply to the solution considered here. The main tool for the proof of the theorem is the following

Lemma. Let x, $y \in C^1(]0, R[)$ be two positive functions of r and such that

$$rx' = \alpha(r)y - x + x\frac{x + \gamma_1(r)y}{1 - \gamma_2(r)x},$$

$$ry' = y\left(c - \beta(r)\frac{x + \gamma_1(r)y}{1 - \gamma_2(r)x}\right),$$

on $]0, R[$, where $c > 0$, α, β, γ_1, $\gamma_2 \in C(]0, R[)$ with $\alpha_0 = \inf_{r\in]0,R[}\alpha(r) > 0$, $\lim_{r\to R}\beta(r) = \beta_0 \in]0, c[$, $\gamma_1, \gamma_2 \geq 0$ and $\lim_{r\to R}\gamma_1(r) = \lim_{r\to R}\gamma_2(r) = 0$. Also let $1 - \gamma_2(r)x(r) > 0$ for $r \in]0, R[$. Then $R < \infty$.

This lemma is proved in [24] as an adaptation of a lemma proved by Makino [16]. Assuming that the lemma has been proved, we consider the solutions of the reduced field equations as discussed before. We may assume that $e^{F(0)} < E_0$; otherwise the solution is trivial. We choose the largest R such that $e^F < E_0$ on $[0, R[$. Then we have proved that there exists some $R \in]0, \infty[$ such that F, H exist on $[0, R[$ with $e^F < E_0$ on this interval, and

$$\lim_{r\to R-} e^{F(r)} = E_0.$$

We can show that this result holds for $R \in]0, \infty]$, i.e., R may be infinity. In fact, by monotonicity, we know that $F_\infty = \lim_{r\to\infty} F(r)$ exists and is not larger than $\log E_0$. In order to prove that it is exactly $\log E_0$, we assume, by contradiction, that it is less than $\log E_0$. This implies (Problem 13.4.3) that, according to (13.45), $F'(r) \geq cr^{2\ell+1}$ for $r > 0$, with $c > 0$. Then F diverges as $r^{2\ell+2}$ when $r \to \infty$, contrary to the fact that the limit is finite.

Having proved that the case $R = \infty$ is included in what we have discussed so far, we must prove that it cannot occur. To this end, we introduce new variables, which bring the system into the form contemplated in the lemma stated above. To this end we define

$$\eta(r) = \log E_0 - F(r)/2, \tag{13.50}$$

and define the two variables $x(r)$ and $y(r)$ as

$$x(r) = \frac{m(r)}{r\eta(r)}, \tag{13.51}$$

$$y(r) = \frac{[m'(r)]^2 e^{(\ell+2)F(r)}}{8\pi^2 c_{\ell,1/2} r^{2\ell+2} h_{\ell+3/2}(e^{F(r)/2})}. \tag{13.52}$$

It is easy to check (Problem 13.4.4) that

$$r\eta' = -\eta\frac{x + \gamma_1(r)y}{1 - \gamma_2(r)x}, \tag{13.53}$$

where

$$\gamma_1 = \frac{[c_{\ell,1/2} r^{2\ell} e^{-(\ell+2)F(r)} h_{\ell+3/2}(e^{F(r)/2})]^2}{\eta[c_{\ell,-1/2} r^{2\ell} H_\ell(e^{F(r)/2})]^2}, \tag{13.54}$$

$$\gamma_2 = 2\eta. \tag{13.55}$$

It is now easy, if tedious, to obtain the equations for $x(r)$ and $y(r)$ which turn out to have the form considered in the lemma with

$$\alpha = \frac{1}{2\ell+3}\frac{h_{\ell+3/2}(e^{F(r)/2})}{\eta h_{\ell+3/2}(e^{F(r)/2}) + e^{F(r)}\eta h_{\ell+1/2}(e^{F(r)/2})}, \tag{13.56}$$

$$\beta = -\eta - \frac{\eta[c_{\ell,-1/2}r^{2\ell}H_\ell(e^{F(r)/2})]}{c_{\ell,1/2}r^{2\ell}e^{-(\ell+2)F(r)}h_{\ell+3/2}(e^{F(r)/2})} + 2\eta\frac{\tilde{H}_\ell(e^{F(r)/2})}{H_\ell(e^{F(r)/2})}, \tag{13.57}$$

where

$$\tilde{H}_\ell(e^{F(r)/2}) = (2\ell+4)e^{-(\ell+2)F(r)}h_{\ell+3/2}(e^{F(r)/2})$$

$$+(4\ell+5)e^{-(\ell+1)F(r)}h_{\ell+1/2}(e^{F(r)/2}) + (2\ell+1)e^{-\ell F(r)}h_{\ell-1/2}(e^{F(r)/2}). \tag{13.58}$$

We can now apply the lemma, provided α, β, γ_1, and γ_2 satisfy the assumptions. This is lengthy but not hard to verify (Problem 13.4.5). Then R is finite (an explicit upper bound for R can be found (Problem 13.4.5)), i.e., the steady state has compact support. We remark that the support in $\mathbf{v}$ also follows because E and F are bounded. The fact that M is bounded follows easily (Problem 13.4.6)).

We remark that particular cases of the solutions delivered by the theorem. They are reviewed in an article by Shapiro and Teukolski [30] and fall into two classes:

a) The distribution function has the form of a truncated equilibrium distribution $f = e^{E-E_0}H(E-E_0)$, where H is the Heaviside step function.

b) $f = (E/E_0)^{-2\delta}(1-(E/E_0)^2)_+^\delta$. This is included in the theorem discussed above for $-1/2 < \delta < 3/2$.

Other cases were discussed by Tooper [33] and Fackerell [15].

Problems

13.4.1 Show that if the metric tensor is given by (13.13), then the scalar R is given by (13.23) and the components of the Einstein tensor G_{00}, G_{11}, G_{22}, G_{33}, are given by (13.24) through (13.27).

13.4.2 Show that the (22) and (33) components of the Vlasov–Einstein system follow from the rest of the system in the case of spherical symmetry.

13.4.3 Show that, according to (13.45), $F'(r) \geq cr^{2\ell+1}$ for $r > 0$, with $c > 0$ (see [24]).

13.4.4 Check that (13.53) holds (see [24]).

13.4.5 Prove that α, β, γ_1, and γ_2 satisfy the assumptions of the lemma used in the main text to prove that the radius R is finite and provide an upper bound for R (see [24]).

13.4.6 Prove that M is bounded.

13.5 The threshold of black hole formation

The threshold of black hole formation, being a central theme in general relativity, has been studied by many authors, especially in the case of spherical symmetry. In spite of this, much remains to be investigated. In the case of the vacuum Einstein field equations, there is no gravitational collapse; this is guaranteed by a theorem proved by Birkhoff [3] (see below). Then it is necessary to introduce the energy-momentum tensor of matter, in order to learn something about the gravitational collapse. This poses the problem of choosing a model for matter. Most of the investigations assume a field model for matter, such as a massless (real, minimally coupled) scalar field. The papers of Christodoulou [7], [8], [9], [10], [11] are worth a short discussion, because his results are rather deep and should be contrasted with the information to be gathered from the spherically symmetric Vlasov–Einstein system.

In the case of small initial data, Christodoulou [7] showed that the fields disperse to space infinity asymptotically in time ($M = 0$). For large data [8], he proved that for any solution it is possible to define a number M with the property that in the region $r > 2M$ the solution approaches the Schwarzschild solution of mass M asymptotically in time. This result is easy to interpret physically: the system has collapsed to form a black hole of mass M. The previous case [7] corresponds to $M = 0$, but this is the only case in which one can easily answer the question whether a set of given data results in dispersion or black hole formation. A sufficient condition on initial data to ensure that $M > 0$ for the corresponding solution is available [9], though the criterion is rather unpractical.

The results for large data discussed so far describe the structure of the solution for sufficiently large values of r, whereas the internal structure of the black hole is left open. There are data leading to the formation of a naked singularity [10]. This, as explained in the introduction, violates the so-called cosmic censorship hypothesis. This negative implication is limited by the fact [11] that this behavior is unstable even in the restricted class of spherically symmetric initial data.

The numerical work of Choptuik [4] brought an important new element into the study of the Einstein field equations coupled with a scalar field. In fact, he considered a family of initial data for the scalar field, parametrized by an arbitrary constant factor A. All the data considered in the numerical simulation provided the same picture. For values of A corresponding to small data the field dispersed, as expected from the rigorous results [7]. For large initial data a black hole was formed. Then we can define a critical value A_* as the lower limit of those values of A for which we witness the formation of a black hole. The plot of the mass of the latter as a function of A shows that $M(A)$ is continuous; thus one passes continuously from the case of dispersed matter to the case of the formation of a black hole.

In a similar simulation, based, however, on a more complicated matter model, provided by the Yang–Mills equations, Choptuik et al [5] found solutions of two kinds: the first kind (to be called type I) shows a discontinuity at $A = A_*$, the second (type II) shows a continuous transition from the case of dispersed matter to the case of the formation of a black hole. The terminology (type I and II) is borrowed from the phase transitions in statistical mechanics.

The deep difference in behavior for the matter models which have been discussed so far in this section is attributed by Choptuik et al. [5] to the circumstance that there are no regular static solutions of Einstein's field equations for the scalar model, whereas the Yang–Mills model admits the so called Bartnik–McKinnon solutions [1], [32], which are static.

The matter models which have been just discussed are field theoretic. The perfect fluid model, mentioned in the introduction, is the simplest phenomenological model and shows just type II behavior. The study of the Vlasov–Einstein system was begun by Rein and Rendall [24] who showed that for sufficiently small initial data the matter disperses to infinity at large times, i.e., the analogue of Christodoulu's result for small data [7]. No analogue of his large data result [8] seems to be available for collisionless matter, the only result in that direction being that there do exist initial data which develop singularities [27]. The proof refers to initial data which contain trapped surfaces (the existence of these data being part of the proof) and is based on Penrose's singularity theorem [17]. A different kind of result for large data, which is relevant to the problem of computing the collapse of collisionless matter in spherical symmetry, is that if initial data give rise to a singular solution after a finite time, then the first singularity must occur at the center of symmetry [25]. This result refers to the Schwarzschild time variable but an analogous result referring to the case where this variable is replaced by maximal slicing is also available [28].

A plausible conjecture is that black holes are usually formed in those solutions which develop singularities. There are no rigorous results supporting this conjecture, but a large body of numerical work due to Shapiro, Teukolski and collaborators [29], [30], [31], [18], [19] is available. The last two papers are mainly devoted to a discussion of the relative merits of different choices of the time and radial coordinates.

Rein et al [26] carried out a numerical experiment analogous to that of Choptuik et al [5] for the spherical collapse of a given amount of collisionless matter. They used Schwarzshild coordinates and started from a suitable smooth function Af_0, of compact support as initial datum for the distribution function f, where A is a positive constant which can be varied while f_0 is kept fixed.

The system (13.28) through (13.30) must be slightly modified in the time dependent case and becomes (see (12.81)):

$$\frac{\partial f}{\partial t} + e^{(F-H)/2} \frac{v^i}{\sqrt{1+|\mathbf{v}|^2}} \frac{\partial f}{\partial x^i}$$

$$-\frac{1}{2}\left(\dot{H}\frac{\mathbf{x}\cdot\mathbf{v}}{r} + e^{(F-H)/2}\sqrt{1+|\mathbf{v}|^2}F'\right)\frac{x^i}{r}\frac{\partial f}{\partial v^i} = 0, \tag{13.59}$$

$$e^{-H}(rH'-1) + 1 - 8\pi r^2 \int \sqrt{1+|\mathbf{v}|^2} f d^3v = 0, \tag{13.60}$$

$$e^{-H}(rF'+1) - 1 - 8\pi r^2 \int \left(\frac{\mathbf{x}\cdot\mathbf{v}}{r}\right)^2 f \frac{d^3v}{\sqrt{1+|\mathbf{v}|^2}} = 0, \tag{13.61}$$

$$\dot{H} + 8\pi r e^{(F+H)/2} \int \frac{\mathbf{x}\cdot\mathbf{v}}{r} f d^3v = 0. \tag{13.62}$$

The last equation corresponds to the (01)-component of Einstein's field equations and was not present before (here and in the following the dot will denote derivatives with respect to $x^0 = t$). In the case when no matter is present this equation becomes $\dot{H} = 0$ and thus H is time independent; since, as we know from the previous chapter, $H + F$ does not depend on r, if we assume that H and F vanish at infinity (asymptotic flatness), then $H + F = 0$ and F does not depend on time as well; this is a quick derivation of the theorem first proved by Birkhoff [3].

The system (13.60) through (13.62) seems to be overdetermined because there are four equations for three unknowns (f, F, H). Using the energy conservation equation, however, we can show that the last equation of the above system is a consequence of the previous ones (Problem 13.5.1). It is, however, useful to keep it in the numerical code to compute the time derivative of H in a direct way.

For the numerical computation it is useful to change the variables in order to exploit the spherical symmetry in a direct fashion. It is convenient to take:

$$r = |\mathbf{x}|, \quad w = \frac{\mathbf{x} \cdot \mathbf{v}}{r}, \quad L = |\mathbf{v}|^2|\mathbf{x}|^2 - (\mathbf{x} \cdot \mathbf{v})^2. \tag{13.63}$$

In these variables the Vlasov equation for $f = f(t, r, w, L)$ becomes:

$$\frac{\partial f}{\partial t} + e^{(F-H)/2} \frac{w}{\sqrt{1 + |\mathbf{v}|^2}} \frac{\partial f}{\partial r} - \frac{1}{2}\Bigg(\dot{H}w + e^{(F-H)/2}\sqrt{1 + |\mathbf{v}|^2}F'$$

$$-e^{(F-H)/2} \frac{2L}{r^3\sqrt{1 + |\mathbf{v}|^2}}\Bigg)\frac{\partial f}{\partial w} = 0. \tag{13.64}$$

The numerical code used by Rein et al [26] is based on a particle method inspired by the techniques used in plasma physics [2]. The initial condition is assumed to be spherically symmetric, to satisfy the condition $m(r) \leq r/2$, where $m(r)$ is defined by (13.37) (this ensures that the initial hypersurface does not contain a trapped surface), and to vanish outside the set $(r, |\mathbf{v}|, \alpha) \in [R_0, R_1] \times [U_0, U_1] \times [\alpha_0, \alpha_1]$, where $\alpha = \arccos(w/|\mathbf{v}|)$. In the space described by the three variables r, $|\mathbf{v}|$, α, one considers the above set subdivided into uniform meshes and assigns initial values for the (expected) number of particles at each grid point. This permits us to calculate the source terms of (13.60) through (13.62) at the same points. Then (13.61) (suitably discretized) is used to compute F even outside the original grid, because the matter tensor vanishes there and $m(r)$ is constant, so that (13.36) can be used to express e^{-H} in an explicit fashion. The condition $\lim_{r\to\infty} F = 0$ can then be easily used. H is computed immediately from (13.36). Accurate expressions of $\dot{H}$ and H' are provided by (13.62) and (13.60).

The Vlasov equation is integrated by using the method of characteristics: the original grid is transformed into a new one by using the equations for particle trajectories to move the grid points and transfer to these points the number of particles.

One time step is now complete and one can repeat the procedure for the next time step; the calculation can proceed.

The code was tested on a steady solution (a very particular case of the class discussed in the previous section) which could be easily computed by an accurate method. The number of particles was taken to be 2550 with 40 grid points in the radial direction, 10 for the variable $|\mathbf{v}|$, and 10 for α (the fact that 2550 is less than $4000{=}40{\times}10 \times 10$ is due to the fact that the support of f is not rectangular). The maximum errors due to space discretization at $t = 0$ are 0.49% and 0.96% for m and F respectively. As expected, the accuracy of the particle method during the time evolution depends heavily on the size of the time step Δt: the three choices $1/\Delta t = 4000, 8000, 16000$ lead to the maximum errors 6.2%, 3.4%, and 2.1% respectively for m, and 5.9%, 2.9%, and 1.3%, respectively, for F. Thus the particle code seems to track the steady state reasonably well, although a rather small time step is needed. The authors attribute this circumstance, at least in part, to numerical difficulties in tracking the motion of particles near $r = 0$ (where the density takes its maximum value for the solution under consideration). As a partial confirmation of this, they state that for other solutions with zero density near $r = 0$, the time step was taken larger without significant change in the results.

As stated before, the initial data have the form Af_0, where A is a positive constant which can be varied while f_0 is kept fixed. In their first example, Rein et al [26] took

$$f_0 = [50000(2.2 - r)(r - 2)(10.2 - |\mathbf{v}|)(|\mathbf{v}| - 10)(3.1 - \alpha)(\alpha - 2.9)]^2, \qquad (13.65)$$

for $2 \leq r \leq 2.2$, $10 \leq |\mathbf{v}| \leq 10.2$, $2.9 \leq \alpha \leq 3.1$, and $f_0 = 0$ otherwise. Thus initially the mass is concentrated in a rather thin layer with a high speed and the motion is inward (the angle between $\mathbf{v}$ and $\mathbf{r}$ is close to π). In most of the simulations the support of f_0 is divided into 40 cells in the radial direction, 20 for the variable $|\mathbf{v}|$, and 20 for α, resulting in 16000 cells and the time step is 0.005.

If A is taken to be 0.69, the enclosed mass $m(r)$ initially falls inward, but then bounces back and at $t = 8$ has almost returned to the initial position. A similar behavior is shown by F, which is negative and first decreases near the origin, and then almost returns to its initial value. As t grows the particles continue to travel outward and disperse, consistent with the small data result [23].

If A is taken to be 0.75, initially the behavior looks the same, but the motion inward continues even for $t = 8$ and the mass seems to be centered near 0.21, whereas F has formed an abrupt transition near 0.235, where H forms a cusp at its maximum. At the same time instant the largest value of the outward momentum is about 9, while the largest value of the inward momentum is about 189. Examination of the radial flow of particles reveals that it is almost entirely inward. At time $t = 16$ these qualitative features remain the same although the largest value of $|F|$ has grown. A run made with a finer resolution ($80{\times}40 \times 40$ particles, $\Delta t = 0.0025$) shows graphs that are qualitatively very similar, except that the transitions are slightly more abrupt and the largest value of $|F|$ is increased by 7%.

When A is taken larger than 0.75 the results are similar. The same situation occurs for $0.70 \leq A \leq 0.74$, except that a small amount of mass escaped (this amount is never more than 10% of the total mass for all examples considered). Thus it seems that for the choice shown in (13.65) the critical value of A is $A_* \simeq 0.70$.

For $A \geq 0.70$ the radius of the maximum of H and the radius of the abrupt transition in F are nearly the same. If one computes the radius r (as a function A) where H (as a function of r and t) is maximum, it turns out that for $A \geq 0.70$ this maximum is attained at the largest time of computation; for $A \leq 0.69$ it is attained earlier.

The conclusion is that a black hole capturing nearly all the mass is formed for $A \geq 0.70$. This mass is given by $M(A) = r(A)/2$ and seems to depend almost linearly on A ($M(A) \simeq 0.11{+}0.14(A{-}0.70)$). Thus $\lim_{A \to A_*+} M(A) = 0.11$, which indicates a type I behavior (discontinuity of the mass at the point of transition between the subcritical and supercritical behavior).

Another family of initial data considered by Rein et al [26] is

$$f_0 = 0.1(1 - r^2)^2(1 - |\mathbf{v}|^2)^2 \tag{13.66}$$

for $r < 1$ and $|\mathbf{v}| < 1$ and zero otherwise. For this example a smaller time step $\Delta t = 0.00125$ was used and the critical value was found to be $A_* = 1.6$.

A third example [26] was

$$f_0 = 0.1(3 - r)^2(2 - r)^2(1 - r)^2(1 - |\mathbf{v}|^2)^2 \tag{13.67}$$

for $1 \leq r \leq 3$, $|\mathbf{v}| \leq 1$, and $f_0 = 0$ otherwise. The critical value was found to be $A_* = 0.76$.

In all these examples the final time was taken large enough that only minor changes were noticed when increasing it. No signs of singularity formation for any value of A is found. This is consistent with the standard picture where the only singularities formed are those of black hole type and they are avoided by the use of a Schwarzschild time coordinate.

In order to check the interpretation of the numerical solutions as describing the collapse to a black hole, radial null geodesics were computed. The results agree well with the expected picture: geodesics starting at $r = 0$ at early times escape to large values of r. Those starting after a certain time T_1 remain within a finite radius. The limit of this radius as t tends to T_1 from above equals $r(A)$. This provides a consistent picture with a black hole, whose event horizon is generated by the null geodesics starting from $r = 0$ at time T_1.

The picture is rather different from the one provided by the only other case of a phenomenological matter model for which critical collapse has been studied, namely a perfect fluid with a linear equation of state. In the case of pure radiation where the pressure is one-third the energy density, the slightly supercritical collapse can be described as follows [14]. The matter splits almost completely into two parts, separated by a near vacuum region. The outer part of the matter contains almost all the mass and escapes to infinity. The inner part collapses to form a black hole with a small amount of mass. Only a transition of type II is possible, because the above picture becomes more and more extreme as the critical parameter is approached and the black hole mass goes to zero.

The fact that no singularities are observed in the numerical computations can be considered as evidence that the weak cosmic censorship conjecture is true for the Vlasov–Einstein model. Indeed it may even be true in a stronger form than in the case of the massless scalar field model for matter. One may speculate that there are no naked singularities formed for any regular initial data rather than just for generic initial data, because the naked singularities of the scalar field collapse appear to be associated with the existence of type II critical collapse.

Numerical investigations of the kind carried out by Rein et al [26] are extremely important. On one hand they add the Vlasov–Einstein system to the class of matter models for which something is known about critical collapse. As for all numerical simulations, it is clearly desirable to have simulations for other types of initial data so as to discover whether the type of behavior found by Rein et al [26] is prevalent or other types may occur. On the other hand, numerical investigations can also help the mathematical study of the model by providing pictures of what is happening. These pictures can in turn suggest which theorems one should try to prove.

Problem

13.5.1 Using the energy conservation equation, show that the (01) component of the Vlasov–Einstein system in spherical symmetry is a consequence of the previous ones.

13.6 Cosmology with the Vlasov–Einstein system

In this section we study the Vlasov–Einstein system in a cosmological setting with plane, spherical or hyperbolic symmetry. This means that we study the case when one can find coordinates such that the space-time metric has the form

$$ds^2 = A(t,r)(dt)^2 - B(t,r)(dr)^2 - t^2[(d\theta)^2 + \sin_k^2\theta(d\phi)^2] \tag{13.68}$$

where, by definition, plane, spherical, hyperbolic symmetry correspond to $k = 0$, $k = 1$, and $k = -1$ respectively. We have let

$$\sin_k\theta = \begin{cases} 1 & \text{for } k = 0, \\ \sin\theta & \text{for } k = 1, \\ \sinh\theta & \text{for } k = -1. \end{cases} \tag{13.69}$$

$t > 0$ denotes a timelike coordinate, $r \in [0,1]$, and the functions A and B are periodic in r. The angular coordinates $\theta \in [0, 2\pi/(1+k)]$ and $\phi \in [0, 2\pi]$ parametrize the surfaces defined by constant values of t and r, which are (flat) tori for $k = 0$, spheres for $k = 1$ and hyperbolic planes in the case of hyperbolic symmetry ($k = -1$). The coordinates which have been introduced will not cover, generally speaking, the entire space-time manifold, but they do cover a neighborhood of the singularity at $t = 0$, which will allow us, following a paper by Rein [22], to investigate the nature of this singularity.

One way to think of the above metric is to consider the Schwarzschild metric found in the previous chapter. If one passes through the event horizon at $r = 2M$ where g_{00} and g_{11} change sign (simultaneously), the radial coordinate becomes timelike and the time becomes spacelike. If one interchanges the notation between t and r, one obtains the above metric for $k = 1$. This metric is local, of course, and the compactification with respect to r can be easily introduced. The cases of plane and hyperbolic symmetry arise when we require the metric to be invariant not with respect to space rotations (rigid motions on the sphere) but to rigid motions on the Euclidean or Lobachevski plane.

In order to simplify the calculations, it is expedient, as before, to let

$$A = e^F; \quad B = e^H \tag{13.70}$$

where F and H are two new functions of r and t. Our coordinates are $x^0 = t$, $x^1 = r$, $x^2 = \theta$, and $x^3 = \phi$ and the metric tensor is diagonal with the components

$$g_{00} = e^F; \quad g_{11} = -e^H; \quad g_{22} = -t^2; \quad g_{33} = -t^2 \sin_k^2 \theta. \tag{13.71}$$

The determinant of the metric tensor $g = -\det((g_{\mu\nu}))$ is the product of the leading diagonal, $g = e^{F+H} t^4 \sin_k^2 \theta$ and the contravariant components are

$$g^{00} = e^{-F}; \quad g^{11} = -e^{-H}; \quad g^{22} = -t^{-2}; \quad g^{33} = -t^{-2} \sin_k^{-2} \theta. \tag{13.72}$$

We first calculate the twelve Christoffel symbols that do not vanish:

$$\Gamma^0_{00} = \frac{1}{2}\dot{F}, \qquad \Gamma^0_{11} = \frac{1}{2}e^{H-F}\dot{H}, \qquad \Gamma^1_{01} = \frac{1}{2}\dot{H}, \tag{13.73}$$

$$\Gamma^0_{10} = \frac{1}{2}F', \qquad \Gamma^1_{00} = \frac{1}{2}e^{F-H}F', \qquad \Gamma^1_{11} = \frac{1}{2}H', \tag{13.74}$$

$$\Gamma^2_{02} = \Gamma^3_{03} = \frac{1}{t}, \qquad \Gamma^0_{22} = te^{-F}, \qquad \Gamma^3_{23} = k^2 \cot_k \theta, \tag{13.75}$$

$$\Gamma^0_{33} = t \sin_k^2 \theta e^{-F}, \qquad \Gamma^2_{33} = -k^2 \sin_k \theta \cos_k \theta. \tag{13.76}$$

Here

$$\cos_k \theta = \begin{cases} 1 & \text{for } k = 0, \\ \cos\theta & \text{for } k = 1, \\ \cosh\theta & \text{for } k = -1, \end{cases} \tag{13.77}$$

and $\cot_k \theta = \cos_k \theta / \sin_k \theta$. The five components of the Einstein tensor that do not vanish are

$$G_{00} = -\frac{1}{t^2}(t\dot{H} + 1 + ke^F), \tag{13.78}$$

$$G_{11} = -\frac{1}{t^2}e^H\left(e^{-F}(t\dot{F}-1)-k\right), \tag{13.79}$$

$$G_{22} = \frac{t^2}{2}\left\{e^{-F}\left[\ddot{H} - \left(\frac{\dot{H}}{2}+\frac{1}{t}\right)(\dot{F}-\dot{H})\right] - e^{-H}\left[F'' + \frac{1}{2}F'(F'-H')\right]\right\}, \tag{13.80}$$

$$G_{33} = \sin_k^2\theta G_{22}, \qquad G_{01} = -\frac{F'}{t}. \tag{13.81}$$

Using variables analogous to those used in the previous section (and letting $w = e^{H/2}p^1$) we obtain the Vlasov–Einstein system in the following form:

$$\frac{\partial f}{\partial t} + e^{(F-H)/2}\frac{w}{\sqrt{1+w^2+L/t^2}}\frac{\partial f}{\partial r}$$

$$-\frac{1}{2}\left(\dot{H}w + e^{(F-H)/2}\sqrt{1+w^2+L/t^2}F'\right)\frac{\partial f}{\partial w} = 0, \tag{13.82}$$

$$e^{-F}(t\dot{H}+1)+k-8\pi^2\int\sqrt{1+w^2+L/t^2}f\,dL\,dw = 0, \tag{13.83}$$

$$e^{-F}(t\dot{F}-1)-k-8\pi^2\int w^2 f\frac{dL\,dw}{\sqrt{1+w^2+L/t^2}} = 0, \tag{13.84}$$

$$F' + \frac{8\pi^2}{t}e^{(F+H)/2}\int wf\,dL\,dw = 0. \tag{13.85}$$

It is remarkable that k does not enter in the Vlasov equation. The (22) component of the Einstein field equations could also be written, but it turns out to be a consequence of those which have already been written. As should be clear by now, this is due to the conservation equations which supply identities for the complete system. Actually (13.85) is also a consequence of the previous equations (under suitable smoothness assumptions) but it is convenient to retain it as a definition of F' appearing in (13.82). In fact, it would be rather unpleasant, from a technical viewpoint, to control F' in the theorems to be proved. It is easy to show that, even if we consider F' to be *defined* by (13.85), it turns out to be the derivative of F with respect to r in a time interval where the solution is smooth, provided it holds at a time instant in this interval (Problem 13.6.1). We also remark that on the mass-shell of the tangent bundle (Problem 13.6.2):

$$p^0 = e^{-F/2}\sqrt{1+w^2+L/t^2}. \tag{13.86}$$

Rein [22] proves the existence of a unique solution in the time interval $]T,1]$ (with $T \in [0,1[$), if we assign continuous, continuously differentiable and periodic data for f, F, H at $t = 1$. In the case of hyperbolic symmetry ($k = -1$) the extra condition $F(1) < 0$ is required. The physical meaning of this extra condition is easily understood. Let us consider a pseudo-Schwarzschild solution

$$ds^2 = \left(1+\frac{2M}{t}\right)^{-1}(dt)^2 - \left(1+\frac{2M}{t}\right)(dr)^2 - t^2[(d\theta)^2+\sinh^2\theta(d\phi)^2]. \tag{13.87}$$

This is a vacuum solution of the Einstein field equations; in fact it is obtained from the Schwarzschild metric by replacing spherical by hyperbolic symmetry, exchanging t and r and changing M into $-M$. The restriction $F > 0$ is equivalent to $M > 0$. For $M = 0$ the space-time is flat and for $M < 0$ it has a coordinate singularity at $t = -2M$. The space-time could then be extended beyond this singularity, something that Rein wanted to exclude from his investigation.

The proof is obtained by an iterative scheme. After showing that the iterates are sufficiently regular, Rein [22] proves that they have a continuous limit in an interval $[1-\delta, 1]$ $(\delta > 0)$, then that this limit is continuously differentiable. This permits us to show that the limit is indeed a solution of the system. The solution is easily proved to be unique. Then we can extend the solution to a maximal interval $]T, 1]$, where T is a time instant where a singularity shows up ($T \geq 0$ because the system becomes meaningless for $t = 0$).

Rein [22] also gives a criterion sufficient to prove that $T = 0$. If $|w|$ is bounded on the support of $f = f(t, r, w, L)$ for $t \in]T, 1]$, then $T = 0$. The proof is by contradiction, i.e., one assumes $T > 0$ and then proves that the solution can be continued a little bit to the left.

A completely different criterion supplied by Rein [22] in order to have $T = 0$ is to assume sufficiently small data.

Finally Rein [22] examines the nature of the singularity at $t = 0$ for those solution for which $T = 0$. The first result is that the so-called Kretschmann scalar $R_{\alpha\beta\gamma\delta}R^{\alpha\beta\gamma\delta}$ grows faster than t^{-6}. The second result is that the mean curvature of the surfaces of constant t $(c = -(2g_{00})^{-1/2}g^{ij}\dot{g}_{ij})$ (where i and j run from 1 to 3) is negative and grows in absolute value as $t^{-3/2}$. It can also be shown that the limits of the diagonal elements of the tensor associated with the second fundamental form, $c_i^j = -(2g_{00})^{-1/2}g^{jk}\dot{g}_{ik}$, to c are finite if the initial data are small. This question is related to the concept of a velocity dominated singularity [13].

The main purpose of Rein's paper [22] is to investigate the behavior of the cosmological solutions when $t \to 0$. Nevertheless he also examines the behavior of the solutions for $t \geq 1$. His main result is that one can establish a bound on the w-support of f and, in spite of that, the solutions need not exist for all $t \geq 1$. This is in sharp contrast to the spherically symmetric, asymptotically flat case [23], discussed in the previous section, and also to well-known results for the Vlasov–Maxwell [12] system.

The fact that such solutions exist in spherical symmetry ($k = 1$) is an easy consequence of (13.84) which can be integrated to yield

$$te^{-F} = e^{-F_1} - (t-1) - 8\pi^2 \int_1^t w^2 f \frac{dLdwds}{\sqrt{1 + w^2 + L/t^2}} \leq e^{-F_1} + 1 - t \qquad (13.88)$$

over the interval of existence $[1, T[$ (Here F_1 is clearly the value of F for $t = 1$). If we divide by t and let t go to infinity, we obtain the absurd inequality $e^{-F} \leq -1$. Then $T < \infty$. Since one can show that the boundedness of e^F ensures that $T = \infty$ [22], this result implies that e^F tends to ∞ when $t \to T$.

Problems

13.6.1 Show that, even if we consider F' to be *defined* by (13.85), it turns out to be the derivative of F with respect to r in a time interval where the solution is smooth, provided this holds at a time instant in this interval. (Hint: Use the relations which follow, by integrating in time and differentiating in space, from the other equations and connect F' and j to f to show that if

$$Z = F' + \frac{8\pi^2}{t} e^{(F+H)/2} \int w f dL dw$$

then, if $Z(1) = 0$:

$$te^{-F} Z(t) = -4\pi^2 \int_1^t \int f(s) \frac{(1 + 2w^2 + L/s^2)}{\sqrt{1 + w^2 + L/s^2}} Z(s) dL dw ds.$$

The result then follows easily by Gronwall's lemma.)

13.6.2 Show that on the mass-shell of the tangent bundle, p^0 can be expressed in terms of w and L as indicated in (13.86).

References

[1] R. Bartnik and J. McKinnon, Particlelike solutions of the Einstein–Yang–Mills equations, *Phys. Rev. Lett.* **61**, 141–143 (1988).

[2] C. K. Birdsall and A. B. Langdon, *Plasma Physics via Computer Simulation* (McGraw-Hill, New York, 1985).

[3] G. D. Birkhoff, *Relativity and Modern Physics*, (Harvard University Press, Cambridge, 1927).

[4] M. W. Choptuik, Universality and scaling in the gravitational collapse of a scalar field, *Phys. Rev. Lett.* **70**, 9–12 (1993).

[5] M. W. Choptuik, T. Chmaj and P. Bizoń, Critical behaviour in gravitational collapse of a Yang–Mills field, *Phys. Rev. Lett.* **77**, 424–427 (1996).

[6] D. Christodoulou, Violation of cosmic censorship in the gravitational collapse of a dust cloud, *Commun. Math. Phys.* **93**, 171–195 (1984).

[7] D. Christodoulou, The problem of a self-gravitating scalar field, *Commun. Math. Phys.* **105**, 337–361 (1986).

[8] D. Christodoulou, A mathematical theory of gravitational collapse, *Commun. Math. Phys.* **105**, 337–361 (1987).

[9] D. Christodoulou, The formation of black holes and singularities in spherically symmetric gravitational collapse, *Commun. Pure Appl. Math.* **44**, 339–373 (1987).

[10] D. Christodoulou, Examples of naked singularity formation in the gravitational collapse of a scalar field, *Ann. Math.* **140**, 607–653 (1994).

[11] D. Christodoulou, The instability of naked singularities in the gravitational collapse of a scalar field, Preprint.

[12] R. J. DiPerna and P.-L. Lions, Global weak solutions of Vlasov–Maxwell systems, *Commun. Pure Appl. Math.* **42**, 729–757 (1989).

[13] D. Eardley, E. Liang and R. Sachs, Velocity dominated singularities in irrotational dust cosmologies, *J. Math. Phys.* **13**, 99–106 (1972).

[14] C. R. Evans and J. S. Coleman, Critical phenomena and self-similarity in gravitational collapse of radiation fluid, *Phys. Rev. Lett.* **72**, 1782–1785 (1994).

[15] E. D. Fackerell, Relativistic stellar dynamics, *Ap. J* **153**, 643–660 (1968).

[16] T. Makino, On spherically symmetric stellar models in general relativity, *J. of Math. of Kyoto Univ.* **38**, 55–69 (1998).

[17] R. Penrose, Gravitational collapse: the role of general relativity, *Nuovo Cimento B* **1**, 252–276 (1969).

[18] L. I. Petrich, S. L. Shapiro and S. A. Teukolsky, Oppenheimer–Snyder collapse with maximal time slicing, *Phys. Rev. D* **31**, 2459–2469 (1985).

[19] L. I. Petrich, S. L. Shapiro and S. A. Teukolsky, Oppenheimer–Snyder collapse in polar time slicing, *Phys. Rev. D* **33**, 2100–2110 (1986).

[20] K. Pfaffelmoser, Global classical solutions of the Vlasov–Poisson system in three-dimensions for general initial data, *J. Diff. Eqns.* **95**, 281–303 (1992).

[21] G. Rein, Stationary and static solutions of the spherically symmetric Vlasov–Einstein system, *Math. Proc. Camb. Phil. Soc.* **115**, 559–570 (1994).

[22] G. Rein, Cosmological solutions of the Vlasov–Einstein system with spherical, plane, and hyperbolic symmetry, *Math. Proc. Camb. Phil. Soc.* **119**, 739–762 (1996).

[23] G. Rein and A. D. Rendall, Global existence of solutions of the spherically symmetric Vlasov–Einstein system with small initial data, *Commun. Math. Phys.* **150**, 561–583 (1992).

[24] G. Rein and A. D. Rendall, Compact support of spherically symmetric equilibria in non-relativistic and relativistic galactic dynamics, *Math. Proc. Camb. Phil. Soc.* **128**, 363–380, 2000.

[25] G. Rein, A. D. Rendall and J. Schaeffer, A regularity theorem for solutions of the spherically symmetric Vlasov–Einstein system, *Commun. Math. Phys.* **168**, 467–478 (1995).

[26] G. Rein, A. D. Rendall and J. Schaeffer, Critical collapse of collisionless matter – a numerical investigation, *Phys. Rev. D* **58**, 44007 (1998).

[27] A. D. Rendall, Cosmic censorship and the Vlasov equation, Class. Quantum Grav. **9** , L99–L104 (1992).

[28] A. D. Rendall, An introduction to the Einstein–Vlasov system, *Banach Center Publications* **41**, 35–68 (1997).

[29] S.L. Shapiro and S. A. Teukolsky, Relativistic stellar dynamics on the computer. I. Motivation and numerical method, *Ap. J* **298**, 34–57 (1985).

[30] S. L. Shapiro and S. A. Teukolsky, Relativistic stellar dynamics on the computer. II. Physical applications, *Ap. J* **298**, 58–79 (1985).

[31] S.L. Shapiro and S. A. Teukolsky, Relativistic stellar dynamics on the computer. IV. Collapse of a stellar cluster to a black hole, *Ap. J.* **307**, 575–592 (1986).

[32] J. A. Smoller, A. G. Wasserman, A. G. Yau and J. B. McLeod, Smooth static solutions of the Einstein–Yang–Mills equations, *Commun. Math. Phys.* **143**, 115–147 (1991).

[33] H. F. Tooper, Adiabatic fluid spheres in general relativity, *Ap. J.* **142**, 1541–1562 (1965).

[34] P. Yodzis, H-J. Seifert and H. Müller zum Hagen, On the occurrence of naked singularities in general relativity, *Commun. Math. Phys.* **34**, 135–148 (1973).

Physical Constants

Speed of light	$c = 299\,792\,458\,\mathrm{m/s}$
Permittivity of free space	$\epsilon_0 = 8.854\,187 \times 10^{-12}\mathrm{C}^2/(\mathrm{N\ m}^2)$
Permeability of free space	$\mu_0 = 4\pi \times 10^{-7}\,\mathrm{N/A}^2$
Boltzmann constant	$k = 1.380\,658 \times 10^{-23}\,\mathrm{J/K}$
Planck constant	$h = 6.626\,075 \times 10^{-34}\,\mathrm{J\,s}$
Elementary charge	$\mathrm{e} = 1.602\,177 \times 10^{-19}\,\mathrm{C}$
Unified atomic mass	$m_u = 1.660\,540 \times 10^{-27}\,\mathrm{kg}$
Electron rest mass	$m_e = 9.109\,389 \times 10^{-31}\,\mathrm{kg}$
	$= 5.485\,799 \times 10^{-4}\,m_u$
Proton rest mass	$m_p = 1.672\,623 \times 10^{-27}\,\mathrm{kg}$
	$= 1.007\,276\,m_u$
Neutron rest mass	$m_n = 1.674\,928 \times 10^{-27}\,\mathrm{kg}$
	$= 1.008\,664\,m_u$
Gravitational constant	$G = 6.672\,598 \times 10^{-11}\,\mathrm{N\,m^2/kg^2}$
Earth equatorial radius	$R_\oplus = 6.378\,140 \times 10^{6}\,\mathrm{m}$
Earth mass	$M_\oplus = 5.973\,707 \times 10^{24}\,\mathrm{kg}$
Sun equatorial radius	$R_\odot = 6.96 \times 10^{8}\,\mathrm{m}$
Sun mass	$M_\odot = 1.989 \times 10^{30}\,\mathrm{kg}$

Modified Bessel Function

Definition and Properties

a) $K_n(\zeta) = \left(\frac{\zeta}{2}\right)^n \frac{\Gamma(\frac{1}{2})}{\Gamma(n+\frac{1}{2})} \int_1^\infty e^{-\zeta y}(y^2-1)^{n-\frac{1}{2}} dy$;

b) $K_{n+1}(\zeta) = K_{n-1}(\zeta) + \frac{2n}{\zeta} K_n(\zeta)$;

c) $\frac{d}{d\zeta}\left(\frac{K_n(\zeta)}{\zeta^n}\right) = -\frac{K_{n+1}(\zeta)}{\zeta^n}, \qquad \frac{d}{d\zeta}\left(\zeta^n K_n(\zeta)\right) = -\zeta^n K_{n-1}(\zeta)$;

d) $\lim_{\zeta\to 0}(K_n(\zeta)\zeta^n) = 2^{n-1}(n-1)!$;

e) For $\zeta \ll 1$

$$K_n(\zeta) = (-1)^{n+1} \sum_{k=0}^{\infty} \frac{\left(\frac{\zeta}{2}\right)^{n+2k}}{k!(n+k)!} \times \left[\ln\frac{\zeta}{2} - \frac{1}{2}\psi(k+1) - \frac{1}{2}\psi(n+k+1)\right]$$

$$+\frac{1}{2}\sum_{k=0}^{n-1}(-1)^k \frac{(n-k-1)!}{k!\left(\frac{\zeta}{2}\right)^{n-2k}}$$

$\psi(n+1) = -\gamma + \sum_{k=1}^{n} \frac{1}{k}, \qquad \psi(1) = -\gamma = -0.577\,215\,664\ldots$;

f) For $\zeta \gg 1$

$$K_n(\zeta) = \sqrt{\frac{\pi}{2\zeta}} \frac{1}{e^\zeta}\left[1 + \frac{4n^2-1}{8\zeta} + \frac{(4n^2-1)(4n^2-9)}{2!(8\zeta)^2}\right.$$

$$\left. + \frac{(4n^2-1)(4n^2-9)(4n^2-25)}{3!(8\zeta)^3} + \ldots\right];$$

g) For $\mu > 0$ and $a > 0$

$$\int_1^\infty x^{-\frac{\nu}{2}}(x-1)^{\mu-1} K_\nu(a\sqrt{x})dx = \Gamma(\mu)\left(\frac{2}{a}\right)^\mu K_{\nu-\mu}(a);$$

h) $K_\nu(x) = K_{-\nu}(x), \qquad K_{\frac{1}{2}}(2\zeta) = \sqrt{\frac{\pi}{4\zeta}} e^{-2\zeta}$;

i) $\mathrm{Ki}_n(\zeta) = \int_\zeta^\infty \mathrm{Ki}_{n-1}(t)dt = \int_0^\infty \frac{e^{-\zeta \cosh t}}{\cosh^n t} dt$;

j) $\mathrm{Ki}_0(\zeta) = K_0(\zeta), \qquad \mathrm{Ki}_{-n} = (-1)^n \frac{d^n K_0(\zeta)}{d\zeta^n}$;

k) $\mathrm{Ki}_{2n}(0) = \frac{\Gamma(n)\Gamma(3/2)}{\Gamma(n+1/2)}, \qquad \mathrm{Ki}_{2n+1}(0) = \frac{\pi}{2}\frac{\Gamma(n+1/2)}{\Gamma(1/2)\Gamma(n+1)}$;

l) For $\zeta \gg 1$

$$\mathrm{Ki}_n(\zeta) = e^{-\zeta}\sqrt{\frac{\pi}{2\zeta}}\left[1 - \frac{4n+1}{8\zeta} + \frac{3(16n^2+24n+3)}{128\zeta^2} \right.$$
$$\left. - \frac{5(64n^3+240n^2+212n+15)}{1024\zeta^3} + \ldots\right].$$

References

[1] M. Abramowitz and I. A. Stegun, *Handbook of mathematical functions* (Dover, New York , 1968).

[2] I. S. Gradshteyn and I. M. Ryzhik, *Tables of integrals, series and products* 5th ed. (Academic Press, San Diego, 1994).

Index

Progress in Mathematical Physics

Previously published titles

Ablamowicz, R., Tennessee Technological University, Cookeville, TN, USA / **Fauser, B.**, Universität Konstanz, Germany (Eds.)

Clifford Algebras and their Applications in Mathematical Physics, Volume I
Algebra and Physics

2000. 488 pages. Hardcover
ISBN 0-8176-4182-3
Vol. 18

Volume I *Algebra and Physics* is devoted to the mathematical aspects of Clifford algebras and their applications in physics. Algebraic geometry, cohomology, non-commutative spaces, q-deformations and the related quantum groups, and projective geometry provide the basis for algebraic topics covered. Physical applications and extensions of physical theories such as the theory of quaternionic spin, a projective theory of hadron transformation laws, and electron scattering are also presented, showing the broad applicability of Clifford geometric algebras in solving physical problems. Treatment of the structure theory of quantum Clifford algebras, the connection to logic, group representations and computational techniques including symbolic calculations and theorem proving rounds out the presentation.

Ryan, J., University of Arkansas, Fayetteville, AR, USA / **Sprössig, W.**, TU-Bergakademie Freiberg, Germany (Eds.)

Clifford Algebras and their Applications in Mathematical Physics, Volume II
Clifford Analysis

2000. 320 pages. Hardcover
ISBN 0-8176-4183-1
Vol. 19

Volume II *Clifford Analysis* is an up-to-date survey of most aspects of modern-day Clifford analysis. Topics range from applications such as complex-distance potential theory, supersymmetry, and fluid dynamics to Fourier analysis, the study of boundary value problems, and applications to mathematical physics and Schwarzian derivatives in Euclidean space. Among the mathematical topics examined are generalized Dirac operators, holonomy groups, monogenic and hypermonogenic functions and their derivatives, quaternionic Beltrami equations, Fourier theory under Möbius transformations, Cauchy-Riemann operators, and Cauchy type integrals.

Vol. 17: **Baylis, W.E.** Electrodynamics. A Modern Geometric Approach (1999) ISBN 0-8176-4025-8

Vol. 16: **Maslov, V.P.** The Complex WKB Method for Nonlinear Equations I. Linear Theory (1994) ISBN 3-7643-5088-1

Vol. 15: **Leznov, A.N. / Saveliev, M.V.** Group-Theoretical Method for Integration of Nonlinear Dynamical Systems (1992) ISBN 3-7643-2615-8

Vol. 14: **Smirnow, V.A.** Renormalization and Asymptotic Expansions (1991)

Vol. 13: **Haba, Z. / Sobczyk, J.** Functional Integration, Geometry and Strings. Proceedings of the XXV Karpacz Winter School of Theoretical Physics, Karpacz, 20 February–5 March 1989 (1989)

Vol. 12: **Piguet, O. / Sibold, K.** Renormalized Supersymmetry. The Perturbation of N=1 Supersymmetric Theories in Flat Space Time (1986)

Vol. 11: **Ceausescu / Costache Georgescu** Critical Phenomena 1983. Brasov School Conference (1998)

Vol. 10: **Fritz / Jaffe / Szasz** Statistical Physics and Dynamical Systems. Rigorous Results

Vol. 9: **Weldon / Langacker / Steinhardt** Fourth Workshop on Grand Unification

Vol. 8: **Milton K.A. / Samuel M.A.** Workshop on Non Pertubative Quantum Chromodynamics ISBN 0-8176-3127-5

Vol. 7: **Fröhlich** Scaling and Self-Similarity in Physics. Renormalization in Statistical Mechanics and Dynamics

Vol. 6: **Frampton P.H. / Glashow S.L. / Van Dam H.** Third Workshop on Grand Unification, 1982 ISBN 0-8176-3105-4

Vol. 5: **Dita / Georgescu / Purice** Gauge Theories. Fundamental Interactions and Rigorous Results

Vol. 4: **Atwood / Bjorken / Brodsky / Stroynowski** Lectures on Lepton Nucleon Scattering and Quantum Chromodynamics

Vol. 3: **Manin** Mathematics and Physics

Vol. 2: **Jaffe / Taubes** Vortices and Monopoles, Structure of Statistic Gauge Theories

Vol. 1: **Collet, P. / Eckmann, J.P.** Iterated Maps on the Interval as Dynamical Systems (1997) ISBN 0-8176-3026-0

For orders originating from all over the world except USA and Canada:

Birkhäuser Verlag AG
c/o Springer GmbH & Co
Haberstrasse 7
D-69126 Heidelberg
Fax: ++49 / 6221 / 345 229
e-mail: birkhauser@springer.de

For orders originating in the USA and Canada:

Birkhäuser
333 Meadowland Parkway
USA-Secaucus
NJ 07094-2491
Fax: +1 201 348 4033
e-mail: orders@birkhauser.com

Birkhäuser

MIX
Papier aus verantwortungsvollen Quellen
Paper from responsible sources
FSC® C105338
FSC
www.fsc.org

If you have any concerns about our products,
you can contact us on
ProductSafety@springernature.com

In case Publisher is established outside the EU,
the EU authorized representative is:
Springer Nature Customer Service Center GmbH
Europaplatz 3, 69115 Heidelberg, Germany

Printed by Libri Plureos GmbH
in Hamburg, Germany